Clay-Containing Polymeric Nanocomposites

Volume 2

L.A. Utracki

Rapra Technology Limited

Shawbury, Shrewsbury, Shropshire, SY4 4NR, United Kingdom
Telephone: +44 (0)1939 250383 Fax: +44 (0)1939 251118
http://www.rapra.net

First Published in 2004 by

Rapra Technology Limited

Shawbury, Shrewsbury, Shropshire, SY4 4NR, UK

A catalogue record for this book is available from the British Library.

ISBN: 1-85957-482-3

Typeset, printed and bound by Rapra Technology Limited
Cover printed by The Printing House, Crewe, UK

Contents Volume 1

Part 3 Fundamental Aspects

Contents Volume 2

Part 4 Technology of Clay-Containing Polymeric Nanocomposites

Part 5 Performance

Part 6 Closing Remarks

Part 7 Appendices

Part 8 References

Part 9 Index

Part 4

Technology of Clay-Containing Polymeric Nanocomposites

Clay-Containing Polymeric Nanocomposites

4.1 Thermoplastic CPNC

As stated in the Preamble, this book is 'constructed' in five main parts: Part 1 provides an introduction to the science and technology of CPNC; Part 2 summarises the basic elements of polymer nanocomposite technology; Part 3 presents the fundamental aspects, pertinent to the new technology; Part 4 is to focus on the technological aspect primarily as described in the patent literature; and the final Part 5 will discuss selected aspects of CPNC performance.

Part 4 is divided into three parts, the first presenting the technology for CPNC with a thermoplastic polymer matrix, the second with thermosets and the last with elastomers. While focusing on the patent information one must not forget what has been discussed in the open literature as described in the preceding parts.

Discussion of thermoplastic CPNC technology in Section 4.1 will start with the two most important types based on PA and PO, and then follow with a description of the general methods used for preparation of CPNC with thermoplastic matrix. CPNC technology indeed evolved in this order (the exception being elastomeric CPNC explored very early on both sides of the Pacific), thus it is instructive to follow the progress made so far. Section 4.1 will then continue with CPNC having as a matrix vinyl polymers, thermoplastic polyesters, and speciality polymers.

4.1.1 Polyamides (PA)

The first patent on polymerisation of ε-caprolactam in the presence of layered silicate sheets focused on activated anionic polymerisation that resulted in polymer with low extractable monomer content [Brassat, 1975]. The patent used non-intercalated layered mineral (2 to 15 wt%) and the degree of dispersion was not considered, thus in spite of direct implications to the CPNC technology it cannot be counted as belonging to it. One year later another step in this direction was taken – Fujiwara and Sakamoto [1976] from Unitika described the preparation of a clay-polyamide composite. The use of clay as an efficient and inexpensive reinforcing filler sparked interest at Toyota Central Research and Development Laboratories (Toyota). In the US patent application (priority date 26.11.1980) for the first time the XRD was used to identify interlayer spacing [Kamigaito *et al.*, 1984]. The document describes intercalated clay composites, prepared by dispersing MMT in water, adding ε-caprolactam and dichloro dimethyl silane (a sizing agent) and thermally polymerising the monomer at *ca.* 200 °C. The patent stated that besides PA-6, several other polymers could be reinforced using this method, *viz.* PS, PA, PI, PVAc, etc. While claims do not specify the concentration range, Example 1 cites 5 to 20 wt% of polymer. In spite of the resulting small expansion of the interlayer galleries ($d_{001} = 1.5\ nm$) significant improvements in mechanical properties, heat resistance and water absorption were reported. The

Kamigaito *et al.* patent is clearly on the borderline between composites and the new class of materials we now call the clay-containing polymeric nanocomposites, CPNC.

4.1.1.1 PA-Type Nanocomposites from Toyota

The first US patent on CPNC with high mechanical strength and high-temperature characteristics was granted to Toyota [Okada *et al.*, 1988]. It contrasts the 'classical' methods of reinforcing polymers with inorganic materials (e.g., calcium carbonate or mica) with the invented process leading to dispersion of individual phyllosilicate platelets with the interlayer distance of at least 2 nm and concentration of 0.5 to 150 pph of resin. The old methods improved mechanical properties at a cost of embrittlement and increased density. Furthermore, the required filler treatment with coupling or 'sizing' agents caused the organic and inorganic components to form separate phases and non-uniform dispersion.

There are three novel aspects in this patent: (1) intercalation can be conducted in a non-aqueous medium; (2) there is no need for the time, labour and energy consuming drying and grinding steps; and (3) the patent stipulates applicability of the technology to a wide variety of polymers.

The new materials comprised a polyamide (PA = PA-6, PA-66 or PA-11) with uniformly dispersed (on the level of molecular dimensions) clay platelets. The silicate layers were specified as being 0.7 to 1.2 nm thick, with the interlayer distance $d_{001} \geq 3.0$ nm, and CEC = 0.5 to 2 meq/g. The molecular weight distribution of the PA matrix was specified as $M_w/M_n \leq 6$. The manufacturing process involved three steps:

1. Intercalation of clay (e.g., montmorillonite, saponite, beidellite, nontronite, hectorite and stevensite, vermiculite or halloysite);
2. Mixing the organoclay complex with a monomer(s) and, if desired, a catalyst or activator;
3. Heating the mixture to a prescribed temperature to effect polymerisation. The document noted that clays have catalytic activity for ring-opening polymerisation of lactams.

For example, intercalation of MMT or vermiculite (CEC = 1.8 meq/g) was performed by reacting the negatively charged silicate layers with ions of ammonium $-NH_3^+$, trimethyl ammonium $-N^+(CH_3)_3$, or with $-NX^+$, in which X is H^+, Cu^{2+} or Al^{3+}. The preferred intercalants were ω-amino acids (e.g., 12-amino-dodecanoic acid, $H_3N^+C_{12}H_{24}COOH$ or ADA) and dodecylammonium ($H_3N^+C_{12}H_{25}$ or DDA). Intercalation was performed by dispersing clay in an aqueous solution containing the intercalant salt, followed by removing the excess ions from the treated complex either by washing with water, or ion-exchange. The authors mentioned that the use of multivalent inorganic cations transformed the onium ions into complex ones, e.g., $CuHN^+$–, which strongly bonded to clay and engendered exfoliation ($d_{001} > 10$ *nm*). The cationic polymerisation proceeded (without catalyst or accelerator) at $T_{polymerisation}$ = 250 to 300 °C for about 10 to 24 h, with the intercalated clay acting as catalyst. Additives inert to the polymerisation reaction, e.g., glass fibre, pigment and antioxidants, could be incorporated.

The patent provides numerous examples using several intercalants and different concentrations of MMT in the final product. For example, the following cations: Cu^{2+}, Al^{3+}, H^+, $H_3N^+\text{-}(CH_2)_n\text{-}COOH$, (where n = 5, 11 or 17) were used as intercalants. For CPNC with MMT concentration not exceeding 25 phr, full conversion of monomer into PA-6 was achieved and for these samples the interlayer spacing $d_{001} > 10$ nm was obtained. Comparative data were provided for injection moulded specimens of PA-6, clay-filled PA-6 and the new CPNC based on PA-6. The improvements of properties over those of matrix polymer were as follows: the tensile strength increased by 68%, tensile modulus by 88%, HDT by 55 °C, dynamic modulus by 257%, water absorption was reduced by 43%. The elongation at break for the neat PA-6 exceeded 210%, for the clay composite it was 6% and for CPNC 10%.

The following Toyota patent [Kawasumi *et al.*, 1989] modified the Okada *et al.* manufacturing technology by replacing the costly and time consuming process of dispersing clay in a large volume of water followed by drying and pulverising, which required a large amount of energy. The new process operated at higher clay concentration and the separation was done by filtration. Next, the filtrate (still containing *ca.* 80 wt% H_2O) was mixed with monomer, and the mixture was heated to the polymerisation temperature while water was driven off. The patent claims listed applicability of the invention to CPNC with numerous polymers as the matrix, *viz.* PA (PA-66, PA-6 and PA-11), PEST, PC, PI, PES, PPS, POM, PPE, PES, fluorinated polymers, vinyls, thermosets, and rubbers.

To prepare PA-based nanocomposites the new process also involves three steps:

1. Intercalating clay to form a complex capable of being swollen by liquid monomer(s) (at $T > T_m$).
2. Mixing the organoclay with a lactam (caprolactam, caprylolactam or dodecanolactam), an aminoacid (e.g., 6-amino-*n*-caproic or 12-amino-dodecanoic acid), or a PA-salt (e.g., hexamethylenediamine adipate). In this step a base catalyst and an activator could be added.
3. Heating the mixture at the polymerisation temperature. The best results were obtained keeping the mixture at a temperature slightly above the T_m of the monomer for a period that ensured even dispersion of organoclay in the monomer. Note: ε-caprolactam has $T_m = 62$ °C and boiling point $T_b = 142$ °C.

The desired clay should have CEC = 0.5-2.0 meq/g and a large surface area, e.g., smectite, vermiculite or halloysite (i.e., one negative charge per area of 0.25-2.0 nm^2). The document noted that clays with CEC > 2.0 meq/g have the interlayer bonding force too strong for exfoliation, while those with CEC < 0.50 meq/g have insufficient intercalation capability (in the intervening years these limits have been moved!). Prior to intercalation the clay should be purified and ground into the desired shape and size as well as ion exchanged to form Na salt, Na-MMT.

According to the document, intercalation could be accomplished by dispersing the clay in an aqueous solution of a positively charged intercalating agent, reacting it, and then washing with water to remove excess ions. Onium salts of strong acid have been used, e.g., these of trimethylamine, triethylamine, hexylamine, cyclohexylamine, dodecylamine, aniline, pyridine, benzylamine, bis(aminomethyl)benzene, amino-phenols, ethylenediamine, hexamethylene-diamine, hexamethylene-tetramine, polyallylamine, alanine, 4-amino-butyric acid, 6-amino-caproic acid, ω-amino-dodecanoic acid, and ω-amino-hexadecanoic acid, etc. The suitable strong acids are HCl, HBr, H_2SO_4 and phosphoric acids. Furthermore, the suitable intercalating ions include other ions, e.g., trimethyl phosphonium, $P^+(CH_3)_3$; dimethyl sulfonium ion, $S^+(CH_3)_2$, etc. The intercalant may have such functional groups as vinyl, carboxyl, hydroxyl, epoxy, or amino. In the case where the matrix is PA, the preferred intercalant is ω-amino-$C_{12\text{-}18}$

acid, *viz.* $H_3N^+C_{12}H_{24}$ COOH. For vinyl polymers, the intercalant may be selected from between the unsaturated compounds, *viz.* methacryloyl-oxyethyl-trimethyl ammonium chloride, $CH{=}C(CH_3)COOCH_2CH_2N^+(CH_3)_3Cl^-$, and/or 4-vinylpyridinium chloride.

Polymerisation of a lactam to PA-6 with narrow *MWD* was carried out in the presence of a base catalyst (e.g., NaOH, $NaOCH_3$, sodium or a potassium salt of lactams) and an activator, for example: *N*-acetylcaprolactam, acetic anhydride, CO_2, phenyl isocyanate or cyanuric chloride at a level 0.01-5 mol% of polymerisable lactam. At T = 120-250 °C the catalysed polymerisation time varied with T from 1 to 300 min. CPNCs with other polymers were prepared following the standard procedure: dispersing a suitable organoclay in a monomer(s) and then following the polymerisation procedure suitable for a given monomer(s). It is important that the organoclay should not retard the polymerisation. The *MWD* of PA-6 in the newly produced CPNC was $M_w/M_n \leq 6$.

The patent formally admitted that the reasons why the PA-based CPNC has distinctive properties were not clear, but the following observations were itemised:

- The chemical bonding between PA molecules and silicate layers firmly resists thermal and mechanical deformation. This is reflected in high modulus and tensile strengths as well as excellent thermal characteristics (e.g., high softening point and high temperature strengths).
- The materials show total exfoliation with the interlayer spacing d_{001} > 10 nm.
- The materials show high dimensional stability, abrasion resistance and surface smoothness.
- The barrier properties and water resistance originate from the uniform dispersion of clay platelets.
- The embrittlement and other problems of conventional composites are eliminated because the clay platelets are dispersed on the molecular scale and are end-tethered to macromolecules.
- The process eliminates several steps of the earlier methods, *viz.* surface treatment and mixing of reinforcements.
- The aspect ratio remains high as the intercalation/exfoliation uses a chemical reaction instead of mechanical forces.

The following example provides the best illustration of the advantages of the new process:

(1) *New 'wet' process.* 100 g of Na-MMT was dispersed in 1.75 L H_2O containing 51.2 g of ADA and 6 ml of HCl, followed by mixing at 80 °C for 1 h. After washing, the mixture was filtered to obtain the intercalated complex (MMT-ADA) containing 80-90 wt% of H_2O, with d_{001} = 1.6 nm. Next, to 100 g of ε-caprolactam was added 2.0, 5.3 or 8.7 phr of MMT-ADA. The monomer was polymerised at 180 °C under N_2 for 5-7 h.

(2) *Old 'dry' process.* For comparison, similar polymerisation was carried out, but with MMT-ADA that was dried and ground.

(3) Pertinent characteristics of these two processes are summarised in **Table 63**. Evidently, there is a small systematic increase in M_n, a reduction of M_w/M_n, a significant improvement in the degree of exfoliation, and a dramatic reduction in the total time required for the process.

Table 63 Characteristics of Toyota CPNC's prepared in the 'wet' and 'dry' process. Data [Kawasumi *et al.*, 1989]

MMT-ADA (parts/100 g lactam)	New 'wet' process				Old 'dry' process			
	M_n (kg/mol)	M_w/M_n	d_{001} (nm)	Prep. time (h)	M_n (kg/mol)	M_w/M_n	d_{001} (nm)	Prep. time (h)
2.0	33	5.0	>10	11	32	5.6	3-10	64
5.3	25	4.8	>10	12	23	5.2	3-10	65
8.7	24	5.3	>10	13	18	6.0	3-10	66

In other examples, ethanol or an aqueous solution of DMF replaced water. CPNC with PMMA or a thermosetting resin was prepared. To produce PMMA-based CPNC the ADA was replaced by an ester of methacrylic acid with ω-terminated vinyl compound, *viz.* methacryloyl-oxy-ethyl-trimethyl ammonium chloride, $CH{=}C(CH_3)COOCH_2CH_2N^+(CH_3)_3Cl^-$. Good exfoliation was reported.

The next patent in this series was deposited by Toyota jointly with Ube Industries Ltd. [Deguchi *et al.*, 1992; 1995]. The intrinsic merits of the original invention were confirmed, but the industrial exploration of the procedure demonstrated three needs for improvement: (1) According to the original patent, the PA-based CPNC was produced by infiltrating organoclay with a lactam, amino acid, or PA-salt, then polymerising the monomer(s). As a result the PA macromolecules were end-tethered to clay platelets through the end amino group. As a result, the PNC showed insufficient dye-affinity, as well as poor coating and printing properties. (2) It was difficult to economically produce CPNC with high *MW* matrix, high clay content and uniform composition. (3) The process was difficult for the production of nanocomposites with di-amine di-carboxylic acid type polyamides, e.g., PA-66.

The new patent is much more narrow in scope, focused on the production of CPNC with PA, in particular with PA-6 as the matrix. The preparation of CPNC is as follows:

1. Dispersing > 1 wt% of a clay (CEC = 0.50-2.0 meq/g) in a medium containing, e.g., ω-amino-dodecanoic acid (ADA) and HCl, where it swells to a degree $S > 5$ ml/g (S is volume per 1 g of dry clay). For example, after 1 h at 80 °C MMT-ADA organoclay was obtained. The dispersion medium could be water, alcohol, DMSO, DMF, acetic or formic acid, pyridine, aniline, phenol, nitrobenzene, acetonitrile, propylene carbonate, ether, MEK, carbon-tetrachloride, *n*-hexane, etc. The preferred silicate was Na-MMT having a CEC = 1.19 meq/g, an occupied area per one negative charge of 1.06 nm^2, and interlayer spacing of d_{001} = 1.25 nm. After reaction and filtering the complex containing 88 wt% H_2O had d_{001} = 1.8 nm. Next, ε-caprolactam was added to obtain the composition with a ratio of MMT-DDA:H_2O: ε-caprolactam = 1:9:9, followed by mixing. The swelling degree (measured by centrifugation at 500G for 4 h) was S = 14.6 ml/g.

2. A TSE was fed with a mixture of 72 wt% PA (e.g., PA-6, PA-66 or other PA) and 28% of the clay/water/monomer complex. From the extruded strands unreacted ε-caprolactam was extracted, then the product was dried under vacuum. Excellent clay dispersion was observed under TEM – mainly individual platelets with few stacks of maximum two parallel platelets.

The new method was economic, the clay platelets were uniformly dispersed, without blocking the amino end groups. The product had good dye-affinity and whitening. The method was more efficient, yielding CPNC with a wide range of viscosity, applicable to any type of PA matrix. CPNCs with 0.05-30 phr of dispersed clay were prepared. The material showed excellent mechanical properties, heat resistance, improved dye-affinity and whitening resistance during stretching.

The key to the Toyota-Ube technology is the use of double intercalant: ω-amino-dodecanoic acid (ADA) and: ε-caprolactam. This acid was selected since –COOH groups are known to catalyse the ring-opening polymerisation of lactams. The C_{12} acid was a compromise between the commercial availability and good intercalating capabilities. As shown in **Figure 25**, an aqueous solution of ω-amino-dodecanoic acid alone engenders insufficient interlayer distance (*viz.* d_{001} = 1.72 nm) for the subsequent melt exfoliation. However, addition of ε-caprolactam increases the spacing to d_{001} = 3.87 nm. At 25 °C the interlayer spacing corresponds to the AA molecules being aligned perpendicularly to the clay platelets surface, with ε-caprolactam inserted between them. However, as the *T* increases to 100 °C the latter component is inserted between the –COOH groups of AA and the opposite clay layer. The interlayer spacing in the CPNC as a function of intercalated MMT content is shown in **Figure 44**. The least-squares hyperbolic predicts exfoliation only at low clay concentration, MMT < 11.3 wt%, i.e., below 5 vol%. By the end of 2002 several major companies (e.g., Bayer, Honeywell Polymers, Mitsubishi Gas, Nanocor, Ube, and Unitika) had licensed the Toyota/Ube technology for the manufacture of PA-based CPNC.

A new method, fascinating by its simplicity, for the production of PA-based nanocomposites involves compounding in a TSE of PA-6 with water slurry containing 2 wt% Na-MMT [Hasegawa *et al.*, 2003]. During compounding the water was removed by vacuum-aided devolatilisation. According to OM and TEM the clay platelets were exfoliated and dispersed homogeneously. The relative viscosity of the CPNC containing 1.6 wt% of clay was almost equal to that of neat resin indicating that PA-6 hydrolysis did not occur during the process. Tensile and flexural moduli of the CPNCs were 28 and 14% higher than those of neat PA-6, respectively, while the tensile and flexural strength were similarly higher (by 28 and 12%, respectively). The Izod impact strength of the new nanocomposites was 12% lower compared to PA-6 and HDT increased from 75 °C for the neat resin to 102 °C. Gas permeability was 31% lower than that of neat PA-6. In short, the properties of the new PA-6/MMT nanocomposites were nearly equal to those of conventional CPNC prepared by compounding PA-6 with organoclay, as described in the preceding patents [Deguchi *et al.*, 1992; 1995], see also **Table 94.**

The new Toyota process opens new vistas for the technology. The use of a diluted aqueous suspension of MMT with molten PA-6 may not be easily adapted to other PA resins, especially those with higher processing temperature (generation of high steam pressure!). However, once MMT is exfoliated in aqueous medium, part of the water can be replaced with a higher boiling point solvent, miscible with molten PA, e.g., glycols or a lactam. Furthermore, the method should also work for other highly polar macromolecules, *viz.* polyesters or polycarbonates.

4.1.1.2 PA-Type Nanocomposites from AlliedSignal Inc.

It is interesting that Toyota being mainly a mechanical engineering company chose the chemical route for the production of PA-based CPNC, whereas AlliedSignal (now Honeywell) having strong roots in the chemical industry selected the mechanical, i.e., by melt exfoliation.

In the first pertinent patent from Honeywell, Maxfield *et al.* [1995] emphasised the dominant formation of γ-crystals of PA-6 during compounding with nanoparticles. The γ-phase was found resistant to conversion to the more thermodynamically stable α-form. The inorganic nanoparticles mentioned in the patent could be either layered or fibrillar. The diameter of the fibres (e.g., imogolite or V_2O_5) should be from 1 to 5 nm, with the average length from 30 to 200 nm. The layered materials (layer thickness of 0.7 to 2.5 nm and aspect ratio p = 1 to 1000) included phyllosilicates; smectite; vermiculite; illite; ledikite; layered double hydroxides; chlorides; chalcogenides; cyanides; oxides; etc. Evidently, the most preferred were the common clays, *viz.* MMT, nontronite, beidellite, hectorite, saponite, magadiite, etc. The amount of nanoparticles added to polymer should be > 5 ppm, homogeneously dispersed in the γ-phase polymer. To be effective, the layered materials should be dispersed into stacks with ≤ 2 platelets. The CPNC may include other additives, *viz.* nucleating agents, fillers, plasticisers, impact modifiers, chain extenders, plasticisers, colorants, UV-stabilisers, thermal stabilisers, mould release lubricants, antistatic agents, pigments, fire retardants, and the like.

As far as CPNC technology is concerned, the novelty was the use of MMT that was treated with alkylammonium cations and a silane coupling agent. However, it should be recognised that silane treatment has been a part of classical composites technology for a good 50 years, and it was cited in the principal claim of the Kamigaito *et al.* patent of 1984. However, the new invention used the silanes to provide sufficient intercalation and affinity between the organoclay and the matrix polymer to exfoliate the clay during melt compounding. The invention is applicable to any PA that has MFR = 0.5 to 10 g/10 min, able to crystallise in the γ-phase and is melt processable, *viz.* PA-4, PA-6, PA-8, PA-9, PA-10, PA-11, PA-12, PA-18, etc. The preferred resins are PA-6 and PA-12 with 30 to 50 wt% of γ-phase crystallinity. The new compositions showed improved rigidity and water resistance while retaining toughness, surface gloss and abrasion resistance.

The process described involves melt blending of PA with intercalated nanoparticles at $T \geq T_m$ in a TSE. The mixture is sheared until the desired level of dispersion or exfoliation has been achieved. For example, pellets of PA-6 and organo-treated MMT were heated to $T > T_m$, then sheared in an extruder, injection moulding machine, or internal mixer. The melt temperature, the residence time and the design of the extruder (single screw, twin screw, number of flights per unit length, channel depth, flight clearance, mixing zone, etc.) were the variables used to control the amount of applied shear stress and the residence time. The shearing should be carried out until about 95 wt% of material exfoliates to the desired extent.

The patent specifies that, prior to compounding, the natural clay must be purified, the metal cations (that engender repulsive interactions with the molten polymer) must be removed, and the clay must be intercalated to increase the interlayer distances to d_{001} > 1.5 nm. The intercalating agents of the invention included any type of ammonium ion, *viz.* $-NH_3R^{1+}$, $-NH_2R^1R^{2+}$, $-NHR^1R^2R^{3+}$, $-NR^1R^2R^3R^{4+}$, where the R^1, R^2, R^3 and R^4 are the same or different organic substituents.

To facilitate exfoliation and to prevent re-aggregation, the patent advises the use of a compatibilising agent, having one part bonding to the clay surface and another able to interact with the matrix. For PA-6 and a smectite the preferred compatibilisers were either neutral organic molecules, zwitterionic or cationic compounds, *viz.* amides, esters, lactams, nitriles, ureas, carbonates, phosphates, phosphonates, sulfates, sulfonates, nitro compounds, etc. Preferred neutral organics are monomers or oligomers. Useful cationic compounds include onium species, historically used as intercalants, *viz.* ammonium [$-N^+(CH_3)_3$, $-N^+(CH_3)_2(CH_2CH_3)$], phosphonium or sulfonium [$-P^+(CH_3)_3$, $-S^+(CH_3)_3$] derivatives of aliphatic, aromatic or arylaliphatic amines, phosphines or sulfides. However, it may be convenient to use a compatibilising agent that is different from the intercalant. For example, alkyl ammonium cations may be used to replace the metal cations of a smectite and in turn be partially replaced by a silane coupling agent. In this case, the alkyl ammonium cation is an intercalant while the silane is a compatibilising agent, specifically selected to interact with the polymer matrix. Thus, the preferred intercalating/compatibilising agents include silane, ammonium, sulfonium or phosphonium derivatives of octadecyl amine, octadecyl phosphine, trimethyl dodecyl sulfide, octadecyl sulfide, dimethyl di-dodecyl amine, octadecyl amine, di-octyl phosphine, methyl-octadecyl amine, di-octyl sulfide, decyl sulfide, etc. Their amount should be about 0.8 to 1.2 mmole/g of clay.

For example, organoclay may be prepared by dispersing MMT (5 to 15 wt% in water) with sodium hexametaphosphate at 80 °C using a high-speed mixer, and then adding a suitable ammonium chloride of, e.g., octadecylamine, 11-aminoundecanoic acid dioctyl amine, dimethyl dodecyl amine, methyl octadecyl amine, dimethyl di-dodecyl amine, etc. The amine-complexed clay may be filtered out or centrifuged, followed by rinsing it with water, and rigorously drying (at 100 to 160 °C under vacuum for 8-24 h in the presence of phosphorous pentoxide). Next, the organoclay should be treated with silanes in an organic solvent (e.g., dioxane, glyme, diglyme, DMSO, MEK). Alternatively, an aqueous suspension of clay may be treated with water-soluble silane, *viz.* trialkoxysilane. For example, 5 wt% MMT-ODA (0.8 mmole/g) was dispersed at 60 °C in a dioxane solution of amino-ethyl amino-propyl trimethoxysilane. The ratio of silane to mineral was about 0.20 mmole/g. The organoclay with 0.6 mmole of ammonium and 0.2 mmole of silane per 1 g of clay was dried, and then ball milled into a powder of about 100 mesh.

The documents also lists several commercial organoclays as suitable, e.g., Gelwhite HNF and Claytone APA® from SCP (MMT-ODA), Volclay from AMCOL (MMT-2M2ODA), Bentone 34 from Rheox Inc. (MMT-2M2ODA), Laponite S, from Laporte Ind. (FH-2M2ODA). Prior to melt compounding with PA, these materials should be treated with silane, e.g., amino-propyl trimethoxy silane or 3,3-epoxy cyclohexyl ethyl trimethoxy silane.

The resulting CPNC is a thermoplastic that may be used to produce articles by conventional methods, *viz.* melt spinning or casting, film blowing or extruding, vacuum, sheet, or injection moulding, etc. Examples of such articles are components for technical equipment, castings, household and sport components, bottles, containers, components for the electrical and electronics industries, car components, circuits, fibres, semi-finished products which can be shaped by machining and the like. These PNC can be fabricated as film laminates (25 to 75 μm thick, with the platelets aligned parallel to the film surface), e.g., for food packaging, powder coating and hot-melt adhesives.

The next patent [Maxfield *et al.*, 1996] stressed the use of the classical 'sizing' agents, *viz.* organosilanes, organotitanates, and organozirconates [Plueddemann, 1982]. As in the previous patent, the process for the preparation of PA-based CPNC consists of two steps:

1. Reacting inorganic layered or fibrillar particles with an organosilane, organotitanate or organozirconate. These species must have groups able to bond covalently with the particles and others that either react with a polymer precursor or are miscible with the product of polymerisation. Insertion of these species may be enhanced by heating, ultrasonication or by the use of microwaves.
2. Polymerising the precursor in the presence of the intercalated MMT to form a CPNC. The precursors may be selected such that the resulting matrix is formed of one or more thermosetting and/or thermoplastic polymers or rubbers. The preferred monomers are caprolactam, caprylactam, laurolactam and the like; and diamines which form PA by reaction with dicarboxylic acids such as hexamethylene-diamine, decamethylene-diamine, etc. The amount of inorganic particles in the polymerisation may vary from about 0.1-10 wt%. The average platelet aspect ratio should be $15 \leq p \leq 300$. The particles should be uniformly dispersed in the matrix polymer with the interlayer distance $d_{001} \geq 5$ nm.

The new process improved properties of PNC, such as yield strength, stiffness, HDT, toughness and impact strength, and it shows superior resistance to diffusion of polar liquids and gases. The products also showed enhanced barrier properties, and yield strength at elevated temperatures. The complexes of organosilanes, organotitanates and/or organozirconates, as well as complexes with primary or secondary ammonium cations showed thermal stability at $T > 300$ °C, below which most thermoplastics are polymerised.

In one example, MMT was dispersed (5-15 wt%) in an aqueous solution of sodium hexametaphosphate at 60 °C. A solution of a caprolactam-blocked isocyanato-propyl-(tri-ethoxy)silane and 1-trimethoxy silyl-2-(*m,p*-dichloro methyl)-phenyl-ethane, was added (24.7 g silane and 11.3 g ε-caprolactam) so that the concentration of organosilanes was 0.01-1.1 mmole/g of clay. The suspension was stirred under N_2 for 2 h at $T = 110$ °C. The intercalated MMT was filtered, dried (at $T = 60$-160 °C in a vacuum for 8-24 h) and ground to about 100 mesh, then combined with caprolactam and aminocaproic acid. The interlayer spacing of the treated MMT should be at least $d_{001} = 1.5$ nm. The mixture was heated at 255 °C for 4 h, then the residual monomer and oligomers were extracted with boiling water for 1.5 h.

This second AlliedSignal patent describes technology similar to that of Toyota; first the clay is intercalated, dispersed in a monomer and then exfoliated during polymerisation. The key difference is the use of an organometallic compound as the reactive compatibiliser.

The third patent in this series describes CPNCs comprising a continuous polymeric phase ($T_m \geq 220$ °C) and clay dispersed in it (pre-intercalated with a secondary or primary ammonium or a quaternary phosphonium cation) extrusion compounded in a TSE [Christiani and Maxfield, 1998]. Since this patent incorporates the technologies described in the previous ones, it is broader, covering melt exfoliation and polymerisation methods. Furthermore, the uses of silanes that may compatibilise clay or organoclay with any polymer make it more general.

As far as PA-based CPNC is concerned, the document emphasises the melt compounding of PA comprising 2 to 5 wt% organomodified MMT or hectorite. For example, PA-6 was melt compounded in a TSE with 4 wt% of MMT, which was pre-intercalated with a dipentyl ammonium cation (0.125 mole per 100 g of clay). **Table 64** summarises results of TGA for several organoclays. The significantly better thermal stability of the organoclay with secondary ammonium (2-ary) is to be expected from the general principles of organic chemistry. The reversed stability of the two others may be due to the aromatic substituent in the 4-ary.

The process yielded CPNC with high specific modulus, strength and DTUL. In addition, the platelet orientation and surface nucleation promoted faster crystallisation and higher crystallinity of PA-6, thus improving the moisture resistance. Recognising these benefits, AlliedSignal investigated the use of CPNC as a matrix for making both short and the long glass fibre (GF) reinforced composites.

Akkapeddi, [1999, 2000] extended the general method of CPNC preparation to the manufacture of GF composites with CPNC as the matrix. These were produced in a TSE by first pre-mixing the organoclay thoroughly into the PA-6 melt and then adding chopped GF through a downstream feed port. By selecting proper intercalant and TSE screw design, a single step process was developed in which the high-shear kneading and mixing elements in the initial zones achieved clay exfoliation and the distributive mixing elements in the final zones ensured uniform incorporation of GF. The processing was carried out in a co-rotating TSE (CORI), either 10-barrel segments, 28 mm Leistritz or 40 mm Werner Pfleiderer machine. A commercial, extra-dry PA-6 (having M_w = 30 kg/mol) was compounded with organoclay using two procedures:

1. Direct compounding with 3-5 wt% of MMT or hectorite intercalated with specially designed organo-quaternary ammonium at 260 °C in a single step, under high shear mixing conditions.
2. The organoclay was masterbatched first into PA-6 (at 25 wt% loading) and then re-extruded in a second step with more PA-6 to dilute the clay content to ≤ 5 wt%.

Conventional chopped GF with diameter, $d \cong 10$ μm and length of about 3 mm was added through a downstream feed port in zone 6 of the extruder. The GF was compounded with the molten, premixed CPNC either as a one-step extrusion process or in a second extrusion step. The extrudate was quenched, pelletised then dried under vacuum at 85 °C, and injection moulded into standard ASTM test specimens.

The organoclays generated had the general structure of an ionic complex in which the silicate anion on the clay surface is bound to an organoquaternary ammonium cation. The structure was designed to engender strong bonding of clay to the organic intercalant.

$$\Big| \!\!-\! \text{Si} - \text{O}^{-+}\text{NR}_1\text{R}_2\text{R}_3\text{R}_4$$

Table 64 Thermogravimetric analysis of intercalated MMT. Data [Maxfield *et al.*, 1996]

Intercalant	Decomposition temperature, T (° C)	Weight loss at T - 100 to 300 °C (% of onium conent)
Di-pentyl ammonium (2-ary)	275	5.6
Di-methyl dodecyl ammonium (3-ary)	190	27.8
Benzyl-tallow-dimethyl ammonium (4-ary; Claytone APA®)	220	38.2

High resolution TEM confirmed clay exfoliation in CPNC containing ≤ 5 wt% - the platelets observed 'edge-on' showed an approximate thickness of about 1-5 nm and a length exceeding 100 nm. At or above 7 wt% clay levels, the platelets aggregated into short stacks. Thus, the nanoclay concentration was kept at an optimum level of 3.5 to 4 wt% (based on PA-6).

The mechanical properties were measured according to the standard ASTM specifications using 'dry as moulded' (DAM) or conditioned at 50%RH (ISO 1110: 1987) specimens. The results are presented in **Table 65**. Dynamic mechanical analysis (DMA) data indicated that CPNC has improved modulus retention with temperature compared to unfilled PA-6 and even mineral-filled PA-6 (see **Figure 139**). The clay high surface area strongly affected PA-6 crystallisation, *viz.* the increased rate translated into > 30% reduction of the cycle time during injection moulding. Furthermore, the platelets' high surface area and their chemical affinity for PA-6 chains induced lamellar orientation. The high crystallinity and the presence of a large number of thin, oriented platelets contributed to the low moisture absorption rate (see **Figure 140**) and improved modulus retention after moisture equilibration.

The aim of this work was to investigate the relative benefit of GF reinforcement in PA and a CPNC matrix. Both short GF and long glass fibre composites were prepared. As the data in **Table 65** and **Figure 139** illustrate, the unique combination of nano- and macro-scale reinforcements gave the opportunity to design PA-6 composites with low filler level, high specific modulus and strength, lighter weight parts with improved surface gloss, reduced mould and tool wear, reduced moisture absorption, etc. It is noteworthy that at any GF level, CPNC showed higher modulus than PA-6. However, the organoclay matrix composites are more brittle than PA-6. The presence of GF contributes to increased crack initiation relative to the unfilled PA-6 or its CPNC, without offering significant resistance to crack propagation.

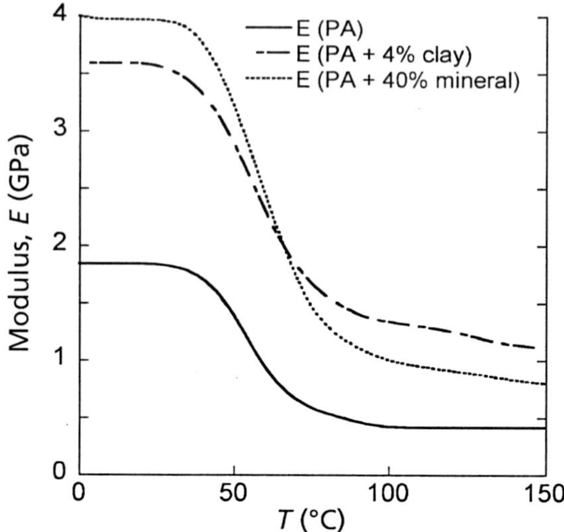

Figure 139 Modulus-temperature behaviour of PA-6, its PNC containing 4 wt% of treated MMT, and its composite with 40 wt% of Wollastonite (mineral calcium silicate). Addition of 4 wt% of treated MMT increased HDT by a similar amount to 40 wt% of a mineral, and increased the high temperature modulus by about one decade (in comparison to PA-6), more than the mineral. After [Akkapeddi, 2000].

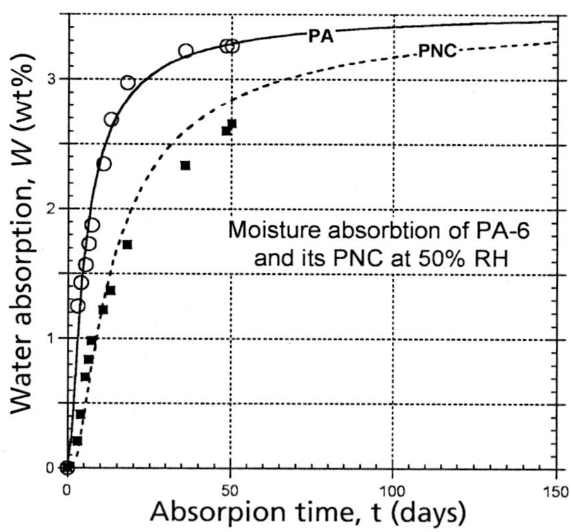

Figure 140 Water absorption versus exposure time to 50% RH for PA-6 and its CPNC with ca. 4 wt% organoclay. The experimental data were fitted to a single a exponential: W = a_o exp{-a/t}. The final moisture (a_o = 3.544) is the same for both materials, but the rate of its absorption for CPNC is about 3× slower than that for the matrix PA-6, viz. a_1 = 3.96 and 11.10 for PA-6 and PNC, respectively. Data [Akkapeddi, 2000].

Table 65 Properties of PNC from AlliedSignal. Data [Akkapeddi, 1999, 2000]

Property	Unfilled	0.01 wt% NC	0.05 wt% NC	0.1 wt% NC
	Dry PA-6			
Flex strength (MPa)	2,770	3,070	3,130	3,160
Flex 5% modulus (MPa)	108	120	124	124
Tensile yield (MPa)	77	83	86	85
Impact strength (J/m)	69	66	64	65
	Dry PA-66			
Flex strength (MPa)	3,070	3,100	3,120	3,160
Flex 5% modulus (MPa)	118	120	120	120
Tensile yield (MPa)	84	84	84	85
Impact strength (J/m)	58	54	44	45
	Wet PA-6			
Flex strength (MPa)	417	507	550	546
Flex 5% modulus (MPa)	18	22	24	24
Tensile yield (MPa)	21	27	29	28
Elongation at break (%)	300	280	270	270
	Wet PA-66			
Flex strength (MPa)	690	678	703	705
Flex 5% modulus (MPa)	28	27	28	28
Tensile yield (MPa)	39	39	39	39
Elongation at break (%)	300	250	250	260

4.1.1.3 AMCOL Technology for PA

Nanocor Inc. is a wholly owned subsidiary of AMCOL International Corporation and a global supplier of nanoclays for CPNC. Commercial production of Nanomer® began in 1998. During the last decade AMCOL has produced the largest volume of patents (some of them have over 60 pages!) on clay intercalation (e.g., since 1996 Nanocor received at least 34 US patents on the topic). The main emphasis of these documents is on clay purification and intercalation for the use of organoclays in several industries. However, several patents either mention the use of hydrophilic polymers as intercalants, the use of patented organoclays in CPNC or even more directly concentrate on the preparation and performance of specific polymeric nanocomposites.

AMCOL patented intercalation of layered phyllosilicates by either oligomers or polymers [Beall *et al.*, 1999]. The principal requirement was that the intercalating polymer must interact with the clay and be sorbed into the interlamellar galleries, increasing the basic spacing to, preferably, d_{001} = 3.5 to 4.0 nm. To accomplish this task, the polymer should have an aromatic ring and/or a polar group (*viz.* a carbonyl; carboxyl; hydroxyl; amine; amide; ether; ester; sulfate; sulfonate; sulfinate; sulfamate;

phosphate; phosphonate; or phosphinate). The intercalation is easiest using water-soluble polymers, e.g., P4VP, PVAl, or their mixtures. As the concentration of the intercalant increases, the d_{001} also increases in steps. For example, at 50 to 80 wt% of P4VP (on dry weight of clay) an interlayer spacing of 3.0 to 4.5 nm has been achieved without the use of a low molecular weight intercalant (e.g., onium or silane). Evidently, polymerisable monomers may be incorporated by mixing with the intercalated clay. These monomers may be acrylics (*viz.* acrylic acid) or a mixture of a diamine and a dicarboxylic acid suitable for the production of a PA.

However, most interesting is the comment that numerous water-insoluble, melt-processable polymers (degree of polymerisation DP = 10 to 100) may also be used as intercalants. These include PA (PA-6 or PA-66), PEST, PC, TPU, polyepoxides, PO, polyalkylamides and their blends. The intercalated clay may be prepared as a concentrate, containing 20-80% of the layered material. The mixture may include other additives, *viz.* nucleating agents, fillers, plasticisers, impact modifiers, chain extenders, colorants, mould release lubricants, antistatic agents, pigments, fire retardants, and the like.

Exfoliation of clay should provide delamination of at least about 90 wt% of the platelets, homogeneously dispersed in the matrix. Some systems may require shearing (e.g., in a TSE), others may exfoliate naturally when heated to $T > T_m$ of the matrix polymer, or by cycling the pressure, e.g., from 0.05 to 6.0 MPa. However, the patent stressed that to accomplish intercalation the clay must have at least 10 to 15 wt% of water (based on the dry weight of the clay). The water can be included in the clay as received, or it may be added prior to or during mixing with a polymer.

As an example the document describes melt-intercalation/exfoliation (under a blanket of dry N_2) of 20 wt% Na-MMT in PA-6. The compounding was carried out 40 to 50 °C above the T_m of PA, e.g., at $T \geq 230$ °C. The compounding resulted in a high degree of exfoliation, evidenced by the XRD diffraction pattern on which the characteristic peaks for Na-MMT virtually disappeared.

Lan *et al.* [2000] continued this work. This patent describes methods of co-intercalation of clay by contacting it with (1) a surface modifier (at least 20% of CEC), which has an alkyl radical and at least 6 carbon atoms (*viz.* an alkyl with a polar group: hydroxyl, carbonyl, carboxylic acid, amine, amide, ester, ether, lactam, lactone, anhydride, nitrile, oxirane, halide, pyridine, etc.), and (2) 16 to 20 wt% (on dry weight of clay) of a polymerisable monomer, a polymerisable oligomer, a polymer (e.g., epoxy, PA, PVAl, P4VP, PC, PEST, etc.), or their mixture. The first intercalation step is usually carried out in aqueous medium – its role is to expand the interlamellar galleries to at least 1 nm. The second intercalant may be incorporated from a solvent or melt – this step should result in exfoliation. The intercalation process follows closely the procedure described in the preceding patent. The intercalated system may be used directly or as a concentrate. All five examples of the patent describe CPNC with MMT intercalated with substituted pyrrolidone (e.g., 1-dodecyl-2- pyrrolidone or 1-octadecyl-2- pyrrolidone as the first intercalant) and epoxy as the second intercalant, which, after curing with Jeffamine D400, formed the matrix.

However, even when the main emphasis of the patent is for the use of organoclays with epoxy, the implications are broader. Thermoplastics (especially PA, PEST, and polymers of α-β-unsaturated monomers) can be used as a matrix for CPNC described in this patent (the most preferred are PA-6, PA-12 and PA-66). Furthermore, the strategy involves a partial intercalation (< 20% of CEC!) with a low molecular weight first intercalant followed by either a polymerisable monomer, oligomer, or polymer, having polar groups as the second intercalant that eventually leads to exfoliation. This strategy is in close agreement with the conclusions based on the self-consistent field theory, reached by Balazs *et al.* [1998]. The authors showed that thermodynamics strongly favours polymeric intercalants over the customary short chain quaternary ammonium ions (for the analysis see Section 3.1 of this book).

More recently, Lan *et al.* [2001a] described an intercalation procedure that reacts (in aqueous medium) clay anions with (at least 25% of CEC) multi-charged onium ions, *viz.* di-ammonium, di-sulfonium, ammonium/phosphonium, etc. The schematic structure of a di-onium is:

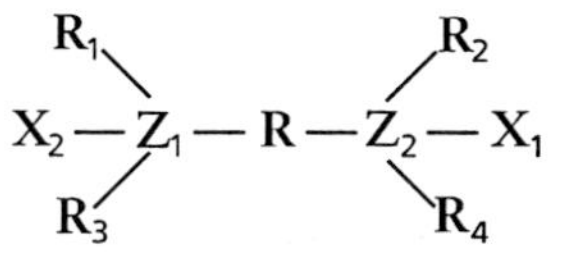

$$\begin{array}{ccccccccc} & R_1 & & & & & R_2 & & \\ & \diagdown & & & & & \diagup & & \\ X_2 - & Z_1 & - & R & - & Z_2 & & - X_1 & \\ & \diagup & & & & & \diagdown & & \\ & R_3 & & & & & R_4 & & \end{array}$$

where Z_1 and Z_2 are cations; R is an organic radical (alkylene, aralkylene or substituted alkylene; C_3 to C_{24}, straight or branched chain) separating the onium ions by 0.5 to 2.4 nm; R_i (i = 1 to 4) are moieties, the same or different, selected from between: H, alkyl, aralkyl, benzyl, substituted benzyl, ethoxylated alkyl; propoxylated alkyl; ethoxylated benzyl; propoxylated benzyl; and finally X_i may be positively charged or not radical.

Next, from 0.05 to 40 wt% of the intercalated clay is extruder-compounded with a polymeric matrix (MW = 1 to 500 kg/mol), which co-intercalates and/or exfoliates the clay. The patent focuses on the polymer or oligomer obtained by polycondensation of *meta*-xylylene diamine and adipic acid (MXD-6), but it may be applied to other polymers, *viz.* PA, epoxy, PVAl, PC, P4VP, PEST, etc.

The invention leads to organoclays with multicharged cationic substances that are deemed to be more suitable for CPNC preparation. The di-amines (e.g., ditallow amine, Duoquad T50, Ethouomeen T13, etc.) are commercially available at a reasonable cost, and may provide complete ion-exchange for the interlayer cations using less organic material, thus leaving more space into which the polymer may diffuse. Dispersing MMT intercalated with tallow di-amine chloride (TADA; 21 wt% organic phase; d_{001} = 1.7 nm) in either PA-6, MXD-6 or PMMA gave excellent quality transparent pellets and cast films. XRD of these products indicated that the interlayer spacing increased to $d_{001} \geq 3.1$ nm. Intercalated clay was also successfully used in the reactive process. CPNC of PA-6 prepared by polycondensation in the presence of MMT intercalated with TADA showed superior barrier properties to the one obtained in the presence of the traditional organoclay intercalated with a monotallow ammonium ion.

In the following patent Lan *et al.* [2001b] reported on a dramatic reduction of O_2 permeability, P_O, through MXD-6 film upon incorporation of MMT-ODA. CPNC was prepared by compounding the two ingredients in a TSE. As shown in **Figure 141**, the data can be approximated with the simplest permeability expression that assumes constant, orthogonal orientation and aspect ratio (p) of uniformly dispersed platelets. The drawn line was computed for a rather high value of $p = 600$. Thus, taking P_O of the neat polymer as a reference, incorporation of 2.5 and 5.0 wt% of organoclay reduced O_2 permeability by 73 and 91%, respectively. However, according to the patent, the highly efficient reduction in oxygen permeability is a result of the organoclay acting (alone, or in combination with the MXD-6 matrix) as an oxygen scavenger.

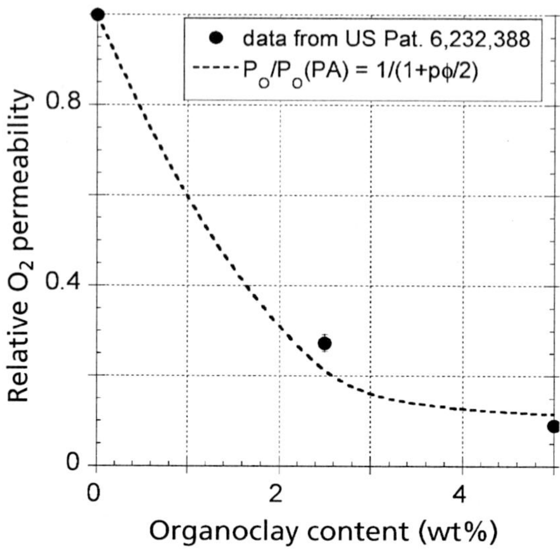

Figure 141 *Relative oxygen permeability through MXD-6 CPNC film versus organoclay content. Data [Lan et al., 2001b]. The line was calculated from Equation 23 with p = 600.*

The latest patent from AMCOL on CPNC technology again focuses on nanocomposites with controlled gas permeability [Lan *et al.*, 2002a]. As before [Lan *et al.*, 2001a] clay intercalation is carried out in the molten state, with a mixture (from 1:3 to 3:1) of at least two organic cations. The process is applicable to most thermoplastics (*ca.* 5 to 85 wt%) to form a matrix, *viz.* PA, PEST, PI, PEI, PAI, PPE, epoxy, PO, PS, PMMA, etc. While any clay can be used, the preferred is a smectite, e.g., MMT, saponite, hectorite, mica, vermiculite, bentonite, with CEC = 0.9 to 1.5 meq/g. The intercalant mixture comprises alkyl or polyalkoxylated ammonium ions (*MW* = 0.6 to 1.1 kg/mol) and alkyl phosphonium ions, *viz.* tetrabutyl, trioctyl octadecyl, tetra-octyl, or octadecyl triphenyl.

Preparation of CPNC is carried out in two steps: (1) full exchange of clay Na^+ with a mixture of at least two organic cations in water at 50 to 80 °C, and (2) melt compounding matrix polymer with the intercalated clay. Alternatively, step (2) can be replaced by dispersing organoclay in a monomer, which then is polymerised. The process should result in exfoliation of clay, so 75 to 90% of the platelets will be individually dispersed or in the form of short stacks, about 10 nm thick.

The authors stated that for successful exfoliation, it is important that the melt compounding of clay with a polymer results in negative free enthalpy of mixing. This they tried to accomplish by pre-treating the clay with suitable intercalants. The use of mixtures of onium ions is aimed at achieving the most suitable balance of polar and non-polar interactions, as it is easier to blend available cations than to design and synthesise new ones. For example, an ethoxylated ammonium ion (EOA, e.g., Ethoquad-18/25 or Jeffamine-506) would reduce the clay content in organoclay below 50 wt% and cause serious de-watering problems. By using a 50:50 molar ratio of the EOA with octadecylammonium (ODA), the system is easier to handle and it leads to better exfoliation when melt compounded with a polymeric matrix.

The preferred polymeric matrices include those used in the multilayer structures, *viz.* PEST, PA, EVAl, etc. The preferred PEST and PA are, respectively, PET and MXD-6 or PA-66. The compounded mixture may also include such additives as colorants, pigments, carbon black, glass fibres, fillers, impact modifiers, antioxidants, stabilisers, flame retardants, reheat aids, crystallisation aids, acetaldehyde reducing compounds, recycling release aids, oxygen scavengers, plasticisers, nucleating agents, mould release agents, compatibilisers, etc. Prepared CPNC may be formed by standard processing methods into film, sheet, pipes, tubes, profiles, thermoformed articles, moulded articles, preforms, and containers, especially bottles.

As an example, Na-MMT was intercalated with a mixture of Jeffamine-506 (EOA) and Jeffamine-505 (oligo oxy propylene amine, POA), which resulted in a small expansion of the interlayer spacing, $d_{001} = 1.40 \pm 0.02$ nm (see **Figure 142**). Next, the dry mixture of PET with 1 wt% of the organoclay was extruder compounded, then formed into *ca.* 0.25 mm thick film. Results of the permeability measurements are shown in **Figure 142** as relative permeability, P_O/P_O(PET), versus POA content in the intercalating mixture. Considering that the interlamellar gallery height in dry organoclay was marginal, $h \cong 0.44 \pm 0.02$ nm, comparable to the diameter of a $-CH_2-$ unit of 0.45 nm, the reduction of the oxygen permeability by 22 to 35% is surprising. Unfortunately, information about the degree of exfoliation or about the effect of higher clay loading is not available. The small reduction of permeability is mainly due to the low clay content, but crystallisability of the matrix resulting in random orientation of the clay platelets may also contribute. Better results would be expected for MXD-6 as the matrix.

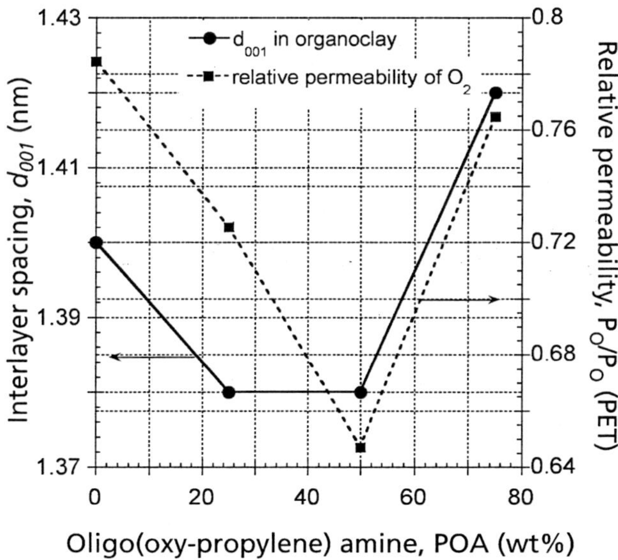

***Figure 142** Interlayer spacing in dry organoclay and relative oxygen permeability through a PET film containing 1 wt% organoclay versus POA content in the mixture with EOA used as intercalants (see text). Data [Lan et al., 2002a].*

Two recent patent applications from AMCOL are of utmost importance, especially for those interested in the use of organoclay for enhanced barrier properties. The patents focus on the preparation of CPNC with MXD-6 as the matrix [Lan *et al.*, 2002b, c]. The first of these specify that the new organoclays must have less than 2 wt% of quartz. Presence of this contaminant causes the formation of voids, which in turn (especially upon stretching) are responsible for haze and increased fluid permeability. In examples, new organoclay from Nanocor containing 0.33 wt% of quartz was used. The document shows that for MDX-6/organoclay nanocomposites the haze (H_a) linearly increases with quartz content (w_q), *viz.* $H_a = 7.1838 + 4.2452w_q$, with the correlation coefficient squared, $r^2 = 0.96$. The patent application is applicable for the preparation of CPNC with any polymeric matrix, but it specifically mentions polymers used to control permeability, *viz.* polyamides, PEST, EVAl, e.g., MXD-6 and PET. The objective of the second application is similar, but here salts (formed during ion exchange with Na-MMT) are extracted with either water or ethanol to reduced haze and gas permeability.

4.1.1.4 Other Technologies for the Production of CPNC with PA Matrix

Yasue *et al.* [1995] developed reinforced PA (PA-6, -66, -46, -12) composites with 0.1 to 20 pph of synthetic fluoromica (FM). The recommended process involves melt polymerisation of a monomer in the presence of non-intercalated mineral filler. When required, this process may be followed by a solid-phase polymerisation step. The stated aim was to prepare reinforced PA, free from the customary reduction of toughness in standard composites, and without the expensive and time consuming pre-intercalation, as is the case for CPNC with organoclays. The new material may also contain pigments, heat stabilisers, antioxidants, weathering agents, flame retardants, plasticisers, mould release agents and other reinforcements, so long as these do not seriously degrade the performance.

A large part of the document describes the preparation and properties of various types of FM. For example, the material may be obtained by heating a mixture of talc with sodium and/or lithium silicofluoride(s) or fluoride(s) under N_2 at T = 700 to 1,200 °C for 5 to 6 h. Thus, talc is the layered matrix material and Na and/or Li ions are intercalated into the silicate layers to engender the desired expanding characteristics. The amount of Na and/or Li silicofluoride(s) and/or fluoride(s) in the mixture with the talc is 10 to 35 wt%. The resulting FM has a general formula:

$$\alpha\text{-(MF)}\ \beta\text{-(aMgF}_2\ \text{bMgO)}\ \gamma\text{-SiO}_2$$

where M represents Na or Li, with α, β, γ, a and b being numerical coefficients provided that

$$0.1 \le \alpha \le 2;\ 2 \le \beta \le 3.5;\ 3 \le \gamma \le 4;\ 0 \le a \le 1;\ 0 \le b \le 1 \text{ and } a + b = 1.$$

The interlayer spacing measured by XRD for the suitable FM should be: d_{001} = 0.9 to 2.0 nm.

The composition of six FMs used in the studies is listed in **Table 66**. The mechanical strength, toughness, heat resistance, and dimensional stability of CPNC containing 5 wt% of these FMs are also given. In **Figure 143** values for dry CPNC containing 5 wt% of one of the six FM (M1 to M6) dispersed in PA-6 are displayed; the respective values for neat PA-6 are shown as well.

Table 66 Composition of fluoromica (FM) and performance of PA-6 and PA-66 comprising 0 or 5 wt% of FM. Data [Yasue *et al.*, 1995]

Polymer	PA-6							PA-66						
FM clay	Nil	M1	M2	M3	M4	M5	M6	Nil	M1	M2	M3	M4	M5	M6
Talc, wt%	-	80	80	80	80	80	80	-	80	80	80	80	80	80
Na_2SiF_6	-	20	-	6	10	16	6	-	20	-	6	10	16	6
Li_2SiF_6	-	-	20	-	10	2	6	-	-	20	-	10	2	6
K_2SiF_6	-	-	-	-	-	2	-	-	-	-	-	-	2	-
NaF	-	-	-	6	-	-	-	-	-	-	6	-	-	-
Al_2O_3	-	-	-	8	-	-	8	-	-	-	8	-	-	8
Performance of dry composition														
η_r	2.6	2.6	2.6	2.7	2.5	2.5	2.4	2.4	2.3	2.4	2.3	2.2	2.2	2.3
σ (MPa)	72.6	79.4	78.4	78.4	78.4	79.4	76.5	78.4	84.3	83.3	83.3	84.3	84.3	83.3
ε_b (%)	180	31	30	28	46	52	68	110	27	31	29	31	27	21
NIRT	3.3	3.5	3.1	3.4	2.9	3.8	2.7	4.6	4.2	4.4	4.3	4.5	4.2	4.4
HDT (°C)	55	121	120	122	116	121	118	74	153	150	150	155	153	154

Table 66 Continued...

Polymer	PA-6							PA-66						
FM clay	Nil	M1	M2	M3	M4	M5	M6	Nil	M1	M2	M3	M4	M5	M6
Performance of wet composition														
H_2O Absorp. (wt%)	**5.8**	3.1	3.0	3.1	3.0	3.1	3.1	**3.8**	2.1	2.0	2.0	2.0	2.1	2.0
Dimens. change (%)	**1.2**	0.7	0.7	0.7	0.6	0.7	0.6	**0.8**	0.5	0.5	0.5	0.4	0.5	0.5
σ(MPa)	33.3	58.8	57.8	57.8	58.8	58.8	59.8	**52.9**	67.7	68.6	69.6	67.7	67.7	66.7
ε_b (%)	>200	56	53	55	95	150	140	180	46	42	40	47	46	42
NIRT	**4.5**	4.5	4.7	4.5	4.8	4.2	4.4	**5.6**	5.2	5.0	5.2	5.1	5.1	5.0

Note: η_r - relative viscosity of PA at 25 °C, c = 1 g/dl in phenol/tetrachloroethane; σ= tensile strength, ε_b = strain at break, NIRT = notched Izod impact strength at room temperature, HDT = heat deflection temperature under load of 18.6 kgcm^{-2}

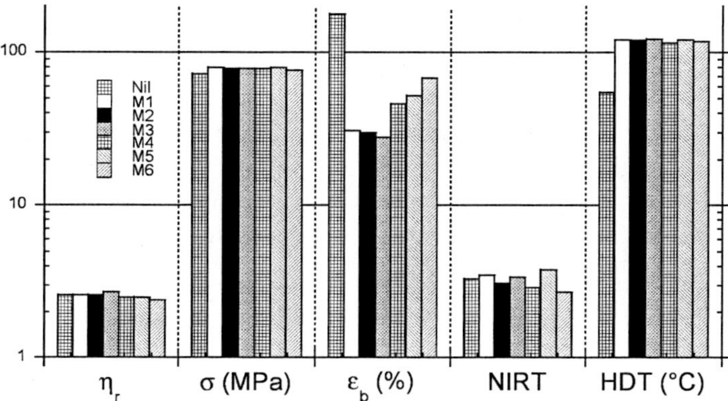

Figure 143 *Properties of PA-6 comprising 5 wt% of FM: M1 to M6. For explanation see text and data in* ***Table 66****. Properties are (from the LHS): relative viscosity (η_r); tensile strength (σ); elongation at break (ε_b); notched Izod impact strength (NIRT); and heat deflection temperature (HDT).*

It is evident that incorporation of FM improved the tensile strength, HDT, dimensional stability and reduced the water absorption. At the same time the relative viscosity (a measure of the PA molecular weight) and impact strength remained virtually unchanged, but the elongation at break was seriously reduced. In the absence of data on the interlayer spacing it is difficult to judge to what extent the polymer managed to expand the interlamellar galleries of FM. However, the cited values show a small improvement of strength and HDT. For 5 wt% clay loading larger effects would be expected for fully exfoliated CPNCs. Thus, most likely the non-intercalated FM is dispersed as aggregates, at best in the form of short stacks of platelets. The remarkable information provided in this patent is the ability of FM to disperse uniformly in the PA matrix without pre-treatment.

It is very unfortunate that information on the aspect ratio of these synthetic clays is missing. Of the six FM compositions tested M5 shows the best overall performance. It seems that incorporation of larger K-atoms has a beneficial effect. It would be interesting to explore these synthetic clays more thoroughly, using the well-established intercalation procedures.

Recently, Showa Denko patented rigid, flame-resistant PA CPNC [Inoue *et al.*, 2002]. The material comprises: (1) PA, (2) clay complexed with triazine, (3) fibrous reinforcements and (4) flame retardant.

Re 1. Any PA may be used, obtained by ring-opening polymerisation of a lactam or from polycondensation of di-acid with di-amine, having either aliphatic or aromatic moiety. While the patent focuses on PA, other thermoplastic resins (e.g., PP, PE, styrenics, PEST, POM, PC, PI, PPS, PES, LCP, fluoropolymers, silicone resin, TPU, etc.) may be blended in. Their amount, however, should not exceed *ca.* 60 pph of the principal resin – the PA.

Re 2. The clay complex is obtained by inserting a triazine ($C_3H_3N_3$) or its derivative into the interlamellar galleries. Suitable natural or synthetic clays include 2:1 and 1:1 type phyllosilicates, such as smectite (i.e., saponite, hectorite, MMT, beidellite, nontronite, stevensite, and the like), vermiculite, mica (phlogopite, biotite, lepidolite, muscovite, palagonite, chlorite, etc.), as well as kaolin, serpentine, and the like. The preferred triazine should also be a flame retardant for PA, e.g., melamine, cyanuric acid, and melamine cyanurate derivatives. The intercalant may be a mixture of a triazine and 0.1 to 2 mol of a Lewis acid per 1 mol of the triazine. The amount of the triazine intercalant is 0.3 to 5 times the equivalent amount of the clay CEC. The complexing takes place in a suitable solvent in which the ingredients are uniformly dispersed then reacted and dried. The organoclay is added to PA in amounts of 0.05 to 5 phr.

Re 3. The fibrous reinforcement may include whiskers (aluminium borate, silicon nitride, potassium titanate, etc.) and/or inorganic fibres (glass or carbon fibre, wollastonite, etc.). They may be surface treated with a coupling or 'sizing' agent. They are added to PA in amounts from 1 to 50 phr.

Re 4. The flame retardant may be based on a triazine, phosphate, metal, organic halide, etc. An auxiliary agent, e.g., Sb_2O_3, Fe_2O_3, and similar may also be added. The amount added to the formulation ranges from *ca.* 1 to about 40 phr.

The four basic components of the formulation may be dry blended (e.g., in a Henschel or ribbon mixer) then melt compounded in an internal mixer or a TSE. Alternatively, the mixing may be accomplished during polymerisation. Various conventional additives, such as antioxidants, UV absorbents, lubricants, antistatic agents, colorants, reinforcement, may also be added.

As an example the following procedure was described. 200 g of synthetic mica (Somasif ME-100, CEC = 0.8 meq/g) was dispersed in 10 L of water at 60 °C, the stoichiometric amount of melamine (1,3,5-triazine-2,4,6-triamine; $C_3H_6N_6$) and HCl were added and the mixture was stirred for 1 h. The precipitate was filtered, washed, dried, and ground to particles with diameter of *ca.* 5 μm. XRD confirmed expansion of the interlayer spacing from the initial (dry clay) value of d_{001} = 0.96 to 1.28 nm. The procedure was repeated with MMT (Kunipia-F; CEC = 1.19 meq/g). The organoclay with 15 wt% of melamine showed d_{001} = 1.30 nm. Larger values of d_{001} were obtained by either replacing melamine with a melamine cyanurate and/or by using excess intercalant. For the sake of comparison ME-100 was also intercalated with 2M2ODA. At 40 wt% of organic content the interlayer spacing was d_{001} = 3.5 nm.

The test results are summarised in **Table 67** for PA-66 with 1 phr organoclay and 15 phr of GF. The first four compositions are based on the invented intercalant (melamine or its derivative), the last three are for the reference. It is interesting that melt compounding of PA-66 with slightly intercalated clays (No. 1 to 4) gives better results that clay with 2M2ODA, the 'classical' quaternary ammonium intercalant (No. 5) – the thermal decomposition may be responsible for this behaviour. The first four compositions also outperformed the reference compositions as far as the flame resistance (V-0 according to the UL-94 standard), mould deposition, bleed-out and appearance are concerned. Owing to good appearance and ease of processing the new compositions may be used, e.g., for the electric and electronic parts, in automotive applications, in house appliances, mechanical parts, etc.

Table 67 Properties of PA-66 containing 1 phr of modified clay and 15 phr of GF. Data [Inoue *et al.*, 2002]

No.	Organoclay or other filler	d_{001} (nm)	Flex modulus, E (MPa)	HDT (°C)	Deformation (mm)	Relative shrinkage (TD/MD)
1	ME + m	1.28	6.3	245	0.5	1.94
2	MMT + m	1.30	6.1	244	0.7	1.86
3	ME + mc	1.50	6.2	244	0.7	1.96
4	ME + 2xm	1.35	5.9	240	1.0	2.12
5	ME + 2M2ODA	3.5	5.8	230	1.2	1.88
6	Talc	0.96	6.0	243	1.0	2.13
7	nil	-	5.6	244	6.2	2.30

Notes: ME = synthetic clay, Somasif ME-100; MMT = Kunipia-F; m = melamine; mc = melamine cyanurate; 2M2ODA = dimethyl-di-octadecyl ammonium chloride; 2x = twice stoichiometric

The similarity of approaches used by Inoue *et al.* [2002] and by Yasue *et al.* [1995] is noteworthy – there is no effort to secure high polymer-clay interaction that may lead to exfoliation. The small enhancement of the mechanical properties for both cases suggests intercalation. The ease of preparation of these materials and their good processability makes them industrially attractive. Finally, the use of melamine as an intercalant by Inoue *et al.* [2002] is an important development. This is the first time the component specifically designed to provide enhanced stability (flame retardance) has been used as intercalant, replacing the notoriously unstable quaternary ammonium intercalants. This promising route is very much worth pursuing.

Melt compounding was also used by Liu *et al.* [1999] to prepare CPNC with PA-6 as matrix. The organoclay was prepared by intercalating Na-MMT (CEC = 1.0 meq/g) with ODA. The interlayer spacing increased from d_{001} = 0.98 to 1.55 nm. Next, dry PA-6 was melt compounded in a TSE (30 rpm, T = 180, 210, 230 and 220 °C), pelletised, dried and injection moulded into test specimens. Processing increased the interlayer spacing to d_{001} = 3.68 nm and the XRD peak was depressed, thus partial exfoliation took place. The mechanical properties of CPNC with 4.2 wt% of organoclay were found significantly improved over those of PA-6 (see **Table 68**). In accord with previous documents the authors also observed increased γ-crystalline PA-6 content upon addition of organoclay. Similar results were reported by Dahman [2000] for melt exfoliated PA-6.

Cheng *et al.* [2000] reported a significant difference in suitability between CPNC and PA-6 for the preparation of microporous membranes. While neat PA-6 generated asymmetric membranes with tight skin and porous sublayer, the intercalated mica/PA-6 system produced skinless microporous membranes with open, bicontinuous structure that could easily be controlled. Hence, while

Table 68 Effect of 4.2 wt% organoclay on the performance of PA-6. Data [Lui *et al.*, 1999]

Property	PA-6	CPNC	Ratio: CPNC/PA-6
Yield strength, σ_Y (MPa)	68.2	91.3	1.34
Tensile modulus, E (GPa)	3.0	4.1	1.37
Flexural strength, σ_f (MPa)	93.5	150	1.60
Flexural modulus, E_f (GPa)	2.4	4.2	1.75
Notched Izod impact strength, *NIRT* (J/m)	28.0	26.0	0.93
HDT (°C/K) at 1.82 MPa	62/335	112/385	1.15

microfiltration membranes produced by the standard method from PA-6 were useless, the CPNC engendered perfectly adjustable, useful products.
Katahira *et al.* [1998a-d] published a series of articles on the use of mica for the production of CPNC based on PA-6. Thus, Na-mica flakes were cleaved and dispersed into hydrolysed and protonated ε-caprolactam (ε-CL). Phosphoric acid has been used as a catalyst for the protonation. The intercalation took place in two steps: a rapid (even at 20 °C) exchange of Na^+ in mica galleries with protonated ε-CL^+, then a significantly slower exchange of Na^+ ions solvated by water. The latter step could be accelerated by heating to $T > 60$ °C. The intercalated mica had the interlayer spacing of d_{001} = 1.47 nm.

During the polymerisation the pre-intercalated mica exfoliated. High bending strength and modulus were obtained for the CPNC as a result of the cleaved thin layers with a high aspect ratio and good dispersion. The CPNC preparation consisted of three steps. In the first, ε-CL was protonated. In the second, mica was intercalated by ion exchange of Na^+ with ε-CL^+. In the third, ε-CL was polymerised at $T > 260$ °C and PA-6 formed within the mica galleries caused exfoliation.

A CPNC layer was used to reduce permeability through a polymer-paper laminate for packaging applications [Shih *et al.*, 2000]. The CPNC was a PA-6/smectite type. Usually, the multilayer laminate is dominated by PO, thus a tie-layer material between PO and PA may be required. A more recent patent from the company [Adur *et al.*, 2002] focuses on the preparation of CPNC with improved barrier properties, useful in the manufacture of liquid food containers with extended shelf life. Since the material is to be used to laminate paperboard the specifications are relatively narrow. Thus, 30 to 50 wt% of clay (MMT, bentonite or kaolinite, either neat or treated with amino-propyl tri-ethoxy silane or vinyl triethoxy silane) was melt compounded with a polymer (EVAc, EVAl, PA, PEST, PO, etc.). Optionally, 1 to 15 wt% of a coupling agent, dispersion aid and/or compatibiliser might be used (e.g., maleic anhydride, ethylene-co-acrylic acid, ethylene-co-methacrylic acid, ethylene-co-maleic anhydride, PO-grafted with acid or anhydride or other acid derivative functional group, etc.). Drying the materials may not be necessary as compounding in a TSE provides for evacuating the volatiles, but dried compositions are easier to process.

The CPNC was directly applied to the paperboard substrate (as a film 1 to 30 μm thick) or with a tie-layer, causing a reduction of O_2 and water vapour permeability and enhancing mechanical properties. While the preferred CPNC of this invention has EVAl as the matrix, examples with PA and LDPE were also cited. One of these specified that 25 wt% of clay was melt compounded into EVAl with ethylene-co-acrylic acid as compatibiliser. The dispersed platelets (1 to 2 nm thick, aspect ratio p = 1000) reduced O_2 permeability by 76% and that of water vapour by 14%. At the same time the tensile modulus increased by 30%. However, judging by the quoted properties in this and other cited examples, exfoliation has not been achieved.

Fischer and Gielgens [2002a] described a new nanocomposite based on anionic clays, known as layered double hydroxides (LDH, with CEC = 0.5 to 6 meq/g) having the chemical formula:

$$\left[M_{l-x}^{2+}M_x^{3+}\left(OH\right)_2\right]\left[A_{x/y}^{y-}\times nH_2O\right]$$

where: M^{2+} = Mg^{2+}, Zn^{2+}, Ni^{2+}, Fe^{2+}, Cu^{2+}, Co^{2+}, Ca^{2+}, or Mn^{2+}; M^{3+} = Al^{3+}, Fe^{3+}, Cr^{3+}, Co^{3+}, Mn^{3+}; x = 0.15 to 0.5; y = 1 to 2; n = 1 to 10; and A = Cl^-, Br^-, NO_3^-, SO_4^{2}, CO_3^{2}. Suitable LDH may be a hydrotalcite or a hydrotalcite-like material. Before compounding with a polymeric matrix (≥ 70 wt%) LDH must be treated with organic intercalants, so at least 20% of the anions are either miscible or reactive with the matrix, and at least 5% of these anions contain a second charge-carrying group. Anions of carboxylic, sulfonic and phosphonic acids as well as hydrogen sulfates may be used. These anions may contain the alkyl or phenyl group with a reactive group, *viz.* hydroxy, amino, epoxy, vinyl, isocyanate, carboxy, hydroxy phenyl and anhydride. The second charge-carrying group is cationic, e.g., ammonium, phosphonium, sulfonium, carboxylate, sulfonate, phosphonate and sulfate. It is expected that the second charge-carrying group repels the LDH sheets, thus helps the exfoliation process. The polymeric matrix may be selected from between: PA, PO, PEST, PC, TPU, vinyl polymers, and polyepoxides. The CPNC of this invention may be prepared by either melt compounding organo-LDH with molten polymer in a TSE or by mixing it with a monomer, which subsequently is polymerised. One of the authors' objectives was to provide a method for preparing CPNC with homogeneous composition, controllable quality and composition, *viz.* superior heat resistance, mechanical and impact strength. Since LDH may be synthesised, reproducibility of performance can readily be accomplished. The prepared CPNC may be shaped by conventional means, into fibres, packaging and construction materials.

The following patent [Fischer and Gielgens, 2002b] provided an example of the process. First, a hydrotalcite: $[Mg_6Al_2(OH)_{16}][CO_3 4H_2O]$ (CEC = 4 meq/g) was dispersed in an aqueous solution of α,ω-amino-undecanoic acid at 80 °C. The components were allowed to react for 3 h while mixing. The resulting precipitate was washed, freeze-dried, and then mixed with 90 g of ε-CL and 10 ml water. To effect polymerisation the mixture was heated to 260 °C and stirred under dry N_2 for 6 h. A transparent melt of a PA-6 nanocomposite was obtained, indicating exfoliation, confirmed by XRD.

One of the oldest companies involved in the intercalation of clay was National Lead Co., transformed into Rheox Inc. and recently into Elementis Specialties, Inc. For example, Carter *et al.* [1950] patented intercalation of clays with ammonium or phosphonium cations, e.g., dodecyl amine, aniline, melamine, pyridinium, etc. A recent patent from Rheox Inc. [Ross and Kaizerman, 2002] describes a method of preparation of CPNC designed primarily as rheological additives for liquid organic systems (e.g., paints and coatings) providing viscosity modification and levelling properties. The new system comprises:

(a) smectite clay,

(b) one or more quaternary ammonium intercalants, and

(c) a polymer (e.g., PA, PEST, PC, PO, TPU, P4VP, PVAl, PEG, PPG, epoxy, etc.) able to intercalate the clay.

The new process is a response to the perceived dilemma of the industry: the existing methods for the preparation of organoclays lead to materials that are either easy to process and isolate, but are difficult to disperse in a matrix, or they have improved dispersing characteristics but are inconvenient to manufacture. For example, organoclays prepared by intercalation of smectite with quaternary ammonium salts are easy to separate and wash, but they are difficult to exfoliate during compounding with polymers. Beall *et al.* [1996] described clays intercalated with *water-soluble polymers* that may be miscible with most polymers. However, to isolate them one must boil off water (expensive), incorporating into the product all contaminating ingredients (*viz.* salts, non-bonded to clay polymer, etc.). Similarly, the reactive method for the production of CPNC may result in well-dispersed clay in the polymeric matrix, but this technique requires that an expensive polymerisation line be dedicated to the process and plant contamination with clay may become an issue. According to Ross and Kaizerman, the hydrophobic organoclay prepared according to the new process may be washed with water to remove reaction salts and excess organic materials then filtered out to give a clean inexpensive organoclay, which may readily be modified to ascertain good compatibility with the matrix.

Re (a) The clays considered by the patent are 2:1 type smectites, having CEC = 0.5 to 1.0 meq/g, e.g., MMT, bentonite, hectorite, saponite, stevensite, beidellite and their mixtures.

Re (b) The low molecular weight intercalant is a quaternary ammonium salt, which may be derived from natural oils such as tallow, soya, coconut, palm, corn, soybean, etc. Aliphatic and aromatic groups may also be incorporated. The following quaternary ammonium salts were found useful with the R_i groups listed in **Table 69**:

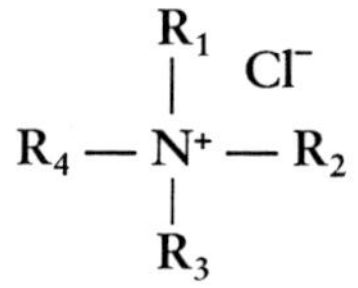

$$\begin{array}{c} R_1 \quad Cl^- \\ | \\ R_4 - N^+ - R_2 \\ | \\ R_3 \end{array}$$

The preferred quaternary ammonium cations are 2M2HTA and MS2EtOH. The nanocomposite comprises 3.5 to 12.0 wt% of the organoclay.

Re (c) The non-anionic organic materials can be either liquid or melt, provided that it is capable of intercalating clay, e.g., PA, TPU, PEST, PC, PO, epoxy, PEG, PPG, P4VP, PVAl, aliphatic polyesters, styrenics. However, the polyester amide copolymers (Thixatrol®) manufactured by Rheox are preferred.

Table 69 Quaternary ammonium cations used for intercalation. Data [Beall *et al.*, 1996]

Quaternary ammonium cation	Code	Radicals			
		R_1	R_2	R_3	R_4
Dimethyl dihydrogenated tallow	2M2HTA*	$-CH_3$	$-CH_3$	HT	HT
Methyl stearyl bis[2-hydroxyethyl]	MS2EtOH	$-CH_3$	$-C_2H_4OH$	$-C_2H_4OH$	$-C_{18}H_{37}$
Dimethyl dibehenyl	2M2BhA	$-CH_3$	$-CH_3$	$-C_{22}H_{45}$	$-C_{22}H_{45}$
Methyl tris[hydrogenated tallow	M3HTA	$-CH_3$	HT	HT	HT

*Note: *HT = hydrogenated tallow; Bh = behenyl ($C_{22}H_{45}$)*

The organoclay may be prepared by a standard method, *viz.* dispersing 2 to 8 wt% of smectite in water, centrifuging to remove contaminants, and then adding non-anionic organic material and mixing until a clay-organic intercalate is formed. Next, the mixture is heated and a quaternary ammonium salt is added to form the final organoclay, which is filtered and washed with water. Alternatively, the clay may be compounded with a non-anionic organic material and some water in a TSE. After this intercalation step, the quaternary ammonium salt is added and then the composition is washed and dried. These procedures may be modified by either reversing the order of addition, or by simultaneously adding the two components to the clay.

The final organoclay can be used with virtually any polymer. However, those preferred have polar functional groups, e.g., hydroxyl, urethane, ester, amide acid, ketone, aldehyde, halide, cyanide and thiol functionality. The listed examples include homopolymers or copolymers of PEST, PA most particularly PA-6, polyethers, PC, POM resins and their blends. While the concentration of organoclay depends on the particular system and application, the most preferred range is 3.5 to 12.0 wt%. The final method of CPNC preparation is direct batch or continuous type compounding, e.g., in a TSE. These CPNC show improved tensile modulus, tensile strength, gas barrier and heat distortion temperature.

As an example, 2.8 wt% of a pre-hydrated aqueous hectorite slurry was mixed with 30 wt% (on dry weight of clay) of a non-anionic organic material at 70-80 °C with stirring. Several polymers or copolymers were used as the non-anionic organic material, e.g., poly(vinylpyrrolidone-co-acrylic acid), poly(vinyl methyl ether-co-maleic anhydride), poly(vinyl formal), PEG, PVAl, PA, TPU, PDMS, polyacrylamide, polyester amide, and a polyester. Next, a solution of alkyl quaternary ammonium compound (MS2EtOH or 2M2HTA) in isopropanol was added to react up to 100% of CEC. The product was filtered and re-dispersed in water, re-filtered, dried and ground into powder. According to XRD the d_{001} spacing varied from 1.93 to 4.39 nm. It has been observed that incorporation of the non-anionic organic materials caused the additional (over that caused by the quaternary ion) expansion of the interlayer spacing in these organoclays by Δd_{001} = 0.28 to 2.55 nm. Some of these organoclays were melt compounded with PETG using a roll mill at 177 °C for 2 min. Excellent dispersion was obtained using hectorite intercalated with 2M2HTA and polyamide or with MS2EtOH and polyester amide. By contrast, the CPNC prepared with traditional organoclays (without non-anionic organic material) or with clays containing a polymer, but not quaternary salt gave poor dispersion.

4.1.1.5 Mechanical Exfoliation of PA-Type CPNC

Several patents discussed above describe two alternative methods for the preparation of CPNC: reactive, involving monomer(s) polymerisation in the presence of organoclay, or direct compounding of PA with the latter compound.

Currently, several companies around the world manufacture CPNC, *viz.* AlliedSignal (now Honeywell), Bayer, BASF, Dow, DuPont/ICI, Eastman, Mitsubishi, Montell, RTP Co., Showa Denko, Solutia, Toyota, Ube Intl., Unitika, etc. However, little is known as far as the actual method of production is concerned. For example, judging by the patent literature one may suspect that AlliedSignal is using a 'hybrid' method, Ube Intl. a reactive process, etc. However, it is fairly certain that RTP Co., being a well-recognised and innovative compounding company, uses mechanical exfoliation.

Several years ago [Jommersbach, 1999] announced the first commercial CPNC produced by compounding. The grades contained 3 to 8 wt% of organoclay. The announcement is skimpy on technical details, but it mentions that the clay surface area exceeds 750 m^2/g, suggesting modified MMT. The increase in HDT over unfilled PA-6, comparable to a 20-30% load of a standard mineral-filled compound was considered as particularly worth attention in view of under-the-hood automotive applications. Furthermore, the reduction in heat release rates and enhanced barrier properties against moisture, gasses, and fragrances were listed as of particular value. The new product was designed for film or sheet extrusion. As shown in **Table 70**, the direct compounding method did indeed significantly enhance the mechanical performance of the PA-6 matrix.

Table 70 The mechanical performance of PA-6, its CPNC and composites with mineral filler and glass fibres. Data: RTP Co.

Properties	Unfilled PA-6	PA-6 + 4 wt% organoclay	PA-6 + 30 wt% mineral	PA-6 + 10 wt% GF
Tensile strength, σ_b (MPa)	69	92	65	96
Flexural modulus, E_f (GPa)	2.9	4.3	4.8	4.1
HDT (°C)	65	102	92	190
Density, ρ (kg/m^3)	1130	1140	1380	1210

Further details were provided by Solutia [Lysek *et al.,* 1999]. The patents describe melt compounding of PA (MW > 40 kg/mol) with organoclay pre-intercalated with a mixture of two or more (quaternary) ammonium ions and optionally treated with γ-amino-propyl tri-ethoxy silane. The compounding may be carried out in diverse processing equipment, *viz.* SSE, TSE, internal mixer, Farrel Continuous Mixer (FCM), etc. The invention may also be applied to copolyamides as well as to PA-blends.

Li *et al.* [2000] developed CPNC comprising a blend of a thermoplastic engineering resin (PA, PI, PEST, PC, PSF, POM, styrenics, PPE, etc.) with 5 to 35 wt% of an elastomeric functionalised copolymer, and 0.5 to 10 wt% of a smectic clay selected from: MMT, nontronite, beidellite, etc. The key to success is the selection of copolymer. The preferred one was a brominated copolymer of isobutylene with 0.5 to 20 mole % of *para*-methylstyrene (Exxpro™):

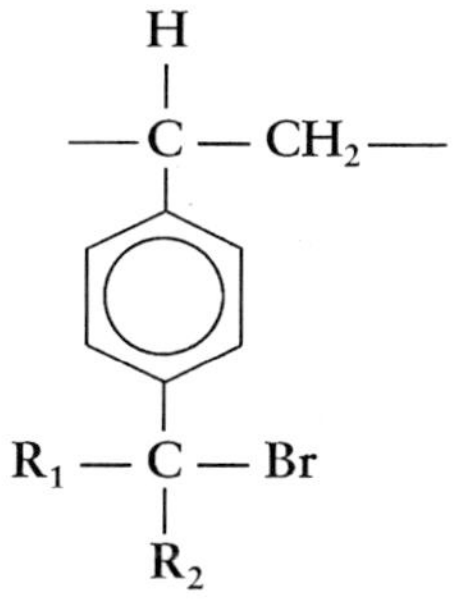

H
|
—C— CH_2—
|
(benzene ring)
|
R_1 — C — Br
|
R_2

The new process provided rubber-toughened thermoplastic CPNC, which had not only superior tensile strength and modulus, but also markedly improved impact strength. The compositions were also relatively thermally stable at processing and moulding temperatures of up to about 300 °C as well as more resistant to oxidative degradation than similar compositions containing conventional elastomeric impact modifier based on polymers with unsaturated double bonds.

The invention particularly favours PA having T_g or T_m = 160-230 °C, thus preferably PA-6 or its copolymer PA-6/66. Furthermore, the focus is on impact modification not exfoliation. For these reasons, the authors used commercially available PAs and CPNCs from Ube Intl., *viz.* PA-6 (PA-1022B), PA-6/2 wt% organoclay (PA-1022C2), PA-6/66 = 80/20 wt% copolymer (PA-5034B) and PA-6/66 with 2 wt% organoclay (PA-5034C2). These materials were extruder-blended (TSE-CORI under a stream of N_2) with 10, 20, or 30 wt% of the Exxpro elastomer. The extruder operated at 400 rpm and at T = 200-230 °C for PA-6/66 and 220-250 °C for PA-6. The extruded pellets were dried then injection moulded into specimens for the mechanical tests. The results are presented in **Table 71**.

The data show that the flexural modulus as well as the tensile yield strength decreased with increasing levels of elastomer. The Izod impact strength (NI) at 18 °C markedly increased with addition of elastomer. At elastomer levels ≥ 15 wt% the impact strength of the rubber-toughened PA-6/66 nanocomposite rapidly approaches values for the rubber-toughened PA. The NI of the rubber-toughened PA-6 nanocomposite followed the values for the rubber-toughened PA-6 blend. Thus, the CPNCs show high impact strength as well as high stiffness, i.e., modulus

Table 71 Effect of incorporation of Exxpro™ into PA-6, PA-6/66 and their CPNCs. Data [Li *et al.*, 2000]

PA-phase content*	*NI* (18 °C) (kJ/m²)	*NI* (0 °C) (kJ/m²)	*NI* (-20 °C) (kJ/m²)	*E* (GPa)	σ_y (MPa)	σ_{max} (MPa)	ε_b (%)
PA-6 100%	0.140	0.0876	0.0700	2.455	60.00	68.97	300.0
PA-6 90%	0.700	0.263	0.2277	2.055	48.28	52.41	209.5
PA-6 80%	3.538	0.525	0.4378	1.669	42.07	47.59	246.8
PA-6 70%	3.853	2.679	0.4203	1.297	32.41	33.79	137.7
PA-6+2% clay 100%	0.070	0.052	0.0700	3.448	79.31	79.31	189.0
PA-6+2% clay 90%	0.420	0.263	0.2977	2.621	57.93	57.93	150.8
PA-6+2% clay 80%	3.082	0.350	0.2452	1.965	42.07	45.52	228.7
PA-6+2% clay 70%	4.116	1.0508	0.5254	1.359	34.48	34.48	148.6
PA-6/66 100%	0.140	0.0700	0.0525	2.014	51.72	55.17	241.3
PA-6/66 90%	2.679	0.1926	0.1226	1.655	40.00	61.38	311.5
PA-6/66 80%	3.835	0.5254	0.4203	1.317	32.41	51.03	283.6
PA-6/66 70%	4.343	3.2750	0.5429	1.076	28.28	46.21	290.3
PA-6/66+2% clay 100%	0.088	0.0525	0.0350	2.186	59.31	71.03	314.5
PA-6/66+2% clay 90%	0.560	0.0700	0.0876	1.972	40.00	60.00	283.0
PA-6/66+2% clay 80%	3.538	0.2802	0.2277	1.400	33.79	51.03	289.2
PA-6/66+2% clay 70%	4.133	0.9107	0.4378	1.152	28.28	40.69	239.1

*Notes: *The composition is given in wt% of the PA-phase - the rest being the elastomeric copolymer, Exxpro™; NI = notched Izod impact strength at 18, 0 and -20 °C, Y = flexural modulus; σ_y = yield stress; σ_{max} = maximum stress; ε_b = maximum strain at break*

and high tensile properties. The impact strength of the nanocomposites with the elastomer essentially retained its value over a broad temperature range.

Singh and Haghighat [2000] patented a new method for the production of high-use temperature organoclays with PA, PEI, PI, PPE, etc. The patent is in two parts, the first describing the new, high temperature intercalant to be used with any anionic clay, the second introduces a new type of synthetic clay, obtained through sol-gel conversion. The new, intercalant is tetraphenyl phosphonium (TPP) chloride shown in **Figure 144**. While the patent describes organoclays with TPP, for some applications modification by the introduction of aliphatic and/or

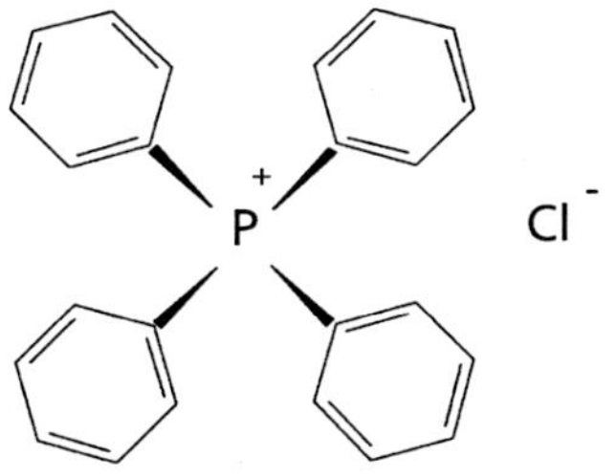

***Figure 144** Tetraphenyl phosphonium (TPP) chloride*

polymerisable organic groups may offer additional advantages, such as reactive end-tethering or improved matrix miscibility.

According to the authors, CPNC with the new intercalant shows increased thermal stability to at least 250 °C. This improvement is achieved without affecting other physical properties. The higher thermal stability facilitates the processing and recycling of CPNC. Furthermore, TPP 'eliminates the poisonous alkyl ammonium' intercalants that limit the applications of CPNC for food packaging. Using standard processing methods the new CPNC may be formed into films, fibres or mouldings for the automotive, aerospace, electronic and food and beverage industries.

For example, CPNC of PEI with organoclay (MMT pre-intercalated with TPP through a cation-exchange) has been prepared by direct melt compounding. Similarly, extrusion-blending of PA-6 with 4 wt% of the new organoclay significantly improved performance (see **Table 72**). For example, incorporation of the new organoclay enhanced the tensile modulus and strength by 91 and 55%, respectively, with slightly (by 22%) improved impact strength. The significant increase in HDT extends the temperature range of these CPNCs to under-the-hood applications in automobiles. A reduction in water adsorption accompanies the improvements in mechanical properties. The key for obtaining superior CPNC properties at low clay loadings is homogeneous exfoliation and the presence of favourable interactions at the organic-inorganic interface.

The second part of the patent deals with the 'organically modified layered alumino-silicates' (or ORMLAS), which combine the layered silicate and the organic intercalant/compatibiliser in a single compound and extends the use-temperature of CPNC to over 250 °C. The ORMLAS are prepared by the sol-gel process where the organic groups are incorporated into the molecular structure through the use of organically modified silicon precursors containing $Si–C_xH_y$ bonds. The organic functionality is directly bonded to the structural Si atom. The standard procedure is to combine an alcohol solution of aluminium chloride with an alcohol solution of an organically functionalised alkoxysilane, e.g., alkyl tri-ethoxysilane. The mixture is crosslinked to form a gel at appropriate pH, the gel is aged, filtered, washed and then vacuum dried. The resulting organophilic material is precipitated as a powder, and screened for incorporation into a polymeric matrix for direct melt intercalation. The ORMLAS may be specifically designed to delaminate in the presence of a specific type of polymer. They yield nanocomposites that extend the range of the long-term use-temperatures to $T > 250$ °C. ORMLAS are dispersible in a variety of polymers, such as PA, PEI (Ultem, $T_g = 215$ °C), PI ($T_g > 275$ °C) or PAE ($T_m > 325$-350 °C).

Table 72 Properties of PA-6 and its CPNC with 4 wt% organoclay. Data [Singh and Haghighat, 2000]

Property	PA-6	CPNC of PA-6
Tensile modulus, E (GPa)	1.1	2.1
Tensile strength, σ_b (MPa)	69	107
HDT (°C)	65	145
Impact strength, $NIRT$ (kJ/m^2)	2.3	2.8
H_2O abosrption, P_{H2O} (%)	0.87	0.51
Thermal expansion, α (μm/m)	13	6.3

In Section 2.3.8 melt compounding of PA-6 with 5 wt% of either Cloisite® 15A or Cloisite® 30B was discussed [Dennis *et al.*, 2000, 2001; Fornes *et al.*, 2001]. The authors remarked that, since melt blending of PA-6 with C30B resulted in easy exfoliation, the studies were conducted with PA-6/C15A as a model system. Evidently, MMT intercalated with methyl tallow bis[2-hydroxyethyl] ammonium can readily be exfoliated during compounding in a TSE. This observation underlines the importance of the thermodynamic interactions.

Another patent that specifically describes the preparation of CPNC by melt exfoliation originates from DSM [Korbee and van Geenen, 2002]. The invention focused on the preparation of PA-6 nanocomposites by mixing molten polymer, 0.05-30 wt% solid anisotropic particles (layered, preferably MMT, or fibrous, for example imogolite or V_2O_5) with a high aspect ratio (p = 100 to 10,000), and a liquid, e.g., 10-40 wt% water. The mixing was carried out in a co-rotating TSE, equipped with a liquid injection gate and degassing port. The pressure within the kneading zone was 1-1.5 MPa. The new process comprises fewer steps, it is more economic as it is not necessary for the clay to be pre-intercalated. Furthermore, the compounding with wet PA takes place at temperatures well below the T_m of dry PA, is facilitated by low viscosity, uses less energy and less shear heat is developed in the melt. As a result, there is a smaller risk of gel formation and degradation; hence a better end product is obtained.

The liquid that is to be used in the process must be miscible with molten polymer, capable of expanding the interlamellar gallery and sufficiently volatile for removal in the terminal part of the compounding TSE. For these reasons, the liquid is chosen from between: water, aliphatic, aromatic and cycloaliphatic hydrocarbons, dioxane, tetrahydrofuran, cyclohexanone, ethyl acetate, acetonitrile, ethanol, etc. Particularly suitable is water or an aqueous solution of the aforementioned solvents. Surprisingly, the process using water causes no substantial degradation of PA. While PA-6 is preferred, the patent stresses its applicability to all polymers with –CONH– bonds between the repeating (aliphatic or aromatic) units. Its molecular weight should be from M_w = 9 to 40 kg/mol. Optionally, the PA may be blended with, e.g., PP, ABS, PPE, PC, PEST, etc. Other additives, for example fillers and reinforcing materials, flame retardants, foaming agents, stabilisers, or pigments may also be incorporated. The CPNC obtained according to this method may be formed by injection moulding into automotive parts that demand particularly good mechanical properties, e.g., for casings and connectors.

As an example, PA-6 was cryogenically ground to a particle size of less than 1 mm, dry blended with 5 wt% of MMT then fed at 4 kg/h to a ZSK TSE (D = 30 mm, L/D = 39) where it was compounded at T = 190 to 215 °C. Water was injected into the melt at 20 g/min and it was discharged via the degassing gate at the end of the extruder at a pressure varying from 3 to 6 kPa. According to TEM, uniform dispersion of the clay platelets was obtained. The HDT of PA-6 and its CPNC with 5 wt% MMT were, respectively, 96 and 152 °C.

Similarly, PA-66/clay nanocomposites were produced by Liu and Wu [2002a,b] by melt exfoliation. First, the organoclay was prepared by co-intercalation of quaternary ammonium and epoxy into Na-MMT. Thus, clay (CEC = 0.80 meq/g) was dispersed in hot water then a solution of trimethyl hexadecyl ammonium bromide (3MHDA) was added. The precipitate was dried, ground and mixed in an internal mixer with a diglycidyl ether of bisphenol-A (DGEBA, MW = 360). The final organoclay composition was: MMT = 67, 3MHDA = 20 and epoxy = 13 wt%. The interlayer spacing of Na-MMT, its adduct with 3MHDA and the organoclay was determined as: d_{001} = 1.24, 1.96 and 3.77, respectively. The epoxy evidently diffused into the interlamellar galleries, thus a chemical reaction or at least some favourable interaction between MMT-3MHDA and epoxy must have occurred. Unfortunately, such a mechanism has not been identified.

The melt exfoliation was carried out in a TSE operated with a screw speed of 180 rpm, at T = 270, 290, 290 and 280 °C from hopper to die, respectively. Well-dried PA-66 was compounded with 0 to 10 wt% of organoclay. The dispersion was studied by means of XRD and TEM. The CPNC containing less that 7 wt% organoclay showed good exfoliation (no XRD peak), but at higher concentration a small, broad peak indicated an intercalation. The tensile strength and modulus of these CPNCs increased with clay loading by up to 37 and 77%, respectively. Similarly HDT increases from 74 to 167 °C (at 10 wt% organoclay). All three dependencies followed the 2nd order polynomial (see **Figure 145** and **Figure 146**). By contrast, the notched Izod impact strength of CPNC reached a local maximum (about 50% higher than that of PA-66) at *ca.* 5 to 6 wt% organoclay loading. The rate of water absorption was drastically reduced. For example, at 10 wt% organoclay loading it was reduced from 7.6 (for PA-66) to 4.7% for CPNC. Evidently, this property is related to the diffusion coefficient, which decreased with clay loading by about 40%. Finally, it was observed that incorporation of clay affected the crystalline structure of PA-66 – besides the α-crystals present in neat polymer, a γ-form was observed in CPNC.

More recently, Uribe-Arocha *et al.* [2003] used a TSE to compound PA-6 with Cloisite® 30B at 250 °C. The authors did not measure the degree of dispersion, but the system is known to easily exfoliate [Dennis *et al.*, 2000, 2001], thus good dispersion may be assumed. The aim of the work was to study the effect of the specimen thickness on mechanical properties. The specimens (0.5, 0.75, 1.0 and 2.0 mm thick) were injection moulded using either neat PA6 or its CPNC with 5 wt% C30B. For PA-6 DMA and tensile tests showed a small influence of the specimen thickness on the performance. By contrast, for CPNC the effect of specimen thickness on the mechanical and thermomechanical properties were severe (see Section 3.51). The authors interpreted this observation on the basis of skin/core morphology: skin formed of well aligned clay platelets and core with randomly oriented ones.

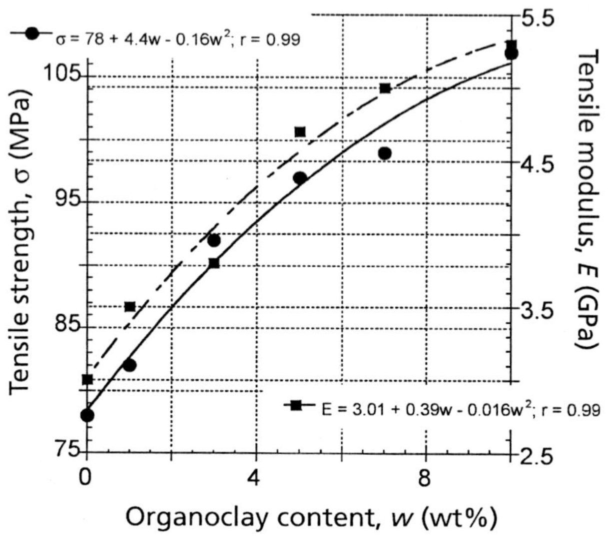

Figure 145 *Tensile strength and modulus of PA-66 as a function of organoclay (MMT-3MHDA-DGEBA) content. Lines are second-order polynomials. Data [Liu and Wu, 2002].*

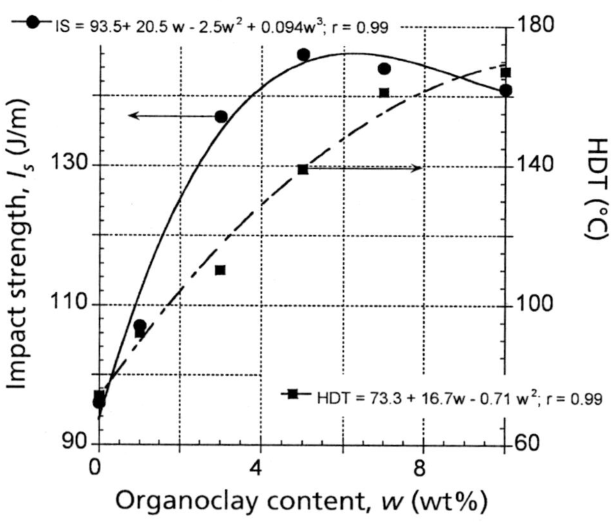

Figure 146 *Izod impact strength and HDT of PA-66 versus organoclay (MMT-3MHDA-DGEBA). The lines are 3rd and 2nd order polynomials. Data [Liu and Wu, 2002].*

After the initial straining, the sample started to neck. SEM of such specimens showed void formation in the core that eventually resulted in failure.

To accommodate a wide range of performance characteristics the plastics industry makes use of blending technology [Utracki, 2002]. With growing frequency this approach is also being used for CPNC. For example, George and Ballard [2003] applied for patent protection for CPNC with PA/poly(ether-*b*-amide) (PEBAX) blend as the matrix. The resulting compositions are to be used for the production of small diameter catheters. The organoclay (I.42) may be incorporated into either phase, *viz.* PA (PA-6, PA-11 or PA-12), or PEBAX as well as into both of them. By proper selection of the polymer and copolymer grades, judicious choice of concentration and location of organoclay, the performance may be tuned to a specific application.

4.1.1.6 PA-6/Kaolinite Nanocomposites

By contrast with the smectite clays most commonly used in CPNC (2:1 type, with two tetrahedral SiO_4 sheets on both sides of an octahedral $AlO_2(OH)_4$ sheet), kaolinite is a 1:1 type clay, characterised by large aspect ratio. It is composed of alternating SiO_4 tetrahedral and $AlO_2(OH)_4$ octahedral sheets. Accordingly, polymer/kaolinite nanocomposites may show different performance from those of polymer/smectite CPNC. Unfortunately, the low CEC = 0.02 meq/g of kaolinite makes the standard method of intercalation difficult.

Komori *et al.* [2000] demonstrated that kaolinite might be intercalated by methanol. In the process the interlayer spacing expanded from d_{001} = 0.72 to 1.11 nm. Drying reduced this value to 0.86 nm and extensive drying under vacuum at 150 °C to 0.82 nm. The latter kaolinite did not contain free water or methanol molecules and the expansion of d_{001} by 0.10 nm resulted from the presence of chemically bonded methoxy groups, proved by the ^{2}H NMR (the location was assigned by ^{13}C CP/MAS NMR). From the difference in the CH_3O- line width the formation of Al-O-C bonds was concluded:

$$= Al - OH + HO - CH_3 \Rightarrow = Al - O - CH_3 + H_2O$$

The number of methoxy groups (x) per $Al_2Si_2O_5(OH)_{4-x}$ was determined as x = 0.36. The chemical formula of dry, methanol-treated kaolinite is: $Al_2Si_2O_5(OH)_{3.64}(OCH_3)_{0.36}{\cdot}0.58H_2O$. The possibility of exchange reactions between the bonded methoxy groups and other organic molecules makes the use of kaolinite in CPNC a reality.

Itagaki *et al.*, [2001] obtained kaolinite/PA-6 intercalated compound (KPA-6) by treating methoxylated kaolinite with ω-amino-hexanoic acid (AHA). The product was compounded in a TSE at 240 °C with neat PA-6 (Ube PA-1015B) to obtain comparable mineral content to that in commercial CPNC (Ube PA-1015C2). After compounding, the interlayer spacing in KPA-6 was slightly reduced to 1.15 nm, thus the system was only intercalated. The tensile and Izod impact tests are presented in **Table 73**. As the data show, incorporation of kaolinite improved the tensile strength and modulus of PA-6. Reinforcement was obtained for all three systems with kaolinite: with unmodified clay, intercalated with AHA or intercalated with PA-6. Evidently, for the latter system the improvements are the largest, but in comparison to the unmodified kaolinite the differences are hardly significant. This may indicate the presence of inherent interaction (chemical bonding?) between PA-6 and the clay. The surface of kaolinite particles is covered

Table 73 Properties of PA-6 and its CPNC containing Kaolinite (KPA-6). Data [Itagaki *et al.*, 2001]

System	Clay content w, (wt%)	Tensile strength, σ, (MPa)	Tensile modulus E, (GPa)	*NIRT** (J/m)
PA-6 (extruded)	0	72.7 (-)**	1.16 (-)	34.5 (-)
PA-6/kaolinite	1.42	79.2 (0.77)	1.29 (0.78)	27.7 (0.57)
PA-6/kaolinite-AHA	1.48	77.0 (0.72)	1.25 (0.73)	21.6 (0.42)
PA-6/KP6	1.38	80.4 (0.80)	1.33 (0.83)	27.7 (0.58)
Ube 1015C2	1.80	89.1 (0.68)	1.36 (0.65)	25.4 (0.41)

*Note: *Notched Izod impact strength at room temperature.*
***The values in parentheses are relative property values per 1 wt% of clay, e.g.* $(0.xy) = \sigma/w\,\sigma_{PA\text{-}6}$

by -OH groups, which may react with PA-6. For comparison, the properties of exfoliated PA-6/MMT system with 2 wt% organoclay are also shown. The reduced properties are shown in parentheses. These were calculated as:

$$P_r \equiv P(CPNC)/[P(PA\text{-}6) \times w(clay)] \qquad (177)$$

where P is a property and w is clay content in wt%. The outcome is quite surprising as the commercial PA-6/MMT system shows the worst performance. Part of the blame is due to overestimation of the clay content (if organoclay content is 2 wt%, the mineral content cannot be higher than about 1.6 wt%), but the *ca.* 10% correction is not sufficient to change the overall conclusion. The improvement of performance engendered by kaolinite is even more surprising when the effect of the degree of dispersion is taken into account – kaolinite was intercalated, whereas MMT was fully exfoliated! The improved performance may be due to the higher aspect ratio of kaolinite than that of MMT, as well as to the difference in bonding (covalent versus ionic) and the structure of the hairy clay platelets ('hairs' on one versus two sides). This is a significant, promising development.

4.1.2 Polyolefins (PO)

In this part only the patents specifically designed for PO nanocomposites will be discussed. The interested reader should also consult Section 4.1.3 where patents of a general nature are presented. Most of these discuss technologies applicable to CPNC with PO as matrix.

Preparation of polyolefin (PO)-based CPNCs is significantly more difficult than that of polar PAs. Thus, as an intermediate step between PO/clay and the PA/clay nanocomposites, a discussion on the elastomeric copolymer/clay system may be informative – see Section 4.3.

Surprisingly, the first patent application on 'organoclay-polyolefin compositions' was deposited in 1961 [Nahin and Backlund, 1963]. The process consisted of three steps: (1) intercalation of MMT with an onium salt, (2) mixing 5 to 80 wt% of organoclay with a polymer (PE, PP, PS, PVC or a similar), and (3) irradiating the mixture with high intensity γ-rays (at 1 to 1000 Mrads) to crosslink the system. The final product showed increased modulus without reduction of tensile strength, thus a clear advantage over the filled/reinforced thermoplastic resin. However, in spite of similar chemical composition, these systems were quite different from the modern CPNC – the authors used pre-intercalated MMT as a filler, without attempting to obtain exfoliation.

Kurokawa *et al.* [1996, 1997] developed a complicated procedure for preparing CPNC with PP as matrix, exploring the possibilities for exfoliating clay in a PO matrix. The process involved solution polymerisation of di-acetone-acryl amide, or 2-pentanone-4-methyl-4-acrylamide (DAAM):

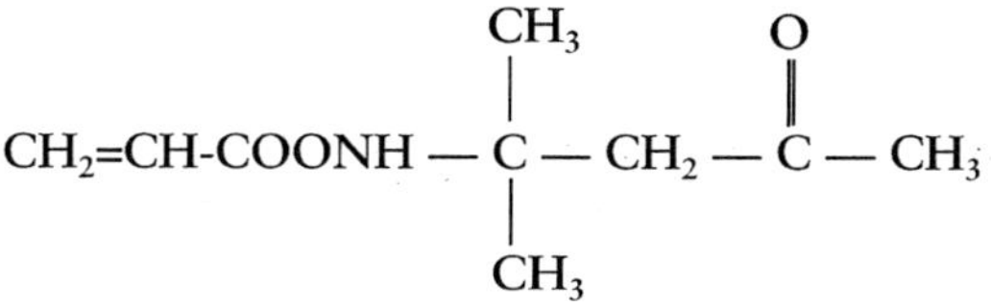

$$CH_2{=}CH\text{-}COONH - \underset{\displaystyle CH_3}{\overset{\displaystyle CH_3}{\underset{|}{\overset{|}{C}}}} - CH_2 - \overset{\displaystyle O}{\overset{\|}{C}} - CH_3$$

in the presence of ammonium-intercalated clay followed by mixing with maleated PP. It is of note that *ca.* 65 wt% of poly-DAAM was extracted by MeOH, without change in the interlayer spacing. Compounding the adduct with PP produced exfoliated CPNC containing: 3 wt% clay, 2 wt% poly-DAAM, 5 wt% PP-MA and 90 wt% PP. The authors did not postulate the mechanism responsible for the good dispersion of individual clay platelets, but the key clearly was the presence of poly-DAAM. The high concentration of polar side groups (amide and ketone) provided an intermediate interacting layer between the clay platelet and PP-MA. Miscibility of long chains of the latter compound with PP also contributed to full delamination.

As the following examples from patents show, simplification of the Kurokawa *et al.* procedure by eliminating the poly-DAAM step is possible, but difficult. When onium-intercalated clay platelets are hidden behind a layer of alkyls the heavily maleated PP has a tendency to form rich in maleic anhydride phase separated domains within the PP-matrix phase – thus no exfoliation. When the concentration of MAH groups is low, the PP-MA is miscible with PP, but the amount of compatibilising interaction of the organic and inorganic phases may be insufficient – thus again no exfoliation. In between there is a narrow region of composition, macromolecular configuration and molecular weight where exfoliation may be possible. The Kurokawa *et al.* procedure provided a mechanism that forced the MAH groups to interact strongly with the intermediate layer, widening the gap of beneficial compositions of PP-MA. In the latter publication it was shown that the degree of exfoliation depends on type of clay – the best was obtained using synthetic hectorite, the worst using mica [Oya *et al.*, 2000].

Jeong *et al.* [1998] provided further evidence for the need of a polar compatibiliser. The authors dispersed MMT-DDA in HDPE or NBR, obtaining short stacks with d_{001} = 1.77 and 2.15 nm, respectively. One interesting conclusion was that HDPE lamellae formed parallel to the clay surface, thus the amorphous surfaces of the lamellae interfaced with the dodecyl intercalant chains.

4.1.2.1 Toyota Patents on PO-Based CPNC

The first patent on CPNC with polymeric matrices that include PO came from Toyota [Usuki *et al.*, 1989]. The matrix resin was described as a polymer or copolymer such as: vinyl, thermoset resin or rubber. Thus, the patent claims are very broad, but the main emphasis is on the development of reinforced rubbers with either organoclay or carbon black (CB). The new nanocomposites comprise a resin (other than PA!) and uniformly dispersed layered silicate.

By contrast with 'classical' composites (which due to poor bond strength between the inorganic particles and the organic matrix are brittle) the new CPNC showed high modulus and good impact strength. Furthermore, the amount of clay that could be incorporated was significantly extended. Thus, superior mechanical performance, as well as high heat, water and chemical resistance characterised the final product. These CPNCs were developed for automotive or aircraft parts, building materials and thickeners for paint or grease.

The patented process comprised three steps:

(1) Intercalation of mineral clay (CEC = 0.5 to 2.0 meq/g) by ion exchange with onium salt.

(2) Mixing the organoclay with a monomer and/or oligomer.

(3) Polymerisation of the monomer and/or oligomer in the mixture, and/or kneading it with solid rubber.

When rubber is to be the matrix, CPNC comprises a clay complex and a rubber. The complex is made of clay intercalated with liquid rubber, having positively charged groups. The interlayer distance in the complex should be $d_{001} > 3.0$ nm. Furthermore, it is important that the liquid rubber of the complex is miscible with the solid rubber. The rubber may be natural, synthetic, a thermoplastic elastomer (e.g., *1,2*-polybutadiene, SBS, SIS) or blends. A diene rubber is preferred for co-vulcanisation with the liquid rubber. This structure is responsible for the composite material having superior mechanical characteristics, oil resistance, fatigue resistance, and processability.

When CPNC with a vinyl polymer matrix is prepared, it is desirable to use onium salt having a terminal vinyl group. In these polymers the unit chain may be: ethylene, propylene, butadiene, isoprene, chloroprene, vinyl chloride, vinylidene chloride, vinyl fluoride, vinylidene fluoride, styrene, α-methylstyrene, divinylbenzene, acrylic acid, alkyl acrylate, methacrylic acid, alkyl methacrylate, acrylamide, alkyl acrylamide, acrylonitrile, vinyl alcohol, norbornadiene, *N*-vinylcarbazole, vinylpyridine, vinylpyrrolidone, 1-butene, isobutene, vinylidene cyanide, 4-methylpentene-1, vinyl acetate, vinyl isobutyl ether, methyl vinyl ketone, phenyl vinyl ketone, methyl vinyl ether, phenyl vinyl ether, phenyl vinyl sulfide, and acrolein.

Mixing the clay with the monomer and/or oligomer is accomplished in a mortar or vibration mill. The polymerisation may be carried out in bulk, in suspension or in solution (e.g., using water, CS_2, CCl_4, glycerol, toluene, aniline, benzene, chloroform, DMF, DMSO). The polymerisation may be radical, cationic, anionic, coordination, or polycondensation. Furthermore, the clay may catalyse the reaction. When the onium salt has a terminal vinyl group that becomes a part of the polymeric matrix the system is end-tethered. The polymerisation must take place in the interlayer space, expanding the interlayer distance. As a result, the clay is exfoliated and the individual platelets bound to the polymer are uniformly dispersed in the matrix.

For example, Na-MMT (CEC = 1.19 meq/g or one negative charge per surface area of 1 nm^2; 1 x 100 x 100 nm layers) was dispersed in water, then added to an aqueous dispersion of liquid polybutadiene (PBD) along with HCl solution. The product was filtered, dried and characterised by pulse NMR. Two spin-spin relaxation times (T_2) were observed, the short T_{2S} = 10 μs for the glassy state and the long T_{2L} > 1 ms for the rubbery. The latter component originates in the part of the system where the molecular mobility is restricted, i.e., where is a strong bond between MMT and PBD - about 20% of the rubber molecules were restrained by the MMT surface. X-ray diffractometry indicated that MMT platelets were exfoliated (d_{001} > 8.8 nm) and uniformly dispersed in the PBD matrix. The complex was cooled with liquid N_2 and hammer milled into particles smaller than 3 mm. The crushed complex was mixed with NBR (33% AN), sulfur (vulcaniser), dibenzothiazyl disulfide (vulcanisation accelerator), ZnO (vulcanisation auxiliary), and stearic acid. The mixture was kneaded in an 8-inch roll mill at 50 °C. In the final product the amount of MMT was 5-10 phr.

In conclusion, the above disclosure suggests preparation of polyolefin-based PNC by polymerisation of olefinic monomers in the presence of intercalated clay platelets. In Japanese Patent Publication No. 198645/1989 it was proposed to incorporate ammonium ions at the chain ends either of the main or a side chain of a PO macromolecule. Unfortunately, incorporation of the ammonium group at a PO chain-end is chemically difficult. Furthermore, since the PO would be inserted into the interlamellar galleries at a single stage, the time for the interlayer expansion may not be sufficient. Nevertheless, these 15 year old ideas clearly predate the theoretical calculations that provide thermodynamic justification for the strategy.

The following patent from Toyota introduced further improvements to the CPNC behaviour [Usuki *et al.*, 1999]. Two articles by Kawasumi *et al.* [1997] and by Kato *et al.* [1997] provided further details. Thus, nanocomposite may be prepared using MMT-ODA and a guest molecule (MW = 0.1 to 100 kg/mol) with a polar group. Alternatively, an oligomeric guest molecule with a polar group may be added to the clay-onium complex, and then the mixture compounded with non-polar polymer. The most important feature of the invention is the use of a compatibiliser, 'a polar guest molecule' that bonds with the pre-intercalated clay and is miscible with the matrix polymer.

As was shown earlier, the interlayer distance depends on the ammonium alkyl chain length, C_nH_{2n+1} (n = 6 to 18). However incorporation of compatibilising 'guest' molecules with a polar group (in **Figure 147** hydroxylated, low molecular weight polybutadiene) expands the interlamellar galleries without limit (for 10 g of hydroxylated PBD per 1 g of MMT-2M2ODA the interlayer spacing d_{001} > 8.8 nm). Thus, the modified clay can be exfoliated and uniformly dispersed. When the 'guest molecules' are miscible with the main polymer, the dispersed clay (enrobed in compatibiliser) will not re-aggregate. However, when the 'guest molecules' are immiscible with the main polymer, the clay-phase (i.e., pre-intercalated clay with polar guest molecules) will form one phase and the main polymer the other. Thus, miscibility is the key for the success of this processing strategy.

As the clay expands, its surface area increases improving the CPNC performance, especially as far as the barrier and mechanical properties are concerned. Since the attachment to clay restricts macromolecular motion, the tensile strength, elastic modulus and creep resistance are all improved.

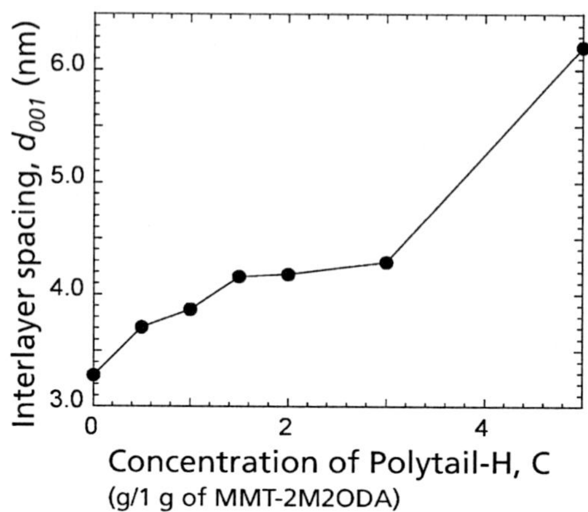

Figure 147 *Interlayer spacing of MMT-2M2ODA as a function of the 'guest molecule' concentration (hydroxylated PBD; Polytail-H). Concentration is expressed in grams of Polytail per gram of MMT-2M2ODA. Data [Usuki et al., 1999].*

For example, the rubber nanocomposite may be used in applications where a general rubber is used, but it is particularly effective for improving dynamic and barrier properties to gas or liquids.

The low *MW* guest molecule must have at least one polar group at its main and/or side chain, *viz.* hydroxyl (-OH), halogen (-F, -Cl, -Br, -I), carboxyl (-COOH), anhydrous-carboxylic acid (MAH), thiol (-SH), epoxy, and primary, secondary or tertiary amine ($-NH_2$, -NH-, =N-). The chain length of the guest molecule should be at least as long as that of the intercalant alkyl, but shorter that that of the main polymer in the matrix. Preferably it is a linear or branched olefinic or paraffinic oligomer or low molecular weight polymer (of the same type as the main polymer in the matrix), with an aromatic ring or not, e.g., PE, PP, PI, or PBD (*MW* = 0.5 to 100 kg/mol), having a polar group to bond with the clay surface. When low molecular weight polymer is used, it is important that it efficiently entangles with the main polymer, thus that its molecular weight $M_n \geq M_e$. At concentrations ≥ 10 wt% the compatibiliser sufficiently expands the galleries.

Thus, the revised method for producing CPNC with PO matrix again includes three steps:

1. Incorporating an organic onium ion into the clay interlamellar gallery. As described before, the organoclay is usually prepared by dispersing well-purified clay in an aqueous solution containing organic onium salt. After reaction the product precipitates, it is filtered out and washed with water to remove the excess of onium ions. Depending on the following step, drying may not be necessary.
2. Contacting the organoclay with a guest molecule having a polar group capable of diffusing into the interlamellar galleries and bonding with the clay. The organoclay may be added to a solution containing the compatibilising guest molecules. Alternatively, it may be melt compounded with the compatibilising 'guest polymer'. However, since a clay-ammonium complex is unstable at $T \geq 200$ °C its exposure to high temperatures must be minimised.

3. Next, the complex of compatibilised organoclay prepared in step two may be blended with the matrix polymer, having the same statistical chain units as the compatibiliser, or others, but miscible. The clay complex may be brought into contact with the first and the second guest molecules in any order. Thus, the difficult step of grafting a polar group onto a high *MW* polymeric chain is omitted.

As an example, Na-MMT (CEC = 1.2 meq/g) was first treated in aqueous medium at 80 °C with dimethyl-distearyl-ammonium chloride (2M2ODA), then (after drying) it was redispersed in toluene and mixed for 6 h with end-hydroxylated PBD (average carbon number = 210; *MW* = 3 kg/mol). The toluene was evaporated under reduced pressure to provide the compatibilised clay with the interlayer spacing d_{001} = 3.87 nm. This complex could be melt blended either with PBD or a polymer that is miscible with PBD.

In another example, CPNC was produced as above, except that MMT was intercalated with ODA and maleated low *MW* PP (Yumex-1010, *MW* = 30,000, softening point = 145 °C) was used as compatibiliser. (*MW* and acid value are used to characterise Yumex resins. The acid value of Yumex 1001 and Yumex 1010 is 26 and 52 mg KOH/g, respectively.) The clay complex was added to molten PP at 200 °C, and then mixed for 30 min. A moderate expansion of the interlayer spacing, d_{001} = 3.82 nm, was achieved.

The Usuki *et al.* patent claims specify:

- The natural or synthetic layered clay with CEC = 0.5-2.0 meq/g, e.g., MMT.
- An organic onium ion selected from between ammonium (hexyl, octadecyl, dimethyl-dioctyl, trioctyl, dimethyl-distearyl, lauric acid, octyl, 2-ethyl-hexyl, dodecyl, 1-hexenyl, 9-dodecenyl, 9-octadecenyl, 9,12-octadecadienyl, or 9,12,15-octadecatrienyl), sulfonium or phosphonium. As the cited examples indicate, the preference seems to be for ODA. The selected onium ion must ionically bond to the clay, expanding the interlayer spacing and making it susceptible for interacting with the compatibilising guest molecules.
- A guest molecule (with molecular length ≥ than that of onium ion radicals) must have a polar group, be able to at least partially incorporate into expanded interlamellar clay galleries and bond to clay. Preferably, it is a linear or branched olefin or paraffin ($M_n \geq M_e$ = 1-100 kg/mol), having at least one polar group selected from between hydroxyl, halogen, carboxyl, anhydrous carboxylic acid, thiol, epoxy and amino groups. At least 10 wt% of the total *MW* must be incorporated into the interlayer section of clay.
- The main polymer is a synthetic resin selected from between: PE, PP, PS, PIB, acrylics, PU, SBS, PBD, liquid polyisoprene, liquid butyl rubber, etc. The molecular weight may be *MW* = 1-500 kg/mol.

The current production of PO-based CPNC is carried out by the 'dry compounding' method described in this patent [Usuki, 2000a,b]. The older method [Usuki *et al.*, 1989] did not produce uniformly dispersed clay platelets in either PP or polyisoprene - initially, the dispersion was excellent, but then the platelets tended to aggregate under stress. In other words, non-polar PE or PP macromolecules did not sufficiently interact with the onium salt intercalated clay to prevent re-aggregation – addition of maleated low *MW* PP changed the balance. As reported by Kato *et al.* [1997], good intercalation/exfoliation may be obtained during melt compounding as long as the acid value of the maleated PP (PP-MA) is reasonably high, for example, PP-MA with acid value of 52 worked quite well, but that with acid value of 7 mg KOH/g not at all.

In another article from the group [Kawasumi *et al.*, 1997] MMT was first intercalated with ODA, and then treated with low *MW* PP-MA. Two critical conditions for good exfoliation were stressed: (1) strong interaction of PP-MA with the organoclay, and (2) miscibility of the PP-MA chains with the main polymer.

Hasegawa *et al.* [1998] (also from Toyota) prepared PP-clay nanocomposites intercalating purified MMT with ODA, then melt compounding the organoclay with PP-MA as a compatibiliser in a TSE. The PP-MA was Yumex 1010 with M_w = 30 kg/mol and acid value 52 mg KOH/g – depending on the composition (the used weight ratios of organoclay to PP-MA were: Ξ = 1:3, 1:2 and 1:1) the interlayer spacing changed from d_{001} = 6.3, 5.7 and 3.4 nm, respectively. Next, PP was melt blended with the compatibilised organoclay. In spite of relatively high clay content (4.7 to 5.0 wt%) the TEM micrographs showed excellent exfoliation for CPNC prepared with compatibilised clay at a ratio Ξ = 1:3, but worsening as that ratio increased to Ξ = 1:2 and 1:1 – in the latter case short stacks of clay were present.

Thus, this work focused on another important aspect of the technology, the concentration of compatibiliser added in the second step of the CPNC preparation. As a corollary conclusion, 'the short-cut' of this method involving single compounding of organoclay, PP-MA and PP must result in inefficient dispersion of clay lamellae, thus loss of performance. The reported dynamic storage modulus versus temperature (T = -50 to 160 °C) was systematically higher than that of PP/PP-MA matrix and higher than that of PP up to 130 °C. For example, at 25 and 80 °C the modulus of the CPNC with 5 wt% clay and 22 wt% PP-MA was, respectively, 1.5 and 1.8 times higher than that of PP. Evidently, addition of such a large quantity of low molecular weight compatibilising PP-MA reduced the reinforcement effect of the clay.

Two years later Hasegawa *et al.* [2000c] obtained a general patent for the preparation of CPNC, including those with PO. Owing to the general nature of the document it will be discussed later, along with similar documents. By the end of 2002, Toyota had licensed the technology for making PP nanocomposites to four major PO-resin manufacturers, two in the USA and two in Japan.

4.1.2.2 Dow Patents on CPNC Technology for PO

Karande *et al.* [1998] described preparation of a polymer foam by dispersing an organoclay into an olefinic or styrenic polymer melt having polar functionality so that at least a portion of the polymer intercalates between layers of the particles, and then expanding the polymer with a blowing agent into a foam. The organoclay may also be dispersed in a monomer, which is then polymerised to form a polymer melt prior to, or during the foaming step. The organoclay used in the examples was Claytone™ HY (MMT with 2M2HT-ammonium chloride). The polymer was Affinity plastomer reacted with vinyl trimethoxy silane (1.5 wt%). The co-reactive crosslinking agent was Sartomer™ 350, a trimethylol-propane trimethacrylate. Compounding was carried out in a W&P ZSK at T = 130 to 211 °C with a chemical foaming agent. In another example PP-MA was used as compatibiliser. The patent does not provide information on the degree of dispersion achieved. The authors claimed better cushioning and elastic recovery properties, improved insulation and ignition resistance.

PO-based CPNCs were prepared from layered materials in three steps [Nichols and Chou, 1999]: (i) intercalation with an organic, polymeric or inorganic intercalant that resulted in d_{001} = 0.5 to 60 nm; (ii) drying or calcinating the filtered out product at T = 50-80 °C or at 450-550 °C, respectively; and (iii) dispersing the intercalated clay in a monomer or melt blending it in an extruder until at least 80 wt% of the layers are exfoliated (clay platelet aspect ratio p = 10-2000). CPNCs were prepared from diverse multilayered materials dispersed in a thermoset, thermoplastic or elastomer. The claims specifically mention LDPE and LLDPE copolymers with density ρ = 850-920 kg/m^3 and MFR = 0.1-10 g/min. The inorganic intercalant was obtained by hydrolysing a metallic alcoholate: $Si(OR)_4$, $Al(OR)_3$, $Ge(OR)_4$, $Si(OC_2H_5)_4$, $Si(OCH_3)_4$, $Ge(OC_3H_7)$, $Ge(OC_2H_5)_4$, etc. To be intercalated, the clay (MMT, nontronite, beidellite, volkonskoite, hectorite, saponite, sauconite, magadiite, or kenyaite) should be swollen in an aqueous medium.

The patent specifically describes CPNC having 'delaminated or exfoliated particles derived from a multilayered inorganic material intercalated with an inorganic intercalant'. When an organic intercalant is employed it can be removed by calcination. Preferably the CPNC is prepared by melt compounding a thermoplastic polymer with expanded clay under suitable conditions. The resulting CPNC showed an excellent balance of properties and superior heat or chemical resistance, ignition resistance, superior resistance to diffusion of polar liquids and of gases, yield strength in the presence of polar solvents such as water, methanol, ethanol and the like, or enhanced stiffness and dimensional stability, as compared to standard composites. The CPNC were found useful in a wide variety of applications including transportation (e.g., automotive and aircraft) parts, electronics, business equipment such as computer housings, building and construction materials, and packaging materials.

It is interesting to note that Dow has been active in the area of modification of aluminosilicates, mainly for catalytic (Zeolite) applications [Millar and Garces, 2000]. The latter patent describes methods for adjusting porosity by reacting amorphous or crystalline silicates with an alkali-aluminate, followed by extraction. Thus, the porosity increases, boosting the adsorbing ability of the compound for organic molecules (e.g., methanol, hexane or benzene). The main goal has been improvement of the catalytic applicability of zeolites in, e.g., alkylation of benzene [Samson *et al.*, 2001]. Another patent described the preparation of 'aerogels' (to be used as components of a catalyst for addition polymerisation) starting with, e.g., NH_4^+ MMT^-, freeze-dried, calcined at 500 °C, and then treated with dimethyl aniline and tripropyl aluminium [Sun *et al.*, 2001]. However, since the patent by Nichols and Chou in 1999, there has been no further information about this promising approach.

The fundamental research of Balazs and her co-workers on the thermodynamics of nanocomposites resulted in a series of recommendations for the preparation of CPNC (see Section 3.1.5). These recommendations form the basis of a patent assigned to Dow [Ladika *et al.*, 2001; Fibiger *et al.*, 2001]. The document specifies that PO-type CPNC should comprise: (1) a non-polar polymer, such as PE or PP; (2) an intercalated multilayered silicate (quaternary ammonium treated MMT or sepiolite); and (3) a functionalised polymer, e.g., amine-terminated PE or PP, miscible with the non-polar polymer and having MW > 3 kg/mol. Melt blending these three components results in CPNC. According to the claims the polymer (1) is either PE, PP, PS, SEBS, elastomer, poly(vinyl cyclohexane), an ethylene-styrene interpolymer, etc. The multilayered silicate (2) is selected from between: MMT, nontronite, beidellite, volkonskoite, hectorite, saponite, sauconite, magadiite, octasilicate, fluoromica, fluorohectorite, etc. The clay should be pre-intercalated. The functionalised polymer (3) should have an organic cation ('sticker') and a polymer chain, miscible with the main polymer. Its concentration should be within the range from 0.1 to 10 wt%. Melt compounding is supposed to result in exfoliation, defined by the patent as dispersion into 'one, two, three, four or more than four layer units', hence rather in intercalation.

The key of the invention is the long-chain intercalating organic cation. Its molecular weight ranges from 5 to 300 kg/mol. It may have one or several cations, such as an amine hydrochloride group, a quaternary ammonium group, a sulfonium group or a phosphonium group. However, the organic cation must have, what the authors call, a 'pendent polymer chain'. The term includes a polymer chain with a cationic group, either on one or both ends (telechelic) or on a side group. Suitable intercalants may be prepared by reacting PP-MA with *N,N*-dimethyl-ethylene diamine in the presence of chlorobenzene at 130 °C for 4 h. Next, the resulting imine-amine functionalised PP may be converted to the amine hydrochloride form. The patent provides several examples that illustrate the new procedure. These are summarised in **Table 74**. The authors provide several comparative examples prepared under similar conditions to those listed in **Table 74**, but without the amine-terminated PP. Invariably, in this case little if any exfoliation was reported and the mechanical performance was significantly lower.

Another patent originating from cooperative research and assigned to Dow describes a CPNC, as a dispersion of layered nanofiller particles [Alexandre *et al.*, 2002b]. The patented process involves preparation of a clay-catalyst complex, which in turn can be dispersed in a monomer to polymerise into a CPNC. Thus, the PNC was prepared in several steps:

1. The clay interlayer spacing was expanded by first swelling it in water, to form a gel-like slurry. For example, untreated clay (Na-MMT, laponite, or hectorite) was dispersed in hot deionised water.
2. Following the work described in an earlier patent by Sun *et al.* [2001], the slurry was centrifuged and then freeze-dried to obtain organophilic clay.
3. Next, a clay/alkyl aluminoxane complex was prepared. About 10 wt% of the dehydrated and expanded clay was dispersed in a non-polar solvent (such as pentane, hexane, heptane, octane, or toluene). Methyl aluminoxane (MAO) was then added in a stoichiometric excess (with respect to active sites in the dehydrated clay) to form a MAO/clay complex. The complex was filtered, washed and dried.
4. The MAO/dehydrated clay complex was redispersed in a solvent along with a catalyst (Ziegler-Natta or metallocene) that promotes polymerisation of α-olefins or styrenics. At this stage the reactive complex comprises: clay with expanded interlayer spacing, MAO chemically attached to it, and a suitable polymerisation catalyst.
5. The reactive complex was dispersed in monomer at a concentration in the range from 1 to 10 vol%. Suitable monomers are: ethylene, propylene, 1-pentene, 1-butene, 4-methyl-1-pentene, 1-hexene, 1-octene, 1-decane, styrene, ethylidene norbornene, 1,4-hexadiene, 1,5-hexadiene, 1,7-octadiene, 1,9-decadiene, dicyclopentadiene or their mixtures.

Table 74 Summary of the examples cited by Ladika *et al.*, 2001

Polymer	Organoclay	Functionalised polymer	Processing	Results
93 g PP-6524	6 g Clayton HY	1 g $PP\text{-}NH_2$	Mix1	E_f = 1.72 GPa
93.9 g PP-H705	6 g 2M2ODA2C	0.1 g $PP\text{-}NH_2$	Mix1	*sf*
93.8 g PP-H705	6 g 2M2ODA2C	0.2 g $PP\text{-}NH_2$	Mix1	*sf*
93.65 g PP-H705	6 g 2M2ODA2C	0.35 g $PP\text{-}NH_2$	Mix1	E_f = 1.42 MPa
93.5 g PP-H705	6 g 2M2ODA2C	0.5 g $PP\text{-}NH_2$	Mix1	*sf*
96 g PP-H705	3 g acid-treated MMT	1 g $PP\text{-}NH_2$	Mix1	E_f = 1.42 GPa *IS* = 1.76 J
95 g PP-H705	3 g acid-treated MMt	2 g $PP\text{-}NH_2$	Mix1	E_f = 1.54 GPa *IS* = 2.06 J
27 g PP-9934	5.4 g 2M2HT + 1.8 g octa-decan amide	1.8 g $PP\text{-}NH_2$	Mix1	*E* = 3.40 GPa
28.6 g PP-9934	7.4 g 2M2HTFM	2 g $PP\text{-}NH_2$ + 2 g PP-MA	Mix2	*E* = 3.69 GPa
28.6 g PP-9934	7.4 g 2M2HTFM	2 g $PP\text{-}NH_2$ + 2 g PP-MA	Mix2	*E* = 3.69 GPa E_f = 2.90 GPa
28.6 g PP-112	7.4 g 2M2HTFM	2 g $PP\text{-}NH_2$ + 2 g PP-MA	Mix2	E_f = 2.24 GPa

Notes: Polypropylene: PP-6524 is Montell grade 6524, PP-H705 is PP Dow grade H705-04Z, PP-9934 is Amoco grade 9934, PP-112 is Dow grade DC 112. Organoclay: Clayton® HY is ammonium intercalated MMT, 2M2ODA2C is di-octadecyl-di-methyl ammonium magadiite, 2M2HTA is sepiolite treated with 2-methyl 2-hydrogeneted tallow ammonium chloride, 2M2HTFM is fluoromica treated with 2-methyl 2-hydrogeneted tallow ammonium chloride. Functionalised polymer: $PP\text{-}NH_2xHCl$ *(MW is 15 kg/mol), PP-MA is Eastman, Epolene Brand, grade G-3003. Processing: Mix1 is Internal mixer @ 180 °C, 60 rpm for 5 min, Mix2 is Internal mixer @ 180 °C, 100 rpm for 5 min. Results:* E_f *= flexural modulus (ASTM D-790); E = tensile modulus (ASTM D-638); IS = impact strength, sf stands for 'significant exfoliation' - reported for all listed examples*

6. Polymerisation transforms the suspension into CPNC with well-dispersed clay platelets. Thus, the nanocomposite can be prepared directly by *in situ* polymerisation of an olefin or the styrene, without the customary clay ion exchange, without the need for polar compatibiliser, and without shear.

The procedure is well illustrated in the following example. MMT was dispersed while stirring in deionised water at 80 °C for 2 h. The resultant gel was centrifuged for 30 min at 2000 rpm. The filtrate was freeze-dried under vacuum, obtaining a fluffy white solid. The degassed solid was dispersed in dry heptane and a solution of MAO in toluene was added. Stirring at 50 °C continued for 1 h, and then the solvents were removed under vacuum. The resultant powder was washed with toluene to remove excess MAO, and re-dispersed in heptane. Next, the suspension was heated to 80 °C and the catalyst was added forming yellowish slurry. After 1 h of ageing at 80 °C, ethylene was under pressure of 172 kPa. During polymerisation, the polymer-loaded clay particles were well dispersed in the reaction solvent. After 6 h of reaction and removal of solvent, 35 g of PE-clay CPNC was obtained as a free-flowing white powder. The powder was compressed into a disc and then into a transparent film. A STEM image showed nanoscale clay particles uniformly dispersed in the PE matrix. The tensile properties of three selected polymerisation products are displayed in **Figure 148**. Only the sample with 3.4% clay is CPNC, the two other are for reference: neat HDPE obtained following the same procedure, and a composite containing 2.3 wt% of calcined clay. The selected nanocomposite showed the best performance. However, there is no significant improvement of properties over the reference HDPE as

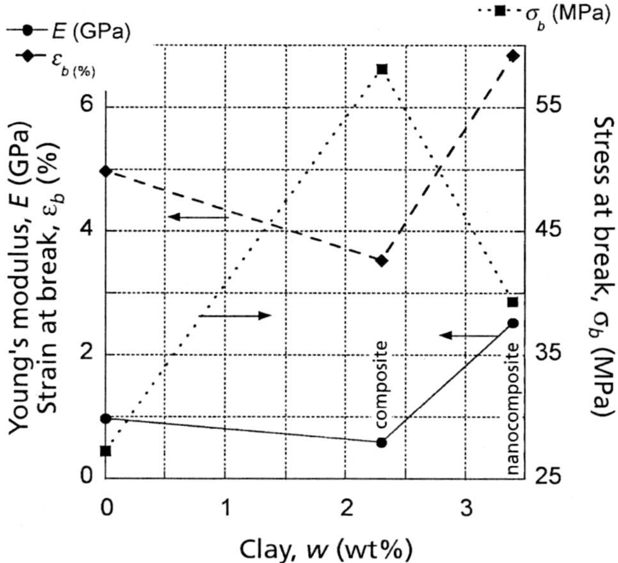

Figure 148 Tensile properties of three reaction products described by Alexandre et al. [2002]: the first sample (0% clay) is neat HDPE, polymerised without any filler; the 'composite' (2.3% clay) comprises calcined clay; the 'nanocomposite' sample (3.4% clay) is an exfoliated polymerisation product. Young's modulus (E), stress at break (σ_b) and strain at break (ε_b) are shown.

well as over the composite with calcined, micron-size clay particles. There is no information regarding the d_{001} spacing. Furthermore, poor stability of the nanocomposite dispersion has frequently been observed during the post-polymerisation forming stage, e.g., Heinemann *et al.* [1999], Manias *et al.* [2001], Jin *et al.* [2002], Alexandre *et al.* [2002a], etc. By the end of 2002 Dow obtained a license for the Toyota technology for the manufacture of PO-based CPNC.

4.1.2.3 Sekisui Chemical Patent on PO-Based CPNC

Fukatani *et al.* [2002] proposed a refreshingly new approach to exfoliation of clay. Their patent describes preparation of PO nanocomposites, but the implications are general. In essence, a standard cation-exchanged organoclay and a polymer are treated with a secondary intercalant, an end-functionalised compound with a group able to interact with the clay platelet edge. Two types of the secondary intercalants were proposed: (i) one able to bond with the clay hydroxyl groups (here the key is to minimise interactions with the clay platelet flat surface), and (ii) another one having anionic end-groups able to react with the clay cations that are located at the platelet edges. In both these cases it is desirable that the secondary intercalant has another reactive group providing a chemical bond between the clay complex and the polymeric matrix. As the schematic below shows, the attachment of the end-functionalised compound to the clay edges is expected to provide good leverage for the mechanical peeling off of clay layers (3) from the stack (4). As everyday experience shows, it is easier to remove the top sheet from a stack taking it by the edge than by the middle (see **Figure 149**).

The patent describes PO-type CPNC with 100 phr PO and 0.1 to 50 phr organoclay in the form of three consecutive inventions. Its validity is general, useful for the preparation of CPNC with a variety of polymers, *viz.* PP, EPR, PE, polymers of isoprene, 1-butene, 1-pentene, 1-hexene, 4-methyl-1-pentene, 1-heptene, 1-octene etc., with M_w = 5 to 5,000 kg/mol (preferably 20-300), and M_w/M_n = 3 to 40. The type of clay specified in the document may be either synthetic or natural, e.g., mica, smectite (MMT, saponite, hectorite, beidellite, stevensite and nontronite), vermiculite, halloysite, etc. The CEC = 0.5 to 2 meq/g, and the aspect ratio (defined as): p = (major diameter + minor diameter)/(2 x thickness) = 50 to 200.

The cationic intercalants are mainly quaternary ammonium or phosphonium salts with alkyls having at least 8 C-atoms, *viz.* ammonium salts of lauryl trimethyl, stearyl trimethyl, trioctyl, distearyl dimethyl, dihydrogenated tallow dimethyl, distearyl diphenyl, etc.

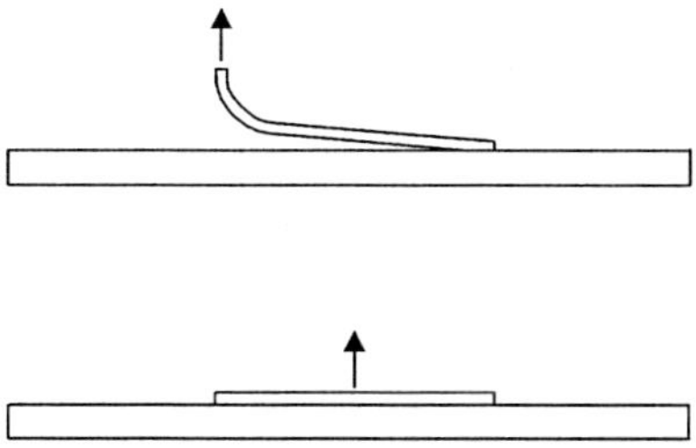

Figure 149 *The basic concept of peeling from the stack one clay platelet after another – by an edge (top) or by the middle (bottom). After [Fukatani et al., 2002].*

According to the first invention, CPNC is prepared in two steps: (1) by ion-exchange of clay with a cationic intercalant, and (2) by compounding the pre-intercalated clay with a PO resin and a secondary intercalant, which has at its molecular end reactive groups able to interact with the clay hydroxyl groups. These functional groups are selected from between: alkoxy, alkoxy-silyl, epoxy, carboxyl, hydroxyl, maleic anhydride, isocyanate and aldehyde groups. The secondary intercalant may be oligomeric or polymeric. Two types of molecules were specifically identified, the well-known 'sizing agents' (*viz.* silanes or titanate compounds) forming one type, and compounds with glycidyl, carboxylic acid, alcohols and other hydrogen-bonding groups forming the second. These compounds may have another reactive group, such as vinyl, amino, epoxy and acryloyl. The latter reactive group should provide adequate bonding between the clay complex and PO matrix. Execution of this invention should result in dispersion of clay platelets with interlayer spacing $d_{001} > 6$ nm.

The idea behind this invention is not exactly new, but the focus on the platelet edges is. The secondary intercalant should form covalent or hydrogen bonds with hydroxyl groups on a clay platelet edge and serve as a physicochemical tie for the mechanical peeling off of the top layers during compounding. Evidently, miscibility of this secondary intercalant with PO is essential. In the case of immiscibility, the secondary intercalant should be able to react with the matrix molecules during compounding. For example, when a vinyl trimethoxy or triethoxy silane is used, the grafting reaction can be initiated by any source of free radicals.

The second invention uses another type of secondary intercalant. To obtain enhanced grafting of organic molecules to the clay edges, an anionic compound (a carboxylate, sulfonate, polyphosphate, and sulfate or phosphate ester) is to be used. The organic anions are to react with the clay cation groups located at the platelet edges. The reaction takes place after the standard cation-exchange pre-intercalation of clay. The anionic compounds having a high chemical affinity for the clay cations are, e.g., a carboxylate (Na-laurate, Na-stearate, Na-oleate), alcohol sulfate (Na-lauryl alcohol sulfate, NH_4-lauryl alcohol sulfate, Na-cetyl alcohol sulfate, etc.), sulfated oil, sulfated fatty acid esters and sulfated olefins, sulfonate (Na-alkyl benzene sulfonate, etc.), phosphate salts and many others. As in the first invention, here also the secondary anionic intercalant should have another reactive functional group (*viz.* vinyl, amino, acryloyl or epoxy), able to react with PO-matrix molecules and provide adequate bonding between the clay complex and the matrix.

The third invention extends the method described in inventions 1 and 2 to diverse types of CPNC. For example, the system may comprise: 100 phr of a thermoplastic resin, and 0.1 to 100 phr of organoclay of the type as described in the first two inventions. Examples of thermoplastic resin include POM, PVC, PVAc, PS, PO, acrylics, PVAl, cellulose derivatives, various rubbers (e.g., NBR, SBR), etc. The composition may also include 2 to 300 phr of a plasticiser, etc. The compounding can be carried out in a batch or continuous mixer, *viz.* a TSE. The CPNCs of this third invention find diverse applications. For example, a CPNC with plasticised polyvinyl butyral as a matrix, may be used for an interlayer of a laminated glass, a sound-insulating interlayer in architectural or automotive applications. CPNC with plasticised PVC is useful in agricultural vinyl sheeting, food packaging, leather, film, damping sheet, wall covering and tubes, coatings and adhesives.

Several examples illustrate the patented inventions. Two clays (MMT and synthetic mica) were cation exchanged with distearyl dimethyl ammonium chloride (2M2ODA).

Invention (1). To stirred organoclay powder a 2 wt% aqueous solution of amino-propyl-trimethoxy silane (octadecyl trimethoxy silane, methacryl-oxytrimethoxy silane, vinyl trimethoxy silane, glycidoxy-propyl trimethoxy silane, or alkoxy-silyl-terminated polyisobutylene with M_w = 20 kg/mol) was added drop-wise. The resulting powder was dried and used with either PP or PE as the matrix.

The selected polymeric matrix and the clay complex were compounded in a TSE at T = 200 °C at a weight ratio of 92.3:7.7. The extruded strands were pelletised then compression moulded into 0.1 or 2 mm thick plates. When the clay complexes contained either methacryl-oxy- or vinyl trimethoxy silane, a plunger pump was used to feed liquid peroxide, the 2,5-dimethyl-2,5-bis(*t*-butyl-peroxy)-hexane. The free radicals caused a grafting reaction between the unsaturated secondary intercalant and the matrix, thus securing good bonding in the system. The CPNC were evaluated for flexural modulus and oxygen permeability. The same resin with or without standard clay-2M2ODA was also extruded, moulded and tested. Thus, for example, the PP flexural modulus F = 1.37 GPa and oxygen permeability P_O = 2.84 ml/m/day was found improved by addition of clay-2M2ODA to F = 1.46 GPa and P_O = 2.41 ml/m/day. However, the composition containing the secondary intercalant of this invention yielded: F = 3.22 GPa and P_O = 1.45 ml/m/day. Similarly, improved performance was obtained when alkoxy-silyl-terminated polyisobutylene was used as the secondary intercalant: F = 3.42 GPa and P_O = 1.2 ml/m/day. The improved stiffness was considered to result from enhanced exfoliation.

Invention (2). The resins and organoclays were the same as in the examples above. Sodium salts of alkyl-benzene sulfonic and oleic acids, acryloyl-containing sulfonate ester and ammonium polyphosphate were used as secondary intercalants. As before, 2 wt% aqueous solution of a secondary intercalant was added drop-wise to clay-2M2ODA powder while stirring. The compound was dried, melt blended with a PO resin, compression moulded and tested for F and P_O. When the clay complex contained either Na-oleate or acryloyl-sulfonate ester, a plunger pump injected liquid 2,5-dimethyl-2,5-bis(*t*-butyl-peroxy) hexane to the extruder, causing the clay complex to bond with the matrix. Again, good improvement of performance was obtained. Where peroxide was used F = 3.22 and 3.12 GPa and P_O = 1.32 and 1.26 ml/m/day were obtained. The best performance in this series was shown by CPNC treated with ammonium polyphosphate: F = 3.42 GPa and P_O = 1.1 ml/m/day.

Invention (3). First, MMT-2M2ODA was blended with plasticiser (triethylene glycol diethyl butyrate) then polyvinyl butyral was added and melt mixed in an internal mixer at T = 140 °C for 5 min. The product was compression moulded into a 0.5 mm thick sheet. The CPNC was found to have finely dispersed clay platelets. The composition showed remarkable improvement of properties, e.g., tensile strength of the resin increased from 3.6 to 103 MPa, increased elastic modulus and reduced bleeding of plasticiser. The most important accomplishment was the good balance of flexibility and strength, which was previously difficult to achieve.

4.1.2.4 Diverse Technologies for the Preparation of CPNC with PO-Matrix

A simple method for the preparation of PP-based PNC was described by Yoo *et al.* [1999]. PP was melt compounded at 200 °C with 5 wt% of organoclay

(Cloisite® 20A; MMT-2M2HTA) in an internal mixer. During compounding the mixture was subjected to ultrasonic (US) irradiation. The modulus was found to increase with the dose of US energy, in spite of the concurrent PP chain scission. Unfortunately, information on the extent of the intercalation achieved is not provided.

Ultrasonic radiation (US) has often been used for dispersing one material in another, *viz.* for preparing emulsions, suspensions (including dispersing such reinforcing particles as carbon nanotubes or organoclay in a monomer or polymer solution), and blends as well as to prepare CPNC. For example, US has been used to accelerate clay intercalation in aqueous media [Beall *et al.*, 1999; Tsipursky *et al.*, 1999], to improve the degree of exfoliation during extrusion compounding of CPNC [Yoo *et al.*, 1999], to monitor extrusion compounding [Tanoue *et al.*, 2003a; 2004], to disperse organoclay in a monomer [Okamoto *et al.*, 2000], or to initiate radical polymerisation of a monomer during the reactive exfoliation process. During emulsion polymerisation, US plays a dual role, improving the organoclay dispersion and generating free radicals, etc. More details are provided in Sections 2.1.3.1; 2.3.5; 2.4.1; 4.1.2.4; and 4.1.4.1.

Recently Lee and Mielewski [2003] applied for patent protection for the use of US during melt compounding in an extruder and/or injection moulding of CPNC with a thermoplastic matrix. The examples focus on the preparation of nanocomposites of PP with such organoclays as Cloisite® 20A (MMT-2M2HTA) or Nanomer I.30E (ODA) – a compatibilising PP-MAH may also be added. The document specifies that sonication is performed at a frequency from 5 to 10^{10} kHz (the frequency used in examples was 20 kHz), at an energy input of 5 to 100 kJ/g (used *ca.* 40 kJ/g), i.e., for about 10 to 600 s. As the examples illustrate, sonication is more effective for systems with lower molecular weight (i.e., low melt viscosity). For example, in the case of low *MW* sonication increased d_{001} from 2.36 to 3.6 nm, while in the case of high *MW* from 3.5 to 3.7 nm. Since the test specimens contained 5 wt% organoclay the processing improved intercalation of the clay, but did not result in exfoliation. Ultrasonics are known to cause thermomechanical degradation of PP hence the overall effect of US can only be judged by measuring the full range of performance characteristics – unfortunately this information is not provided.

Heinemann *et al.* [1999] reported on the preparation of PO-based CPNC by melt and reactive exfoliation of hectorite, intercalated with dimethyl benzyl stearyl ammonium ions (2MBODA). The intercalation was followed by XRD measurements. Melt compounding with HDPE reduced the intensity of the XRD peak and simultaneously the interlayer spacing from d_{001} = 1.96 (of organoclay) to 1.41 nm. Furthermore, when HDPE was polymerised in the presence of the organoclay, the XRD also indicated reduction of the peak intensity and the interlayer spacing. The authors speculated that compressive stresses during compounding were responsible for the reduction of interlayer spacing. Since a similar (but less intensive) reduction was obtained as a result of polymerisation, this mechanism seems unlikely. Instead, one may propose two processes, one that involves a change of molecular configuration of the intercalating 2MBODA, e.g., caused by phase separation in the presence of the paraffinic monomer or polymer. The second involves selective exfoliation - it can be postulated that exfoliation starts with the intercalated clay sandwiches that have the largest interlayer spacing. Exfoliating these makes them invisible to X-rays, hence only the intercalated sandwiches with low intercalated space will remain. The compounding method resulted in poorer dispersion of clay platelets than the polymerisation one. Better results were obtained with short-branch PO macromolecules (e.g., LLDPE-type ethylene-octene copolymer) than with linear HDPE. Furthermore, it was reported that some organoclay interfered with the metallocene-catalysed polymerisation.

Performance of CPNCs prepared by the polymerisation and compounding methods depended on the molecular weight and the branching index (N, defined as number of branching C per 1000 C-atoms), which varied significantly. For this reason, the original results were fitted to the relation (see **Figure 150**):

$$\log_{10} E = a_o + a_1 \log_{10} M_w - a_2 N \qquad (178)$$

Results of interpolation for M_w = 100 kg/mol are plotted as Young's modulus E versus N. Evidently, systematically the lowest modulus was obtained for neat resin (without clay), the highest for CPNC prepared by synthesis and intermediate for compounded CPNC. It is worth noting that the efficiency of the compounding method decreases with N (as well as with M_w). This is reasonable, since in this case exfoliation that is controlled by diffusion requires longer mixing time, not provided in the comparative TSE process.

In another publication from Mülhaupt's laboratory, CPNC with PP as the matrix were prepared by melt compounding at 210 °C [Reichert *et al.*, 2000]. First, synthetic hectorite (FM; Somasif ME100 sodium fluoromica, CEC = 0.7-0.8 meq/g) was intercalated by ion exchange with alkyl amines: C_nH_{2n+1}-NH_2, where n = 4, 6, 8, 12, 16 or 18. As shown in **Figure 151**, the interlayer spacing linearly increased with the alkyl chain length: $d_{001} \leq 1.98$. Next, the organoclay (5 or 10 wt%) was dry-blended with PP-MA (20 wt%) and PP. Interlayer distance remained unchanged for n = 4, 6 or 8, but increased for the longer alkyl chins, $n \geq 12$, with increasing content and anhydride functionality of PP-MA. The performance was sensitive to the type of PP-MA used – good results were obtained for Hostaprime

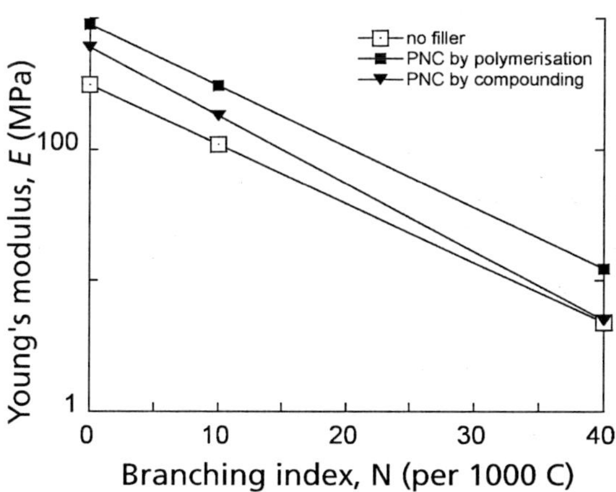

Figure 150 Young's modulus versus branching index for LLDPE, as well as its PNCs, prepared either by metallocene-catalysed polymerisation in the presence of organoclay, or by melt compounding in a twin-screw extruder. The data [Heinemann et al., 1999] were analysed by means of Equation 178. The plot represents the results interpolated to M_w = 100 kg/mol.

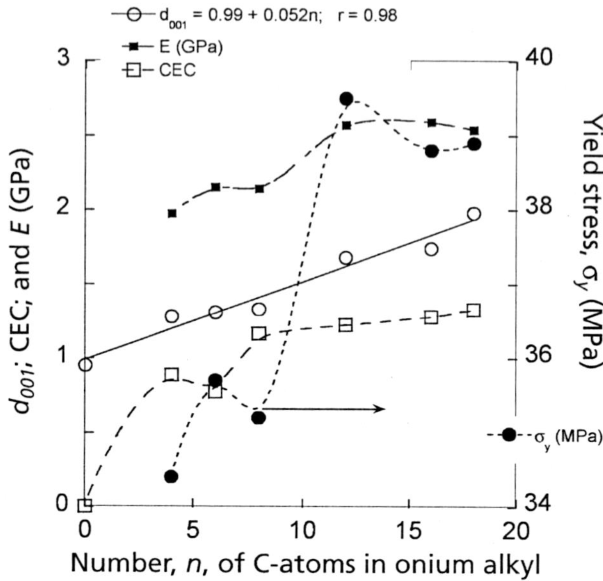

Figure 151 *PP/hectorite CPNC prepared by melt compounding of PP + PP-MA + clay intercalated with C_nH_{2n+1}-NH_3^+. Results for 5 wt% clay and Hostaprime HC5 compatibiliser. Data [Reichert et al., 2000].*

HC5 (M_w = 32 kg/mol; MAH = 4.2 wt%), poor for Epolene E43 (M_w = 23 kg/mol; MAH = 2.9 wt%). For FM pre-intercalated with C_{16}-ammonium, the values of Young's modulus increased from 1.49 (for PP) to 2.59 and 3.46 GPa (for 5 and 10 wt% clay, respectively). Similarly, the yield stress increased from 33 (for PP) to 39 and 44 MPa for the same CPNC compositions. TEM and AFM indicated that full exfoliation had not been achieved.

In this context, a patent from Exxon is of interest [Elspass and Peiffer, 2000]. It describes the formation of CPNC using a very much standard MMT intercalated with di-alkyl ammonium salt. The polymeric matrix is composed of two miscible, 'nonpolar', amorphous elastomers, one of lower *MW* than the other (*MW* = 50 kg/mol is the boundary). High barrier and mechanical properties confirmed stable exfoliation (detected by XRD and TEM). Judging by the listed examples, it seems that there are two possible mechanisms that prevent re-aggregation. The first example describes CPNC with two fractions of brominated isobutylene-co-*p*-methyl styrene. These molecules have strong dipole moments, which apparently are sufficient to obtain good immobilisation of macromolecules by the clay surface forces, in between the onium-clay complexes. The second example describes the formation of CPNC with low and high *MW* PBD, then crosslinking and compression moulding the specimens. Here, platelet immobilisation by crosslinking and low stresses during compression moulding may be responsible for the absence of re-aggregation.

Lan and Quian [2000] modified the method of preparation of PO-based CPNC developed by Toyota. The authors used a silane coupling agent, *viz.* $(R_1)_3{\equiv}Si$-R_2, where R_1 is either CH_3O- or C_2H_5O-; and R_2 is either CH_3-, C_8H_{17}-, NH_2-C_3H_6-

or NH_2-C_3H_6-(NH-C_3H_6)$_2$-. Thus, for example, PP-based CPNC was prepared by mixing PP with 2-5 wt% PP-MA and 2-6 wt% of silane-treated organoclay (Nanomer® I.31PS). Alternatively, when PP is pre-blended with 0.5 to 4 wt% of a silane coupling agent a standard onium-intercalated MMT (e.g., Nanomer® I.30P) may be used. The methods were tested using either PP with MFR = 4 g/10 min or TPO with MFR = 12 g/10 min. The process was conducted in two stages: (1) preparation of a masterbatch containing 50-60 wt% Nanomer, and (2) dilution of the batch with the matrix resin to 2.6-6 wt% of the organoclay. Excellent exfoliation was obtained. The pre-set goal (40% increased tensile modulus, 20% improved HDT and 10% improved Izod impact strength) was achieved at 6 wt% organoclay loading.

Zhang *et al.* [2000b] achieved high dispersion by melt mixing in a TSE a MMT (pre-intercalated with 2M2ODA-ammonium bromide), PP-MA (MFR = 1.2 g/10 min; 1% MAH) and PP (MFR = 1.5 g/10 min). The clay platelets formed short stacks. As shown in **Figure 33** (Part 2) addition of only 0.1 wt% of clay caused a dramatic increase of the notched Izod impact strength (from 9 to 26 kJ/m^2) and more moderate increase of the impact strength (from 29 to 30 MPa).

Nanocor, the manufacturer of onium ion-modified MMT (Nanomers® I.30P, I.31PS, and I.44PA for PO matrix) developed two processes for the preparation of CPNC with PO as a matrix [Qian *et al.*, 2001]:

1. First step – preparation of a concentrate (feed: PP, compatibiliser and Nanomer), followed by second step – dilution to the desired clay concentration (feed: concentrate and PP). Vacuum line is connected to zone 16D.
2. Single pass process (process as schematically shown in **Figure 30**, but with side-feeder where the first vent was located).

For either process the authors used a Leistritz co-rotating TSE, with diameter *D* = 27 mm and *L/D* = 36:1, at screw speed = 500 rpm. For both processes the screw configuration is the same (see **Table 75**). The configuration is identical to that shown in **Figure 30**, except that instead of the first vent a side-feeder was used. The TSE is fed with PP/Nanomer/compatibiliser (a concentrate) diluted downstream with PP fed from the side-feeder CPNC [Qian et al., 2001].

Table 75 Functions and temperatures for the Leistritz TSE barrel zones. Data [Qian *et al.*, 2001]

Zone/T (°C)	Function	Zone/T (°C)	Function
4D/RT	Feed	24D/170	Kneading/dispersing
8D/160	Conveying	28D/170	Vacuum devolatilisation
12D/165	Melting/dispersing	32D/170	Conveying
16D/165	Conveying/side-feeding	36D/170	Conveying and building pressure
20D/170	Kneading/dispersing	Die/170	Strand pelletising

When the first process is used, the concentrate contains 50-60 wt% organoclay and the rest is made of a compatibiliser (PP-MA) and standard PO. The first passage through TSE pre-disperses the organoclay, facilitating full CPNC formation during the subsequent dilution to *ca.* 2-7 wt% of organoclay. When the second (single pass process) is used, the first part of the TSE is used for the preparation of a concentrate (*ca.* 50 wt% organoclay), which in turn is diluted by side-feeding PO in zone 16D, and then homogenised in the second part of the extruder. The concentrate must also contain a compatibiliser. However, preparation of CPNC with low density organoclay powder (bulk density = 0.3 to 0.5 g/ml), at concentrations above even 20 wt% is a challenge.

Thus, for example, using the first process a modest clay dispersion in PO matrix was obtained – XRD indicated increased organoclay interlayer spacing from d_{001} = 2.2 to 3.2 nm, while TEM showed that organoclay was randomly dispersed as short stacks. Nevertheless, the reported properties for CPNC with organoclay loading of 6 to 7 wt% are quite impressive, *viz.* barrier improvement by a factor of 1.5 to 2.5, tensile strength by 13 to 33%, flexural modulus by 24 to 57%, and HDT by 20.5 to 20.8 °C.

Starting from the thermodynamic analysis of clay intercalation, Manias *et al.* [2001] examined two possible preparation methods for PO-type CPNC: (1) using a functionalised PP and common organo-MMT, and (2) using neat PP and clay that was modified by a semifluorinated organic compound. The former method is based on the need for improving the interactions between the polymer and the clay, to make them more favourable than those between the intercalant and clay. The aim of the second method is the same, but the approach is reversed – to reduce the interactions between the intercalant and clay so as to make those between polymer and clay more favourable. Both methods were experimentally examined by incorporating *ca.* 6 wt% of dimethyl-dioctadecyl ammonium intercalated MMT (MMT-2M2ODA) in the matrix.

1. Re Method (1). In principle, this approach has already been explored in several patents and articles. However, the authors introduced a new twist, using as a matrix a random copolymer of PP with functionalised monomer. The copolymers were prepared by copolymerising propylene with *ca.* 1 mol% *p*-methylstyrene (*p*MS) using a metallocene polymerisation method. Subsequently, the *p*-methyl groups were reacted to attach either a hydroxyl (-OH) or maleic anhydride functionality (-MA). Even with such a low functionalisation level the copolymers promoted nanocomposite formation with MMT-2M2ODA.
2. Re Method (2). MMT-2M2ODA was reacted with a semifluorinated alkyl-trichloro silane, CF_3-$(CF_2)_5$-$(CH_2)_2$-Si-Cl_3 (FS). This second intercalant tethered to the clay surface through a reaction of the trichloro silane with hydroxyl. Thus, the MMT-2M2ODA + FS organoclay was fully ion-exchanged with 2M2ODA and it contained *ca.* 60 wt% of FS.

The CPNCs were formed by annealing the physical mixture in a vacuum oven at 180 °C, unassisted by shear or solvent. In the case of Method (2) systems, extrusion was also used. The XRD data are summarised in **Table 76**. It is evident that all cases intercalated structures were obtained, indicating favourable thermodynamics for the nanocomposite formation. TEM showed that the intercalated tactoids coexisted with exfoliated platelets.

Table 76 Summary of XRD and TEM data for melt-intercalated CPNCs. Data [Manias *et al.*, 2001]

No.	Organoclay	Matrix	d_{001} (nm)	Exfoliation (%)	Method
0	MMT-2M2ODA	(nil)	1.98	0	–
1	MMT-2M2ODA	PP-co-pMS	2.9	25	Static
2	MMT-2M2ODA	PP-co-pMS-MA	2.9	40	Static
3	MMT-2M2ODA	PP-co-pMS-OH	3.0	30	Static
4	MMT-2M2ODA	PP-b-PMMA	2.76	20	Static
A	MMT-2M2ODA + FS	(nil)	2.4	0	–
B	MMT-2M2ODA + FS	PP	3.6	25	Static
C	MMT-2M2ODA + FS	PP	4.0	40	Extruded

Notes: 2M2ODA = dimethyl-dioctadecyl ammonium; FS = CF_3-$(CF_2)_5$-$(CH_2)_2$-Si-Cl_3; PP-co-pMS = copolymer of propylene ca. 1 mol% p-methylstyrene; for other symbols - see text

Manias *et al.* [2001] also examined the stability of the degree of dispersion during melt processing at 180 °C. In systems where neat PP was the matrix (e.g., PP/MMT-2M2ODA or PP/MMT-2M2ODA + FS) the clay platelets re-aggregated within minutes. However, systems with functionalised PP as the matrix (e.g., PP-co-*p*MS-MA with MMT-2M2ODA + FS) showed good stability. This seems to indicate that direct or indirect tethering (e.g., by means of end-functionalised compatibiliser) is required for good and stable performance of CPNC. The authors also reported several material properties, *viz.* mechanical performance, heat deflection temperature, transparency and flame retardancy. Examples of the results are shown in **Figure 152**. They also mentioned that Method (2) (MMT co-intercalated with semifluorinated alkyl-trichloro silane) is applicable to other polymers, e.g., polycarbonates and polydienes.

PP was melt compounded with clay in three stages of modification: (1) Bentonite comprising 30 wt% quartz crushed and screened into powder with diameter $d < 90$ μm. (2) The powder from (1) was dispersed in NaCl aqueous solution, where quartz sedimented and Ca^{2+} was exchanged by Na^{+}. (3) The Na-MMT from (2) was quantitatively ion exchanged with *N*-cetyl pyridinium chloride (CPC) [Pozsgay *et al.*, 2001]. CPNC was prepared by melt blending one of these clays with PP in a TSE at *ca.* 200 °C, without any compatibiliser (e.g., functionalised PP). Evidently clays (1) or (2) did not exfoliate, while (3) showed moderate intercalation with $d_{001} \approx 2$ nm. The strongest effect on the crystallisation temperature (T_c increased by *ca.* 15 °C) was obtained for clay (1) with quartz; organoclay (3) was virtually neutral and Na-MMT showed intermediate performance. By contrast, yield stress improved only with organoclay (3), especially at a low loading of *ca.* 2 wt%. At 10 wt% clay loading the tensile

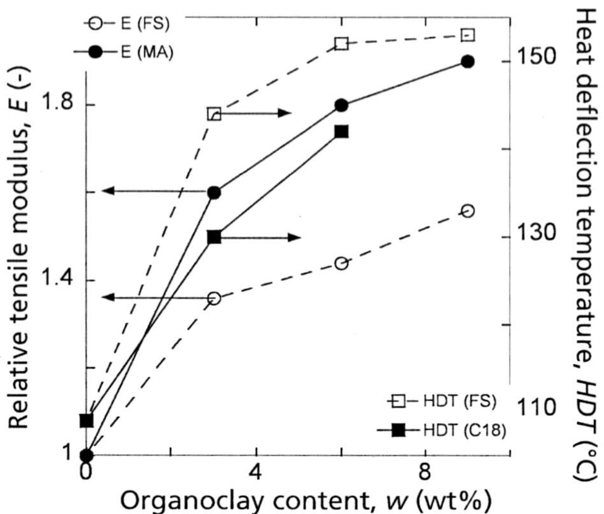

***Figure 152** Relative tensile modulus and heat deflection temperature versus organoclay content for PP with MMT-2M2ODA + FS (FS) or with MMT-2M2ODA (C18), as well as for PP-co-pMS-MA with MMT-2M2ODA (MA). Data [Manias et al., 2001].*

strength decreased by a factor of *ca.* 3 for organoclay and by *ca.* 1.5 for the two other clays. The poor performance of the PP/MMT-CPC most likely results from the thermodynamic immiscibility of these two components.

Hambir *et al.* [2001; 2002] prepared PP/organoclay CPNC by melt intercalation, using different PP resins, compatibilisers and clays. After drying the ingredients at 65 °C for 8 h, the CPNCs were prepared in an internal mixer at 200 °C and 60 rpm. A single composition: PP/PP-MA/clay = 85/15/5 (in wt%) for each combination was prepared. The CPNC ingredients with codes are listed in **Table 77**. XRD of Cloisite® 6A (C6A = MMT-2M2HTA) showed three well-defined peaks at d_{001} = 3.3, 1.8, 1.1 nm. Compounding PP1 with C6A and compatibiliser PB or VB/6A the XRD indicated expansion; for PB $d_{001} \geq 4.4$ nm with 56% exfoliation, while for VB d_{001} = 3.4 nm and 21% exfoliation. Smaller spacing was reported for systems with ODA-clay. Similarly, for PP3 with C6A and PB-compatibiliser (i.e., replacing PP1 with more viscous PP3) a smaller gallery expansion (by *ca.* 0.5 nm) was obtained. Surprisingly, still smaller gain (by 0.2 nm) was found for PP2 having the lowest M_w. As the data in **Figure 153** show, there is a direct correlation between the degree of dispersion and the mechanical behaviour. The only system that had significant expansion of the interlayer spacing, PP1/PB/C6A, shows outstanding improvement of the tensile modulus by a factor of 6.8 to 4 (a decrease with *T*), whereas for all the other systems the ratio varied from 0.8 to 1.5.

Thus, it seems that this series of measurements indicates that: (1) a sufficient MAH content in compatibiliser is important – PB with 1 wt% outperformed VB with 0.65 wt% MAH; and (2) for the melt compounding method the matrix viscosity is important – for low viscosity there is not sufficient stress transfer to

Table 77 Polymers, compatibilisers and clays. Data [Hambir *et al.*, 2002]

No.	Sample	Code	Supplier	Grade/MFI/MAH	M_w (kg/mole)
1	Polypropylene	PP1	IPCL	Koylene M0030/10/-	241.0
		PP2	Hüls	Vestolen 2000/70/-	164.0
		PP3	Hüls	Vestolen 7000/4/-	362.0
2	Compatibiliser	PB	BP Chem.	Polybond PB 3150/-/1 wt%	188.0
		VB	Vin Enterprise	Vinbond VB100/-/0.65 wt%	151.0
3	Organoclay	C6A	SCP-Cloisite® 6A 2M2HTA	(dimethyl dihydrogenated tallow ammonium chloride)	-
		ODA	Cloisite® -NA+	(octadecyl amine)	-

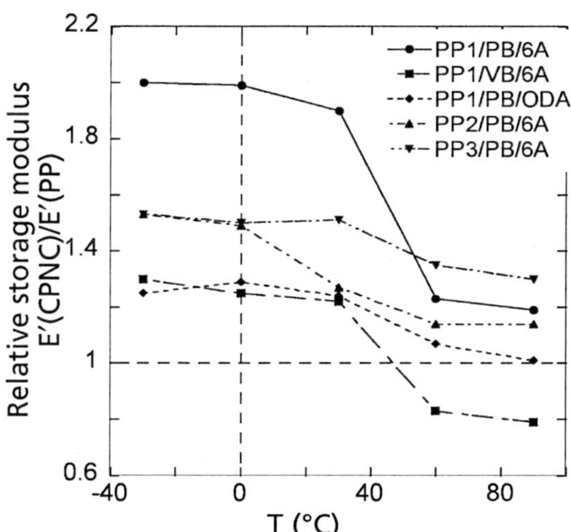

Figure 153 Relative tensile storage modulus versus temperature for five CPNCs of the type: PP/compatibiliser/organoclay – see text. Data [Hambir et al., 2002].

clay stacks to peel off individual platelets, for too high viscosity the matrix diffusion into the interlamellar galleries is too slow. Improvement of rigidity was particularly significant at $T < T_g$. Thus, an optimum molecular weight exists, since owing to the thermal instability of ammonium intercalants there is little freedom in modifying the rheology by the processing temperature.

A new route for the preparation of CPNC with PP as matrix involves the use of MAH as an intercalant and reactive reagent that grafts to the polymeric matrix [Tjong *et al.*, 2002]. It is important to recognise that to make the problem more difficult the authors used vermiculite. PP/vermiculite CPNC with an intercalated or exfoliated structure was prepared by melt compounding. As it will be apparent from the following discussion, the method is applicable to a variety of polymers.

Vermiculite (VA) is a phyllosilicate with the general formula: $[(Mg, Fe^{3+}, Al)_3 Mg_{0,33}]Al\ Si_3O_{10}\cdot(OH)_2\ (H_2O)_4$ (hardness = 2.5; density = 2.3 g/ml; crystal = octahedrally coordinated, monoclinic with perfect [001] cleavage). As MMT, vermiculite layers of about 1 nm thickness are composed of octahedral alumina or magnesia sandwiched between two tetrahedral silicate sheets. The basal spacing d_{001} = 1.4-1.5 nm depends on the interlayer cations and the degree of hydration. Its CEC = 1.8 meq/g is relatively high, thus VA is more difficult to intercalate then MMT. However, since vermiculites have high aspect ratio, $p \leq 2{,}500$, they are of great interest for use in packaging applications with high barrier properties.

The Tjong *et al.* method for the preparation of PO/VA nanocomposites consists of three steps:

1. Acid treatment of vermiculite (VA). An aqueous suspension of VA with HCl (pH = 3.0-4.0) was stirred at RT for 8 h. After washing (pH = 7.0), the filtrate was dried overnight at 160 °C. The final VA product was ground into a fine powder.
2. Intercalation of VA with MAH. The VA powder (50 g) was mixed for 12 h with MAH (50 g), and 100 ml of acetic acid (a carrier-solvent of MAH). The resulting MAV slurry was dried in a vacuum oven at $T \leq 70$ °C for 24 h, and kept under vacuum until needed.
3. Compounding MAV with a polymer. In the final step MAV was compounded in a TSE at 15 or 35 rpm with PP in the presence of dicumyl peroxide (DCP). The TSE barrel temperature profile was: 200-220-230-180 °C. The weight ratio of PP to MAV to DCP was either 96:4:0.3 or 90:10:0.3. The extrudates were pelletised, dried and then injection moulded into standard dogbone tensile bars.

XRD showed that the interlayer spacing of VA before treatment was d_{001} = 1.183 nm, after HCl-treatment it increased to 2.608 nm. However, XRD of MAV as well as that of the CPNCs containing 2 and 5 wt% of VA was featureless, indicating full exfoliation. MAH is a flat molecule with the longest length of about 0.513 nm (see the schematics below), thus it can easily enter the interlamellar galleries of VA.

CO
O
CO

The mechanical properties of PP/MAV nanocomposites are summarised in **Table 78**. There is a significant improvement of the tensile strength and moduli with a dramatic reduction of toughness. One may suspect that the peroxide not only caused MAH to graft, but also induced '*vis-breaking*' of PP macromolecules. Clearly, the method is intriguing by its generality and simplicity, but it needs optimisation.

The organoclays, commercial or prepared in a laboratory, are contaminated by residual solid particles (e.g., up to 3 wt% of quartz) and reaction byproducts:

Table 78 Relative (to PP) mechanical properties of PP/Vermiculite CPNCs. Data [Tjong *et al.*, 2002]

Composition	VA (wt%)	σ (MPa)	E (MPa)	G′ (GPa)	ε_b (%)	E_b (J)
PP	0	1.00	1.00	1.00	1.00	1.00
PP(MAV)-2/15	2	1.18	1.20	3.04	0.0282	0.057
PP(MAV-5/35	5	1.30	1.54	4.26	0.0153	0.056

Notes: σ= tensile strength; E = tensile modulus; G′= storage shear modulus; ε_b = elongation at break; E_b = energy at break

inorganic salts and excess of intercalant. The presence of quartz and inorganic salts is detrimental to mechanical and barrier properties, but the effect of excess intercalant is less known. Recently, Morgan and Harris [2003] investigated the effects of excess organoclay on the degree of clay dispersion in a PP matrix, as well as on the mechanical properties, and flammability. The authors prepared their own organoclay by intercalating fluoromica (FM; Somasif ME-100) with dimethyl dihydrogenated tallow ammonium chloride (2M2HTA) in a CEC ratio of 1 to 1. The precipitate was repetitively washed with water, filtered, dried, ground into fine powder, and then particles with diameter d < 120 μm were subjected to extraction.

The quaternary ammonium ions have high pK values and strongly bond to clay anions. Theoretically, with a stoichiometric amount of 2M2HT added, the organoclay should not have any intercalant excess. However, since physical adsorption of organic molecules on high energy crystalline solid surface is well known, not all the ammonium ions might have reacted. Thus, the organoclay powder was subjected to a Soxhlet extraction with ethanol for up to 4 days, then dried in a vacuum oven, re-ground, and melt compounded for 10 min with PP and PP-MA in an internal mixer, at 190 °C and 100 rpm. XRD scans of the organoclays indicated a sharp change of diffraction after one day of extraction and relatively small changes thereafter. Thus, the original spectrum displayed multiple peaks with spacing ranging from *ca.* 1.2 to 5.6 nm, with the principal peak at about d_{001} = 3.6 nm. After extraction the main peak moved to lower spacing (2.8 nm), became sharper and reduced in intensity. The major peak before extraction at 3.6 nm decreased in intensity to a shoulder. The harmonics of these two were also apparent. Thus, the removal of adsorbed intercalant reduced the interlayer spacing and it became more regular. The authors did not present corresponding XRD spectra for CPNC. Instead, they indicated a small change in spacing of the organoclay (ranging from +0.2 nm for not extracted to –0.06 nm for 4 days extracted) and evaluated the TEM dispersion in qualitative terms, with a clear preference for the extracted organoclays.

More interesting is the effect of these organoclays on performance. Probably owing to the thermal instability of ammonium ions, the onset of degradation decreased from 280 °C for PP with PP-MA, to 265 °C for all CPNCs. However, there was very significant reduction of flammability from peak heat release rate (PHRR) = 1435 for PP + PP-MA to 498 for CPNC with not extracted organoclay, to PHRR = 519 to

491 kW/m^2 for systems containing organoclay extracted for 1 to 4 days. The effect of intercalant extraction on flexural modulus and Izod impact strength at room temperature (NIRT) are shown in **Figure 154**. Thus, incorporation of about 8.5 wt% of not extracted organoclay into PP + PP-MA matrix increased the flexural modulus by *ca.* 54%, whereas addition of 7.9 wt% of extracted organoclay resulted in a 61 to 66% increase. The increased stiffness is accompanied by enhanced brittleness – NIRT is reduced by 35 to 45%. Unfortunately there is no information on crystallinity that may in part explain these results.

An interesting approach to dispersion of organoclay in PP melt was imposition of a dynamic flow field on CPNC within the injection moulding cavity [Zhang *et al.*, 2003]. This method of melt treatment is relatively common in the plastics industry as it is able to erase weld lines and impose orientation and morphology in multiphase polymeric system (e.g., see [Bevis, 2000]).

Nanocomposites were prepared by melt compounding of PP with PP-MA (0.6 wt% MAH) and (0, 1, 3, 5, or 10 wt%) MMT-2M2ODA in a TSE at 200 °C and screw speed of 110 rpm. The extrudates were injected into the mould cavity where the melt was forced to oscillate (at a frequency of 0.2, 0.33, or 1.0 Hz) by two pistons during the cooling stage. This dynamic packing was expected to engender exfoliation and orientation of the MMT platelets. For the sake of comparison, static (or normal) injection moulding was also carried out. According to XRD d_{001} = 3.3 nm was obtained (5 or 10 wt% organoclay; static or dynamic moulding). Thus, the melt compounding engendered moderate intercalation. However, the intensity of X-ray diffraction in the dynamic specimens was greatly reduced, indicating reduced ordering of the clay platelets. TEM showed that dynamic moulding produced a more homogeneous dispersion. Furthermore, bending and distorting MMT stacks could explain the reduction in XRD intensity.

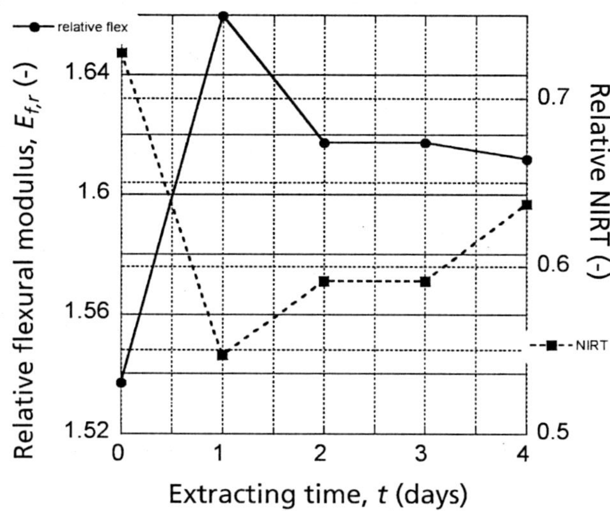

Figure 154 Relative flexural modulus and Izod impact strength for CPNC containing extracted organoclay. The relative magnitude refers to the corresponding performance of the matrix, viz. PP + PP-MA. Data [Morgan and Harris, 2003].

The data from SEM, XRD, and TEM show that dynamic packing resulted in better dispersion of MMT agglomerates. For the specimen without organoclay, dynamic packing imposed a strong orientation of PP macromolecules in the shear direction. By contrast, in CPNCs the matrix orientation was less evident (restricted molecular motion of the PP chains), but the field also weakened the alignment of MMT stacks, created during static injection moulding. The conclusion about randomisation of orientation in injection moulded specimens is confirmed by the relative tensile strength data (relative to the matrix performance), presented in **Figure 155**. As a result of the standard injection the platelets were oriented in the flow direction, strengthening the specimen – the opposite is true for the oscillatory packing. However, at 10 wt% loading the absolute magnitude of the tensile strength for the static specimens reached the maximum value of 56.2 MPa, thus about the same as was obtained for the dynamically injection moulded matrix (PP/PP-MA) specimen, in which the oscillation engendered macromolecular orientation. In the absence of experimental data, one may only speculate that dynamic packing generates mouldings with more uniform mechanical properties, *viz.* modulus and strength hence it may be used where this aspect of performance is important.

Xu *et al.* [2003] prepared PP-based CPNC by milling PP with PP-MA and organoclay (MMT-3MHDA). XRD of the home-prepared organoclay showed a peak corresponding to d_{001} = 1.9 nm. Compounding it with PP virtually did not affect the interlayer spacing. Two types of PP-MA were used, with either 0.6 or 0.9 wt% of MAH. Incorporation of 6 wt% of the former PP-MA did not change d_{001}, but addition of 10, 20 or 30 wt% of the latter increased it to 3.39, 3.68 or 4.01 nm, respectively. The mechanical properties were slightly improved (tensile

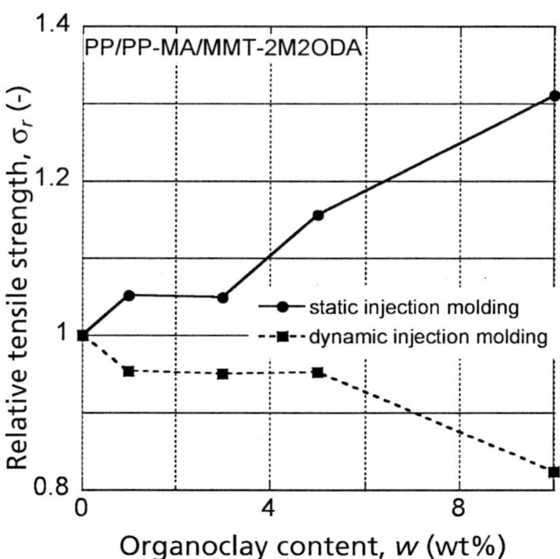

Figure 155 *Relative tensile strength versus organoclay content for injection moulded specimens of PP/PP-MA/MMT-2M2ODA. Two modes of injection moulding were used: a standard or static, and oscillatory or dynamic. Data [Zhang et al., 2003].*

and impact strength up by 25 and 33%, respectively), reaching a maximum for *ca.* 15 wt% PP-MA (at 2 wt% organoclay).

A similar method was used by Gorrasi *et al.* [2003]. Thus, PP was melt compounded with PP-MA and from 2.5 to 10 wt% of synthetic fluorohectorite pre-intercalated with ODA. XRD analysis showed that the organoclay was either delaminated or it formed short stacks with d_{001} = 3.1 to 3.2 nm. The Young's modulus increased with clay loading. Sorption and diffusion was measured for dichloromethane and *n*-pentane vapours. For both solvents, the sorption coefficient was similar to that of PP, but the diffusion (and as a result the permeability) strongly decreased with clay content. The cited data lead to (see Section 5.3) values of the average aspect ratio of p = 826, and p = 107 calculated from Equation 191, for *n*-pentane, and dichloromethane, respectively. Since for both these penetrants the film structure was the same, the apparent change of the aspect ratio must be related to the relative insolubility of the two solvents in the overlapping domains of the ODA-coated clay platelets.

As the previous discussion indicates, the patent and open literature on the preparation of CPNC with PO as the matrix is dominated by melt extrusion or melt compounding processes. However, even for POs there are several publications where the reactive route has been taken. One of these focuses on the preparation of CPNC during Ziegler-Natta polymerisation of ethylene [Jin *et al.*, 2002]. Na-MMT (Cloisite® Na) or MMT-OH (Cloisite® 30B, MMT-MT2EtOH) was used. Using the latter organoclay, in the first step the catalyst, $TiCl_4$, was reacted in toluene with the organoclay –OH groups. Next, after addition of triethyl aluminium, Et_3Al, ethylene was polymerised in a high pressure glass reactor at P = 0.4 MPa and T = 30 to 50 °C. The polymerisation was quenched with dilute HCl solution in methanol, the polymer precipitated, filtered and dried. A similar procedure was used with Na-MMT, but here the amount of –OH groups was smaller, and consequently so was the amount of catalyst that could be fixed to clay (*ca.* 1/3). As a result, during the polymerisation of ethylene full exfoliation was readily achieved with MMT-MT2EtOH, but only a modest intercalation (d_{001} = 1 to 1.4 nm) with Na-MMT. Hydroxyl groups within Cloisite® 30B provided reactive sites for anchoring catalysts in between the silicate layers. Comparison of exfoliation characteristics between these two CPNCs shows that the feasibility of exfoliation during ethylene polymerisation depends on catalyst fixation.

Unfortunately, extrusion of the *in situ* polymerised, fully exfoliated CPNC with Cloisite® 30B, resulted in re-aggregation of the MMT-platelets. XRD of the extrudates showed a shallow peak at $2\theta = 6°$ corresponding to d_{001} = 1.47 nm. The interlayer spacing of Cloisite® 30B is d_{001} = 1.86 nm, suggesting that the intercalant-clay bonding has been destroyed, possibly by the Hofmann elimination mechanism (see Section 3.2 of this book).

Similarly, good exfoliation was observed in CPNCs of *in situ* polymerised PE with metallocene constrained geometry catalyst (CGC) [Alexandre *et al.*, 2002]. Three non-modified clays: synthetic hectorite, natural MMT and kaolin were used. First, the dried clays were treated by trimethyl-aluminium-depleted methyl aluminoxane (MAO) before being contacted by *tert*-butyl-amidodimethyl (tetra methyl-η^5-cyclopentadienyl) silane titanium dimethyl (a CGC catalyst from Dow). Since the co-catalyst was chemically bonded to the clay platelet, the polymerisation of ethylene to UHMWPE (absence of a chain transfer) took place within the interlamellar galleries, pushing the platelets apart. XRD analysis showed that upon polymerisation the diffraction peaks of the original clays disappeared, thus the clays in reactor powders were exfoliated. This conclusion was confirmed by TEM. However, as the authors stress, the PE/clay system is immiscible and the exfoliated structure of CPNC is thermodynamically unstable. As a result, even mild processing (e.g., compression moulding) caused the platelets to re-aggregate into stacks with interlayer spacing ranging from *ca.* 1 to 2 nm. Several sets of properties were determined for these systems.

The mechanical properties were measured using compression moulded specimens. The performance was poor, e.g., the tensile modulus of the CPNC was mostly below the level obtained for neat PE. The specimens were exceedingly brittle. For the materials where hydrogen addition reduced the molecular weight of PE, the performance was significantly better.

In reactor powder the formed macromolecules crystallise virtually non-entangled. Using a Rheometrics Mechanical Spectrometer (RMS-605) it was determined that it takes *ca.* 18 h at 180 °C for the low frequency moduli to increase to a steady-state plateau [Masson and Utracki, 1987]. The time to reach plateau and the magnitude of the effect depended on M_w. Since for each sample its value remained constant and the effect on storage modulus was significantly higher than that on loss, it was concluded that the mechanism responsible was diffusion-controlled entanglement – essential for good mechanical performance. However, in the case of nanocomposites from Alexandre *et al.* [2002], the concurrent re-aggregation of clay platelets would have a negative effect. As expected, addition of clay significantly increased the T_m and crystallinity, suggesting enhanced nucleation.

For the preparation of CPNC with PO as the matrix, Bishop and Niyogi [2002; 2003] patented an oligomeric intercalant of the type:

$$(CH_3)_2\text{-}CH\text{-}(CH_2)_2\text{-}[CH_2\text{-}CH(CH_3)\text{-}CH_2\text{-}CH_2]_n\text{-}NR'_2\bullet HCl$$

where: n = 2 to 17, and R´ is a radical selected from: H or an alkyl C_1 to C_4. The patent provides detailed information on how to prepare the intercalant [e.g., *n*-butyl(3,7,11-trimethyl-dodecyl) amine hydrochloride], an organoclay with smectites, and melt compounded CPNC with PP or a poly(propylene-co-ethylene). In the former case, additional compatibiliser (PP-MAH) may have to be used. Bafna *et al.* [2003] prepared CPNC with HDPE as the matrix using 6 wt% of organoclay (probably Cloisite® 20A, MMT-2M2HTA) and 3 or 12 wt% of maleated polyethylene (PE-MA; 2 wt% MAH) as a compatibiliser. The compounding was carried out in a TSE at 180 to 190 °C and 250 rpm. The orientation and other structural features in the CPNC were studied using two-dimensional (2D) SAXS and 2D WAXS. The authors identified dispersion and 3D orientations of six structural features:

1. Clay clusters/tactoids (120 nm),
2. Modified clay (002) (2.4-3.1 nm),
3. Unmodified clay (002) (1.3 nm),
4. Clay (110) and (020) planes normal to (b) and (c),
5. Polymer crystalline lamellae (001) (19-26 nm), and
6. Polymer unit cell (110) and (200) planes.

Values of these parameters were determined from three scattering projections for each sample through calculation of the major axis direction cosines and through a ternary, direction-cosine plot. The crystallinity in CPNC was slightly lower than that in neat HDPE (78 versus 80%), and T_m was reduced by 1 °C. Consequently, the HDPE lamellae in CPNC were more robust than those in the neat resin. The organoclay interlayer spacing of 2.42 increased upon compounding with 3 and 12 wt% PE-MA to 2.73 and 3.14 nm, respectively; a modest intercalation was obtained. Analysis indicated that in the organoclay used 33% of the clay was not modified. This unmodified clay showed the most direct orientation with the polymer crystalline lamellae, being responsible for the increased lamellar thickness. By contrast, the expanded organoclay was associated with tactoid structures, about 120 nm large.

4.1.3 General Methods of CPNC Preparation

Patents discussed in this part are of a general nature – they are formulated not for a specific type of resin, but claim applicability of the method to diverse polymers. Placing this section after PA and PO patents is deliberate. On the one hand, the literature of PA-nanoclays is voluminous, on the other the general patents frequently provide methods for the preparation of CPNC with PO as matrix, thus are a direct extension of the PO-specific patents discussed in the preceding Sections 4.1.1 and 4.1.2.

In the previously discussed AlliedSignal patent LDPE, PP, P4MP and chlorinated-LDPE were mentioned as polymers that could profit from the invention, but no example was given [Christiani and Maxfield, 1998]. However, the described method of exfoliating the clay by treating them with well-known silanes, titanates or zirconates and then melt compounding in a high stress field may indeed be economically advantageous (more so for PEs than for the readily degradable PP).

Similarly, in AMCOL patents POs have been mentioned (see for example [Beall *et al.*, 1999]). The invention focuses on water soluble polymers, with undocumented extension to water insoluble ones. The CPNC are manufactured by combining a host material, such as a dissolved or molten polymer with intercalate. Contacting clay with a polar polymer that is adsorbed between the platelets forms the intercalate, increasing the interlamellar gallery height to ≥ 1 nm. The intercalated clay is then exfoliated, e.g., when mixed with an organic solvent or a polymer melt, to provide a carrier for chemical delivery, or a PNC.

The best intercalate for exfoliation should contain > 10 wt% of polymer. It is prepared by dispersing clay in a solution of polymer using an intercalating carrier (preferably water, with or without an organic solvent, e.g., ethanol). Alternatively, the dry polymer and clay with ≥ 4 wt% water, may be dry-blended, then the intercalating carrier added to the blend. Sorption may be aided by exposure to heat, pressure, ultrasonic cavitation, or microwaves. The layered material, containing at least 4-15 wt% H_2O, may also be blended with a dissolved polymer (with a carrier) in an extruder at $T \geq T_m + 50$ °C. It is noteworthy that intercalation does not take place in the absence of water! The water-insoluble polymers that may profit from this technology include PA, PEST, PC, PE, epoxies, PO (polymers and copolymers based on ethylene, propylene and other α-olefin monomers); polyalkylamides; and their mixtures. For example, 80 wt% PA was mechanically blended under N_2 with 20 wt% Na-MMT at 230 °C. According to XRD, the PA intercalated and exfoliated the platelets. Similarly, 90 wt% of PET was melt blended with 10 wt% of Na-MMT (containing 8 wt% moisture), causing intercalation and exfoliation of clay. Similarly, melt blending at 280 °C of PC with 50 wt% Na-MMT (containing 8 wt% moisture) resulted in exfoliation.

In the Dow patent already discussed, CPNC was prepared using diverse multilayered materials pre-intercalated with organic and/or inorganic intercalants, then dispersed into thermosets, thermoplastics or elastomers [Nichols and Chou, 1999]. The inorganic intercalant can be a substance obtained by hydrolysing a metallic alcoholate: $Si(OR)_4$, $Al(OR)_3$, $Ge(OR)_4$, $Si(OC_2H_5)_4$, $Si(OCH_3)_4$, $Ge(OC_3H_7)$, $Ge(OC_2H_5)_4$, etc. To be intercalated the clay should be swollen, then dried at 50-80 °C. The process increases the interlamellar gallery from ≤ 0.4 to 60 nm. Melt compounding may be used for the preparation of CPNC. The claims specifically mention LDPE, LLDPE and other linear ethylene copolymers with density ρ = 850-920 kg/m^3 and MFR = 0.1-10 g/min.

4.1.3.1 Hudson's Clay Grafting Method

Hudson [1998, 1999] described a three-step method for the production of PNC:

1. Nanometre-size fillers are functionalised with an aminosilane.
2. A carboxylated or maleated-PO is grafted to the filler through an amine-carboxyl reaction.
3. The modified filler is dispersed in a semicrystalline PO (e.g., PE or PP). Co-crystallisation between the carboxylated or maleated PO-chain and that of the semicrystalline PO matrix improves the interactions between the filler and the matrix, thus performance.

A semi-crystalline CPNC may comprise: (1) 50 to 99 wt% of a semi-crystalline PO, (2) 0.1 to 50 wt% of clay, having dispersible platelets in stacks (the platelets are 1 to 30 nm thick and are swollen with intercalated polymer), (3) 0.001 to 3 wt% of an amino functional silane, reacted with the clay, and (4) 0.1 to 30 wt% of a carboxylated or maleated semicrystalline PO, reacted with the aforementioned amino-functional silane. According to the patent, the nanosize fillers include clays, both natural (e.g., MMT) or synthetic, precipitated, fumed, an aerogel or mined products rich in SiO_2, e.g., diatomaceous earth, mica, wollastonite, etc.

The preferred aminosilane has the formula: H_2NR_1-$Si(R_2)(R_3)R_4$, where R_1 is a C_{3-12} alkylene (butylene, hexylene, or dodecylene), R_2 and R_3 independently are a hydroxyl, a halogen, an alkyl, alkoxy, phenoxy, or aromatic group having from 1-12 carbon atoms, R_4 is a halogen, an alkoxy or phenoxy having from 1-12 carbon atoms. The aminosilane can be added to the filler by any method that exposes its surface to the aminosilane for a sufficiently long time for the reaction to take place. The aminosilane can be added as a gas, liquid, dissolved or dispersed in gas or liquid. Depending on temperatures and catalysts present, the reaction at RT may require seconds, hours or days. For example, clay was mixed with amino ethyl-dimethyl ethoxysilane in either ethanol and water, or dimethyl acetamide, and thereafter reacted for 2-4 hours at 25-90 °C.

The reaction of maleated or carboxylated PO with the aminosilane-functionalised filler can be conducted under a variety of conditions. For economic reasons, reaction with molten PO is preferred. Thus, MA-PP was dissolved in xylene (1-5 wt% MA-PP; M_n = 20 kg/mol; 1 wt% MA-groups), and then aminosilane-functionalised filler was added. The reaction at 120 °C was completed in 30 min. Dewatering drives the reaction between the amine and the carboxyl groups, thus it is faster at higher temperatures and under reduced pressure. The carboxylated or maleated polyolefin should constitute 0.3 to 20 wt% of the filler. The principal polymer of this invention is 7-90 wt% of a semicrystalline PO, with ≥ 15 wt% crystallinity. The term 'principal polymer' is used because small amounts of polymeric additives may be needed to impart impact strength, ease of processability, etc. The carboxylated or maleated PO can be a homopolymer or a copolymer, provided that it is able to co-crystallise with the principal polymer.

Two type of benefits are anticipated from this process. On the one hand it ascertain good bonding between the clay platelets and the matrix, and on the other it generates more tie chains between the crystals in the matrix itself. These interconnect the crystals by crystallising in multiple ones. When a stress is applied, the tie chains are expected to stretch, increasing the modulus and tensile strength. Since the crack growth requires breaking the tie chains between crystalline domains on both sides of a crack, an increase in the relative amount of tie chains should slow the crack growth. In these systems clay mostly remains in short, intercalated stacks [Jeong *et al.*, 1998]. Nevertheless, the mechanical properties of the new CPNC (see **Table 79**) are excellent. With just 1 wt% of nanofiller loading the Young's modulus increased nearly 7-fold to $E = 13$ GPa. At higher loading, further increases were modest, reaching $E = 25$ GPa at 30% loading. The tensile strength also increased with addition of nanofiller, but *ca.* 10% loading was required to achieve a significant increase by a factor of 2 to $\sigma = 60$ MPa. Incorporation of organoclay had a moderately detrimental effect on the Izod impact strength, *viz.* 10% loading reduced it by 18%, and 30% loading by 27%. Finally, the rate of slow crack growth was retarded by addition of chain-grafted filler particles. At 10% filler content, the time to failure increased 4.5 times. It was observed that filler particles do not initiate failure, but rather they act as reinforcement. In conclusion, the new PO-based CPNCs are useful for moulding grades. In particular they may be used to replace PO when decreased crack growth rates are desired.

4.1.3.2 Hasegawa et al. Method with Functionalised Compatibilisers

Hasegawa *et al.* [2000] deposited a general patent on CPNC with a polymer blend as a matrix. The system comprises: (1) organophilic clay and (2) a mixture of two or more polymers, one of which is polyphenylene oxide (PPE), or a copolymer with oxazoline functional groups. Evidently, the main system of interest to the authors is CPNC with PPE/PS as the matrix. However the patent claims are broad, covering virtually any polymer that can be chemically modified by attaching functional groups.

The organoclay is a clay with CEC = 0.5 to 2 meq/g (e.g., MMT, saponite, hectorite, beidellite, nontronite, vermiculite, halloysite, mica), formed by ion exchange with organic ammonium ions, e.g., hexyl-, octyl-, 2-ethylhexyl-, dodecyl-, octadecyl-, dimethyl-di-octyl-, trioctyl-, dimethyl-dioctadecyl-, trimethyl-octadecyl-, etc. The onium content is 0.3 to 3 equivalents of the clay CEC.

The patent focuses on PS/PPE blend nanocomposites, with the styrene copolymer having oxazoline functionality. Thus, the role of the functionalised copolymer is to compatibilise the organoclay-PPE system by reacting with clay and being miscible with PPE. The invention resembles that of Fibiger *et al.* [2001] discussed in Section 4.1.2.2. The essence of the invention is the presence of a functional group having a high affinity for organoclay. When the functional group is able to interact with the clay, the latter becomes exfoliated and the physical properties are significantly improved.

Table 79 Physical properties of PNC with modified clay. Data [Hudson, 1999]

1. Modified clay = 10 wt% silane-treated clay (containing 2 wt% silane) and 90 wt% MA-PP							
Modified clay (wt%)	0	1	2	5	10	20	30
Isotactic PP (wt%)	100	99	98	95	90	80	70
Tensile modulus, *E* (GPa)	1.9	13	14	16	16	17	18
Strength, σ(MPa)	31	32	32	33	43	50	53
Izod impact, *NI* (J/m)	33	31	31	30	31	29	28
2. modified clay = 60 wt% silane-treated clay (containing 2 wt% silane) and 40 wt% MA-PP							
Modified clay (wt%)	0	1	2	5	10	20	30
Isotactic PP (wt%)	100	99	98	95	90	80	70
Tensile modulus, *E* (GPa)	1.9	16	16	18	19	20	23
Strength, σ(MPa)	31	34	35	35	57	64	68
Izod impact, *NI* (J/m)	33	31	29	29	27	25	22
3. Modified clay = 10 wt% silane-treated clay (without silane) and 90 wt% MA-PP							
Modified clay (wt%)	0	1	2	5	10	20	30
Isotactic PP (wt%)	100	99	98	95	90	80	70
Izod impact, *NI* (J/m)	33	31	29	28	27	24	22
4. Modified clay = 60 wt% silane-treated clay (without silane) and 40 wt% MA-PP							
Modified clay (wt%)	0	1	2	5	10	20	30
Isotactic PP (wt%)	100	99	98	95	90	80	70
Izod impact, *NI* (J/m)	33	29	28	27	25	22	21

Blends of any two or more polymers may be used. The simplest case is when the blend is miscible. However, in the case of immiscibility, the polymers should have functional groups facilitating compatibilisation. For example, the resin may be a copolymer with functional groups, *viz.* modified PE, PP, PB, polypentene, EPR, EPDM, ethylene-butene copolymers, polybutadiene, polyisoprene, hydrogenated polybutadiene, hydrogenated polyisoprene, butyl rubber, PS, SBR, SEBS, PA, PC, POM, PEST, PPE, PPS, PES, PEEK, PAr, PMP, PPA, polyethernitrile, polybenzimidazole, polycarbodiimide, PTFE, PAI, PEI, LCP, polysiloxanes, as well as epoxy, melamine, urea, di-allyl phthalate, phenolic, silicone and urethane resins with M_n = 5 to 10,000 kg/mol. The functional monomer (0.01 to 50 mol%) may be selected from between: acrylic monomers, acrylamides, compounds having an unsaturated carbon atom, and monomers having aromatic rings (see below).

When the component polymers are miscible with each other, only compatibilisation between clay and matrix is required, which can be accomplished by introducing the functional groups into one macromolecular component. Examples of miscible blends include PPE/PS, PS/PVME, SMA-MMA/SAN, PVC/SAN, PVC/PCL, PMMA/PVDF, and PC with MMA-copolymers, etc. [Utracki, 1989, 1998, 2002, 2004]. When the polymers are immiscible a compatibiliser must be used to ascertain good interactions between all components of the system. For example, if A and B polymers are immiscible, addition of A-*b*-B block copolymer may serve as a compatibiliser between them. When such a blend is used as a CPNC matrix, it suffices to functionalise the A-*b*-B copolymer to engender exfoliation in the compatibilised blend and assure a high level of exfoliation. However, the phase equilibrium in multicomponent system depends on a precarious balance of entropic end enthalpic contributions, thus the type and extent of functionalisation must be carefully adjusted to the process parameters. Blends of PO (e.g., of PE with PP) are often compatibilised by EPR, thus grafting the latter with functionalised monomer may lead to exfoliated CPNC. For example, good performance of a PP/organoclay system was obtained after compatibilisation with either functionalised EPR or modified butyl rubber and modified EPR.

Virtually any known functional group may be used, *viz.* acid anhydride, carboxyl, hydroxyl, thiol, epoxy, halide, ester, amide, urea, urethane, ether, thioether, sulfonic acid, phosphonic acid, nitro, amino, oxazoline, imide and isocyanate. Furthermore aromatic rings such as benzene, pyridine, pyrrole, furan, and thiophene may also provide sufficient interactions with clay. Monomers having these functional groups may be copolymerised with, e.g., ethylene, propylene, butene, pentene, acetylene, butadiene or isoprene; acrylics and methacrylics, acrylamides and methacrylamides, styrene and methylstyrene, etc.

Hasegawa *et al.* [2000] described diverse methods for the production of CPNC, e.g., in a solvent or in the molten state. The preferred method is melt compounding in an extruder. The patent provides the following example for PO-based CPNC. First, Na-MMT was pre-intercalated with 32 wt% ODA (d_{001} = 2.2 nm). Maleated EPR (MAH content = 0.04 mmol/g) was used as the functionalised polymer and PP as the matrix. The ingredients were compounded at 200 °C in a TSE. Analysis of the product showed that modified EPR formed micelles (containing the organoclay) within the PP matrix. The PP/EPR blend was immiscible and the nanometre-scale clay dispersion was evident only within the functionalised copolymer phase. Similar results were obtained using EPR and maleated-PP. However, when modified PP (MAH content = 0.02 mmol/g) and modified EPR (MAH = 0.04 mmol/g) were melt compounded with organoclay, the latter was found dispersed on a nanometre scale in both phases, the dispersed EPR and the PP-matrix. A similar situation was found when PP was replaced by butyl rubber. The example illustrates another aspect of CPNC technology – directing the rigid clay particles to a specific phase. The mechanical performance of the system very much depends on where the reinforcement is located. For example, improved modulus can be expected for clay located in the crystalline phase – incorporation in an elastomeric phase is to be avoided.

The document provides several other CPNC examples with matrices formed by blends of PO, PS, PPE, acrylics and copolymers (e.g., EPR, SMA, ethylene-co-methyl methacrylate; styrene-co-vinyl-oxazoline).

4.1.3.3 CPNC with Amino-Aryl Lactam Clays

One piece of research describes CPNC preparation in two steps [Liao *et al.*, 2000a,b]: (1) Intercalation of clay (smectite, vermiculite, halloysite, or fluoromica; CEC = 0.5 to 2 meq/g) with a polymerisable amino-aryl lactam, e.g., *N*-(*p*-amino-benzoyl)caprolactam. (2) Mixing 1-30 wt% of the intercalated clay with a thermoplastic polymer. During this step the polymer melts and the intercalated lactam polymerises, which causes the clay to exfoliate, forming a thermoplastic CPNC. According to the patent claims, the method is general, suitable for the manufacture of CPNC with any thermoplastic polymer, *viz.* 'crystalline polar thermoplastic polymers, crystalline nonpolar thermoplastic polymers, non-crystalline non-polar thermoplastic polymers, and non-crystalline polar thermoplastic polymers'.

In this invention, the preferred clay is MMT intercalated with ω-amino dodecyl acid (MMT-ADA). The organoclay was dispersed in a pyridine solution of *N*-(*p*-aminobenzoyl)caprolactam where the latter compound partially ion-exchanges with ADA and partially is physicosorbed by the clay. After vacuum drying, the complex of lactam intercalated MMT-ADA was compounded in an internal mixer with PA-6, PP, PS, or PC, to produce four thermoplastic CPNCs with concentration of the MMT-ADA-lactam of 1, 2, 5, and 10 wt%. The products were hot pressed into films. The XRD spectra for these nanocomposites indicate that good dispersion has been invariably obtained at low clay loading, *viz.* at 1 and 2 wt% of organoclay. The exception is CPNC with PC as the matrix, which remained exfoliated at 5 wt%. In the case of concentrated CPNC with PA-6, PP or PS as the matrix, the characteristic organoclay peak was evident at 5 and 10 wt% organoclay loading.

4.1.3.4 Ishida's Method

Ishida [2001] patented another general method of preparation of CPNC. The method is suitable for a variety of polymeric matrices (55 to 94 wt% of such resin as, e.g., PO, PA PEST, PC, POM, vinyls, styrenics, etc.), diverse clays (5 to 30 wt% of, e.g., MMT, kaolinite, illite, bentonite, mica or halloysite), and (1 to 15 wt%) intercalant. The preferred intercalants include epoxy monomers, PDMS, caprolactam, vinyl monomers and/or oligomers. By varying the type, amount, and mixing sequence one may form intercalated or exfoliated CPNC. The process does not require any special equipment as intercalation and exfoliation occur during compounding.

The first intercalation step may involve treating clay with amines or polyamines (having 1 to 100 carbon atoms), or with silanes (that have at least one silicon atom and at least two functional groups). For example, Na-MMT was intercalated with ADA or hexadecyl amine (HDA) in acidified aqueous medium at 60 °C, then dried under vacuum. The CPNC was prepared by mixing polymer, organoclay, and epoxy (Epon 825) at a temperature above the polymer softening or melting point. Mixing was manual (30 min) or in an internal mixer (e.g., PE with 2 wt% epoxy and 5 wt% of organoclay at 40 or 60 rpm for 5 to 240 min). After 10 min of mixing at 40 rpm the clay diffraction peak was still evident, but it was absent when higher mixing speed (60 rpm) was used. According to the patent, when all three components are added in the beginning of the process, at $T \leq 75$ °C the epoxy monomer solvates and swells the organoclay galleries. At higher temperature the intercalated epoxy polymerises. The reaction (which is catalysed by alkylammonium protons) results in gel formation. Additional diffusion of epoxy monomers into the galleries and progressive expansion of the interlayer spacing results in formation of phase-separated powdered nanocomposite is obtained. Consequently, the epoxy coating prevents the matrix polymer from entering the clay galleries, thus the resulting CPNC is a two-phase mixture of matrix polymer and epoxy-clay dispersion.

To prevent such undesirable phase separation, the matrix polymer should first be heated above its T_g or T_m. Next, while stirring, organoclay and epoxy should be added. CPNC formation depends on the mass transport of molten polymer into clay galleries. The epoxy role is to increase the interlayer spacing thus facilitating polymer diffusion. Two factors contribute to the exfoliation and homogeneous dispersion of clay layers: (1) the ability of epoxy to diffuse into the clay galleries, and (2) the epoxy/polymer miscibility. It is noteworthy that epoxy alone does not exfoliate the clay – a modest expansion of d_{001} by *ca.* 0.3 nm may be obtained.

A summary of data was presented earlier in this book, *viz.* **Table 5**, in Section 1.2.10. Evidently, intercalated or exfoliated CPNCs have been prepared using one specific organoclay (MMT-ADA), one organoclay loading (10 wt%) and one concentration of one type of epoxy (2 wt% of Epon 825). The results are often unexpected, e.g., showing fully exfoliated PVC, but intercalated PEG, SAN or polybutyl methacrylate. Extended studies and optimisation is required. Take, for example, preparation of CPNC with either PE or PP as the matrix. When the epoxy concentration was 2 wt%, the compounds showed two XRD peaks indicating only a partial intercalation. However, when the concentration was increased to 5 wt%, fully exfoliated CPNC with either a PE or PP matrix was obtained. It is noteworthy that these polymers are immiscible with epoxy. According to the author, when 2 wt% epoxy was used it only swelled the clay, while when the concentration was increased to 5 wt%, the clay galleries were greatly expanded, thus the macromolecules could complete the exfoliation process. However, independently of the epoxy concentration, there is no bonding between the dispersed epoxy-covered platelets and the matrix.

Further information on the process may be found in an article [Ishida *et al.*, 2000]. CPNCs with twenty-four polymers were prepared by melt mixing of the polymer, clay and intercalant. As in the patent, Na-MMT intercalated with either ADA or HDA was used. The compatibilising intercalant was either the diglycidyl ether of bisphenol-A (Epon 825) or PDMS. The matrix polymer was selected from between: PE, PS, PC, PCL, PEG, PTFE, PP, PA-6, PVC, SAN, PA-12, PVAl, PVAc, POM, PMMA, poly(1-butadiene), poly(1-butene), polychloroprene, polyisoprene, polyisobutylene, polyethyl methacrylate, polybutyl methacrylate, polyoctadecyl methacrylate, and polyvinyl imidazole. CPNC was prepared by melting the polymer, then adding clay (10 wt%) and either epoxy or PDMS. Epoxy was chosen because it is known to exfoliate clay and has a high boiling point. PDMS was selected because it has been found useful in the preparation of CPNCs (reduction of the clay/polymer interfacial tension coefficient). The time required for reaching equilibrium interlayer spacing depended on the polymer and the preparation method used. While 5 and 10 min was required for PS and PE, respectively, samples with polyisobutylene, polybutadiene, poly(1-butene), and polyisoprene were mixed for 30 min then kept overnight at 140 °C under vacuum.

At an epoxy concentration of 2 and 5 wt% intercalation and exfoliation of clay, respectively, were obtained. However, 5 wt% epoxy is a large quantity considering that even at 10 wt% loading of the organoclay the clay concentration is *ca.* 7 to 7.5 wt%, thus epoxy constitutes *ca.* 40 wt% of the clay/epoxy complex. Considering the density difference, the volume of the inorganic part is significantly smaller than that of the organic one – clay is fully covered in epoxy. To explain the diverse degree of exfoliation the solubility parameter (δ) explanation is again used. The authors postulate that highest miscibility between epoxy and polymer requires that their solubility parameters match, $\delta_{polymer} \cong \delta_{epoxy}$ = 9-11. This explanation is hardly supported by the reported data. One problem may be the omission of the Hansen contribution to the solubility parameter concept – the importance of matching the dispersive, polar and hydrogen bonding parts within the solubility sphere [Hansen, 1967, 1994, 1995]. However, the recent statistical thermodynamic calculations of the solubility parameters of molten polymers indicate that the listed values obtained from the experiments in solutions are quite different from those computed for the melt [Utracki and Simha, 2003b]. The difference originates in the large difference between the free volume present in solution and in the melt. For example, good correlation between the tabulated δ-values and the computed ones was obtained when computations were carried out for T_g + 300 K. Thus, the use of the solubility parameters as listed in the tables is correct for solutions, but not for melts.

4.1.3.5 Edge Reactions of Clay Platelets

As already discussed in Section 4.1.2.3, Fukatani *et al.* [2002] developed a three-step exfoliation procedure: (1) use a standard cation-exchanged organoclay, (2) reaction with end-functionalised compound able to interact with the clay platelet edge surface, and (3) melt mixing of the modified clay with a polymer. By obtaining strong interactions between the secondary intercalant and matrix polymer, the method may be applied to prepare CPNC with any matrix resin. The authors proposed two types of secondary intercalant: (A) one able to bond with the clay hydroxyl groups, and (B) one able to react with the clay cations located at the platelet edges. It is essential that the secondary intercalant should have another moiety that provides strong interactions between the clay complex and the matrix. SCP [Powell, 2001a,b] and later Dow [Chou and Garcia-Meitin, 2001] disclosed similar processes. Thus in Powell's patents, a purified clay is to be intercalated with a branched chain quaternary ammonium ion, then edge treated with a negatively charged organic molecule (e.g., a polyanionic polymer, polyacrylates) and melt compounded with a matrix polymer. More specifically, polyacrylate (PAC for short, e.g., Alcogum SL-76, or JARCO M-25B) was added to an aqueous slurry of purified MMT at *ca.* 00.5 wt% level, prior to a high shearing step in a Manton-Gaulin mill. Following the shearing the clay was treated with quaternary ammonium ion. The invention is pertinent to any polymeric matrix, but illustrated with examples of PA-66/MMT. The sheared clay was either edge-treated with PAC, surface-treated with a dimethyl hydrogenated tallow-2-ethyl hexyl ammonium methyl sulfate (2MHTL8, as in Cloisite® 25A) or both. The treated slurries were dewatered, and the resulting cake was dried, ground and mixed with polymer pellets, and then compounded in a TSE. The degree of dispersion was determined by XRD. As evident from the summary in **Table 80**, the additional

Table 80 Assessment of the degree of clay disperson. Data [Powell, 2001]

Clay (wt%)	Edge treatment	Surface treatment	Comments
7.3	No	Yes	Incomplete exfoliation
4.2	No	Yes	$d_{001} \equiv 7$ nm
6.4	Yes	No	Poor exfoliation
4.7	Yes	No	Poor exfoliation
5.4	No	Yes	Exfoliation
3.4	Yes	Yes	Excellent exfoliation
4.6	No	Yes	Exfoliation
4.6	Yes	Yes	Excellent exfoliation

treatment with PAC improved the dispersion. For a clay loading ≤ 4.6 wt%, after the surface and edge clay treatment, excellent exfoliation was obtained. It seems that surface treatment of MMT with 2MHTL8 is more important than edge treatment, but a combination of both leads to the best results.

MMT was dispersed in water, then first treated with anionic poly(ethylene-co-acrylic acid) (MW = 3.4 kg/mol), and then with mixed quaternary ammonium cations (e.g., MDD2EtOH and MDD2HOH). The resulting organoclay might be incorporated into a variety of thermoplastic or thermoset polymers. In the latter case, it is preferred to mix the organoclay with a monomer or pre-polymer, and then cure the system [Chou and Garcia-Meitin, 2001].

The Rheox patent [Ross and Kaizerman, 2002] was discussed in Section 4.1.1.4. The invention describes a new, doubly intercalated organoclay. Thus, clay was intercalated in aqueous medium with a non-ionic, water-soluble (or water dispersible) polymer and a quaternary ammonium ion. The authors have shown that incorporation of polymers significantly improves the interlayer spacing, e.g., from *ca.* d_{001} = 2 to 4.4 nm. The role of the polymer is to hydrogen bond to –OH groups on the clay platelet surface – these are mainly available on the platelet edges, away from the flat surfaces crowded with quaternary onium cations. The doubly intercalated organoclay may be dispersed into a thermoplastic resin using a TSE or an internal mixer. Polyolefins were specified as resins suitable for CPNC with new organoclay, but no examples were provided.

4.1.4 Vinyl Polymers and Copolymers

To this category of CPNCs belong nanocomposites with polymeric matrices, which contain vinyl monomers, *viz.* –[CH_2-CHX]–, where X is an atom or group (excluding H and C_nH_{2n+1}, assigned to PO). Owing to commercial importance, three main resin types may be distinguished: (1) styrenics, (2) acrylics, and (3) vinyl halides (e.g., PVC). The CPNC of a vinyl polymer widely used for the control of permeability, polyvinyl alcohol (PVAl), will be discussed along with other water soluble resins.

Vinyl-type CPNCs have been prepared using chemical means (polymerisation in the presence of clay), or physical means, i.e., dispersing organoclay either in a polymer solution or in a molten resin.

4.1.4.1 Polymerisation in the Presence of Clay

Vinyl monomers are relatively easy to polymerise by different reaction mechanisms and in different media. Polymerisation in the presence of organoclay has been most frequently used. The preferred mechanism is free radical, historically in bulk, but today mainly in emulsion or suspension, although solution polymerisation has also been used. Since clay intercalation is usually performed in water, the emulsion or suspension polymerisation is quite natural, especially when the resulting CPNC latex can be directly used, e.g. as paints, adhesives, sealants or laminations.

It is important to ascertain that clay is wetted by the monomer. In emulsion polymerisation one may start with neat clay, intercalate it by means of pH control and mechanical or ultrasonic mixing, then intercalate with water soluble monomer or oligomer and carry on polymerisation starting in the aqueous phase. Thus, in this procedure, the costly pre-intercalation with onium is avoided.

However, when the monomer forms its own 'oil' phase, the clay must be treated to reduce its hydrophilic character. If the clay is not transferred to the monomer phase, it cannot be exfoliated during polymerisation. In this case, a limited degree of intercalation may be expected, rather resulting from high stress melt compounding during the forming stage than from polymerisation. These CPNCs will show behaviour similar to filled composites.

The classical thermodynamic description of three-phase morphology is given by the well-known 'Neumann's triangle' equation:

$$\cos\Theta_2 = \left(v_{12}^2 + v_{23}^2 - v_{31}^2\right)/\left(2v_{12}v_{23}\right) \qquad (179)$$

The relation is general, predicting relative placement of three liquid phases, characterised by three interfacial tension coefficients, v_{ij}. For the emulsion polymerisation of vinyl monomer(s) to produce well dispersed CPNC the following index assignment may be made (see **Figure 156**) [Inoue, 2002]: (1) = organoclay; (2) = 'oil' phase containing vinyl monomer(s); (3) = aqueous phase. Thus, to have the organoclay dispersed within the 'oil' phase, Equation 179 requires that:

$$\nu_{13} > \nu_{12} + \nu_{23} \qquad (180)$$

Since the lattice theory predicts that the interfacial tension coefficient is proportional to the square root of the binary interaction parameter [Utracki, 2002a]:

$$\nu_{ij} \propto \sqrt{\chi_{ij}} \Rightarrow \sqrt{\chi_{13}} > \sqrt{\chi_{12}} + \sqrt{\chi_{23}} \qquad (181)$$

In other words, the positive (repulsive) interaction coefficient between the aqueous phase and organoclay must be larger than the sum of those between the organoclay and the oil phase and the oil phase and the aqueous medium.

The theory was derived for three mutually immiscible phases where $\chi_{ij} \geq 0$. As a result it does not require that the organic shell of well-intercalated organoclay be miscible with the 'oil' phase. However, miscibility between these two, $\chi_{12} < 0$, would facilitate the process and may lead to better CPNCs.

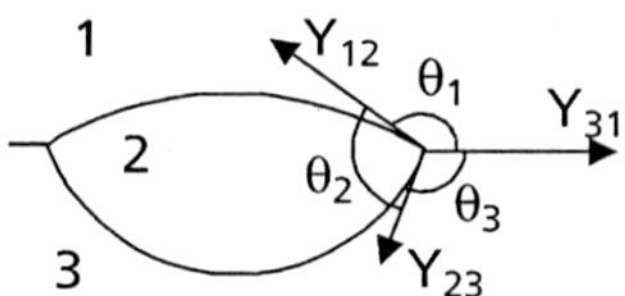

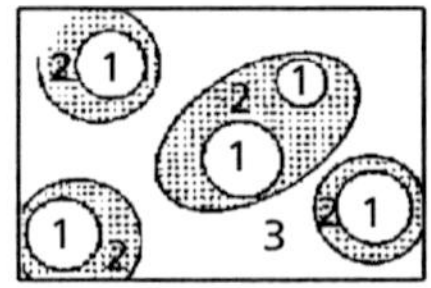

Figure 156 Neumann's triangle and the desired morphology for the three phases in emulsion or suspension polymerisation: phase (1) = organoclay; phase (2) = 'oil' phase containing vinyl monomer(s); phase (3) = aqueous phase. The depicted structure is expected for $v_{13} > v_{12} + v_{23}$. After [Inoue, 2002].

The polymerisation methods may be divided into subcategories by the type of polymerisation processes involved, e.g.:

1. Bulk polymerisation by the free radical and coordination methods
2. Emulsion and suspension methods
3. Solution polymerisation methods.

4.1.4.1.1 Bulk Polymerisation by the Free Radical and Coordination Methods

The earliest patent on the polymerisation of vinyl monomer(s) in the presence of clay belongs to Toyota [Kamigaito *et al.*, 1984]. The document describes preparation of intercalated clay composites by contacting MMT with a monomer (e.g., styrene, vinyl acetate) at room temperature, then mixing with dichloro dimethyl silane and in a few minutes effecting the polymerisation. In the cited example, compositions containing 78% intercalated clay, were hot pressed into parts that showed high heat, abrasion and flame resistance. The interlayer spacing of these materials was $d_{001} \cong 1.5$ *nm*. This patent is on the borderline between highly filled clay composites and CPNCs.

Another patent from the company (priority 1987) also used bulk polymerisation for the preparation of CPNC [Kawasumi *et al.*, 1989]. CPNCs with a diversity of clays and matrices, e.g., thermoplastics (including vinyls), thermosets and rubbers, could be prepared in three steps:

1. Intercalating clay to facilitate its swelling by molten monomer (at $T > T_m$). When vinyl polymer was to be the matrix, an ammonium salt having a terminal vinyl group, *viz.* chloride of methacrylic acid-*N,N*-dimethyl ethanolamine, may be used as the intercalant.
2. Mixing the organoclay with a monomer, e.g., styrene or methyl methacrylate (MMA). In this step a catalyst or an initiator, e.g., potassium persulfate, is usually added.
3. Heating the mixture at the polymerisation temperature. For example, polymerisation of MMA was carried out for 5 h at $T = 60$ °C with continuous mixing.

As XRD showed, the polymerisation of MMA in the presence of vinyl-terminated MMT resulted in exfoliation; $d_{001} > 10$ nm. The CPNC contained 5.7 wt% of MMT dispersed in a PMMA matrix with $M_w = 91.5$ kg/mol, was suitable for injection moulding. The tensile test showed that the modulus was $E = 5.68$ GPa, to be compared with $E = 3.20$ GPa obtained for a composite with non-exfoliated clay.

In a related publication, Na-smectite (CEC = 0.866 meq/g) was ion exchanged with either oligo(oxypropylene)-diethyl-methyl-ammonium chloride, $(C_2H_5)_2(CH_3)N^+(O\text{-}i\text{Pr})_{25}\ Cl^-$, or methyl-trioctil-ammonium chloride, $CH_3(C_8H_{17})_3N^+\ Cl^-$, labelled SPN and STN, respectively [Okamoto *et al.*, 2000]. The organoclays were dispersed in either MMA or styrene (St) using ultrasonics for 7 h at 25 °C. Next, a peroxy initiator was added and free radical polymerisation was carried out at 80 °C for 5 h (for MMA) or at 100 °C for 16 h (for St). XRD for solid organoclays SPN and STN, indicated that d_{001} = 4.20 and 1.81 nm, respectively. Dispersing 10 wt% of these in a monomer and polymerising it gave the following interlayer spacings:

1. MMA/STN d_{001} = 2.96 nm, and then PMMA/STN reduced spacing: d_{001} = 2.66 nm
2. MMA/SPN exfoliation, and then PMMA/SPN small shoulder at d_{001} = 4.55 nm
3. St/SPN exfoliation, and then PS/SPN strong diffraction at d_{001} = 3.65 nm.

Thus, only in the case of PMMA/SPN was a high degree of dispersion (at 10 wt% loading!) achieved. Considering the immiscibility of PS with PEG, a low degree of exfoliation in PS/SPN nanocomposite was to be expected. At the same time, the hydrogen bonding between the oligopropylene chain and PMMA would provide better dispersion than that for PS.

TEM bright field images for PMMA/STN showed the presence of aggregates containing *ca.* 10-20 parallel clay platelets with a basal spacing between them of about 3 nm – this is consistent with the XRD result. For PMMA/SPN, about 200 nm long stacks of *ca.* 10 clay platelets were found randomly oriented in the PMMA matrix. For PS/SPN, short stacks of only 2-3 platelets at a distance of ≤ 4 nm were evident. However, randomisation of the clay stacks seems to be better than that observed for PMMA/SPN. This illustrates the fact that XRD patterns must be interpreted with caution, preferably with a number of TEM images of the dispersion. The temperature dependence of the torsional storage modulus, G', at a frequency of 6.28 rad/s and strain of $\gamma = 0.05\%$ was carried out at T = 25 to 135 °C for PMMA/SPN, PMMA/STN and PS/SPN below the glass transition ($T_g \cong$ 25-90 °C). The measured value of G' was respectively, 20%, 34% and 57% higher than that of systems without clays.

In the following paper from this laboratory, the authors examined the correlation between the internal structure and the ionic conductivity of SPN/polymer systems [Okamoto *et al.*, 2001b, c]. As polymers, PS and copolymers of MMA with 1 mol% of either acrylamide (AA), *N,N*-dimethyl aminoethyl acrylate (AEA), or *N,N*-dimethyl aminopropyl acrylamide (PAA) were used. Specimens had a constant organoclay content of 10 wt%. Preparation of the CPNC followed the method described in the preceding paper [Okamoto *et al.*, 2000]. Test specimens for the electrical conductivity were compression moulded at 5 MPa in a hot press kept for 100 s at 170-200 °C, quenched to room temperature and then cut into squares of 28×28×0.8 mm. The sheets were dried before use. The electrical conductivity was measured with an impedance analyser at T = 90-150 °C under a constant potential of 100 mV.

TEM showed that in PMMA/AA/SPN short stacks with 4-5 platelets were dispersed by about 5 nm one from another. In PMMA/AEA/SPN the stacking was more extensive and the thickness a bit larger. The tests showed generally low conductivity with several conductivity relaxation times. The absolute magnitude of conductivity (see **Figure 157**) seems to correlate with the degree of dispersion

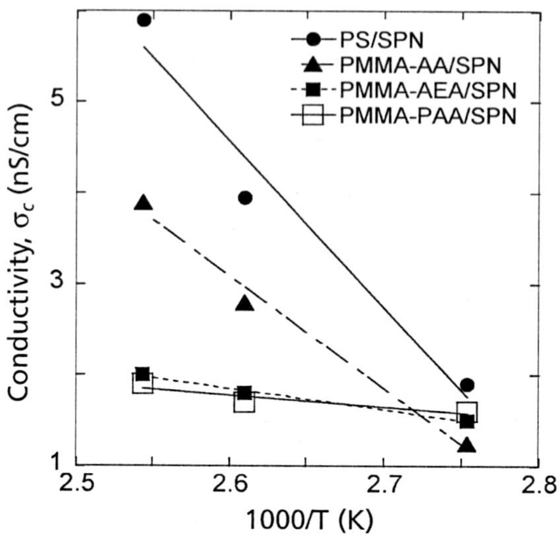

***Figure 157** Conductivity at 10^5 Hz of CPNC containing 10 wt% of SPN organoclay in PS and PMMA copolymers containing 1 mol% of second co-monomer (see text). Data [Okamoto et al., 2001].*

(as well as with clay content, not shown). For the same matrix polymer, the finer the clay dispersion the higher the conductivity.

Fu and Qutubuddin, [2000; 2001] and Qutubuddin *et al.* [2002] used a bulk polymerisation method to prepare end-tethered, PS-based CPNC. The authors provided a concise review of the topical literature. The preparation of their CPNC started with the synthesis of vinyl-benzyl dimethyl dodecyl ammonium chloride (VDAC), which was then used to intercalate Na-MMT by cationic exchange in aqueous medium. XRD indicated that intercalation by onium ions caused the interlayer spacing to increase from d_{001} = 0.99 nm (measured for Na^+-MMT) to d_{001} = 1.92 nm. Owing to good miscibility of VDAC with styrene, uniform dispersion of the intercalated platelets in monomer was obtained. The copolymerisation of VDAC with styrene was initiated by azobisisobutyronitrile (AIBN) or $K_2S_2O_8$ and conducted for 48 h at 60-70 °C. XRD and TEM showed that exfoliation depends on the degree of substitution of Na^+ by VDAC. When only 50% of CEC ions were exchanged, the exfoliation was partial. Full exfoliation was obtained for high degrees of substitution. At a clay loading of 7.6 wt% the dynamic modulus of CPNC was 60 to 70% higher than that of PS. The thermal degradation of CPNC started at a temperature about 15 °C higher than that of neat PS.

The exfoliation reported by Fu and Qutubuddin originates from good miscibility of the intercalated MMT with styrene and the end-tethering of the PS macromolecules. This is in contrast with several explored routes where either the organoclay was not reactive or where the reactive group was attached to a compound immiscible with a PS matrix – in either of these types only modest intercalation was obtained.

Wilkie and his collaborators have also studied the effect of miscibility on the degree of exfoliation and the performance of CPNC. Thus, three onium salts were prepared (see **Figure 158**) and used to intercalate Na^+-MMT following a standard procedure [Zhu *et al.*, 2001a]. The CPNCs were prepared by a bulk polymerisation method, mixing together 3 g of organoclay, 1 g of AIBN and 100 g of styrene. The polymerisation was conducted for 48 h at 60 and 80 °C, then unreacted styrene was vacuum distilled off, the PS not bounded to the clay was extracted by toluene, and clay unattached to polymer was removed by filtration. The viscosity average molecular weight of the extracted PS and that prepared as a control was $M_v = 100 \pm 20$ kg/mol. The results of XRD and TEM analyses are summarised in **Table 81**. Clearly the best dispersion was obtained for MMT intercalated with reactive *N,N*-dimethyl-*n*-hexadecyl-(4-vinylbenzyl) ammonium (VB16). As was the case for Fu and Qutubuddin, copolymerisation of the intercalated clay with styrene resulted in full exfoliation. Note that the difference between the two reactive intercalants (VDAC and VB16) was the additional 4 CH_2 segments in the latter intercalant. Looking at the tensile test data in **Table 82**, it seems that their role was beneficial, but more detailed studies are needed to confirm this point.

Using TGA/FTIR, the authors observed that MMT intercalated with either ammonium or phosphonium ions degrade by the Hofmann elimination mechanism (see **Figure 88**). The data indicated that phosphonium treatment leads to marginally more thermally stable CPNC than either one of the two prepared with ammonium salts at a cost of tensile performance. The CPNCs were also tested for thermal stability, flammability and mechanical performance. Abbreviated results for PS and its CPNCs with *ca.* 3 wt% organoclay are presented in **Table 82**. Comparing the CPNC data with those of PS, one notes that as far as flammability and thermal stability are concerned there is a serious improvement of performance upon incorporation of 3 wt% organoclay. However, the data

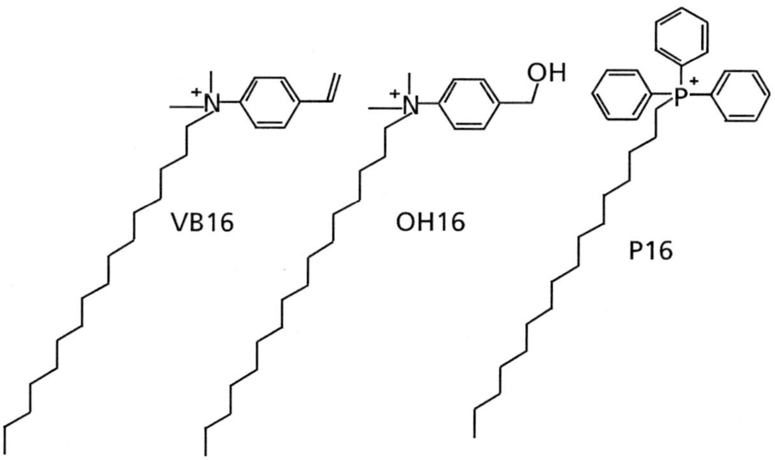

***Figure 158** The intercalating cations: VB16 = N,N-dimethyl-n-hexadecyl-(4-vinylbenzyl) ammonium; OH-16 = N,N-dimethyl-n-hexadecyl-(4-hydroxymethylbenzyl) ammonium; and P16 = n-hexadecyl triphenyl phosphonium. Reprinted with permission from Zhu et al. [2001a]. Copyright American Chemical Society.*

Table 81 Interlayer spacing of organoclays before and after *in-situ* polymerisation of styrene. Data [Zhu *et al.*, 2001a]

Clay or organoclay	Interlayer spacing, d_{001} (nm)		TEM image analysis of CPNC
	organoclay	CPNC	
Na-MMT	0.96	Not applicable	Not applicable
VB-16	2.87	No peak	Isolated clay layers; exfoliation
OH-16	1.96	3.53	Intercalated stacks *ca.* 30 nm thick
P-16	3.72	4.06	Isolated platelets and short stacks

Table 82 Test results for PS and CPNC based on it comprising *ca.* 3 wt% of organoclay. Data [Zhu *et al.*, 2001a]

Material	Thermal stability		Flammability		Tensile tests	
	Onset T (°)	10% weight loss T (°C)	PHRR* kW/m^2	Time to burn out (s)	Strength at break (%)	Strain at break (%)
PS	282	355	1024	190	100	100
PS/VB-16	378	408	584	226	300	145
PS/OH-16	363	400	502	227	120	100
PS/--16	378	400	586	225	90	100

**PHRR = peak heat release rate*

show that there is very little difference between the performances of the three organoclays – flammability and thermal stability seem insensitive to the degree of dispersion (see **Table 82**). Gilman *et al.* [2000] reported similar observations, stressing that while all CPNCs with MMT show reduced flammability, to be effective clay should be nanodispersed, but not necessarily exfoliated. Apparently the degree of intercalation for PS/OH-16 and for nPS/P-16 was sufficient to be effective.

The tensile tests are much more discriminating. There is little doubt that the exfoliating and end-tethering by VB-16 leads to a superior product – three-fold improvement in tensile strength paralleled by increase of the strain at break by 45% is most impressive, especially considering the relatively small amount of clay. The data show dramatic differences in mechanical performance between the three CPNC specimens. The PS/OH-16 specimen contains a benzyl alcohol functionality, which may undergo a Friedel-Crafts reaction with hydrogen of the PS aromatic ring at $T \cong 250$ to 300 °C, end-tethering the polymer chain in a similar manner as in the case for PS/VB-16. Evidently, the method is less efficient than direct reaction with a vinyl group attached to MMT platelets. Surprisingly, the phosphonium salt offered a small improvement in thermal stability and flammability, and rather poor mechanical performance. Note that all polymerisations resulted in similar M_v, thus the differences are not related to matrix.

In another publication from this laboratory [Zhu *et al.*, 2001b] PS-based CPNC were prepared using synthetic mica-montmorillonite clay (SMM; CEC = 1.4 meq/g), intercalated with ammonium and phosphonium salts. The ammonium intercalants were VB-16 and dimethyl benzyl hydrogenated tallow ammonium chloride (M2BHTA), the phosphonium was stearyl-tributyl phosphonium bromide (P18). Previous studies have shown that even when the clay content is as low as *ca.* 0.1 wt% the peak heat release rate in a cone calorimeter is reduced by 40%, thus not much different from that observed at higher amounts of clay. Hence, the goal of this work was to determine whether the presence of paramagnetic iron in the CPNC can result in radical trapping, which in turn would enhance the thermal stability. The results showed that indeed the presence of structural iron (rather than specially added) significantly increased the onset temperature of thermal degradation. The effect was particularly significant for CPNC with intercalated clay, but less so for the exfoliated systems.

Again in this work CPNCs were prepared by bulk polymerisation, then the degree of dispersion was determined using XRD and TEM. The interlayer spacing for PS/M2BHTA and PS/P-18 was, respectively, d_{001} = 5.88 and 5.95 nm. The intercalation was confirmed by TEM. XRD of PS/VB-16 had no diffraction peak, but TEM micrographs showed the presence of individual platelets and short stacks. Thus, again the polymerisation of styrene onto the cation with a vinyl group (as for SMM/VB-16) resulted in exfoliation, while bulk polymerisation in the presence of SMM/M2BHTA resulted in intercalation with stacks *ca.* 50 nm thick. It is noteworthy that before polymerisation the intercalated SMM/VB-16 had an interlayer spacing smaller than that of SMM/M2BHTA, *viz.* d_{001} = 4.53 and 5.04 nm, respectively. One may conclude that end-tethering leads to better dispersion than straight bulk polymerisation.

Zeng and Lee [2001] tried to prepare exfoliated CPNC with either PMMA or PS as the matrix using the bulk polymerisation procedure. The effects of monomer (MMA or styrene), initiator (either benzoyl peroxide (BPO) or 2,2′-azobisisobutyronitrile (AIBN)), and organoclay on the structure of nanocomposites were investigated. As organoclay, Na-MMT was intercalated with either non-reactive dimethyl dihydrogenated tallow ammonium chloride (Cloisite® 20A, C20A), or with reactive 2-methacryloyl-oxyethylhexadecyldimethyl ammonium bromide (MHA).

To prepare CPNC, the organoclay was dispersed in a monomer and the mixture was exposed to US irradiation for 6 h before initiator was added, then sonicated for another 1 h. Polymerisation was carried out under isothermal conditions for 4 to 20 h. The materials were post-cured at 105 °C, and then injection moulded into tensile bars. The CPNCs were characterised by XRD, TEM and evaluated for dimensional stability at T = 50 to 135 °C for up to 4 h. The interlayer spacing of the prepared CPNCs is summarised in **Table 83**. As the data show, mutual miscibility of initiator, monomer and the intercalant determines the clay dispersion. AIBN with two polar nitrile groups has higher polarity and is more hydrophilic than BPO. Thus, the MMA/organoclay + AIBN system shows a better clay dispersion than that with BPO. However, full exfoliation was only obtained when reactive MHA was used. Surprisingly, better dispersion was obtained with styrene than with MMA – the latter has a stronger affinity to the polar clay surface than

Table 83 MMT dispersion in bulk polymerised MMT/PMMA and MMT/PS nanocomposites containing 5 wt% of organoclay. Data [Zeng and Lee, 2001]

Material	Reaction T (°C)	d_{001} (nm)	Comments based on XRD and TEM data
Cloisite® 20A	-	2.3	
C20A+MMA+BPO	70	3.6	Independent of BPO conc.; large clay stacks
C20A+MMA+AIBN	50	4.9	Individual platelets and short stacks
C20A+styrene	60	3.4	Same spacing for BPO and AIBN
Na^+-MMT	–	0.95	
MHA	–	1.95	
MHA+MMA+AIBN	50	No peak	Mainly exfoliated platelets and short stacks
MHA+styrene+AIBN	60	No peak	Exfoliation

styrene. The authors also reported large differences in dimensional stability. In the absence of clay, the dogbone specimen greatly deformed. Addition of 5 wt% of clay significantly improved the dimensional stability – the exfoliated PS/MHA provided the best performance.

The importance of miscibility between the intercalant and polymeric matrix is well illustrated by another publication from the same laboratory [Zhu *et al.*, 2002]. Three CPNCs with PMMA as the matrix were prepared by bulk polymerisation following the same procedure as described above for PS-type CPNC. To start with, Na^+-MMT was intercalated with ammonium salts, *N,N*-dimethyl-*n*-hexadecyl allyl ammonium chloride (Allyl-16), *N,N*-dimethyl-*n*-hexadecyl benzyl ammonium chloride (Bz-16) and *N,N*-dimethyl-*n*-hexadecyl-(*p*-vinyl benzyl) ammonium chloride (VB-16). By contrast with the PS nanocomposites discussed above, the ones with PMMA as matrix showed poor clay dispersion. The results of XRD and TEM analysis are summarised in **Table 84**. Even end-tethering of PMMA macromolecules did not result in exfoliation. Two possible explanations can be offered: (1) poor miscibility of the intercalants with PMMA; and (2) high polymerisation rate of PMMA, resulting in vitrification that hinders diffusion of monomer into the interlamellar galleries. Surprisingly, the best dispersion (VB-16/MMT) resulted in the lowest value of tensile strength – reduction of PMMA tensile strength by 20%. For the PMMA nanocomposites Allyl-16/MMT gave better overall performance than VB-16/MMT.

Translucent acrylic nanocomposites were described by Dietsche *et al.* [1999; 2000]. The new material was prepared by copolymerising methyl methacrylate with dodecyl methacrylate in the presence of 2-10 wt% bentonite intercalated with *N,N,N,N*-dioctadecyl dimethyl ammonium ions (2M2ODA). Addition of

Table 84 Dispersion of MMT platelets in PMMA matrix. Data [Zhu *et al.*, 2002]

System	Interlayer spacing, d_{001} (nm)		TEM observations
	Organoclay	CPNC	
Na-MMT	(1.20)	–	–
MMT-Allyl-16	2.08	3.40	Stacks ≥ 50 nm thick, no single plates
MMT-Bz-16	1.80	3.27	Stacks ≥ 50 nm thick, no single plates
MMT-VB-16	2.52	4.65	Stacks ≥ 50 nm thick, single plates

n-dodecyl methacrylate improved interactions between clay and matrix. As determined by XRD, TEM and AFM the bentonite formed anisotropic nanoparticles with average thickness of *ca.* 18 nm, average length of 450 nm, and interlayer distance of d_{001} = 4.8 nm. Addition of clay accounted for improved stiffness-to-toughness balance, higher T_g and thermal stability in comparison to the corresponding copolymer. The reported data also include Izod impact strength, light transmittance coefficient, elongation at break, TGA, and flammability.

Fischer and Gielgens [2002a] used layered double hydroxide (LDH) in their CPNC. Irreproducibility of performance of CPNC containing natural clays was the main reason for using synthetic material of the general formula:

$$\left[M_{l-x}^{2+}M_x^{3+}(OH)_2\right]\left[A_{x/y}^{y-}\cdot nH_2O\right]$$

where M^{2+} is a bivalent cation, e.g., of Mg, Zn, Ca, etc.; M^{3+} is a trivalent cation, e.g., of Al, Cr, Fe, etc.; *x* is a number between 0.15 and 0.5; *y* is 1 or 2; *n* is a number from 1 to 10; and *A* is an anion, e.g., Cl^-, Br^-, NO_3^-, SO_4^{2-} or CO_3^{2-}. Thus, LDHs are anionic clays that are readily modifiable and can be homogeneously dispersed in a polymeric matrix. Furthermore, the clay quality and composition is controllable. The preferred LDH is a hydrotalcite or a hydrotalcite-like material, with a large surface area and CEC = 0.5 to 6 meq/g. It is desirable that at least 95% of all LDH anions are reactive and/or compatible with the polymeric matrix. Such a modification is carried out by ion exchange by, e.g., replacing NO_3^- by an anion of carboxylic, sulfonic, phosphonic or sulfate acid, which in addition contains an alkyl or an alkyl phenyl group having 6 to 22 carbon atoms.

When reactive anions are desired, the ion and the reactive group should be attached to opposite terminal methylene groups. It is advantageous that at least 20% of these anions contain a second charge-carrying group, which may provide a repellant energy that forces the clay platelets apart. These repulsive groups may be cationic, such as amino. LDH may be used to prepare CPNC with ≥ 70 wt% of either polyadduct or polycondensate matrix. However, instead of mixing the LDH with a polymer it may be preferred to disperse it in a monomer, then polymerise the latter. Accordingly, in the following patent [Fischer and Gielgens, 2002b] the method of CPNC preparation has been specified to follow two steps: (1) mixing LDH with at least 50 wt% of a monomer, and (2) bulk polymerising the monomer. As in the preceding patent, LDH should have at least

20% of all anions compatible and/or reactive with the polymer, *viz.* R′–RCOO$^-$, R′–ROSO$_3^-$ or R'–RSO$_3^-$, where R is a straight or branched alkyl or an alkyl phenyl group having 6 to 22 carbon atoms and R' is a reactive group. In the non-reactive anions R′ = H. For example R′ = –OH, –NH$_2$, –CO$_3^{2-}$, epoxy, vinyl, isocyanate, hydroxyphenyl and anhydride group – the choice depends on the matrix. CPNCs based on LDH with reactive anions are exceptionally stable.

4.1.4.1.2 Emulsion and Suspension Methods

Since intercalation of clay in aqueous media under controlled conditions (clay concentration, time, temperature, pH and pK), can lead to exfoliation, emulsion and suspension polymerisation seems to be the simplest route to CPNC. This certainly is the case when monomers show miscibility to water. Emulsion polymerisation of MMA in the presence of Na-MMT resulted in a moderate expansion of the interlayer spacing [Lee and Jang, 1996]. Owing to the ion-dipole bonding, the PMMA macromolecules were found oriented parallel to the clay platelet surface. The thermal stability and tensile properties of CPNC were found substantially enhanced. As it will be discussed later in the text, intercalation was also obtained by Noh *et al.* [1999a] for CPNC with SAN as matrix – these authors also started with Na-MMT!

The emulsion polymerisation of styrene in the presence of reactive organoclay (Na$^+$-MMT with CEC = 0.534 meq/g intercalated with methylstyrene ammonium chloride, MSA) was used to prepare CPNC with PS as the matrix [Laus *et al.*, 1998]. The organoclay + styrene was stirred for 24 h before water, sodium lauryl sulfate: $CH_3(CH_2)_{10}CH_2OSO_3Na$ or SLS, and $K_2S_2O_8$ were added to start the emulsion polymerisation. The recovered CPNC was extracted with $CHCl_3$ and the residual PS-MMT was ion exchanged with trimethyl hexadecyl ammonium bromide (3MHDA). The molecular weights of the two polymers, the extracted PS (hence not attached to MMT; marked PS) and the one recovered after ion exchange (hence end-tethered to clay; marked PS-MSA) were measured. The SEC data indicated (see **Figure 159**) that the molecular weight exponentially decreased with increasing organoclay content. The decrease originated in the increasing probability of chain termination or chain transfer reactions. As the figure shows, independently of the decreasing M_w, the glass transition temperature, T_g, increased with clay loading by up to 11 °C. Evidently, the increase is related to immobilisation of PS chains on the clay surface.

Dynamic mechanical analysis at 1 Hz and T-scanning rate of 4 °C/min showed that the storage modulus, G', in the glassy region increased with clay loading, but the increase was modest, *ca.* 60% for 25 wt% of organoclay. XRD showed rather poor intercalation; for MSA-intercalated clay as well as for polymerised CPNC the interlayer spacing was: d_{001} = 1.53 to 1.55 nm. On a geometrical basis one may argue that independently of end-tethering the macromolecules, they form a single layer between two adjacent MMT platelets, with slightly tilted aromatic rings.

Elspass *et al.* [1997; 1999] patented CPNC with improved mechanical properties and reduced permeability to small molecules (barrier properties). These materials are particularly useful as tyre inner liners and inner tubes, as barrier layers in films or coatings, etc. The proposed process involves standard steps: dispersing clay in water, adding intercalant, monomer(s), free radical initiator, and then polymerising. The novelty is the use of onium intercalant that also serves as the latex emulsifier. The monomers are to be selected from two types of vinyls – those that result in rigid thermoplastics and the others that are elastomeric

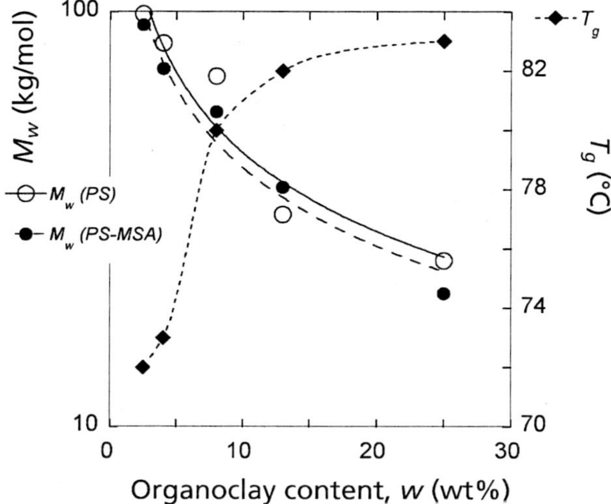

Figure 159 Weight-average molecular weight of free PS and that attached to MMT platelets as functions of the organoclay loading. Variation of the glass transition temperature is also shown. Data [Laus et al., 1998].

after polymerisation, *viz.*: styrene, *p*-methyl styrene, vinyl chloride, vinyl acetate, acrylonitrile, maleic anhydride, succinic anhydride, acrylates; and butadiene, isoprene, chloroprene, ethylene, propylene, butene, hexene, heptene, isobutylene, and octene. Clay loading ranges from 0.2 to 4 wt%. However, the process was slow, requiring *ca.* two days to complete.

A non-reactive process was also described. Thus, an emulsion or microemulsion was prepared by dispersing polymer and surfactant in a polar liquid, and then adding organoclay and either mechanically shearing the mixture or using US to produce latex containing a polymer intercalated clay. Furthermore, the authors also describe preparation of nanocomposites by using a polymer having functional groups to intercalate and at the same time to form a matrix, thus bypassing the whole reactive process. The groups of interest may be selected from between: quaternary ammonium, phosphonium, maleate, succinate, carboxyl containing groups, acrylate, benzylic hydrogens, benzylic halogens, aliphatic halogens, and oxazoline.

The patent especially focuses on styrene-rubber and styrene-acrylonitrile-rubber copolymers with $-50 \leq T_g\,(°C) \leq 100$. The desired *MW* of these copolymers is $M_n \geq 15$ kg/mol, and some unsaturation or other reactive sites for vulcanisation or crosslinking must be provided. As an example, MMT was dispersed in water overnight (at a concentration below 1.5 wt% the platelets became fully exfoliated). To the dispersion trimethyl dodecyl ammonium bromide (3MDDA), and then the monomers (isoprene, and styrene), and initiator (AIBN) were added. The mixture was polymerised for 46 h. A solid CPNC prepared from the latex had $d_{001} \cong 3.6$ nm, the average number of silicate layers in a stack was about 260. The oxygen permeability at 26.3 wt% clay loading was modestly reduced by a factor of 3.

The claims are quite broad, *viz.* the three components for intercalation are described as:

1. At least one liquid selected from the group consisting of water, dimethyl formamide, aliphatic alcohols, aliphatic alkanes, esters, and ethers.
2. 0.1-70 wt% of intercalant/surfactant selected from between: quaternary ammonium, phosphonium, maleate, succinate, surfactants having carboxyl containing groups, acrylate, benzylic hydrogens, benzylic halogens, aliphatic halogens, and oxazoline. The intercalant is used at the ratio of 1 to 20 by weight of clay. It either performs the role of intercalant and emulsifier-surfactant, or another surfactant anionic, cationic, and nonionic may be added.
3. 0.2-4 wt% of clay (MMT, saponite, beidellite, hectorite, stevensite, vermiculite, etc.), preferably MMT. As a result of *in situ* emulsion or microemulsion polymerisation of selected monomer(s), the clay becomes intercalated with 0.1 to 63 wt% of a polymer.

The Elspass *et al.* patent of 1999 described the use of US for enhanced dispersion during emulsion or microemulsion polymerisation. By contrast, Wang *et al.* [2001] patented the use of high energy US irradiation for the reactive preparation of PS/MMT systems. First, MMT was intercalated with a long chain quaternary ammonium salt, and then US was used to disperse organoclay in styrene as well as to generate free radicals that cause polymerisation of PS macromolecules directly attached to the clay surface. The resulting CPNC was stable after separation and drying. DSC indicated that incorporation of MMT increased PS T_g.

More details about this process were provided in the following article [Zhang *et al.*, 2002a]. US was used to effect microemulsion polymerisation of styrene, with minimal contamination by surfactants and reaction byproducts. During the US irradiation cavitation generated local temperature and pressure up to $T \leq 5000$ K and $P \leq 50$ MPa with a heating and cooling rate greater than 10^9 K/s. Thus, the role of US was for dispersing, emulsifying (at low surfactant concentration), and initiating the chemical reaction. The equipment used was a 20 kHz US generator with titanium horn of diameter 25 mm and power output from 150 to 1500 W. The isothermal reaction was conducted with the US horn immersed directly in the reaction system with N_2 purging. The reaction was conducted during continuous irradiation. After 15 min of US irradiation the average particle size was about 30 nm increasing (and broadening the size distribution) to about 37 nm after 60 min. The conversion was *ca.* 70%/h. The molecular weight of PS was very high: $M_w \cong 2{,}500$ kg/mol and $M_w/M_n \cong 1.06$, which may be a sign of branching.

Noh *et al.* [1999a] also used US during the emulsion polymerisation of SAN/MMT in the presence of clay. In the process, US replaced mechanical stirring for dispersing Na-MMT (CEC = 0.9 meq/g) in water. To this suspension styrene, acrylonitrile and SLS-emulsifier were added. The reaction (initiated by $K_2S_2O_8$) took 5 h. After coagulation, the product was washed, dried and extracted with hot tetrahydrofuran (THF) for up to 5 clays. The nominal clay content varied from 5 to 40 wt%, while that measured after extraction was 62 to 73 wt%, thus the extent of intercalation by SAN was quite low! XRD analysis of the SAN nanocomposites indicated that the emulsion copolymerisation resulted in low interlayer spacing, *viz.* d_{001} = 1.9 to 1.4 nm. Such a low interlayer spacing is quite understandable, considering the high clay concentration in the 'oil' phase. Lee and Jang [1996] (from the same laboratory) also obtained poor dispersion during emulsion polymerisation of MMA, starting with Na-MMT.

In the following article, the properties of CPNC, prepared either by emulsion or by solution methods were compared [Noh and Lee, 1999b]. The emulsion copolymerisation was carried out as described above. However, different types of clay were used: Na-MMT (Cloisite® NA+) in emulsion polymerisation, and Cloisite® 6A (C6A) in solution polymerisation in cyclohexanone. As a result, in these SAN CPNCs the interlayer spacing was: in emulsion d_{001} = 1.72 nm and in solution d_{001} = 1.64 nm, hence both processes resulted in a modest degree of intercalation. The molecular weight of the copolymer from the emulsion and solution processes was M_w = 53 and 4.8 kg/mol, respectively. The most dramatic differences between SAN, and CPNCs based on it, were observed in thermogravimetric analysis. Whereas at 500 °C only about 5 wt% of neat SAN remained, the solution-type PNC retained 20 and emulsion-type 70 wt%. Improvement of the mechanical properties was modest, but again the emulsion-type CPNC was superior. Optical microscopy of the solution-type CPNCs showed the presence of large clay aggregates (several tens of μm in diameter!). This suggests that C6A phase separated during the solution polymerisation, forming its own phase dispersed by vigorous mixing in the polymer solution.

In the more recent publication from this laboratory, a one-step emulsion polymerisation in the presence of Na-MMT was used to prepare CPNC with ABS as matrix [Jang *et al.*, 2001]. The emulsion terpolymerisation started with latex of 1,4-polybutadiene (PBD, 30 wt% solids, particle size = 100 nm) mixed with Na-MMT and co-monomers at the weight ratio of styrene/acrylonitrile/PBD = 55/25/20. As emulsifier a rosin soap and as initiator cumene hydroperoxide were used. Polymerisation was conducted for 5 h at 80 °C, while stirring at 400 rpm. The coagulated product was washed with water and methanol, dried and pulverised. To remove free SAN and ABS, the product was extracted with boiling acetone, and then with boiling toluene, for 48 h and 120 h, respectively. Since it was impossible to form CPNC with more than 50 wt% of clay, nominal compositions containing 10 to 50 phr of MMT were prepared. After extraction of free polymers, the ABS content in these CPNCs was *ca.* 65 to 25 wt%, indicating limited intercalation by SAN.

XRD data of the extracted CPNC and Na-MMT are presented in **Table 85**. The emulsion polymerisation increased the MMT interlayer spacing by *ca.* 0.6 to 0.4 nm, thus a partial intercalation has been achieved. As expected, the intensity of the diffraction peak increases with the MMT loading. These experimental data can be used to fit to a general curve of $d_{001} = f(w)$ with the correlation coefficient squared, r^2 = 0.99914. The dependence predicts that at infinite dilution (w = 0) the interlayer spacing should be: d_{001} = 2.05 nm. Thus, unless there is an opposite trend at low clay content, the data suggest that inherently the system is non-exfoliating. Either during polymerisation the macromolecular chains are bridging the interlamellar galleries (the polymer near the clay surface is partially crosslinked) or the monomers dissolve in the PBD phases, too large to enter the interlamellar galleries.

After extraction of the unbounded polymer from the CPNC, the DSC does not show any thermal transition, indicating the absence of a discrete polymeric phase. In the not extracted CPNCs the well defined $T_g \cong$ 102 to 124 °C for increasing MMT loading from 0 to 50 wt% (see **Table 85**). Considering the poor intercalation, this increase originates in confinement of SAN copolymer not

Table 85 Interlayer spacing in extracted ABS/MMT nanocomposites, and T_g in their non extracted homologues. Data [Jang *et al.*, 2001]

Material	MMT* w (wt%)	d_{001}* (nm)	T_g (°C)
ABS	0	(2.05)##	102
ABS10	35.2	1.75	106
ABS20	47.2	1.70	108
ABS30	60.9	1.66	118
ABS40	71.7	1.56	121
Na+-MMTa	100	1.15	–

*Notes: *d_{001} in extracted CPNC; **Data for not extracted CPNC; ##value obtained from nonlinear extrapolation of the experimental data to infinite dilution*

between individual MMT layers (no transition observed after extraction), but rather between stacks of lightly intercalated clay particles. The TGA data show that, independently of MMT content, the CPNC decomposes at $T_d \cong 430$ °C, hence due to poor dispersion the thermal-insulation effect of MMT was not significant.

To prepare an electrorheological (ER) fluid, Na-MMT was emulsion polymerised with SAN in the presence of SLS and initiated by $K_2S_2O_8$ at 82 °C [Kim *et al.*, 2000]. The coagulated CPNC was washed, vacuum dried and extracted with hot tetrahydrofuran (THF) for 5 days. The weight-average molecular weight of SAN was M_w = 530 kg/mol. The XRD patterns of the clay and SAN/clay system gave the d_{001} = 0.98 and 1.74 nm, respectively. This confirmed the insertion of SAN polymer into the interlamellar galleries. The ER fluids were prepared by suspending the SAN/clay particles in silicone oil. When exposed to the externally applied electric field, the fluid showed typical ER behaviour with pseudoNewtonian flow at high shear rates. Evidently, for this specific application exfoliation is not critical. Similar results were obtained by [Kim *et al.*, 2002]. The authors emulsion polymerised styrene in the presence of 2 to 10 wt% of Na-MMT – the interlayer spacing of Na-MMT, d_{001} = 1.195 nm increased to 1.511 nm. Thus, limited intercalation was achieved.

On the basis of these repeated, internally consistent experimental data, one may postulate that as a rule when starting with Na-MMT in emulsion or suspension polymerisation, lightly intercalated clay is obtained with a monolayer of polymer in the interlamellar galleries, yielding d_{001} = 1.5 to 1.7 nm.

Chen *et al.* [2000] started the emulsion polymerisation of PS by intercalating Na-MMT (CEC = 0.92 meq/g) with trimethyl cetyl ammonium bromide [$C_{16}H_{33}(CH_3)_3N^+ Br^-$] (3MHDA; at a concentration of 1.5×CEC; giving d_{001} = 1.92 nm). Next, 5 wt% of the organoclay was dispersed in styrene as a stable transparent dispersion. Then an aqueous solution of $(NH_4)_2S_2O_8$ initiator and an emulsifier were added and the mixture polymerised at 70-80 °C for 5 h under a blanket of N_2. The molecular weight and its distribution were determined by size exclusion chromatography (SEC) as: M_w = 336 kg/mol and M_w/M_n = 4.3. The glass transition temperature, T_g = 103 °C, was determined by DSC. Wide angle XRD gave the interlayer spacing as d_{001} = 5.45 nm in stacks *ca.* 50 nm thick.

The extruded CPNC film was sheared at T = 200 ± 10 °C at a shear rate of *ca.* 100 s^{-1}. After quenching, the specimen was scanned at room temperature (XRD, TEM and FTIR), then heated to a desired temperature (and kept there for 1.5 min), scanned and then heated to a higher temperature and measured, and so on. TEM showed that initially the clay stacks were well aligned in the shear direction. As the temperature increased the stacks became disoriented, then re-aligned at *ca.* 95 °C and randomised at higher *T*. FTIR dichroism indicated that initially the phenyl rings were parallel to the shearing direction and the MMT stacks. However, after heating at 95 °C, the phenyl rings turn perpendicular to the MMT surface. There was no apparent orientation at $T \geq 110$ °C – randomisation in the molten PS at $T > T_g$ is to be expected.

In the following publications [Chen *et al.*, 2001a, b; Fan *et al.*, 2002], PS/MMT nanocomposites were also prepared using 5 wt% of the same MMT-3MHDA (d_{001} = 1.92 nm) and homogenising the dispersion by US. After polymerisation no distinct diffraction peaks were discernible, suggesting exfoliation. However, stable exfoliation in a system where MMT is intercalated with paraffinic, hence immiscible with PS, onium cations should not be expected. Indeed, TEM showed that polymerisation caused MMT particles to break into small stacks, *ca.* 40 nm thick, randomly dispersed in the PS matrix. Within the stacks the interlayer distance was in the range of 5-18 nm. The system was intercalated. The CPNC decomposition onset temperature was at T_{deg} = 417 °C, i.e., higher by 11 °C than that of PS. Similarly, the T_g increased from 100 to 103 °C. The samples were soluble in good solvents for PS, *viz.* toluene and chloroform, partly in benzene, and not soluble in the non-solvent, acetone.

The CPNC emulsion with monodispersed spherical particles, *ca.* 200 nm in diameter, could be directly formed as a film. Alternatively, it could be coagulated and extruded. The pellets fractured under liquid N_2 showed fibrils rather than the smooth surface observed for neat PS extrudate. After chemical etching, MMT flakes ('the diameter of the planar montmorillonite primary particles is less than 350 nm') became visible. It is a bit puzzling how clay platelets having diameter of 350 nm can fit inside 200 nm emulsion particles. The authors also do not mention the stability of the clay dispersion during processing. Considering the immiscibility of alkyl intercalant with PS, the system is not expected to be thermodynamically stable when sheared in the molten state. The authors applied for patent protection in China.

In the Kuo *et al.* [2001] patent, the first step is also the intercalation of clay (MMT, mica, or talc; loading = 0.05 to 30 wt%) with a cation such as a pyridinium or a quaternary ammonium salt, e.g., hexadecyl-pyridinium or 3MHDA chloride. Optionally, the organoclay may also be treated with organosilane, having a vinyl, epoxy, or acrylic group, e.g., vinyl-triethoxysilane, 3-methacryloxy-propyl-trimethoxysilane, or 3-glycidyloxy-propyl-trimethoxysilane. In the second step, the modified clay is dispersed in a vinyl monomer (styrene, acrylonitrile, or acrylic monomer) and bulk polymerised at T = 50 to 100 °C in the presence of a free radical initiator, e.g., BPO. In the third step, when the bulk polymerisation reaches 10 to 50% conversion, an aqueous suspension of PVAl was added and the polymerisation was completed by the suspension mechanism. According to XRD analysis, intercalation increased MMT interlayer spacing to $d_{001} \geq 1.7$ nm. Polymerisation in the presence of organoclay further increased it to $3.1 \leq d_{001}$ (nm) ≤ 3.6. It is interesting that the maximum gallery expansion was reached after 1 h of reaction and the d_{001} level remained nearly constant for the remaining 15 h. The spacing was about the same with or without the organosilane. Thus, intercalated CPNCs with PS, SAN, HIPS, ABS and PMMA were prepared.

An inverted intercalation method was recently proposed in a patent application from the University of Akron. Fully exfoliated CPNC was obtained by first emulsion polymerising a vinyl monomer in the presence of a polymerisable intercalant/surfactant, e.g., 2-methyacryloyl-oxyethyl-trimethyl ammonium chloride, then by adding a suspension of unmodified clay [Brittain and Huang, 2002]. The identified monomers are acrylates, methacrylates, styrene, vinyl chloride, acrylonitrile, butadiene, isoprene and their mixtures (preferably MMA). The unmodified clay is either Na^+-MMT or synthetic laponite. These fully exfoliated during the process and remained exfoliated after melt processing. The CPNC showed a modest increase of T_g (by 5 to 15 °C) and a significant increase (by *ca.* 62 °C) of the thermal decomposition temperature. The invented process is based on the idea that after polymerisation the ammonium groups attached to the polymeric chain will form an ionic bilayer around each drop. For this reason, slow addition of unmodified clay suspension causes a spontaneous ion exchange between Na^+ (of Na-MMT) and the ammonium ion of the polymer terminal group. According to the patent, the ion exchange is rapid and efficient. The selected type of polymerisable intercalant/surfactant, 2-methyacryloyl-oxyethyl-trimethyl ammonium chloride, is well miscible with the PMMA matrix and the CPNC is thermodynamically stable.

Recently Rohm and Haas filed four patent applications for the reactive preparation of CPNCs to be used as thickeners, dispersants, plastics additives, adhesives, coatings, sealants, thermoplastic resins, etc. The first of these describes a process for the preparation of an aqueous polymer/clay dispersion with a monomer mixture containing 10 to 25 wt% of an acid containing monomer and, optionally, a surfactant. After polymerisation of a portion of the monomers the clay becomes intercalated or exfoliated [Lorah and Slone, 2002a]. The monomer may be selected from between: alkyl methacrylate, 2-ethylhexyl methacrylate, lauryl methacrylate, stearyl methacrylate, acrylonitrile, methacrylonitrile, ethylene, butadiene, vinyl acetate, styrene, vinyl aromatic monomers, etc. The acid containing monomer may be, methacrylic anhydride, maleic anhydride, methacrylic acid, maleic acid, fumaric acid, styrene sulfonic acid, ethylmethacrylate-2-sulphonic acid, phospho-ethyl methacrylate, etc. The monomeric mixture may also comprise a polyunsaturated monomer, *viz.* divinylbenzene, divinyl pyridine, diallyl phthalate, etc. The clay is, for example, MMT, saponite, beidellite, hectorite, vermiculite, kaolinite, synthetic phyllosilicate, etc. A chain transfer agent may also be added. Characteristically, the clays do not need to be pre-intercalated for this process. The presence of ≥ 10 wt% of acid containing monomer(s) (in all used) affords sufficient intercalation.

According to this invention, preparation of CPNC follows seed emulsion polymerisation with surfactant(s), where the CPNC constitutes the core seed. The ratio of the polymer core to the polymer shell may vary from 90:10 to 10:90. The core polymer should have $T_g \geq 30$ °C. The resulting latex may be directly used in a variety of applications, such as coatings, thickeners, dispersants, plastics additives, adhesives, coatings, print pastes, personal care products, household or industrial cleaners, or flexographic inks. Nanocomposites containing 0.1 to 20 wt% of clay show significant improvements of, e.g., tensile strength, barrier properties, flexibility, film forming ability, etc. Owing to a wide range of requirements imposed on the final products, the patent application describes many additives and processing steps to be selected for specific applications, *viz.* pre-intercalated clay or neat clay, chain transfer agents, crosslinkers, varnishes, etc.

Clay dispersions may be prepared by subjecting an aqueous clay suspension to a shearing force (e.g., US irradiation, grinding, high speed blending, milling), which partially exfoliates the clay without intercalant. Since dispersions of exfoliated clay can be quite viscous, it is preferable that the clay is only partially exfoliated, causing the shear viscosity of the suspension not to exceed 5 Pas. Depending on the application, the emulsion polymerisation may require addition of a surfactant either prior to or after polymerisation. Surfactants may be used to react with the clay and/or to stabilise the monomer droplets. Thus, usually these have a hydrophilic head group and at least one oleophilic tail with 4 to 30 carbon atoms. In addition, a low molecular weight, water insoluble co-stabiliser, for example, cetyl alcohol, hexane, or hexadecanal, may be used. The initiator, a redox type, should be active at T = 55 to 90 °C.

For many applications, a multistage emulsion polymerisation is desired as it provides opportunity to form particles with diverse structures, e.g., core/shell or core/sheath particles, multilayer core/shell particles, interpenetrating network particles, etc. When this type of material is to be reinforced with nanoparticles, the physics dictates that the clay should be incorporated into the rigid phase. For example, when CPNC with a rubbery first stage polymer core and a second stage rigid polymer shell is to be prepared, the clay should only be incorporated into the shell. For use as a thermoplastic resin the CPNC may be recovered either by spray drying or coagulation and drying. Following the 70-odd years old I. G. Farbenindustrie approach, to fine-tune performance these CPNCs can easily be blended as latices or as recovered powders [e.g., see Utracki, 1998].

The patent provides numerous examples illustrating various processes for diverse applications. However, whether the clay was fully exfoliated or not was of limited interest – only the emulsion viscosity was used as an estimate of exfoliation. For example, since higher viscosity of the aqueous clay suspension was obtained at pH = 2 to 6 than at $pH \geq 9$ the conclusion was that exfoliation is greater in acidic than in neutral or basic medium. Distribution of clay between aqueous and oil phases after polymerisation also was not discussed.

As an example of the new process, a reactor was charged while stirring, with water, an anionic surfactant and Na-MMT. (While the anionic surfactant was not specifically identified, this category includes, alkyl carboxylic acids, sulfonic acids, sulfuric acid, phosphate acid ester or salt, e.g., sodium dodecyl-benzene sulfonate). The reactor was heated to 85 °C then a solution of Na-acetate in water was added. In a separate vessel, a monomer emulsion was prepared by mixing water, anionic surfactant, styrene, MMA, methacrylic acid and an alkyl mercaptan (chain transfer agent to control the molecular weight). A small quantity of this emulsion was added to the reactor, initiated with $Na_2S_2O_8$ and polymerised into polymer seeds. Next, the monomer emulsion was gradually fed into the reactor, then at T = 65 °C aqueous solutions of ferrous sulfate, *tert*-butyl hydroperoxide and sodium sulfoxylate formaldehyde were added to start the polymerisation. The resulting latex had a 22% solids content, pH = 4.2 and particle size of 156 nm. It was neutralised by ammonia to a pH = 9.5. Its application was as alkali soluble resin containing 4.5 wt% clay. A viscous gel was obtained when the clay content reached 21 wt%. Thus, it seems that after the polymerisation a relatively large portion of unmodified clay remained dispersed in the aqueous phase – the Na-MMT being hydrophilic would tend to stay in this phase.

However, to prepare CPNC with high strength and good barrier properties it was necessary to incorporate exfoliated clay into the polymeric matrix. To achieve this goal the above procedure was modified. The reactor was charged with water, an emulsion polymer (containing 25 wt% clay) and $(NH_4)_2S_2O_8$, and then T increased to 85 °C. As before, in a separate vessel, a monomer emulsion was prepared, fed to the reactor and free radical polymerisation was initiated. The recovered CPNC showed higher tensile strength, low permeability, greater fire retardancy and less surface tack than material without clay. The latex retained its ability for film formation. The above process could be further modified by additionally charging the reactor with an ammonium intercalant. The polymerisation was conducted as already described. The resulting latex showed good performance and was less water sensitive than those prepared without hydrophobic clay intercalant.

The patent by Lorah and Slone [2002b] used the technology described in the preceding document. However, as the claims specify, this time the clay ought to be lightly modified by an intercalant, either before or during the polymerisation. The term lightly modified refers to clays that have been intercalated to the stage where they are slightly hydrophobic, but not enough to destabilise the emulsion. The amount of intercalant used ranges from about 10 to 100% on the clay CEC. The patent lists a variety of intercalating cationic surfactants, for example, the classic onium salts. Furthermore, the intercalant may be polymerisable, i.e., capable of copolymerising with selected monomer(s). These compounds include, e.g., the ethylenically unsaturated amine salts of alkyl-benzene sulfonic acids, alkyl olefin sulfonic acids, alkyl alcohol sulfuric acid esters, alkoxylated alkyl alcohol sulfuric acid esters, fatty acids, fatty phosphate acid esters, etc. One example of a suitable salt is:

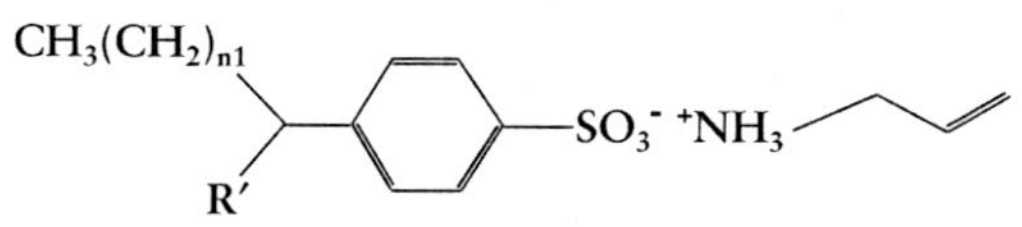

$CH_3(CH_2)_{n1}$ / R' –(benzene ring)– $SO_3^- \, {}^+NH_3$ (allyl)

The patent provides several examples. Thus, reactor was charged with water, polymerisable surfactant (allyl-ammonium salt), Na-MMT, methacrylic acid, and Na_2CO_3. Separately, a monomer emulsion was prepared by mixing together water, allyl-ammonium, butyl acrylate, methyl methacrylate, and methacrylic acid. The reactor was heated to 85 °C, then a small amount of the monomer emulsion was added to form a polymer seed. Next, $(NH_4)_2S_2O_8$ was added to initiate polymerisation and the remaining part of the monomer emulsion was fed, then the temperature was reduced to 65 °C and ferrous sulfate + *tert*-butyl hydroperoxide + *iso*-ascorbic acid were added to the reactor. After completion of the reaction the pH was adjusted to 7.5. The CPNC latex was cast into film and tested for the maximum tensile strength at T = 22 °C and the relative humidity, RH = 50%. Three batches were prepared with 0, 2, and 5 wt% of clay. Their relative tensile strength was, respectively, 100, 127 and 141%. The clay was at least partially exfoliated. The latex was used for coatings, adhesives, caulks, sealants, plastics additives, and thermoplastic resins.

The third patent application in this series describes preparation of CPNC by suspension polymerisation of monomer(s) in the presence of hydrophobically modified clay, dispersed within the monomer phase [Lorah *et al.*, 2002]. The process involved dispersing monomers in water, then adding emulsifiers and initiators while mixing. The water soluble emulsifier may be: gelatin, methylcellulose, PVAl, alkali salts of polymethacrylic acid, etc. The applications of the resulting CPNC range from coatings, sealants, caulks, adhesives, binders, caulks, traffic paint, to plastics additives.

Many examples use Cloisite® 15A (C15A with 2M2HTA) and Cloisite® Na-MMT as the hydrophobically modified and unmodified clay, respectively. Since the data are not provided, one must be satisfied with a statement that 'the use of pristine clay will lead to higher viscosity in the aqueous nanocomposite dispersion and less enhancement in property'. For example, to prepare CPNC with PS as the matrix the reactor was charged with an aqueous solution of SLS and $HNaCO_3$. In a separate vessel a mixture of styrene, polystyrene, a surfactant and C15A was prepared by stirring for *ca.* 20 min and then homogenised using US for 30 min. The resulting dispersion was gradually added to the reactor, homogenised for 30 min and then polymerised at 75 °C for 2 h.

The fourth patent application also describes the preparation of an aqueous CPNC dispersion [Lorah and Slone, 2002c]. While emulsion polymerisation seems to be the main objective, the patent also mentions suspension, solution and even bulk polymerisation methods as suitable. The claimed compositions are similar to those of the preceding documents. The clay may be pre-intercalated. The aim of this patent application is to protect modifications of the process, which may improve the physical properties of CPNC. For example, the addition of a second, multivalent cation (Ca, Mg, Cu, Mg, Fe, or Zr) provides stronger bonding between the clay and the matrix. The modification is applicable to any polymerisation method that employs acid containing monomer, oligomer, or stabiliser, which may be attracted to the second cation before or after polymerisation. Surprisingly, the cation may be added to the clay dispersion at any point during or after polymerisation as $Ca(OH)_2$, $Mg(OH)_2$, or $MgSO_4$, in the amount of 0.1 to 1 wt% (on dry clay). At least a portion of the second cation exchanges with the first cation. In systems where the first cation is either Na^+ or a cationic intercalant, Ca^{2+} will first exchange with the former and then with the latter compound. Upon addition of the second cation the aqueous CPNC dispersion may show reduction of viscosity by one order of magnitude. The process is illustrated by the following example.

The reactor was charged with water, a polar surfactant and Na-MMT, heated to 85 °C, and then $Ca(OH)_2$, methacrylic acid, and Na_2CO_3 were added. In a separate vessel, a monomer emulsion was prepared from water, polar surfactant, butyl acrylate, methyl methacrylate and methacrylic acid. A small quantity of the monomer emulsion and $(NH_4)_2S_2O_8$ was added to the reactor (to form a polymer seed). The remaining monomer emulsion was fed into the reactor at T = 85 °C, then T was reduced to 65 °C and ferrous sulfate + *tert*-butyl hydroperoxide + *iso*-ascorbic acid was added. After polymerisation the temperature was reduced to $T < 45$ °C and pH was increased to 7.5. Finally, a bactericide was added. The CPNC was cast into film and tested for the maximum tensile strength and elongation at break at T = 22 °C and RH = 50%. Results of the test are shown in the first row of **Table 86** The data indicate that homogenisation is not as important as addition of $Ca(OH)_2$ and the order of its addition. Surprisingly, the addition

Table 86 Tensile strength (σ_{max}) and strain at break (ε_b) for acrylic-based CPNC. Data [Lorah and Slone, 2002c]

No.	Clay (wt%)	Homogenised	Ca^{2+}	σ_{max} (MPa)	ε_b (%)	Comments
1	2	N	Y	2.87	1045	Process described in the text
2	2	Y	Y	2.85	1059	Same as (1), but homogenised for 20 min
3	2	Y	Y	1.74	1110	Same as (1), but methacrylic acid added before $Ca(OH)_2$
4	2	N	Y	1.74	1099	Same as (3), but homogenised for 20 min
5	2	N	Y	0.886	1294	Same as (1), but $Ca(OH)_2$ added to finished latex
6	0	N	Y	0.654	1272	Same as (1), but no clay added
7	0	N	N	0.407	1040	Same as (1), but neither clay nor $Ca(OH)_2$ added
8	2	Y	N	0.562	1239	Same as (1), but no $Ca(OH)_2$ added

of $Ca(OH)_2$ seems to have a greater effect than that of clay (compare (6) with (8) in **Table 86**). This may indicate that Ca^{2+} partially crosslinks the acrylic resin, either through randomly placed methacrylic acid co-monomer units, or through ion clusters.

Exfoliated PS/MMT nanocomposites were obtained by emulsion polymerisation carried out in three steps: (1) pre-swelling a clay (Na-MMT, saponite, hectorite, or fluorohectorite) in H_2O; (2) combining the suspension with styrene (or styrene and methyl methacrylate), emulsifying agents, initiator and a reactive intercalant having affinity for the layered silicate; and (3) emulsion polymerising the system to form exfoliated CPNC [Choi *et al.*, 2003]. The reactive intercalant of choice was 2-acrylamido-2-methyl-1-propane sulfonic acid (AMPS), but several other compounds are also named in the document. The properties of the exfoliated CPNC (containing up to 3 wt% MMT) were superior to these of the matrix polymer, *viz.* the temperature for 50% weight loss increased by *ca.* 50 °C, the elastic modulus increased by 95% (PS matrix) or by 660% (PS-PMMA matrix). The new materials are expected to be used where heat resistance, stiffness, and/or dimensional stability are required.

4.1.4.1.3 Solution Polymerisation Methods

Moet and Akelah [1993] prepared PS-based nanocomposite by intercalating Na-MMT with vinyl-benzyl-trimethyl ammonium chloride (VM), which increased the interlayer spacing from d_{001} = 0.96 to 1.5 nm. Next, the purified and dried intercalate (VM-MMT) was dissolved in acetonitrile, and mixed with styrene and AIBN initiator. After solution polymerisation and extraction of PS not bounded to clay, d_{001} = 2.45 nm and T_g = 99.5 °C was determined.

In the following paper [Akelah and Moet, 1996] the process was repeated with two clays: Na-MMT and Ca-MMT. After intercalation with VM the clays showed virtually identical interlayer spacing, *viz.* d_{001} = 1.50 and 1.49 nm, respectively. The solution polymerisation (in either toluene or THF) yielded CPNC. The reactor powder was in the form of spherical particles, *ca.* 150-400 nm in diameter. For CPNC with 5 to 50% organoclay the amount of extractable PS decreased from 80 to 10 wt%, respectively. The extraction significantly changed the composition, as instead of the nominal clay content of 5, 10, 25 and 50 wt%, when measured after extraction it was, respectively, 25, 43, 43 and 56 wt%. The interlayer spacing (as measured by XRD) showed smaller variation, *viz.*: $d_{001} \cong$ 2.3, 2.1, 1.7 and 2.1 nm, respectively. However, the interlayer spacing of not extracted, compression moulded CPNC with 5, 10 and 25 wt% organoclay showed d_{001} = 20, 20 and 14.6 nm, respectively. These excellent values (especially considering the high clay content) indicate that inside the interlamellar galleries the not-tethered PS chains formed a solution with their tethered homologues. Neither the processing nor the performance of these intercalated CPNCs was reported, thus it is not known whether the dispersion remains stable under conditions of industrial forming processes.

The method of CPNC preparation developed by Moet and Akelah was adopted for the preparation of CPNC with syndiotactic sPS during metallocene polymerisation of styrene at 60 °C [Geprägs *et al.*, 1998]. The patent specifies that the resulting thermoplastic moulding composition contained: (A) 10 to 99 wt% of a syndiotactic vinyl aromatic polymer, (B) 0.1 to 20 wt% of a delaminated clay, and (C) 0 to 50 wt% of other additives and processing aids. However, the examples focus on sPS/MMT nanocomposites showing good flow, heat resistance, mechanical and surface properties at low density. The materials were obtained by first pre-intercalating MMT (CEC = 0.9 meq/g) with vinyl benzyl trimethyl ammonium chloride (VM). First, 3.13 g of organoclay was dispersed under N_2 in 104.2 g of styrene for 1 h at 60 °C. Next, methyl aluminium oxide (MAO), di-isobutyl aluminium hydride (DIBAH) and pentamethyl cyclopentadienyl trimethyl titanium (Cp.TiMe3) in toluene were added. After 2 h of polymerisation at 60 °C, the product was precipitated with methanol, washed and vacuum dried. The product contained 97 wt% of sPS (M_w = 343 kg/mol; > 98% syndiotactic pentads) and 3 wt% of organoclay. At a later date the process was broadened to elastomer toughened CPNC of virtually any thermoplastic [Grutke *et al.*, 2002]. Thus, the material contained three ingredients: (1) thermoplastic resin, (2) delaminated phyllosilicate dispersed in it, and (3) rubber or rubber mixtures (with particles ranging in diameter from 10 to 1000 nm). The thermoplastic can be sPS, PO as well as PA or another commodity or engineering polymer.

Biasci *et al.* [1994] compared the properties of CPNC with PMMA as a matrix prepared using reactive and non-reactive routes. The nanocomposites were prepared by: (1) intercalating Na-MMT with 2-(*N*-methyl-*N,N*-diethyl ammonium iodide) ethyl acrylate (QD1) or 2-(*N*-butyl-*N,N*-diethyl ammonium bromide) ethyl acrylate (QD4), then copolymerising the organoclay with MMA, and (2) intercalating Na-MMT with a copolymer of MMA and the ammonium salts. Using the first method, intercalation increased the interlayer spacing from d_{001} = 0.95 to 1.42 and 1.49 nm for QD1 and QD4, respectively. Copolymerisation in acetonitrile solution resulted in minor additional expansion (to d_{001} = 1.57 nm). However, intercalation with macromolecular ammonium salts (the second method) gave larger interlayer spacing (up to d_{001} = 2.96 nm for inorganic content of 58.8 wt%), providing significant improvement of thermal stability. The structure of the CPNCs was later studied by solid state NMR [Forte *et al.*, 1998].

Muzny *et al.* [1996] used a synthetic clay ($Si_8[Mg_{5.54}Li_{0.46}H_4O_{24}]^{0.46-}$ $Na^{0.46+}$, Laponite RD; resembling hectorite) with layered hydrous magnesium silicate platelets 25-30 nm diameter and 1 nm thick, having CEC = 0.55 to 0.73 meq/g. The intercalant with a reactive vinyl group was dimethyl hexadecyl allyl ammonium bromide (modified common surfactant: cetyl trimethyl ammonium bromide, CTAB). The authors found that once the level of the intercalant reaches 100 of CEC the organoclay precipitates. However for high intercalant/surfactant content of *ca.* 7 to 100 fold of CEC, a milky suspension of double-layer colloidal dispersion, stable for at least one week, was obtained. The stability was confirmed by dynamic light scattering. The intensity-time correlation function indicated a monodispersed micellar dispersion. The hydrodynamic volume of the 0.1 wt% Laponite suspension significantly increased (peak position moved from 22 to 138 nm), but the system remained stable. The intercalation increased the interlayer spacing from d_{001} = 1.2 to 2.4 nm.

The CPNC was obtained by dissolving acrylamide in the aqueous suspension of Laponite, deoxygenating the system and adding $K_2S_2O_8$. Starting with a stable, monodispersed micellar suspension of intercalated clay the free radical polymerisation of acrylamide resulted in well-exfoliated CPNC, without regular spacing below 60 nm (SANS data). Incorporation of 0.1 wt% of the synthetic clay increased the transition temperature of polyacrylamide (PAM) from 134.7 to 139 °C. CPNC with higher clay content had the transition above 148 °C. Other properties were not measured. It would be interesting to repeat this work using a high aspect ratio Na-MMT.

Since PVC is mainly produced by suspension or emulsion polymerisation in aqueous medium, the preparation of PVC-based PNC would seem easy. However, there are complicated reactions between the intercalating ammonium ion and macroradicals, thus not much information on the formation of PVC-based PNC during the polymerisation is available. There is also little known about the effects of clay addition on PVC performance.

More recently exfoliated CPNCs of PVC with Cloisite® –6A, –10A or –30B were obtained by the suspension polymerisation or melt intercalation method [Hu *et al.*, 2004]. It was found that while the reactive method resulted in exfoliation for any of the three organoclays, melt compounding resulted in intercalated systems – the best dispersion was obtained with Cloisite® 10A and the worst with Cloisite® 6A (no longer commercially available).

Tabtiang *et al.* [2000a] used a bypassing method for the preparation of CPNC with PVC as the matrix – by blending the resin with PMMA-clay complexes. The PMMA/clay CPNC developed was considered as an additive for PVC when used at elevated temperatures. The CPNC was prepared through solution polymerisation of methyl methacrylate in the presence of clay pre-intercalated with either dodecyl- or hexadecyl-trimethyl-ammonium ions (3MDDA or 3MHDA, respectively). The process was conducted at 80 °C for 60 min, and then the complex was filtered, washed, dried and ball-milled. The polymerisation of MMA (added to a hexane dispersion of organoclay) was initiated with AIBN. The PMMA-based PNC was obtained in 5 h long radical polymerisation at 68 °C.

In the product the organoclay content was: 10, 20, 50, or 100 phr. After drying the products were either melt fluxed then compression moulded at 180 °C for 5 min, or statically annealed at 175 °C. GPC data showed that up to 7 vol% of 3MDDA the molecular weight or the polydispersity was independent of clay content, but at > 15 vol% the chain transfer resulted in a reduction of M_w and increased M_w/M_n. A significant amount of chain branching was also observed.

The 3MDDA contained 17.7 wt% of dodecyl ammonium. As a result of intercalation the clay interlayer spacing increased from d_{001} = 1.29 to 1.79 nm. Similar values were obtained for 3MHDA. XRD of the polymerisation products showed absence of peaks from 2θ = 1 to 10°, thus the interlayer spacing was ≥ 8.8 nm. However, after shearing the specimens with 3MDDA or 3MHDA showed XRD peaks corresponding, respectively, to d_{001} = 2.9 and 3.88 nm, the same for all the organoclay contents. Thus, there is a significant re-aggregation of the clay platelets induced by shear stress. Comparing to the static annealing effects, shearing greatly speeds up the re-agglomeration. Thus, the clay platelets that were well exfoliated and randomly distributed in the matrix during the melt flow re-assembled into closely packed short stacks. TEM of sheared 3MDDA systems showed that the short stacks had 3 to 5 platelets and that the distances between aggregates was about 8.5 nm, but isolated platelets were also present. Similar, although less pronounced, re-aggregation was observed for the statically annealed specimens. In the latter case, the clay platelets were aligned under pressure parallel to the mould surface. These PMMA-clay nanocomposites were melt blended with PVC, increasing the modulus and HDT of its rigid formulations without affecting clarity.

In a parallel publication, the properties of PMMA/clay CPNC prepared by solution polymerisation and melt compounding were compared [Tabtiang *et al.*, 2000b]. Both systems were fluxed in an internal mixer to give comparable melt processing histories. As in the preceding publication, the reactive route gave d_{001} > 8.8 nm, which, after melt mixing, decreased to d_{001} = 2.92 nm, independent of the organoclay content. The corresponding value for the melt compounded CPNC was d_{001} = 3.02 nm. These values indicate expansion of the organoclay interlamellar galleries by Δd_{001} = 1.13 and 1.23 nm, respectively. Depending on the assumed PMMA chain conformation, either one or two macromolecules reside within the gallery.

The dependence of the glass transition temperature on organoclay content (T_g versus w) for the two types of CPNCs was quite different. While $T_g \cong$ 106 °C of melt compounding systems was virtually independent of composition, that of the polymerised system increased from 106 °C for neat PMMA (w = 0), to 121, 121 and 124 for w = 3.8, 6.7 and 14.5 vol% of organoclay, respectively. This indicates that the reactive route causes a reduction of PMMA chain mobility. Since the intercalant does not participate in the polymerisation, the reduction must be caused by adsorption of the newly formed macromolecules on the clay surface. Some of these macromolecules remain on the surface after melt compounding. By contrast, the melt compounding route resulted in expansion of the organoclay interlayer spacing from d_{001} = 1.79 to 3.02 nm, but apparently here the macromolecules are isolated by the onium ion from the clay surface.

Nakada and Kanaida [2001] patented a reactive method for the production of exfoliated CPNC. For example, onium intercalated smectite was dispersed in a solution of a monomer in organic solvent, e.g., styrene in xylene solution, and then polymerised. The resulting CPNC was found to be exfoliated.

Owing to the hydrophilic nature of clay the Ziegler-Natta (Z-N) or metallocene polymerisation methods that would lead to CPNC are inefficient [Zhang *et al.*, 2002b]. To change this unhealthy situation, Zhang *et al.* [2002b] deposited a patent, which simultaneously solves the basic problems: (1) sPS manufacture becomes efficient. (2) The reactor fouling during sPS polymerisation in a stirred reactor is eliminated. (3) The production of exfoliated CPNC with sPS as the matrix is reliable. Basically, the process aims to cover the clay platelets with an insulating copolymer. Once the clay surface is covered, the Z-N or metallocene catalyst can safely be deposited, and then the monomer polymerised. The clay may be natural or synthetic smectite, e.g., Na-MMT. For example, clay was dried for 6 h at 400 °C then dispersed in toluene solution of SAN (ratio clay/SAN = 20:1). After 2 h of mixing, the solvent was removed and a co-catalyst, methyl aluminoxane (MAO), and a catalyst, $Cp^*Ti(OMe)_3$ and styrene were added. The polymerisation was quenched after 1 h – conversion was 53.5%, and the activity = 4.86 kg/mmol-Ti-h (tripling the MAO content increased conversion and activity to 78.6% and 7.14 kg/mmol-Ti-h, respectively). In spite of the high MMT content (4.1 wt%) XRD showed full exfoliation, the flexural modulus and Izod impact strength were 380 MPa and 80 kJ/m^2, respectively. Thus, incorporation of MMT increased the performance of sPS by 41% (flexural modulus) and 19% (Izod impact strength). The claims specify that the method is general, applicable to a variety of substituted styrene monomers.

Recently, Zerda *et al.* [2003] prepared CPNC with PMMA as matrix, using supercritical carbon dioxide ($scCO_2$) as the solvent for MMA polymerisation. Four types of Cloisite® were used: Na^+-MMT, C15A, C20A and C25A. An attempt was made to produce intercalated nanocomposites containing a high level of organoclay. Below 40 wt% of the nanofiller the interlayer spacing depended on the length of the fully extended intercalant chains. At higher clay content, the d_{001} spacing decreased. A simple model for estimating this transitional concentration resulted in a relation for the polymer-to-organoclay mass ratio:

$$R = m_p/m_{organoclay} = \rho_P (d_{CPNC} - d_{organoclay}) S_{organoclay} \qquad (182)$$

where ρ_P is the polymer density, $d_{organoclay}$ and d_{CPNC} are the interlayer spacing in the organoclay and in CPNC, respectively, and $S_{organoclay}$ is the clay surface area in m^2/g. To derive Equation 182, the authors assumed that the increase of the interlayer spacing is solely due to incorporation of polymer between clay platelets, without a significant change in the total free volume, with the polymer density the same in the bulk as within the interlamellar galleries.

4.1.4.2 Other CPNC Prepared by Solution Method

The solution methods discussed in this part involve the use of a common solvent for the matrix polymer and organoclay. The increased macromolecular mobility accelerates diffusion of macromolecules into the interlamellar galleries. Thus, in principle, the solution method with a well-chosen common solvent is equivalent to melt exfoliation, but at lower temperature and over longer time. Considering the high cost of solvent and its removal, the method has been only used to prepare model CPNCs for diverse experimental studies.

Ren *et al.* [2000] prepared a series of intercalated CPNCs based on a disordered polystyrene-*b*-1,4-polyisoprene diblock copolymer (PS-*b*-IR) with varying amounts of organoclay. The molecular weight of the copolymer was M_w = 17.7 kg/mol and polydispersity M_w/M_n < 1.07. The PS content was 44 wt%, and the copolymer order-disorder transition temperature was below 80 °C. The organoclay was MMT (CEC = 0.90 meq/g) intercalated with 2M2ODA. After intercalation, the excess intercalant was extracted with boiling ethanol. The remaining alkylammonium (26.8 wt%) corresponded to the clay CEC.

CPNCs were prepared at RT by dispersing appropriate quantities of organoclay in a toluene solution of the block copolymer. XRD gave the interlayer spacing of organoclay as $d_{001} \cong 2.3$ nm. Five compositions were prepared with clay content of 0.7, 2.1, 3.5, 6.7, and 9.5 wt%. The solutions were homogeneous. The recovered nanocomposites were dried under vacuum. Solution blending of the organoclay with PS (M_w = 30 kg/mol), or with various quantities of block copolymer resulted in further intercalation to $d_{001} \cong 3.1$ or 3.5 nm (independent of the clay loading), respectively. By contrast, the same process using IR (M_w = 17 kg/mol) did not affect the organoclay spacing ($d_{001} \cong 2.3$ nm remained). This result is consistent with IR/organoclay immiscibility. In all cases the XRD peak was sharp, indicating a highly regular layered clay structure.

Apparently, toluene caused the organoclay to expand – the intercalation by PS or the PS-*b*-IR block copolymer took place by progressive replacement of toluene molecules inside the interlamellar galleries by PS segments. The immiscibility of the IR-block with organoclay suggests that it is the PS-block that diffuses into the galleries, leaving most of the PI-block outside. Considering that the gallery height is *ca.* h = 1.5 nm, while the transverse diameter of the polymer iso-diametrical alkyl chain (adopted after Flory *et al.* [1984]) is about 3.5 $-CH_2-$ units with the diameter of *ca.* 0.45 nm on both gallery walls, there is not much space left for the polymer to diffuse into. Because of the gallery restriction the PS-block must adopt a 2D random walk. Statistics indicate that the radius of gyration in 2D is about 50% larger than that in 3D hence the PS block is large enough to fill the gallery between the MMT platelets (in 2D the unperturbed end-to-end distance for the PS-block is about 90 nm). The slightly higher expansion obtained for the block copolymer than for PS may indicate that the IR block may be partially dragged into the gallery. Thus, solution intercalation of either PS or its block copolymer into pre-intercalated MMT failed to lead to exfoliated CPNC. Clearly, MMT-2M2ODA is immiscible with PS homopolymer or PS-block, as well as with IR and IR-block.

For studies of the mechanism of flammability reduction by CPNC several nanocomposites with PP and PS matrix were prepared using pre-intercalated MMT or fluorohectorite (FH), *viz.* tetradecyl ammonium-FH (FH-TDA), octadecyl ammonium-MMT (MMT-ODA) and dimethyl dioctadecyl ammonium-MMT, (MMT-2M2ODA) [Gilman *et al.*, 2000a,b]. The synthetic FH (unit cell: $Z^+_{1.6}[Li_{1.6}Mg_{4.4}(Si_{8.0})O_{20}F_4]$, where Z^+ is the exchange cation) had a higher aspect ratio, p = 500 to 4000, compared to MMT (unit cell: $Z^+_{0.86}[Mg_{0.86}Al_{3.14}(Si_{8.0})O_{20}OH_4]$); with p = 100 to 1000. However, there are some indications that FH platelets are more fragile, prone to extensive attrition during melt compounding.

For comparison the CPNCs were prepared using one of these procedures:

(1) *Static melt intercalation,* where powders of PS and organoclay were ground together, and then heated at 170 °C for 2-6 h under vacuum.

(2) *Solvent intercalation,* where organoclay was dispersed in a toluene solution of PS (M_n = 100 kg/mol) using ultrasonication for 5 min then removing the solvent.

(3) *Extrusion melt intercalation,* where PS and organoclay were dry-blended, and then extruded using a DSM recirculating mini-extruder under N_2 at 150-170 °C for 2-4 min.

The CPNC were characterised by XRD, TEM and cone calorimetry. It was found that all three methods detected intercalated and partially exfoliated nanostructures. However, at the temperatures required for PS extrusion thermal degradation could take place. When the static method was used, the PS with 3 wt% of FH-TDA was intercalated, whereas PS with the same amount of MMT-2M2ODA was *ca.* 25% exfoliated with rest of the clay forming short (2-3 layers), intercalated stacks.

Cone calorimetry was used to measure such properties as heat release rate (HRR), peak HRR (PHRR), as well as smoke and carbon monoxide yield. Four compositions were studied: neat PS and PS mixed with Na-MMT as controls, and two CPNC: PS/MMT-2M2ODA and PS/FH-TDA. Surprisingly, FH-TDA had no effect on PHRR (similar to the not intercalated Na-MMT), whereas MMT-2M2ODA lowered the PHRR by *ca.* 60%. The authors observed reduced flammability for CPNC containing either intercalated or exfoliated MMT.

Kim *et al.* [2003] dispersed Cloisite® 25A (C25A; MMT-2MHTL8) in a $CHCl_3$ solution of PS by stirring for two days at 25 °C. After evaporation of solvent, the dried CPNC showed increased interlayer spacing from $d_{001} \cong 1.94$ nm determined for C25A to $d_{001} \cong 3.27$ nm; hence exfoliation was not obtained. The shear viscosity at 200 °C indicated a systematic increase of the zero-shear viscosity from $\eta_o = 9.3$ to 11.7, 16.9 and 33.3 kPas for clay content 0, 2, 5, and 10 wt% of C25A, respectively. The dynamic measurements also showed a systematic increase of the shear modulus versus frequency dependence in the log-log plot. Thus, even in the absence of exfoliation incorporation of organoclay significantly affected PS behaviour.

Wu *et al.* [2001] prepared CPNC with sPS as matrix by dispersing MMT (CEC = 1.19 meq/g) and MMT pre-intercalated with cetyl pyridinium chloride, CPC (MMT:CPC weight ratio 1:1) in a dichlorobenzene (DCB) solution of sPS ($M_w = 100$ kg/mol) at $T = 140$ °C for 24 h. The CPNCs, both containing 5 wt% clay, were characterised by XRD, TEM, and FTIR. The not pre-intercalated MMT was found to form stacks few tenths to 100 nm thick with $d_{001} = 1.2$ nm, whereas the intercalated MMT was mostly exfoliated with short stacks and $d_{001} > 8.8$ nm. The morphology of the samples did not change after keeping them at 320 °C for 20 min. The authors observed that as a result of clay incorporation, the sPS chain conformation tended to convert from the helical *TTGG* to planar all-*trans* *TTTT* conformation. The clay facilitated formation of the β-crystalline form, particularly in thin sPS films.

Tseng *et al.* [2001] prepared CPNC using sPS and a CPC organoclay, similar to the system used by Wu *et al.* [2001]. However, now five intercalant-to-clay weight ratios were used, namely CPC:MMT = 0, 0.25, 0.5, 1, and 2. The interlayer spacing increased with CPC content, *viz.* $d_{001} = 1.2$ to 1.4, 1.6, 2.1 or 2.1 nm, respectively. The CPNCs were prepared either in DCB solution or in a CORI-TSE at 270 to 290 °C, with screw speed of 250 rpm. The CPNC was characterised by XRD, TEM, FTIR, and DSC. The main purpose was the study of the sPS crystallisation kinetics by fitting the isothermal data to Avrami's equation. For CPNC, the information on the interlayer spacing was not comprehensive – XRD and TEM indicated exfoliation in CPNC containing 5 wt% of CPC:MMT = 1. As shown in **Table 87**, the sPS crystallisation rate depended on the composition as well as on the method of CPNC preparation. The MMT is a nucleating agent, thus neat clay causes more rapid crystallisation than does organoclay. Surprisingly,

Table 87 Avrami's kinetic parameters, *n* and *k*, for sPS/MMT/CPC systems. Data [Tseng *et al.*, 2001]

sPS/MMT (wt%)	MMT/CPC ratio	Process	n	k
95/5	1/1	Solution	2.66	7.8×10^{-2}
95/5	1/0	Solution	2.47	0.12
95/5	1/1	Melt	2.28	0.63
95/5	1/0	Melt	2.64	4.90
95/5	1/0.25	Melt	1.96	1.86
95.5	1/0.5	Melt	2.83	0.81
95/5	1/1	Melt	2.28	0.63
95/5	1/2	Melt	2.68	0.58
97.5/2.5	1/0	Melt	2.44	1.48
99.0/1.0	1/0	Melt	2.92	1.41
99.5/0.5	1/0	Melt	2.78	1.95
100/0	1/0	Melt	2.54	0.11

CPNC prepared in solution shows the crystallisation constant, $k = 0.078$, whereas the same composition prepared in melt has $k = 0.63$. Such passivation of the nucleating agent may indicate that traces of adsorbed solvent remain in the system. The Avrami exponent, n, does not change systematically.

Voulgaris and Petridis [2002] explored the 'co-solvent' route for the preparation of CPNC with hectorite pre-intercalated with 2M2ODA and dispersed in PS/PEMA blends. When the blend was dissolved in THF, either polymer could diffuse into the galleries, but when individual components were used, only PEMA entered the alkyl-filled clay cavities. It is of note that the PS/PEMA blends are immiscible and cast from THF they form large domains with average diameter of about 10 μm. Addition of the organoclay reduces the size of the PS domains, following a typical emulsification curve. For an organoclay content > 15 wt% the size of the PS domains was constant at 1.6 μm and the interlayer spacing, d_{001} = 3.5 nm. On the basis of the DSC data, the authors postulated that 2M2ODA acts like a compatibiliser for the PS/PEMA blend.

4.1.4.3 Vinyl-Type CPNC Prepared by Melt Compounding

The melt intercalation methods are discussed in Section 2.3.8, whereas general principles of melt exfoliation for individual matrix polymers are summarised in Section 2.4.3. Furthermore, the methods are also part of the melt exfoliation of PA or PO, discussed in Sections 4.1.1 and 4.1.2, respectively. Here the focus is on the preparation of CPNC with PS or its copolymers as the matrix via melt compounding.

Static melt intercalation in a PS matrix traces its origin to work by Vaia *et al.* [1993; 1995a,b; 1996]. The process involved mixing organoclay with a polymer powder, pressing the mixture into a pellet, and heating under vacuum at $T_b > T_g$ (e.g., 155 to 180 °C) for $t_b \leq 400$ min. Two organoclays (MMT or FH pre-intercalated with dimethyl dihydrogenated tallow-, dodecyl- or octadecyl-ammonium) and mono- or poly-dispersed PS resins (M_w = 30 to 400 kg/mol) were used. The XRD results showed that the d_{001} peak position changes with t_b from the original value of the organoclay (e.g., d_{001} = 2.13 nm) to new spacing, characteristic for the PS intercalated CPNC (e.g., d_{001} = 3.13 nm). The intensity of the intercalated peak is a measure of the intercalated structure content. It increases with T_b and decreases with M_w. The same behaviour was found for the polymer with high molecular weight, indicating a diffusion-controlled process (see also Section 3.1.8). The intercalation by PS macromolecules was found to be reversible – the PS could be extracted by toluene, replacing the macromolecules by solvent. Sikka *et al.* [1996] reported similar results.

Some specimens were prepared by mixing PS with MMT-organoclay in a laboratory-size TSE at 250 °C. The extrusion, with the residence time $t_r \cong 4$ min, resulted in fully intercalated systems. Thus, the high temperature and shearing reduced the intercalation time from $t_b > 100$ min to less than 4 min, without changing the interlayer spacing. The prohibitively slow static intercalation depends on the Brownian motion of the macromolecules. The results indicated that mechanical mixing provided by ultrasonication, shearing in a rheometer, or compounding in a TSE, accelerates the diffusion and increases the uniformity of structure.

Noteworthy, when PS was replaced by poly-3-bromostyrene (PS3Br; M_w = 55 kg/mol) the MMT-DDA exfoliated under the static process. This resin being more polar than PS, strongly interacted with the MMT surface (even after intercalation with DDA). This underlines the importance of thermodynamic considerations for the preparation of CPNC. These observations were confirmed in the following publication from the laboratory [Vaia and Giannelis, 1997a,b]. Interestingly, FH pre-intercalated with DDA was exfoliated in PS3Br, but only intercalated in poly-2-vinyl pyridine.

Starting in 1999 the use of melt compounding for the preparation of CPNC with PS as a matrix became more frequent. Thus, Fischer *et al.* [1999], Fischer and Gilgens [1999a,b] and Fischer [1999] patented a hybrid method. In the first stage Na-hectorite, Na-saponite and synthetic MMT were statically melt intercalated with poly(styrene-*b*-ethylene oxide) or poly(styrene-*b*-2-vinylpiridine) and then the pre-intercalated clay (clay content: 5 wt%) was compounded with PS in a mini-TSE at 160 °C for *ca.* 5 min. During the first stage the intercalation was as rapid as that by neat PEG. Depending on the block length of the hydrogen-bonding block, either exfoliation or intercalation was obtained. Since a controllable level of intercalation/exfoliation was already achieved in the first stage, the second stage compounding was mainly used to uniformly disperse the exfoliated organoclays in the matrix. The final product contained exfoliated clay platelets, with only few short stacks. Good mechanical performance was reported. The adopted strategy resembles the blend strategy of compatibilisation by addition of a bock copolymer, where one block is miscible with one immiscible polymer and the second with the other. By extension, using appropriate copolymers, the Fischer *et al.* method should be applicable to any clay/polymer system.

A recent patent describes a three-step method for CPNC preparation [Choe *et al.*, 2002]. Similar to Fischer *et al.*, these authors also use a compatibiliser.

However, now the compatibiliser is a homopolymer, thus the flexibility of its application to diverse polymeric matrices is more limited. In particular, the useful polymer should have oxygen atoms in the main chain and be miscible with the matrix. For example, poly-ε-caprolactone (PCL; MW = 10 to 100 kg/mol), owing to its thermodynamic miscibility with SAN, ABS, and PVC, has a wide applicability. Since PCL easily diffuses into interlamellar galleries, it aids the matrix to exfoliate the clay. To prepare CPNC first one must pre-intercalate clay with onium ions; next, prepare exfoliated organoclay/PCL nanocomposite; and then compound the exfoliated organoclay/PCL nanocomposite with the matrix resin. The first step may by bypassed by purchasing a commercial organoclay, e.g., Cloisite® 30A (MMT-MT2EtOH – same as Cloisite® 30B) favoured by the patent. The second step is diffusion controlled – as shown in **Figure 160**, it requires $t > 20$ min to achieve exfoliation at $T = 100$ °C. However, with adequate shearing, full exfoliation could be obtained by compounding in a TSE. The use of TSE for step 3 has some inherent danger, as here the temperature must increase. The reported data show that as temperature increases (T = 140 to 160 °C), the degree of dispersion decreases (see **Figure 161**). The effect may not be entirely caused by the thermal decomposition of the onium ion. Polymer blends usually show a lower critical solution temperature (LCST) hence the effect may be related to phase separation in the intercalant/PCL/SAN system. One of the examples describes that extrusion, using a SSE for mixing 5 wt% Cloisite® 30A with 30 wt% PCL and 65 wt% PVC at 160 °C, resulted in fully exfoliated CPNC.

Wang *et al.* [2001c] melt blended PVC/Cloisite® 30B/di(2-ethylhexyl) phthalate (DOP) in an internal mixer at 170-180 °C. Next, the specimens were annealed under vacuum at 104 °C for 24 h. A major problem here was the thermal degradation of PVC. It was found that organoclay may replace DOP in controlling the degradation time (judged by PVC discoloration), t_d. The reported data may be approximated by the linear equation:

$$t_d = a_o + a_1 w_c + a_2 w_p \tag{183}$$

where w_c is organoclay content (wt%), and w_p is the DOP content (wt%). Least squares fit gives the following equation parameters: $a_o = 5.42 \pm 0.36$; $a_1 = 0.10 \pm 0.02$; and $a_2 = 0.27 \pm 0.02$. The computed standard deviation of data was $\sigma = 0.51$, and the correlation coefficient squared $r^2 = 0.998$. Thus, C30B is about 1/3 as efficient at reducing t_d as DOP. The interlayer spacing of C30B was found to be: d_{001} = 1.9 nm. Melt compounding increased this value, but changing the organoclay content (from 1 to 30 wt%), DOP concentration (from 0 to 35 wt%), or the PVC resins (two resins with M_w = 62 or 233 kg/mol) introduced small changes: d_{001} = 3.7 to 4.4 nm. Evidently, there is no great desire for the PVC segments to be near the aliphatic intercalant groups – only a moderate intercalation has been achieved. The tensile strength of the CPNC increased as the fraction of clay increased from zero to 2 wt%, and then it remained stable, independent of DOP content, provided that the latter did not exceed 10 wt%. Du *et al.* [2003] reported on the thermal degradation and charring of these systems by XPS. By contrast with CPNC with either PS or PMMA as a matrix, in these PVC-based nanocomposites the surface at higher temperatures was dominated by carbon, not by clay oxygen. The presence of clay was found to retard PVC degradation and enhance char formation.

In another paper [Wang and Wilkie, 2002] PVC-based CPNC have been prepared by mixing together a PVC solution in THF with suspension in the solvent of either Na-MMT or Cloisite® 30B – one sample was prepared with addition of

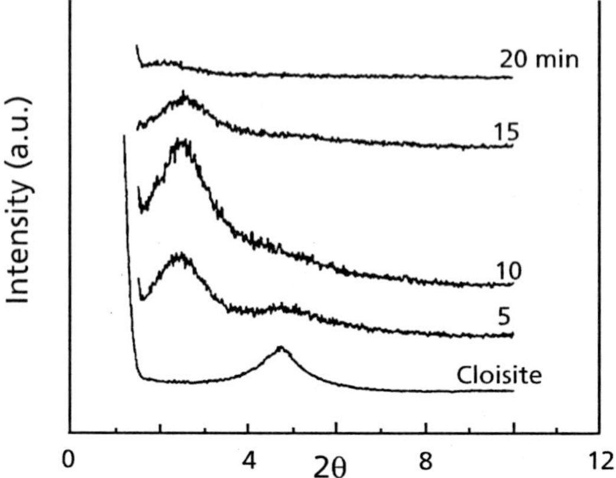

Figure 160 Time dependent XRD spectra indicating progressive exfoliation of Cloisite® 30A/PCL = 5:95 with time. The system was compounded in a mini-max at T = 100 °C. After [Choe et al., 2002].

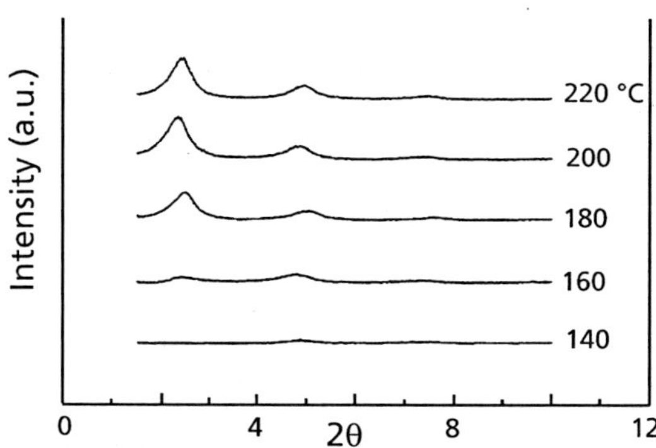

Figure 161 Effect of extrusion temperature on the XRD spectra showing that the interlayer spacing in a Cloisite® 30A/PCL/SAN = 5:30:65 system decreases with increase of the extrusion temperature. After [Choe et al., 2002].

DOP, another without it. Intercalation was conducted under reflux condition (T_b = 65 °C) for 5 h, and then THF was evaporated. The specimens were also annealed at T_a = 90 or 104 °C for 24 or 48 h. XRD show that the as-prepared specimens with DOP had $d_{001} \cong 5$ nm, but during annealing (especially at higher temperature) a highly regular intercalated clay structure developed with $d_{001} \cong 4$ nm. In the absence of DOP, the interlamellar galleries showed little expansion, *viz.* $d_{001} \cong 1.14$ nm. Low magnification TEM demonstrated the presence of micron-size stacks of clay particles.

PVC/Cloisite® 30B nanocomposites with either DOP, DOA or PLA were prepared in a Buss Ko-kneader at T = 130 to 160 °C [Simonik *et al.*, 2002]. The CPNCs were intercalated with d_{001} = 3.48 nm. In spite of the lack of exfoliation good mechanical properties were obtained, *viz.* 4-fold and 8-fold increased tensile modulus at 5 and 10 wt% organoclay loading. The flame resistance was also enhanced.

PS/MMT nanocomposites were prepared by first intercalating MMT with 3MODA, and then, in a TSE at 180 °C, dispersing the organoclay either in PS or in a styrene-methyl vinyl oxazoline copolymer (PS-VO containing 5 wt% VO; see schematic below) [Hasegawa *et al.*, 1999].

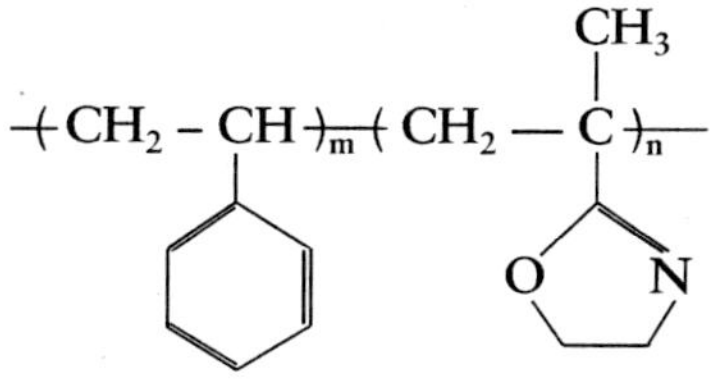

$$-\!\!\left(CH_2 - CH\right)_m\!\!-\!\!\left(CH_2 - \overset{CH_3}{\overset{|}{C}}\right)_n\!\!-$$

O N

The degree of dispersion was characterised using XRD and TEM. At 7 wt% organoclay loading, the interlayer spacing increased from the organoclay value of d_{001} = 2.2 nm, to 3.1 nm (in PS), or to 3.9 nm (in PS-VO). For the latter CPNC the diffraction peak was small, indicating extensive exfoliation. This was confirmed by TEM. The tensile tests gave rather ambiguous results. In **Table 88** the relative values (in relation to the matrix) of modulus, tensile strength and elongation at break are presented. The results are disappointing – independent of whether the system is intercalated or exfoliated the improvement of modulus is small (8 or 38%, for PS or PS-VO, respectively) and 'costly' in increased brittleness. Incorporation of 1.6 wt% of clay in a PA-6 matrix resulted in a 73% increase of modulus. Since the total crystallinity in PA-6 based clay does not significantly change (see Section 3.3.3), the difference between the PS and PA-6 CPNC behaviour is not related to the amorphous/crystalline matrix, but rather to the non-tethering and tethering of macromolecules to the clay platelets. The influence of thermo-oxidative degradation in the melt compounded PS-systems also cannot be ruled out.

PS (M_w = 347 and M_n = 131 kg/mol) was compounded with Cloisite® 6A (MMT-2M2HTA) in an internal mixer at T = 200 °C, for t = 10 min, with the shaft speed N = 45 rpm [Lim and Park, 2000]. Another series of samples was prepared by compression moulding and annealing at the same temperature for 25, 80, and 110 min under N_2. The interlayer spacing was found to increase with annealing time from d_{001} = 3.21 to 3.45 nm. At the same time the storage shear modulus, G', increased by about 70%. In the following paper, Lim and Park [2001] prepared three series of CPNC by compounding PS, SMA or maleated PE with 3, 5 or 10 wt% of Cloisite® 6A in an internal mixer at T = 210 °C, for t = 10 min, with the shaft

Table 88 Matrix polymer property change (in %) upon incorporation of 7 wt% MMT-3MODA. Data [Hasegawa *et al.*, 1999]

Matrix	Modulus	Tensile strength	Elongation at break
PS	8	-10	-27
PS-VO	38	-18	-42

Note: The values were calcualted as: $\Delta P(\%) = 100[P(CPNC)/P(Polymer)-1]$

speed N = 45 rpm. In a preliminary test, the interlayer spacing of the C6A (d_{001} = 3.21 nm) decreased upon heating at T = 210 °C for t = 10 min, to about 2.96 nm. Compounding the organoclay with PS or SMA (7 wt% MAH) resulted in a slight expansion, *viz.* d_{001} = 3.45 or 3.37 nm, respectively, but compounding it with maleated PE resulted in exfoliation. The authors rationalised the difference as caused by the difference in molecular structure of the maleated polymers – while SMA has randomly placed MAH groups in the main chain, PE is only slightly grafted (0.8 wt% MAH). Thus, SMA is bridging the interlamellar gallery gap preventing exfoliation, while PE-MAH is able to graft only to one platelet and cause exfoliation by entropic effects. The rheological measurements showed much higher enhancement of low frequency G' values for SMA than those observed for PS – note that the interlayer spacing is reversed. Thus, even poorly intercalated CPNCs with tethered matrix macromolecules show higher enhancement of properties (HCP behaviour) than a passive matrix in which clay platelets are dispersed.

Wang *et al.* [2001a] prepared CPNC by melt intercalation in a TSE (D = 27 mm; L/D = 40). The organoclay was MMT pre-intercalated with ODA (Nanomer I.30 TC from Nanocor). Compounding 5 wt% I30 with PS resulted in expansion of the interlayer spacing from d_{001} = 2.17 to 3.3 nm. Noteworthy, the process resulted in a modest intercalation not only in PS and PP matrices, but also in PA-6. Incorporation of 10 wt% of the organoclay into PS increased the elastic modulus by a factor of 2.56, but it reduced the tensile strength by 39% and that of the impact strength by 55%. The dynamic viscosity versus frequency plot showed a uniform increase of viscosity for the CPNC (by a factor of *ca.* 1.7 at 10 wt% organoclay loading) without any trace of interaction at the lowest frequency – an expected non-HCP behaviour for the intercalated, non-tethered system.

A dramatic difference in CPNC behaviour was reported by Hoffmann *et al.* [2000] for end-tethered and non-tethered nanocomposites. Synthetic fluoromica (FM; CEC 0.7-0.8 meq/g; d_{001} = 0.95 nm) was intercalated with either 2-phenylethylamine (PEA, M_n = 121 g/mol) or with specially synthesised, amine-terminated PS (ATPS; M_n = 5.8 kg/mol; M_w/M_n = 1.33):

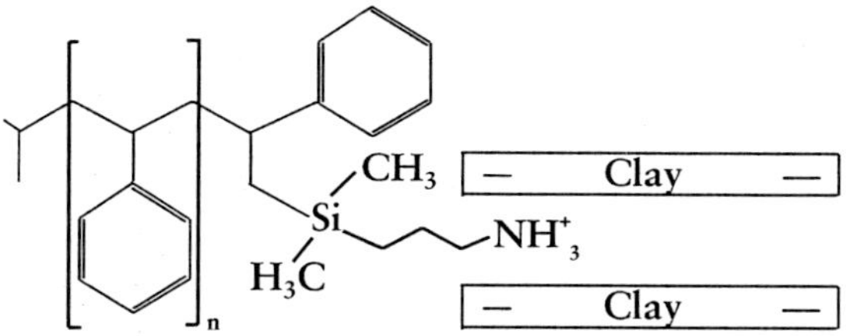

The intercalation was carried out in an acidified THF/water (4 vol%/l) mixture at 40 °C. The white precipitate of the formed organoclay was washed with a hot aqueous solution of THF and with water, and then dried under vacuum at 45 °C for 72 h. Next, CPNCs were prepared by compounding each of the organoclays (5 wt% of clay) with PS (M_n = 106 kg/mol; M_w/M_n = 1.07) at 200 °C in a microcompounder for 5 min.

The products were characterised by XRD, TEM and dynamic melt flow at strain γ = 10% and T = 140 to 220 °C. According to XRD, the CPNCs with FM-PEA did not show increased interlayer spacing over the organoclay, *viz.* d_{001} = 1.4 nm (i.e., the interlamellar gallery space of Δd_{001} = 0.45 nm). By contrast, the one prepared with FM-ATPS was exfoliated (d_{001} > 4 nm); these results were confirmed by TEM. In principle, by varying the *MW* of the ATPS intercalant, diverse morphologies could be obtained, starting (at low *MW*) with an intercalated system, then an exfoliated non-entangled system ($MW < M_e$), and then onto CPNC that not only is exfoliated, but also shows a high degree of association between the hairy clay platelets (HCP). The plot of the dynamic storage modulus, G', versus frequency for the FM-ATPS nanocomposite showed a similar tendency to that observed for exfoliated PA-6 CPNC from Ube, namely a low value of the initial slope, $g' = (d\ lnG'/d\ ln\omega)_{\omega < 0.1} \cong 0.46$. It would be interesting to see whether the mechanical properties of these two end-tethered systems, one amorphous the other crystalline, are also relatively similar.

Park *et al.* [2001] reported on the preparation of CPNC with sPS (M_w = 313 kg/mol; T_m = 270 °C) and Cloisite® 15A (MMT with CEC = 1.25 meq/g, intercalated with 2M2HTA). Owing to the thermal instability of the organoclay at sPS processing temperatures (tests showed that the d_{001} of C15A after 10 min at 240 °C decreased from 2.97 to 1.40 nm), a two-step process was used, i.e., first compounding the organoclay with an amorphous styrenic polymer or copolymer (e.g., PS, SMA, or SEBS-MAH) at $T \cong 200$ °C, cooling and crushing the mixture, and then blending it with sPS at $T \cong 280$ °C. An internal roller mixer was used with $N = 50$ rpm. Simultaneous compounding of all components at $T \cong$ 280 °C was also tried. The products were characterised by XRD, TEM and mechanical test methods.

The XRD spacings for CPNC prepared by compounding C15A with sPS or SMA indicated intercalation (increased d_{001} from 2.97 to 3.4 nm), while compounding with SEBS-MA resulted in exfoliation. Next, these intermediate CPNCs were compounded with 70 wt% of sPS. After compounding C15A/PS or C15A/SMA with sPS the interlayer spacing did not change much, i.e., values of d_{001} = 3.44 to 3.60 nm were obtained for the former mixtures (C15A/PS/sPS) and d_{001} = 3.25 to 3.33 nm for the latter (C15A/SMA/sPS). The numbers correlated neither with C15A content (3 to 9 wt%) nor with the procedure (simultaneous versus step-wise). Fully exfoliated CPNC was only obtained for C15A/SEBS-MAH/sPS containing 3 wt% C15A – at higher C15A content (e.g., 6 and 9 wt%) the system was intercalated with d_{001} = 3.37 and 3.38 nm, respectively.

The two preparation methods yielded CPNC with different microstructures and related properties. The sequential method resulted in more uniform intercalated structures, whereas the simultaneous mixing resulted in poor dispersion of clay. One reason for this may be that the system was only mixed once, whereas during the sequential process it was mixed twice. Another reason may be the degradation of alkylammonium followed by a decrease of interlayer spacing at T = 280 °C. Only the exfoliated nanocomposites showed similar structures and mechanical properties. The enhancement of the flexural modulus (relative to the sPS/amorphous polymer 70:30 blend) at 6 wt% loading of C15A for PS, SMA, and SEBS was, respectively, 20, 23, and 19%. In other words, the method failed to reach the level of stiffness expected for CPNC.

More recently, Yoon *et al.* [2001] reported on the effects of polar co-monomers and shear on the performance of CPNC with PS as the matrix. PS and three commercial styrenic copolymers composed of styrene (St), acrylonitrile (AN) and methyl-vinyl oxazoline (OZ) were used. The selected organoclay was Cloisite® 10A (C10A is MMT-2MBHTA; d_{001} = 1.91 nm). CPNC with 5 wt% of C10A was prepared using either static annealing or compounding in an internal mixer, both at T = 210 °C. Mixing for a specified mixing time was carried out at a shaft speed of N = 50 rpm (equivalent to a volume-average shear rate of $\dot{\gamma} = 65\ s^{-1}$). The products, characterised by means of XRD and FTIR, were only intercalated. Owing to the thermal instability of C10A the static intercalation produced maximum interlayer spacing after a short time ($d_{001} \cong 2.9$ to 3.5 nm), and then the interlamellar galleries stated to collapse back to $d_{001} \cong 1.9$ nm. The strongest effect was obtained for PS, the smallest for the matrix containing AN and/or OZ in its composition. The OZ mers may interact with hydroxyl groups on the clay surface while that of AN with N^+ of the ammonium cation. During compounding, within the first minute of mixing the organoclay intercalation reached its equilibrium value – thereafter the XRD peak positions did not change. Similar spacing and shift of spacing was observed for both static and shearing processes. For the PS matrix the shearing accelerated the d_{001} collapse – after 20 min of compounding the gallery could only accommodate single macromolecules.

Similar behaviour was recently reported by Tanoue *et al.* [2003a,b]. The authors also compounded PS with up to 20 wt% of C10A in a TSE at 200 °C. The compounding resulted in a collapse of the interlayer spacing from that obtained for C10A, $d_{001} \cong 1.92$ to about 1.65 nm, independent of PS molecular weight and clay content. The thermal decomposition of the organoclay was studied using FTIR – the data agreed with the Hofmann decomposition mechanism. The compounding was carried out without addition of stabiliser and without blanketing with N_2. It has been found that the PS molecular weight decreased stepwise upon each re-pass through an extruder (e.g., reduction in one pass from M_w = 294 to 228 kg/mol, i.e., by 22%), but it did not depend on the residence time inside the extruder. Furthermore, the extent of the PS molecular weight decrease was lower in the presence of C10A (e.g., one pass reduction was about 4%, i.e., within the standard error of these measurements: ±5%). There are two possible mechanisms for the C10A stabilising effects. Since the process is oxidative (confirmed in rheological time sweeps with and without a blanket of N_2) the degrading organoclay may act as an oxygen scavenger – formation of =CO groups was detected by FTIR. Another possible mechanism, oxygen diffusion control by the dispersed clay platelets, is more difficult to confirm.

Salahuddin and Shehata [2001] addressed the problem of PMMA polymerisation shrinkage in dentistry. To reduce the shrinkage, up to 1.24 wt% of Claytone® APA (MMT-2MBHTA) was added to a commercial heat-curable acrylic resin formulation (PMMA; M_w = 954 kg/mol). After mechanically mixing the ingredients the specimens were processed by a conventional heat curing method. The XRD data indicated a marginal expansion of the interlayer spacing in the organoclay (d_{001} = 1.8 nm). Since the intensity of this peak progressively decreased with the clay concentration, at least a partial exfoliation may be suspected. A significant decrease of PMMA warpage (by 84%) and linear dimensional changes (by a factor of 2.9) was achieved by incorporating up to 1 wt% of clay. Continuation of this work involved polymerisation of [oligo(oxyethylene) methacrylates] in the presence of montmorillonite. The systems were exfoliated [Salahuddin and Rehab, 2003].

Nanocomposites are replacing neat polymers in multicomponent systems requiring enhancement of properties, *viz.* high modulus, low shrinkage, permeability, or flammability. However, only a few papers have been dedicated to the use of CPNC in structural foams. For example, Wilson [2000] described foam preparation for moulding automotive trim. The patent specifies three components: a thermoplastic (including styrenics), 2 to 15 vol% of layered nanoparticles (> 50% of which are in stacks containing not more than 20 layers; aspect ratio: p = 50 to 300) and a blowing agent. It seems that the clay does not have to be pre-intercalated, provided that it can be dispersed in the selected matrix polymer. The dispersion is engendered during regular extrusion foaming. Thus, the clay replaces glass fibre, improving the surface finish, tensile strength and elongation at break, especially of thin moulded parts.

Similarly, Fibiger *et al.* [2000] patented processing of structural foams with diverse polymers, commercial organoclay (containing quaternary ammonium, e.g., Claytone HY), a compatibiliser (e.g., polyethylene-co-acrylic acid) and a chemical foaming agent. The technology is claimed to be general, applicable for formation of films, sheets or tubes, by the methods of extrusion, pultrusion, moulding or SCORIM forming plastics articles. The process should orient the clay platelets within 30°. Such orientation improves the product performance (e.g., barrier to permeation by gases or vapours) without loss of transparency. The amount of clay is claimed to range from 1 to 20 wt%.

Direct mixing of PS with 1-10 wt% of either Na-MMT or C10A in an internal mixer was carried out at T = 180-200 °C, shaft speed N = 50 or 100 rpm, for t = 3, 5 or 10 min [Uribe *et al.*, 2002]. The interlayer spacing in the product with Na-MMT and C10A was determined as, respectively, $d_{001} \cong 1.1$ and 3.6 nm, virtually independent of the mixing conditions. Tensile measurements indicated that the presence of 5 wt% of Na-MMT or C10A enhanced the modulus by *ca.* 14 or 70%, respectively. However, while Na-MMT slightly improved the modulus and the tensile strength, the latter property in the mixtures with C10A was strongly reduced.

Barbee *et al.* [2002] patented the melt compounding preparation of polymer/clay nanocomposites with 'sufficient exfoliation'. The new CPNC is to be used for films, bottles, and other containers suitable for foods, soft drinks, beers, medicines, etc. The claimed compositions are broad. Thus, the CPNC comprises: (i) a melt-processable polymer, (ii) clay, and (iii) an oligomer or polymer, which is miscible with the matrix polymer and has an onium functional group.

Re (i) The matrix polymer may be selected from between: PEST, PA, PO, PS, EVAl, PU, PI, PEI, PAI, PPE, phenoxy, epoxy, polyacrylate, polyetherester, polyesteramide, their copolymers or blends. However, since it seems that the principal aim of the patent is the reduction of permeability, the resins of interest are those that have been used to enhance barrier properties, *viz.* aromatic polyamides, PARA, EVAl, PET, etc.

Re (ii). The claims are equally broad for the clays – any clay with CEC = 0.9 to 1.5 meq/g will do, e.g., MMT, hectorite, mica, vermiculite, bentonite, nontronite, beidellite, etc. The amount of clay in the CPNC ranges from 0.5 to 15 wt%. The term: 'sufficient exfoliation' is defined as a dispersion where ≥ 50% of the clay is in the form of individual platelets and tactoids with thickness of ≤ 2 nm having a diameter $d \cong 10$ to 3000 nm.

Re (iii). The third ingredient of the patented formulation is a functionalised oligomer or polymer, which has the same monomer units as the matrix polymer. The functionalisation basically means introduction of an organic cation, *viz.* ammonium or phosphonium ions such as, e.g., alkyl ammonium, alkyl phosphonium, polyalkoxylated ammonium, or their mixture. For example, the functionalised oligomer is the hydrochloride salt of octadecyl bis(polyoxyethylene)amine having 15 ethylene oxide units in a chain (commercially available as, e.g., Jeffamine, Ethomeen or Ethoquad), or that of dimethyl-ethanol amine.

The CPNC is prepared in two steps: (a) intercalation of clay in aqueous medium (i.e., water or its solution with dioxane, THF, MeOH, EtOH, acetic acid, acetonitrile, etc.) with 20 to 99.5 wt% of a functionalised oligomer or polymer, and (b) compounding the product of step (a) with the matrix polymer. The composition of the resulting CPNC is: 50 to 99 wt% matrix polymer, 0.5 to 25 wt% clay, and 0.5 to 25 wt% of the functionalised polymer or oligomer. For example, 120 g of a dimethyl amine-terminated oligostyrene (prepared by anionic polymerisation) was dry blended with 8 g of a MMT (pre-intercalated with octadecyl-trimethyl ammonium, 3MODA) and 872 g of PS, and then extruded on a Leistritz Micro-18 TSE at 200 °C. The extruded film showed a significant reduction of O_2 permeability, compared to a clay-free control. There is no information on the expansion of the interlayer spacing or homogeneity of dispersion.

It is gratifying to see how well the invention supports the theoretical calculation by Balazs and her colleagues, e.g., see [Balazs et al., 1998; 1999; Ginzburg and Balazs, 1999, 2000; Zhulina *et al.*, 1999; Kuznetsov and Balazs, 2000; Ladika *et al.*, 2001]. As discussed in Part 3 of this book, these authors using numerical and analytical methods demonstrated that to achieve a high degree of dispersion clay should be intercalated with end-functionalised macromolecular species, miscible with the matrix polymer. A similar approach has been used by Hoffmann *et al.* [2000] to produce exfoliated CPNC with PS as the matrix.

A recent patent from Eastman Kodak demonstrates that multi-branched star-type polymers may be used as CPNC compatibilisers [Robello *et al.*, 2004]. First, the document describes the preparation of one-arm (linear), 5-arm, and 10-arm star polystyrenes. Next, these were mixed with a commercial PS and either Cloisite® 10A or -15A (MMT-2MBHT or MMT-2M2HT, respectively). The mechanical mixture was annealed (without shearing) at T = 185 to 225 °C for up to 24 h. XRD and TEM indicated that while the mixture of organoclay with linear PS yielded intercalated systems, those comprising a small amount of star-branched PS (e.g., PS:star-PS:Cloisite® = 90:8:2) resulted in exfoliation. It was also noted that the rate of diffusion of 10-arm star-PS being lower than that of 5-arm, the former required longer annealing time to exfoliate. While the examples focus on PS as the matrix, the approach is more general. One must expect that successful CPNC can be generated for any miscible blend with PS, *viz.* PPE, PVME, etc. Furthermore, the strategy should be also valid for systems with a suitable set of binary miscibilities, i.e., between the intercalant and star-branched polymer, and between that and the matrix polymer (*nota bene,* the alkyl chains in Cloisite® 15A are immiscible with star-PS, thus good performance is most likely achieved through direct interaction of aromatic rings with the clay surface).

4.1.4.4 Vinyl Polymer Matrix – A Summary

In Section 4.1.4 a large and diverse group of polymers have been examined. Even eliminating polyolefins, the rest, *viz.* styrenics, acrylics and vinyl halides show quite different interaction with silicates, and thus behaviour. The difference may not be so apparent in CPNC produced by the *in situ* polymerisation method, but it will most likely show during the subsequent melt processing step or during the melt compounding preparation method.

A clear illustration of the difference between PS and PMMA interactions with organoclay may be found in a recent work by Hu *et al.* [2003]. The authors used dynamic secondary ion mass spectrometry (DSIMS) to measure the tracer diffusion coefficient, D, in PS and PMMA in the presence and absence of 5 vol% Cloisite® 6A (MMT-2M2HTA in about 40% excess over CEC). As tracers, deuterated PS or PMMA of similar M_w to that of the matrix resin were used. It was found that in PS, D was unaffected by the addition of clay, but it was reduced by a factor of 3 in PMMA. Similar effects were also observed for the zero shear viscosity of PS and PMMA melt-mixed composites with the same volume fraction of clay. Evidently, the macromolecular dynamics of PS are hardly affected by the presence of MMT-2M2HTA, while PMMA is. The authors assigned the observed difference to preferential interactions between the PMMA and the clay platelet surface.

CPNC with vinyl polymer as the matrix may be prepared using numerous methods summarised in **Table 89**. These are segregated into five types. Of these, bulk polymerisation in the presence of clay pre-intercalated with a reactive (and miscible with the polymer) compound leads to end-tethered, exfoliated CPNC, with the HCP type of behaviour. However, as some publications indicate, during subsequent processing clay platelets in these CPNCs may re-aggregate into intercalated structures. For this reason, the second method, by melt compounding, which also may lead to end-tethered structures, is advantageous. The difficulty rests in preparation of a vinyl polymer with reactive end-groups.

4.1.5 CPNC in Water-Soluble Polymeric Matrix

Several US patents from AMCOL for CPNC with water-soluble polymers (e.g., 5,552,469; 5,578,672; 5,698,624; 5,721,306; 5,760,121; 5,837,762; 5,844,032; 5,955,094; 5,998,528; 6,126,734) have been discussed in Part 2 of this book. The term 'water soluble' encompasses such polymers as P4VP, EVAl, PVAl, PEG, polyacrylic acid; polyvinyl oxazoline; polyacrylamide; their chemical modifications, copolymers, blends, etc. These CPNCs are used primarily as replacements for water-soluble resins, especially to provide enhanced gas barrier properties in, e.g., films for food wrap, drink containers, automotive gas tank liners, etc. They are also used as rheology modifiers in, e.g., cosmetics, oil-well drilling fluids, paints, lubricants, in the manufacture of oils and grease, and the like. Another important use of these materials is for the delivery of diverse functional chemicals in, e.g., agricultural or medical applications. In the following text only a few, more recent publications on these materials will be discussed.

Vaia *et al.* [1995] reported direct intercalation (no solvent) of PS and PEG using organically modified silicates. Thus, PEG was directly intercalated into Na-MMT or Li-MMT by heating to 80 °C for 2-6 h, d_{001} = 1.77 nm was obtained. During the process water molecules were replaced by the hydrophilic macromolecules. However, the clay could not be exfoliated.

Early efforts to intercalate MMT with P4VP, PVAl or PEG had little success. Attempts to intercalate MMT with P4VP (MW = 40 kg/mol) produced reasonable results for Na-MMT dispersed in a P4VP/ethanol/H_2O solution [Levy and Francis, 1975]. Apparently, the presence of ethanol was essential. The interlayer spacing d_{001} = 2.6 to 3.2 nm was determined by XRD. PVAl (containing 12% residual acetyl groups) was found to increase the gallery height by only Δd_{001} = 1.0 nm [Greenland, 1963]. Furthermore, when the PVAl concentration increased from 0.25 to 4%, the amount of absorbed polymer was substantially reduced.

Since EVAl has been used as a barrier layer in many packaging applications, incorporation of clay is used to further reduce permeability to fluids.

Table 89 Summary of the information on CPNCs with a vinyl polymer matrix

Clay*/wt%	Intercalant	d_{001} (nm)	Polymer and comments	Reference
Method 1: Bulk polymerisation				
MMT/78	none	1.5	Filled PS	Kamigaito *et al.*, 1984
MMT	methacrylic acid-N,N-dimethyl-ethanol amine	>10	End-tethering, PS or PMMA exfoliated	Kawasumi *et al.*, 1989
MMT/10	oligo(oxypropylene)-diethyl-methyl-ammonium	4.55	Near exfoliation in PMMA; 3.65 in PS	Okamoto *et al.*, 2000
MMT/10	methyl-trioctil-ammonium	2.66	PS or PMMA	Okamoto *et al.*, 2000
MMT/10	oligo(oxypropylene)-diethyl-methyl-ammonium	Short stacks	Intercalation in PS and copolymers of MMA with polar monomers	Okamoto *et al.*, 2001
MMT/7.6	vinyl-benzyl dimethyl dodecyl ammonium chloride (VDAC)	No peaks	Exfoliation	Qutubuddin *et al.*, 2000 - 2002
MMT/?	long chain quaternary ammonium	?	Ultrasonication	Wang *et al.*, 2001 Zhang *et al.*, 2002a
MMT≤30	cetyl-trimethyl-ammonium	3.1-3.6	For styrenics or acrylics	Kuo *et al.*, 2001
MMT/3 or SM*	*N*,*N*-dimethyl-*n*-hexadecyl-(4-vinylbenzyl) ammonium	>10	End tethering, exfoliation in PS	Zhu *et al.*, 2001a
MMT/3	*N*,*N*-dimethyl-*n*-hexadecyl-(4-hydoxy-methylbenzyl) ammonium	3.53	PS	Zhu *et al.*, 2001a

Table 89 Continued...

Clay*/wt%	Intercalant	d_{001} (nm)	Polymer and comments	Reference
Method 1: Continued...				
MMT/3	*n*-hexadecyl triphenyl phosphonium	4.06	PS	Zhu *et al.*, 2001a
SM*/3	dimethyl hydrogenated tallow benzyl ammonium chloride (2MBHTA)	5.88	PS	Zhu *et al.*, 2001b
SM*/3	stearyl-tributyl phosphonium bromide	5.95	PS	Zhu *et al.*, 2001b
MMT/5	dimethyl dihydrogenated tallow ammonium (Cloisite® 20A; 2M2HTA)	3.4-4.9	PS or PMMA	Zeng and Lee, 2001
MMT/5	2-methacryloyl-oxyethyl hexadecyl dimethyl ammonium	No peaks	End tethering. PS or PMMA exfoliated	Zeng and Lee, 2001
MMT/5	dimethyl-*n*-hexadecyl allyl- or dimethyl-*n*-hexadecyl benzyl-ammonium	3.40 3.27	End tethering, but no exfoliation in PMMA. Intercalation in PMMA	Zhu *et al.*, 2002
MMT/5	*N,N*-dimethyl-*n*-hexadecyl-(*p*-vinyl benzyl) ammonium	4.65	End tethering, but no exfoliation in PMMA	Zhu *et al.*, 2002
Bentonite/ 2-10	dioctadecyl dimethyl ammonium	4.8	Poly(MMT-co-dodecyl methacrylate)	Dietsche *et al.*, 1999-2000
LDH*	reactive anions	?	Diverse polymes	Fischer and Gielgens, 2002

Table 89 Continued...

Clay*/wt%	Intercalant	d_{001} (nm)	Polymer and comments	Reference
Method 2: Emulsion polymerisation				
Na-MMT	none	?	Low expansion in PMMA	Lee and Lee, 1996
MMT/25	methyl-styrene ammonium	1.53-1.55	Modest improvements	Laus *et al.*, 1998
MMT/0.2-4	dodecyl trimethyl ammonium	?	Modest improvements	Elspass *et al.*, 1997; 1999
Na-MMT/5-40	none	1.9-1.4	High loading of clay into SAN	Noh *et al.*, 1999a
Na-MMT	none	1.72	Intercalation in SAN	Noh and Lee, 1999b
Na-MMT/35-72	none	1.56-1.75	Intercalation in highly loaded ABS	Jang *et al.*, 2001
Na-MMT	none	1.74	Intercalation in SAN	Kim *et al.*, 2000
MMT	cetyl-trimethyl ammonium	?	Exfoliation after polymerisation, then intercalation on heating	Chen *et al.*, 2000, 2001
MMT or Laponite	2-methyacryloyl-oxyethyl-trimethyl ammonium	No peak	End-tethered, PMMA exfoliated and stable	Brittain and Huang, 2002
A clay/0.1-20	intercalant's role plays 10-20 wt% acid containing co-monomer	?	Variety of emulsions prepared	Lorah and Slone, 2002a
MMT/5	allyl-ammonium/10-100% CEC	?	Tensile strength of the acrylic copolymer increased by 41%	Lorah and Slone, 2002b

Table 89 Continued...

Clay*/wt%	Intercalant	d_{001} (nm)	Polymer and comments	Reference
Method 3: Solution polymerisation				
MMT/5-50	vinyl-benzyl-trimethyl ammonium	1.7-2.3	After extraction of PS clay content was 56%	Moet and Akelah, 1993; Akelah and Moet, 1996
MMT/3	vinyl-benzyl-trimethyl ammonium	?	Syndiotactic PS (sPS)	Gepraegs *et al.* 1998
MMT/59	2-(*N*-methyl-*N*,*N*-diethyl ammonium iodide) ethyl acrylate	1.42	Copolymerisation with MMA	Biasci *et al.* 1994
MMT/59	2-(*N*-butyl-*N*,*N*-diethyl ammonium bromide) ethyl acrylate/MMA	1.49 (2.94)	Clay intercalated with copolymer	Biasci *et al.* 1994
Laponite/0.1	hexadecyl dimethyl allyl ammonium (<100% CEC)	?	Increase of T_g by 5 °C	Muzny *et al.* 1996
MMT/18	dodecyl-(DDA) or hexadecyl-trimethyl-ammonium (HDA)	2.90 or 3.88	Exfoliation after polymerisation; re-aggregation on shearing (d_{001} = 3.02 nm)	Tabtiang *et al.*, 2000
Method 4: Solution mixing				
MMT/9.1	dimethyl-dioctadecyl ammonium	3.5	Polystyrene-*b*-1,4-polyisoprene	Ren *et al.* 2000
MMT or FH*	tetradecyl-, octadecyl- or dioctadecyl-dimethyl-ammonium	?	Intercalation in toluene solution of PS	Gilman *et al.* 2000
MMT/5	cetyl pyridinium	>8.8 1.2-2.1	Low level intercalation in sPS after extrusion	Wu *et al.*, 2001; Tseng *et al.*, 2001

Table 89 Continued...

Clay*/wt%	Intercalant	d_{001} (nm)	Polymer and comments	Reference
Method 5: Melt intercalation; static or while shearing in an internal mixer or a TSE				
MMT or FM	dimethyl dihydrogenated tallow-, dodecyl- or octadecyl-ammonium	3.13	Static intercalation of PS at 155-180 °C, $t \leq 400$ min	Vaia *et al.*, 1993, 1995, 1996; Sikka *et al.*, 1966
MMT	dodcyl ammonium	No peak	Exfoliation in poly-3-bromostyrene	Vaia and Giannelis, 1997
A clay	statically melt intercalated with poly(ethylene oxide-*b*-styrene) or poly(styrene-*b*-2vinylpiridine)	-	Exfoliation or intercalation in PS depends on copolymer	Fischer *et al.*, 1999
MMT/7	octadecyl-trimethyl ammonium	3.1-3.9	In PS or poly(styrene-co-methyl vinyl oxazoline)	Hasegawa *et al.*, 1999
MMT/10	dimethyl dihydrogenated tallow (Cloisite® 6A)	3.45, or 3.37	Intercalated in PS, or SMA (bridging in SMA)	Lim and Park, 2001
Fluoromica	amine-terminated polystyrene	>4	Exfoliated HCP after compounding with PS	Hoffmann *et al.*, 2000
MMT/5-10	octadecyl ammonium	3.3	Compounding with PS	Wang *et al.*, 2001
MMT	dihydrogenated tallow alkyl ammonium chloride (Cloisite® 15A)	2.97-3.4	SPS + amorphous PS, SMA, or SEBS-MAH	Park *et al.*, 2001
MMT	dimethyl benzyl hydrogenated tallow ammonium (Cloisete® 10A; 2MBHTA)	1.9-3.5	Intercalation in poly(styrene-co-acrylonitrile-co-methyl-vinyl) oxazoline	Yoon *et al.*, 2001
MMT/5	dimethyl-benzyl-hydrogenated tallow ammonium (Cloisite® 10A; 2MBHTA)	3.6	Intercalated in PS; 70% improved modulus	Uribe *et al.*, 2002
MMT/0 to 20	dimethyl-benzyl-hydrogenated tallow ammonium (Cloisite® 10A; 2MBHTA)	1.92 to 1.65	Intercalated in three PS; decomposition of organoclay; oxidative degradation of PS	Tanoue *et al.*, 2003

**Notes: FH = fluorohectorite: LDH = layered double hydroxide*

In the earlier patents from AMCOL it was proposed that CPNC based on EVAl would be manufactured by first intercalating a Na-MMT with EVAl oligomer or polymer, and then adding the intercalate to an EVAl matrix [Beall *et al.*, 1998; 1999a]. However, it was found that Na^+ causes degradation of the EVAl complexed to the clay surface, thereby reducing the physical properties of the CPNC. The more recent patents from the company [Serrano *et al.,* 1998a,b] revised the technology. The new CPNCs comprise 40-99.95 wt% EVAl, and about 0.05-60 wt% (preferably > 2 wt%) exfoliated clay. No onium ion or silane coupling agent is used. It has been found that to prepare an EVAl-based CPNC, the pre-intercalate should be formed with a non-EVAl intercalant, e.g., with another water-soluble polymer, such as P4VP, PVAl, a copolymer of vinyl acetate and vinyl pyrrolidone, etc. These intercalating, *water-soluble polymers* are able to complex to the platelet surfaces, covering the Na^+ and shielding the EVAl from its degrading influence. The best results were achieved using 50-80 wt% of an oligomeric (DP = 2-15) or a polymeric (DP > 15) intercalant, which complexes to the clay platelets increasing its interlayer spacing by Δd_{001} = 0.5 to 10 nm. Theoretically, clay may be intercalated with a compound that has a dipole moment greater than that of H_2O (1.85 D). Numerous groups may induce such an effect, *viz.* carbonyl, hydroxyl, carboxyl, amine, amide, ether, ester, sulfate, sulfonate, phosphate, phosphonate, phosphinate, aromatic rings (including lactams, lactones, anhydrides, nitriles, *n*-alkyl halides, pyridines), etc. Depending on T, pH and concentration, thixotropic suspensions can be formulated to have the desired viscosity, *viz.* η = 0.5 to 1.0 Pas.

It has been found that when the concentration of the intercalating polymer in clay is 10-15 wt% the interlamellar gallery expands by Δd_{001} = 0.5 to 1.0 nm, indicating that there is a monolayer of the adsorbed intercalant complexed to the adjacent platelets. When the amount increases to 16-35 wt%, 35-55 wt% or to 55-80 wt%, the gallery expansion (1.0-1.6, 2.0-2.5 or 3.0-3.5 nm, respectively) is sufficient to accommodate 2-, 3- or 4-5 layers, respectively. In **Figure 28** of Part 2 the interlayer spacing versus intercalant content is presented for MMT intercalated with either PVAl or P4VP. According to the patent, the preferred amount of intercalant ranges from 20 to 60 wt%, providing d_{001} = 3 to 4 nm.

The preferred method involves mixing clay (e.g., in a TSE) with the intercalating polymer or its aqueous solution at $T = T_m + 50$ °C. Intercalation could be carried out using monomer(s), e.g., a mixture of a diamine and a dicarboxylic acid. However, the preferred intercalants are *water-soluble polymers*, especially P4VP (M_w = 1-40 kg/mol) that can be transformed into Na-P4VP or K-P4VP, or PVAl, with ≤ 5% acetyl groups and M_w = 2-10 kg/mol. The water content in such a slurry should be > 30 wt%. The main claim of the patent describes strictly a two-component CPNC, comprising 40 to 99.95 wt% of EVAl and the rest being clay (without pre-treatment with onium or silane compounds). The resulting CPNC was fully exfoliated and the individual clay platelets were homogeneously and randomly distributed in the matrix.

Claims of the following patent from AMCOL are very broad, covering many different types of polymeric matrices, thermoplastic as well as thermoset [Beall *et al.*, 1999b]·. These CPNCs are to be used for, e.g., external or heat-resistant body parts by the automotive industry; tyre cords; food containers having improved gas impermeability; electrical components; etc. The strategy is quite similar to that described in the Serrano *et al.* [1998] patents. Thus, the clay is to be pre-intercalated with a polymer having affinity to the phyllosilicate. Polymers with DP = 2-15 having an aromatic ring and/or a polar group (e.g., carbonyl;

carboxyl; hydroxyl; amine; amide; ether; ester; sulfate; sulfonate; phosphate; phosphonate, etc.) may be used. The patent lists virtually all types of water-soluble or highly polar polymers, but again the preference is given to P4VP, PVAl (having < 1% residual acetyl groups), or their mixtures. To engender the desired interlayer spacing (see **Figure 28** in Part 2), the concentration of polymer in organoclay should be *ca.* 20-60 wt%. The intercalation is conducted at $T = T_m + 50$ °C in an extruder using a clay slurry in a solution of the selected intercalating polymer or oligomer containing 30-50 wt% of solvent. The recommended use of oligomers with low *DP*, and relatively high mixing *T* indicates recognition of the critical role diffusion plays in the intercalation process. The resulting masterbatches of organoclay contain *ca.* 20 to 80 wt% of clay. Next, to produce well-exfoliated CPNC the organoclay obtained in the preceding step must be compounded with matrix polymer having *MFR* = 0.01-12 g/10 min. The patent claims are general, but the key word for success is 'miscibility' between the intercalating oligomer and the matrix polymer. The list of suitable matrix polymers is long, including PA, PEST, PC, PPE, PPS, PEI, PSF, PO, TPU, vinyl or acrylic polymers and copolymers, ionomers, polyepichlorohydrin, cellulose derivatives (e.g., CA, CAB, CP), siloxanes, protein plastics, rubbers, diverse elastomers, a variety of thermosets, as well as blends of two or more of these resins.

One of the examples describes compounding PA with 20 wt% Na-MMT (under N_2 at $T = T_m + 50$ °C), which resulted in exfoliated organoclay masterbatch. Other examples concluded that compounding at *T* = 280 °C of either PET with 10 wt% of Na-MMT, or PC with 50 wt% of Na-MMT resulted in exfoliation of the clay platelets, provided that the clay contained 10-15 wt% water. On face value, the patent is surprising, as the presence of water is the *sine-qua-non* exfoliation requirement. Since 'good practice' for processing of the polycondensation type polymers is good drying, the required moisture presence for exfoliation causes an automatic negative reaction – one expects hydrolytic degradation of the macromolecules and reduction of performance. However, when the compounding is conducted under devolatilisation conditions, a re-condensation at the later extrusion stage may at least partially eliminate the degradative effects. Apparently, the authors did not observe negative effects of water on the performance of PA or PEST. One may speculate why the presence of water is essential for efficient exfoliation. At the compounding temperature, $T_m + 50$ °C (e.g., near 300 °C), the water pressure is about 8.5 MPa. Thus, if prior to compounding the water was absorbed between clay platelets, during mixing it had to be converted into high-pressure steam that expanded the clay interlamellar galleries, hence facilitated ingress of the intercalating macromolecules.

Alexandre *et al.* [2001] dispersed Na-MMT (CEC = 0.92 meq/g; d_{001} = 1.26 nm) in EVAc (containing 27 wt% VAc) matrix. Na-MMT was pre-intercalated with 46.2 wt% 2M2ODA. Five EVAc-based samples were prepared on a two-roll mill at 130 °C with: (1) MMT-2M2ODA; (2) Na-MMT; (3) a mixture of Na-MMT and an equivalent amount of 2M2ODA ('one-pot', sequential process: melting PVAc + 2M2ODA, mixing 3 min + Na-MMT); (4) the amount of 2M2ODA equivalent to that present in (1); (5) EVAc alone. The clay content in CPNC samples (1)-(3) was about 3.5 wt%. The XRD analysis of samples (1) and (3) showed the presence of a strong peak corresponding to d_{001} = 4.02 nm, whereas sample (2) showed a weak one corresponding to d_{001} = 1.24 nm, comparable to that of Na-MMT. The TEM micrographs of sample (3) indicated the presence of *ca.* 20 nm-thick stacks (hence of *ca.* 6 MMT platelets).

The properties of these samples are summarised in **Table 90**. The best performance was obtained using the pre-intercalated clay (sample (1); the modulus was enhanced by 55% over that of PVAc + 2M2ODA matrix). However, the sequential addition of the same amount of ingredients also produced CPNC with respectable performance, especially when the low degree of dispersion is taken into account (sample (3); modulus enhancement by 26%). It is noteworthy that direct incorporation of Na-MMT into an EVAc matrix did not affect the mechanical properties – the data are within the range of experimental error.

Shen *et al.* [2002] compared solution and static melt intercalation of PEG (M_n = 173 kg/mol and M_w/M_n = 2.58) with either Na-MMT or an organoclay. The clay (CEC = 0.8 meq/g) was used after drying at 250 °C. The organoclay (B34 from Rheox) is a bentonite pre-intercalated with dimethyl-ditallow ammonium (2M2TA) with a maximum use temperature of 220 °C. The solution intercalation was conducted in a suitable solvent, *viz.* water for Na-MMT and toluene or chloroform for the organoclay. A selected amount of the two components (PEG and clay) was dispersed in a solvent and stirred for 1 day. After drying, the *d*-spacing was determined by XRD. Melt intercalation was performed by mechanically mixing the two components in a mortar then forming pellets in a hydraulic press at 70 MPa. The intercalation took place during annealing at either 85 or 95 °C for about 8 h. Since in the CPNC the polymer content did not exceed 40 wt%, no exfoliation could be expected. Thus, the results must be judged by the relative extent of intercalation at a similar level of inorganic clay content. The interlayer spacing in Na-MMT, d_{001} = 0.96 nm, was found to expand to about 1.4-1.5 nm after incorporation of up to 15 wt% of PEG, then to *ca.* 1.8 nm for higher polymer content (using either method). After incorporation of 15 wt% PEG the interlayer spacing in B34, d_{001} = 3.0 nm, expanded to 3.7 nm from chloroform solution or using a static melt intercalation method (the clay content here is *ca.* 59 wt%). However, intercalation from toluene solution resulted in poor intercalation giving d_{001} = 3.3 nm. Evidently the extent of intercalation in the three-component systems is controlled by miscibility of all

Table 90 Tensile properties of EVAc compositions containing *ca.* 3.5 wt% clay. Data [Alexandre *et al.*, 2001]

No.	Clay	2M2ODA	Young's modulus E (MPa)	Stress at break, σ_b (MPa)	Strain at break, ε_b (%)
1	Na-MMT/ 2M2ODA	NA*	21.42±1.75	28.77±0.47	1365.2±11.2
2	Na-MMt	nil	12.61±0.68	27.74±2.37	1423±34.6
3	Na-MMT	yes	17.44±0.74	29.10±1.15	1367±10.9
4	nil	yes	13.86±0.70	25.36±0.39	1419.1±16.2
5	nil	nil	12.41±1.31	28.66±0.73	1406±28.2

*Note: *The intercalant was incorporated into MMT prior to compounding*

three ingredients: solvent, PEG and intercalant. The thermodynamic interactions of PEG with chloroform are more favourable than those with toluene. The interlamellar gallery height of 0.3 to 0.5 nm may accommodate PEG macromolecules in a planar zigzag conformation, whereas that of 0.6 to 0.8 nm (as in MMT/PEG and B34/PEG systems) may possibly house helical structures.

Koo *et al.* [2003] prepared P4VP/MMT nanocomposites by first dispersing Na-MMT in water for 1 to 2 weeks and then adding P4VP. Two methods were used to disperse clay: (1) simple stirring, and (2) attrition ball milling (using zirconium balls with a diameter of 1 mm in a zirconium jar at 300 rpm). CPNC films were obtained by casting the P4VP/MMT/H_2O suspensions. At a MMT content below 20 wt% the nanocomposites were reported exfoliated and intercalated at higher concentrations. In **Figure 162** the lines represent least-squares fit to a three-parameter Equation 9 (see Section 1.3.2). The values of d_{001} extrapolated to infinite dilution are: 7.73 and 8.56 nm, respectively. However, within the range of composition explored by the authors the two dispersion methods resulted in virtually identical interlayer spacing. Since ball milling may cause attrition of clay platelets the milder method should lead to better performance. The CPNC had good optical clarity and increased thermal resistance with MMT content. The hydrogen bonding interactions between P4VP and MMT may be responsible for the enhanced properties, but it is uncertain whether the system obtained would remain stable under processing condition. The nanocomposites prepared by ball milling showed better transparency, especially at higher clay content (w > 20 wt%) where the degree of intercalation is very similar. Thus, the improvement may originate in the smaller platelet diameter hence reduced aspect ratio, p.

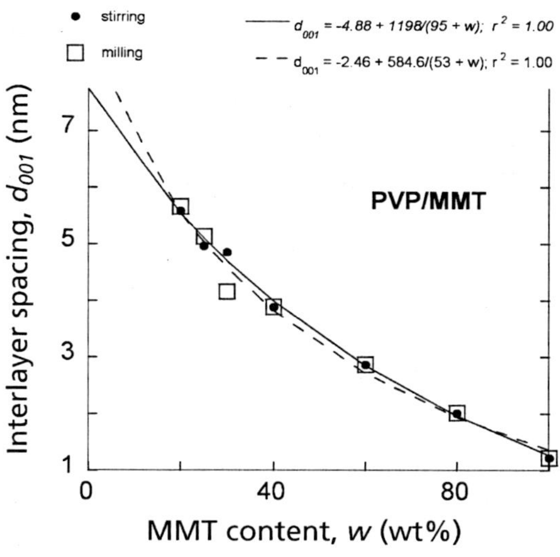

Figure 162 Interlayer spacing of MMT dispersed within a P4VP matrix by either stirring aqueous suspension or ball-milling it. Data [Koo et al., 2003].

In US patent and other patent applications Chaiko explored diverse methods for the preparation of organoclays containing water-soluble polymer [Chaiko, 2003]. The principal process involves dispersing clay in water, and then adding a water-soluble polymer, e.g., PPG, PEG, their mixture or neutralised poly(l-vinylpyrrolidone-co-acrylic acid). Furthermore, to the organoclay an inorganic salt may be added, *viz.* sodium phosphate, sodium hexametaphosphate, tetrasodium pyrophosphate, and polyfunctional phosphonic acid salts, as well as onium salts and/or alcohols. The process expands the interlayer spacing to at least 1.78 nm. For a high concentration of well-intercalated clay (e.g., 10 to 70 wt%) the system behaves as liquid crystal. Dispersing such a mixture in thermoplastic prior to film blowing, results in a significant reduction of permeability to gases or vapours. The author also describes organoclays with platelet edges modified by phosphonic or phosphinic acids, their combinations or derivatives. Furthermore, the advantages of using mixed clays with different CEC values have been described. The main application for these systems is the control of rheology and/or permeability of industrial coatings or packing. A variety of waxes have been found suitable, e.g., polyethylene-block-polyethylene glycol or polyethylene mono-alcohol. Some of these systems may be used with paper, while others may be formed into films or packing containers.

One of the results of this work is the development of potentially efficient, flexible and economic technology for the production of CPNC [Chaiko *et al.*, 2003]. The process combines the preparation of organoclay, followed by dewatering by means of a pressurised extruder into which the matrix polymer is also added. The process yields masterbatches of PO containing 50 to 80% organoclay at a cost of 2.2 US$/kg. Compounding the masterbatch with the matrix polymer leads to full exfoliation without a need for compatibiliser. In a sense, the process resembles that described in 2003 by Hasegawa *et al.*, and by Kato *et al.* The key to the success is on the one hand the thermodynamic miscibility between organoclay and the polymeric matrix, and on the other hand the presence of a substantial amount of water during mixing in a dewatering, pressure extruder. (These extruders are well known in the industry for dewatering rubber emulsions or aqueous slurries by squeezing and draining out water, instead of evaporating it).

4.1.6 Thermoplastic Polyesters (PEST)

Owing to the need for improved rigidity (especially for hot-fill containers) and barrier properties there has been significant effort in the development of technology for the preparation of CPNC with thermoplastic polyesters (PEST) as the matrix. The task is far from simple as the thermal degradation of organoclay at the PEST processing temperatures, as well as hydrolytic degradation of the matrix pose serious complications.

An early work on CPNC at General Electric aimed at upgrading the performance of such engineering resins as PC, PEST or PPE [Takekoshi *et al.*, 1996]. The patented composition contains two components: (1) intercalated layered mineral and (2) polymeric matrix. The layered mineral is a kaolinite, MMT or illite, intercalated with either standard onium substituted pyridinium cation, *viz.* $^{+}N-C_{10-18}$ pyridinium, guanidinium or amidinium. The preferred polymeric system can be selected from between macrocyclic oligomers (of, e.g., polycarbonates, polyesters, polyimides, polyetherimides, polyphenylene ether-polycarbonate and their blends), linear polymers or branched polymers, e.g., PC, PET, PBT, PEN, PPE, PI, PO, PEI, PA, PPS, PSF, PEEK, ABS, styrenics, etc. The ranges of composition have not been identified. The examples focus on MMT intercalated either with primary ammonium chlorides, DDA or HDA, or with *N*-hexadecyl pyridinium (HDP) chloride. The polymers are either PBT (prepared in the presence of organoclay from cyclomer or by direct melt compounding) or PPE. The main method of preparation was melt compounding in a TSE operated at 400 rpm with feed rate 2.4 kg/h at T = 295 (for PPE) or 260 °C (for PBT). Characteristically, compounding PBT with 5 wt% MMT-DDA or MMT-HDP caused a reduction of MW by 29 or 12%, respectively. However, at the same time the shear modulus at 60 °C increased by 97 or 80%, respectively. Evidently, the difference is related to the amount of intercalant present. The data suggest good dispersion.

Kaneka Corp. has been developing CPNC with PEST as the matrix for nearly ten years [Suzuki and Oohara, 1997, 2001; Suzuki, 1999; Oohara and Suzuki, 1999; Suzuki *et al.*, 2000; Kourogi *et al.*, 2000]. The aim has been to provide CPNC containing clay (MMT or fluoromica = FM) dispersed (layer thickness ≤ 5 nm) in a thermoplastic resin (PEST or PC). Furthermore, the organosilane treated clay is to be incorporated in amounts ranging from 0.1 to 50 wt%. The resulting nanocomposites show excellent mechanical properties, heat resistance, and surface appearance. The organosilane used is a Y_nSiX_{4-n}, where n = 0 to 3, Y is a hydrocarbon C_{1-25} group, and X is a hydrolysable or a hydroxyl group. For example, Y may be methacryl oxypropyl, a (polyoxyethylene)propyl, a glycidoxypropyl, an aminopropyl or a γ-(2-aminoethyl)aminopropyl group; and X may be an alkoxy, an alkenyloxy, a ketoxime, an amino group, an aminoxy, an amido, or a hydroxyl group. In particular R_i-amino-propyl trimethoxy silanes were used, where: R_1 is γ-(2-aminoethyl)-, R_2 is γ-glycidoxy-, and R_3 is γ-(polyoxyethylene)-. For example, MMT or FM was dispersed in water by agitating at 5000 rpm for 3 min, then R_1-amino-propyl trimethoxy silane was added, and the mixture agitated for 2 h. The interlayer spacing in the dry precipitate with MMT, (*a*), was d_{001} = 2.6 nm, whereas that with FM, (*d*), 1.8 nm. FT-IR showed absorption bands for a primary amino group, a secondary amino group, and an ethylene group. Similarly treated MMT with R_2-amino-propyl trimethoxy silane, (*b*), had d_{001} = 2.0 nm, and the FTIR showed absorption bands for an epoxy ring, an ether group, and a methylene group. Next, MMT was treated with R_3-amino-propyl trimethoxy silane, (*c*), causing the spacing to increase to d_{001} = 2.4 nm; FT-IR showed absorption bands for ether and ethylene groups. An alternative method of silane treatment involved spraying MMT with 10 wt% of silane, followed by 1 h of mixing – there was no expansion of the interlayer spacing (d_{001} = 1.3 nm). The treated clays were dispersed in monomers prior to polymerisation.

The CPNCs were produced by mixing intercalated clay slurry with a monomer, allowing the mixture to be mixed at $T \cong 120$ °C for about 3 h, before evaporating water and starting the polycondensation. For example, slurry (*a*) was mixed with bishydroxyethyl terephthalate (BHET). After mixing and distilling of water, a stabiliser and Sb_2O_3 polymerisation catalyst were added, and T was raised to 280 °C to produce PET under vacuum. TEM observation and XRD measurements indicated exfoliation with d_{001} > 10 nm. Similar processes were carried out to prepare CPNC with PET, PBT, and PC as matrix in which the treated clays (*a*) to (*d*) were used. At higher clay loading the polymerisation resulted in low MW PEST. To improve the performance it was necessary to increase the MW by solid-state polymerisation. For PET the process was conducted under vacuum at T = 200-210 °C for 9 h. As a result, the same molecular weight of the matrix could be achieved in any CPNC as that of a standard PEST resin. The performance of the CPNC obtained was evaluated in terms of HDT and flexural tests. An example of the results is shown in **Figure 163** for PET and PBT nanocomposites with treated clay (*a*). It is interesting that all three functions, flexural modulus, E_f, flexural strength, σ_f, and HDT show similar increases with increasing clay (inorganic part) content. Furthermore, at 10 wt% loading the E_f increase in PET and PBT matrix is virtually identical: 153 and 155%, respectively. The data also indicate that at a clay concentration exceeding 5 wt% the improvements begin to be less pronounced.

Several patents from Tetra Laval focused on CPNC prepared via the reactive route and used for polymeric containers. Thus Frisk and Laurent patents [1998; 1999a,b] describe CPNCs based on PEST, e.g., PET, its copolymer comprising dihydroxyl compounds other than diethyl glycol (COPET) or their blends. The PET containers must be able to retain CO_2 and to prevent O_2 ingress (especially for beer and wine containers). At present, multilayered bottles are being produced, with layers of EVAl, PA or PEN. Alternatively, either a PET/PEN blend or a PET bottles may be coated with PVDC film.

The containers developed (with 0.1-10 wt% clay) should be able to control the permeability with a minimal amount of exfoliated clay. The barrier efficiency is related to the aspect ratio – clays with p = 50-2000 are known. Even a small amount of clay (the preferred is MMT) reduces the permeability and improves other container properties, *viz.* HDT for hot fill, strength and stiffness, without affecting transparency. The CPNC may be produced using diverse methods, e.g., *in situ* polymerisation, solution intercalation, or melt exfoliation. Any of the three distinct methods may produce CPNC of PET or COPET with nanoparticles

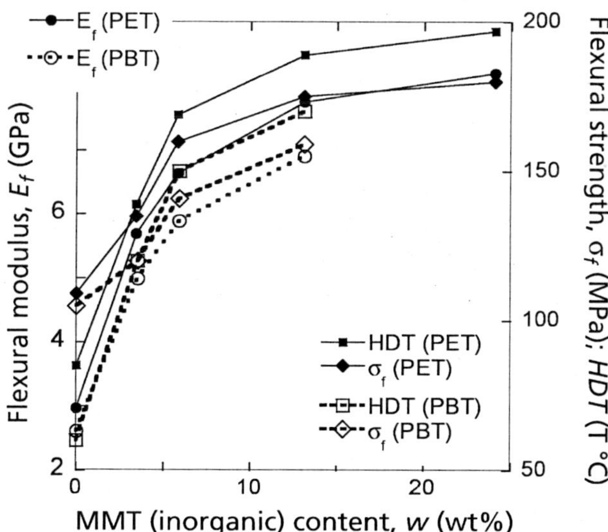

Figure 163 Flexural modulus, strength, and heat deflection temperature for a series of CPNC with PET or PBT as matrix. MMT was treated with γ-(2-aminoethyl) amino-propyl trimethoxy silane. Data [Suzuki and Oohara, 2001].

preserving their high aspect ratio. There is little information on the actual procedure used to obtain a high degree of dispersion for clays with high aspect ratio. The description focuses rather on general methods of polycondensation starting either with terephthalic acid (TA) or its dimethyl ester (DMT). In either case, high vacuum and $T \leq 290$ °C are used. There is no mention of solid-state polycondensation. The three methods are:

1. Clay (preferably Na-MMT) is dispersed in a solvent, and then swelling agent (its role is to bind polymer to the clay surface; hence a 'surfactant' or a compatibiliser) is added to form a clay complex. Next, PET or COPET monomers are added and polymerised *in situ* to form CPNC.
2. As in method (1), the clay suspension is prepared with a swelling agent. Next, the mixture is added to PET or COPET. As a result of trans-reaction the components intercalate into the clay galleries to form a CPNC.
3. Melt exfoliation is the most interesting, as (according to the patent description) it does not require addition of 'a swelling agent'. Thus, PET or COPET is mixed with clay then melted while shearing. During the melt mixing, the polymer exfoliates the clay.

As shown in **Figure 164**, even a small amount of clay provides a substantial increase in the barrier properties. In addition, the clay platelets also enhance the heat stability and mechanical properties, important to allow for the 'hot-fill' applications. The stiffness of the container increased without affecting transparency.

More recent patents from Tetra Laval describe multilayer transparent packaging structures comprising at least two layers: the interior one made of a PE and the exterior one made of CPNC with PET or COPET as a matrix [Frisk, 2000; 2001].

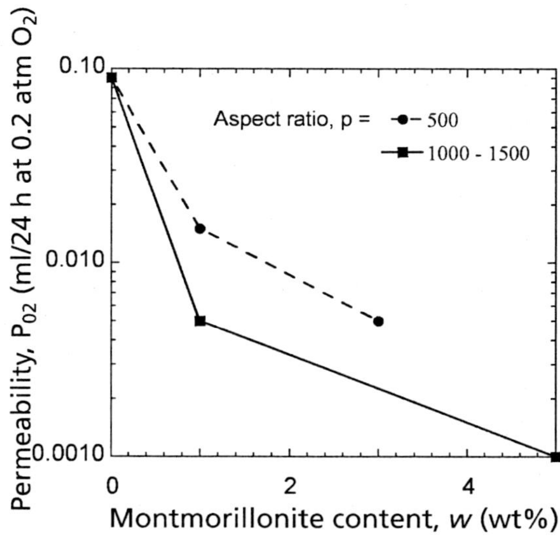

***Figure 164** Permeability of oxygen through a PET container comprising 0-5 wt% of exfoliated clay. Two grades of MMT (differing in aspect ratio, viz. p = 500 or 1000 to 1500) were used. Data [Frisk and Laurent 1998].*

The PE may also be melt compounded in a TSE with organoclay (SCPX from SCP) at 250 °C. Furthermore, an adhesive layer (e.g., EVAl, EMA, EBA, EAA) may have to be used.

PET-based nanocomposites were prepared by *in situ* polymerisation in the presence of clay [Ke *et al.*, 1999]. Since the intercalation method is being patented (in PR China), the details are missing. The published XRD pattern was puzzling, as the raw data showed smaller basic spacing for the intercalated clay than for the not intercalated one! When the clay concentration increased from 3 to 5 wt% the interlamellar spacing decreased from d_{001} = 3.5 to 2.4 nm. Thus, the majority of clay platelets are intercalated. By contrast to neat PET, in CPNC the spherulites were difficult to see. Instead 'a stripe-belt' morphology indicated that PET chains were intercalated into the interlamellar space. The crystallisation rate in CPNC was 3-times faster than that in neat PET. The HDT of PNC containing 5 wt% of clay was higher by 20-50 °C and the moduli 3-times larger than that of PET. The data for T_m, T_g, HDT, heat of fusion, decomposition temperature, and tensile modulus are also provided.

A similar method was recently used by Chang *et al.* [2004], dispersing MMT (pre-intercalated with dodecyl triphenyl phosphonium chloride) in ethylene glycol, then adding dimethyl terephthalate, and polymerising the mixture. The CPNC containing 0 to 3 wt% organoclay were intercalated. After melt spinning the ultimate tensile strength and the initial modulus of the CPNC showed the opposite tendency to that observed for neat PET – while these properties for PET increased with the draw ratio (DR), for CPNC they decreased. However, values of these parameters were found to significantly increase with clay loading (especially at low *DR*).

In 1999 Eastman reached agreement with Nanocor to use nanocomposite technology for improving the barrier properties and heat stability of PET packaging. The announced technology stipulates that owing to the detrimental effects of clay on PET, the simplest solution may be to use PA- or PARA-based CPNC (with 1 to 3 wt% clay) as an inner layer in multilayer PET containers. These CPNC have high gas barrier properties, similar processing characteristics to PET and good adhesion to it. The improved performance makes it possible to reduce wall thickness. However, at the same time work on PET/clay was being carried out in Nanocor laboratories to reduce the gas diffusion rate through the wall of PET containers [Kamena, 1999]. The following year Eastman Chemical deposited a plethora of patent applications for CPNC with improved gas barrier properties [Gilmer *et al.*, 2000a-d; Barbee *et al.*, 2000a-d; Barbee and Matayabas, 2000; and Matayabas *et al.*, 2000]. Considering that most of the authors are employed by Nanocor, evidently the work has been collaborative.

Thus, Gilmer *et al.* [2000a] described a general method of melt compounding a polymer (e.g., PET) with organoclay. The latter was Na-MMT ion exchanged with Jeffamine, engendering a small expansion of the interlayer spacing to d_{001} = 1.4 nm. Better results were obtained using a mixture of Ethoquad and ODA, or commercially available quaternary amines. The quoted improvements of the barrier properties are modest. Another patent [Gilmer *et al.*, 2000b] indicates that with 52 wt% inorganic content the organoclay with Ethoquad had respectable spacing: d_{001} = 3.4 nm. Compounding it with oligo(ethylene terephthalate), OET; M_n = 382 g/mol, in a TSE at 220 to 230 °C resulted in exfoliation with residual tactoids. To increase the *MW* of the PET matrix solid-state polycondensation was used. The next paper [Gilmer *et al.*, 2000c,d] described the use of two intercalants, e.g., 3MODA and tetramethyl ammonium chloride (4MA). The resulting interlayer spacing was modest, d_{001} = 1.5 to 1.9 nm, but then melt compounding with PET resulted in a small reduction of its molecular weight (*ca.* 10% for up to 35% of 3MODA in its mixture with 4MA).

Barbee *et al.* [2000a,b] used mixtures of organoclays, *viz.* either two clays and the same intercalant or the same clay and different intercalants. Even with a modest value of the interlayer spacing, d_{001} = 1.5 to 1.9 nm, the melt compounded CPNC with PET as matrix showed a reduction of O_2 permeability by 61% (inorganic clay content 4.5 wt%). The patent also discusses significant differences in performance between natural clays coming from different geographic locations, as well as between synthetic and natural clays. Incorporation of organoclay into a polymer usually induces a yellow-brownish colour. To hide the discoloration, suitable dyes may be used. The invention described in the following patent elegantly solves this problem by using cationic dyes (e.g., optical, fluorescent brighteners) that serve also as thermally stable intercalants [Barbee *et al.*, 2000c]. The method of CPNC preparation is a standard one: melt compounding of a pre-intercalated clay with molten polymer. The clay is pre-intercalated with a dye, having the cation group (a quaternary ammonium) separated from the chromophore (e.g., stilbenes, coumarins, anthraquinones, anthrapyridones, benzathrones, etc.) by at least two C-groups. For example, basic red Rhodamine 6G (see **Figure 165**) was added in a mixer to an aqueous suspension of Na-MMT. The intercalation took place overnight at 60 °C. Colourless CPNC was prepared by dry-blending this red pigment with a similarly prepared blue one and Claytone· APA with PET. Well-dried mixture was extruded at 275 °C. Unfortunately, there is no information on the extent of intercalation/exfoliation in the CPNC produced, or its properties.

Another Eastman patent [Barbee *et al.*, 2000d] presents a general method for the manufacture of CPNC. Like several others, it also uses three components: (1) matrix polymer, (2) layered clay, and (3) a binder that promotes interactions between (1) and (2), mainly a functionalised oligomer or polymer. Component (1) can be virtually any polymer, but the preference is for PARA (e.g., MXD-6) and PEST (e.g., PET or PETG). Component (2) (used in amounts from 0.5 to 15 wt%) may be any suitable clay with CEC = 0.9 to 1.5 meq/g (*viz.* MMT,

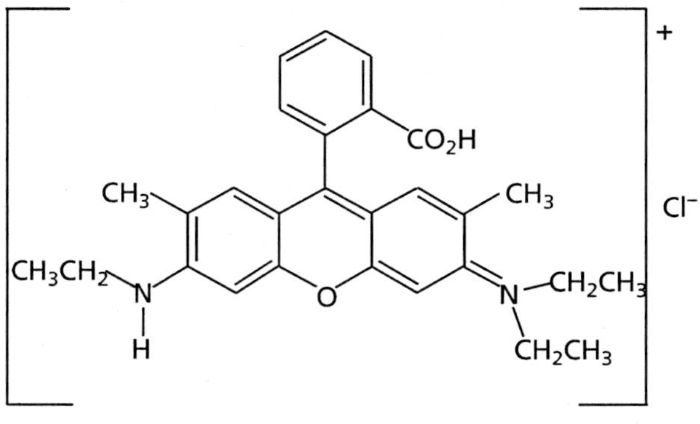

Figure 165 *Basic red Rhodamine 6G used as a cationic (ammonium) intercalant of MMT [Barbee et al., 2000].*

hectorite, mica) which can be dispersed to form individual platelets (at least 50%) and short stacks. The clay should be pre-intercalated with an ammonium or phosphonium compound. Component (3) should have the same monomer units as (1). The process can be either a single step (direct compounding of the desired composition) or two-step. In the latter case, the first step would prepare a concentrate with 0.5 to 80 wt% clay and 20 to 99.5 wt% compatibiliser. For example, first ammonium-functionalised compound (e.g., PCL or low molecular weight PETG) was prepared to intercalate Na-MMT (d_{001} = 1.3 to 3.3 nm). In the second step the concentrate was compounded with PET in a TSE at T = 280 °C – a 'significant' reduction of O_2 permeability was achieved. According to the claims, the quantifying word 'significant' should mean reduction of permeability by at least 10%.

It is noteworthy that a PCL/clay system alone shows good barrier properties. For example, Tortora *et al.* [2002] prepared a series of PCL/MMT-DDA nanocomposites via ring-opening polymerisation of ε-caprolactone in the presence of 0 to 44 wt% of MMT-DDA complex. According to XRD, exfoliation was achieved for compositions containing up to 16 wt% of MMT-DDA. The permeability of the CPNC with water or CH_2Cl_2 as permeants decreased with increasing clay content. For example, the relative (to neat PCL) permeability of CPNC with 30 wt% of MMT-DDA for H_2O and CH_2Cl_2 was 0.09 and 0.03, respectively.

Barbee *et al.*, [2000e] continued the method of the preceding patent only to the pre-intercalation step. For example, MMT was intercalated with Ethoquad T/12 (MT2EtOH, similar to Cloisite 30B). Preparation of a concentrate with PEG was done in methylene chloride (CH_2Cl_2). The resulting suspension was used to coat the PET pellets, which after drying were extruded in a TSE at 275 °C, 200 rpm. The permeability reduction was a respectable 50% at 2 wt% of clay.

Barbee and Matayabas [2000] patented another CPNC with enhanced barrier properties, to be used for packaging carbonated beverages, fruit juices, food products and medicines. The claims specified application of the technology to a variety of polymeric matrices, *viz.* PET, COPET, PCL, PU, linear long chain diols, polyether diols, PSF, PEK, PA, PEST, PEA, PPE, PPS, PEI, vinyl polymers and copolymers, acrylics, EVAl, ABS, MBS, PO, cellulosics, etc. The key is the modified intercalation method.

First, the clay (preferably MMT) is pre-intercalated with an alkyl quaternary onium compound: $MR_1R_2R_3R_4^+X^-$, where M is either N or P; X is a halogen, R_1 is a straight or branched chain alkyl group having 8-20 carbon atoms, R_2, R_3, and R_4 are independently straight or branched alkyl groups: C_{1-4}. Next, the intercalated clay is treated with an expanding agent that separates the platelets into individual ones or small tactoids. The expanding agent is either a polymer, or vitamin-E miscible with the matrix polymer. The compositions containing 0.5-20 wt% of clay showed improved platelet separation. For example, films were prepared from:

a. Neat PET; P_{O2} = 12,
b. PET-based PNC containing 2 wt% MMT pre-intercalated with MT2EtOH; P_{O2} = 11,
c. The same as (b), but pre-swelled with PEG (MW = 3,350 g/mol); P_{O2} = 6.

The oxygen permeability, P_{O2}, is expressed in ml-mil/100 in^2-24 hours-atm.

The preferred clay has CEC = 0.8 to 1.5 meq/g, e.g., smectite, Na-MMT, etc. MMT may be pre-intercalated with alkyl ammonium or phosphonium ions, *viz.* octadecyl-triphenyl phosphonium. The compatibilising (expanding) agents can be polymers, organic reagents or monomers, silane compounds, metals or organometallics, organic cations to effect cation exchange, and their combinations. The preferred agents are PEG, PCL, and polyesters of a dibasic acid and glycol. They can be dissolved in, e.g., CH_2Cl_2 or toluene and then the organoclay dispersed in the solution. The interlayer spacing of the expanded organoclay varies from d_{001} = 2.2 to 4.2 nm. The expanded organoclay may be incorporated into PET either by polycondensation or by melt extrusion, both followed by solid-state polymerisation to increase the *MW*.

Polymerisation of PET in the presence of organoclay usually results in polymer with shear viscosity η = 0.5-1.2 kPas, not useful for making bottles [Matayabas *et al.*, 2000]. CPNC with PET as the matrix, containing 0.5 to 10 wt% clay (aspect ratio p = 10-2000), was prepared by melt polymerisation, followed by solid-state polycondensation. The composition had an inherent viscosity (I.V.) > 0.55 dl/g, η > 3 kPas and gas permeability at least 10% lower than that of neat resin. The product could be formed into films, sheets, containers (manufactured by extrusion or stretch blow moulding), tubes and pipes. The novelty of this patent is the solid-state polymerisation that leads to high *MW* and viscosity. Melt blending of PET with > 2.0 wt% clay may cause severe degradation. Bottles made from such a material would have poor properties. Clay also has nucleating effects that cause rapid crystallisation of a preform, which makes bottle blowing virtually impossible. The solution to this problem is solid-state polymerisation able to increase *I.V.* from 0.1 to 0.6-1.2 dl/g. The process involves heating pelletised polymer at $T_g < T < T_m$. The polymerisation time should be > 12 h under a blanket of an inert gas (Ar or N_2) or under vacuum. A wide range of catalysts (based on titanium, manganese, antimony, cobalt, germanium, tin, zinc, calcium, etc.) may be used. Because of the improved barrier properties, it is surprising that the solid-state polycondensation is so efficient.

For example, Claytone· APA was compounded with PET (*I.V.* = 0.72 dL/g) in a TSE at 280 °C. As the amount of organoclay increased from 0.4 to 6.8 wt% the *I.V.* decreased to 0.68 to 0.50 dl/g, respectively. TEM showed the presence of tactoids with thickness ≤ 50 nm, some larger aggregates, and few if any individual platelets. By contrast, MMT-MT2EtOH was treated with a mixture of OET (M_n = 377 g/mole) and 1,4-cyclohexane dimethanol, and then polymerised at T = 220 to 280 °C under vacuum. The product had *I.V.* = 0.31 dl/g and the viscosity at 280 °C was η = 8.73 kPas. The dried precursor was placed in a solid-state polymerisation unit with a N_2 purge and heated to 218 °C. After 16 h heating was discontinued and the material was allowed to cool. The composite had *I.V.* = 0.70 dl/g, η = 26.1 kPas, and T_m = 239 °C. TEM imaging of this PNC showed the presence of mostly individual platelets and few tactoids and aggregates. Thus, extrusion compounding alone does not provide CPNC with the desired performance, but extrusion followed by solid-state polymerisation does.

Owing to a large reduction of gas permeability these CPNCs can be used as an inner 0.1-2 mm thick layer of a PET container wall [Anonymous, 1999b]. Forming the container may be accomplished through injection stretch blow moulding, injection moulding, extrusion blow moulding or thermoforming. The preferred method of fabricating the nanocomposite polymer container is through two-stage injection stretch blow moulding. The new clay-reinforced PEST exhibits dramatic improvements of mechanical, thermal, barrier, and flame retardant properties without significant loss of toughness or clarity.

The hydrolytic degradation of condensation-type polymers has been a well known problem in the polymer blending industry. Thus, two solutions have been used: (1) the aforementioned solid-state polycondensation, and (2) the molecular repair methods. The latter process is general, based on the principle of coupling two macromolecules in a reaction that involves their chain-end group and a coupling molecule. It takes place during reactive extrusion, and with an appropriate coupler, it can be applied to virtually any polymer. However, the easiest to handle are the hydrolytically damaged condensation polymers. In this case, the reactive additives are usually of the diglycidyl type, e.g., ethylene-glycidyl methacrylate, triglycidyl-isocyanurate, or polyetherimine [Utracki, 2002c].

Li *et al.* [2001b] prepared nanocomposites of PBT organoclay by melt intercalation. The organoclays were Cloisite®-6A (C6A), -10A (C10A) and –30B (C30B). In addition, poly(bisphenol-A-co-epichlorohydrin) was used. Pre-dry blending the powders and mixing them in an internal mixer at 230 °C and 50 rpm resulted in the CPNC. For two-component systems (PBT with organoclay) the best dispersion (a partial exfoliation with short stacks remaining) was obtained using C10A. For three-component systems the results were mixed – the best improvement was observed for C30B, the worst for C10A. The epoxy compound was added at the beginning of the mixing process, thus it affected not only the matrix but the organoclay as well. Neither the molecular weight, the thermal stability of the organoclays nor its performance were measured.

Imai *et al.* [2002] developed a new intercalating compound, which may bind to PEST macromolecules and act as a molecular repair agent. The logic for selecting this route was the observation that the key to high mechanical performance of a composite is the stress transfer from the matrix to the reinforcement. In the case of nanocomposites, the clay must be exfoliated and strongly interacting with the matrix. In the case of PA-based CPNC, there is sufficient ionic and hydrogen bonding between PA and a silicate. PET being less polar may require tethering. The new intercalant to be used with PET should satisfy three conditions: (1) have groups that can react with PET; (2) have a cationic group that can react with clay; and (3) be stable at the polymerisation temperature of PET (275 °C). The selected compound was a dimethyl isophthalate substituted with a triphenyl-phosphonium group (DIP):

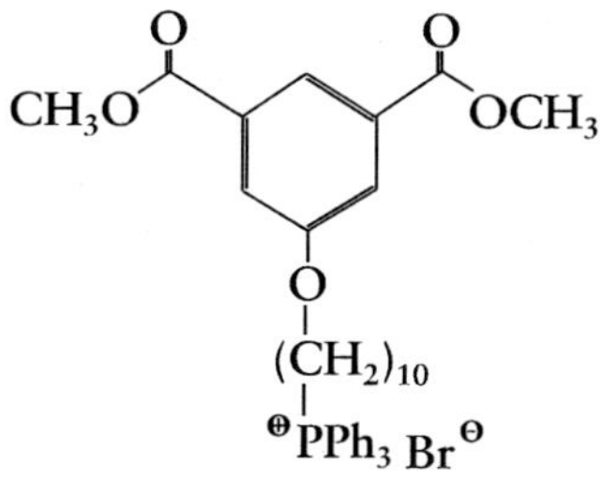

O O
CH_3O OCH_3
O
$(CH_2)_{10}$
$^{\oplus}PPh_3\ Br^{\ominus}$

For the study, fluoromica (FM) was quantitatively intercalated with DIP in a methanol-water solution. XRD showed that the peak corresponding to the interlayer spacing of FM, d_{001} = 0.96 increased to 3.28 nm. The PET nanocomposites were prepared by the polymerisation of bis-hydroxyethyl terephthalate (BHET) in the presence of FM-DIP, using Sb_2O_3 as a catalyst at

275 °C under vacuum for 3 h. As a result, CPNC was obtained showing a residual peak at d_{001} = 1.5-1.6 nm, indicating intercalation. NMR data confirmed bonding between the organoclay and the matrix. The flexural modulus increased with clay loading from 3.5 GPa (PET) to about 6.3 GPa at 8 wt%. This improvement confirms dispersion of the clay in PET and the strong interaction between the two components. However, there is still room for improvements.

Conroy *et al.* [2002] explored a similar approach. To increase the thermal stability of CPNC from about 150 to 450 °C the authors dispersed layered phyllosilicate (e.g., a smectite) in a (molten or dissolved) polymerisable chemical that contains nitrile groups, e.g., a bis(3,4,-dicyano-phenoxy) 4,4´-biphenyl having at least one amine or hydroxyl group. The formed adduct is either readily exfoliated during high temperature curing with at least 10 wt% clay, or it may replace standard organoclays during the preparation of CPNC with, e.g., PC, PA, PEI, PI, PPE, PU, epoxy, phenolic resin, etc. The proposed technology is general, useful for melt compounding with thermoplastics or reactive exfoliation with monomers or pre-polymers. Evidently, for the successful application of the method adequate miscibility between the intercalant and the matrix must be ascertained. However, the new technology (at a cost) provides CPNC with the thermal stability and performance desired for applications in the automotive, aerospace, electronic, and marine industries. Another, general method for the reduction of gas permeability is by coating. Anderson *et al.*, [1998] described CPNC based on melt-processable hydroxy-polyester or polyether. The claims are quite broad: for clays - any layered silicate material, for the intercalant - any onium salt, for the matrix polymer - great variety, and for the method of preparation - melt compounding or reactive exfoliation. The polymer matrix is preferably derived from diglycidyl ether or ester or an epihalohydrin and a di-nucleophile, such as a dicarboxylic acid, a di-functional amine, a bisphenol or a sulfonamide. The preferred concentration of the organophilic multilayered inorganic material is 4-40 wt%. The resulting CPNCs are to be used for coatings, e.g., films, foams, mouldings, laminates, fibres, and hot melt adhesives. The CPNCs may also be reinforced with traditional fibres or mats. The performance can be enhanced by heat treatment, orientation or annealing at *T* - 80-220 °C.

The process is illustrated in examples, where commercial bisphenol-A-epichlorohydrin phenoxy or ethanolamine-diglycidyl ether of bisphenol-A polymers were melt-compounded in an internal mixer with 10 wt% of MMT pre-intercalated with 2MBHTA or 2M2HTA (Cloisite® 10A and 6A?) at *T* = 187-204 °C. Compression moulded plaques showed increased modulus (by up to 81%) and barrier properties to O_2 (by up to 33%), but reduced tensile strength (by up to 27%) and elongation at break (by up to 97%). Better performance was achieved with 2MBHTA than with 2M2HTA, especially in phenoxy matrix. The results were judged acceptable. The relatively small improvements of rigidity and barrier properties (at 10 wt% organo-clay loading!), as well as the presence of micron-sized particles seem to indicate poor clay dispersion. Better performance would be expected from CPNC even prepared within the constrains of the patent.

The coating method has also been favoured by Toray Industries [Harada *et al.*, 1999; 2000]. The developed technology involves preparation under intensive mixing of an aqueous suspension of clay in a solution of water-soluble (or dispersible) polymer and optionally an alkyl amine. The preferred intercalating polymer is PVAl having a high degree of polymerisation, *DP* = 100 to 5,000, and a degree of saponification exceeding 80%. The preferred clay is a smectite, *viz.* MMT, hectorite, saponite, etc.

The suspension is used to coat a polymeric substrate, improving the gas barrier properties and surface finish. To improve the stability, the coated layer may be crosslinked by incorporation of a crosslinkable, soluble component, e.g., epoxy or oxazoline. A dispersion of clay in an isopropyl alcohol solution of PVAl (the clay to PVAl weight ratio = 2:3) was used as the coating solution. Since the barrier properties depend on platelet orientation, coating of the substrate should be 'in-line', providing a coating with a 0.3 to 6 μm thick layer. Furthermore, the extruded and coated PET film may be biaxially stretched 3.2 in each direction. Addition of Mg^{+2} and Al^{+3} ions improves the interaction between clay platelets, thus the overall quality of the coated layer. While any polymeric substrate may be used, some may require surface treatments, e.g., corona discharge for PO. For this reason particularly polar, polycondensation-type polymers may profit from this technology. An electrostatic casting was used for PET films, followed by hot air drying. Reduction of gas permeability by up to two orders of magnitude was achieved.

While melt compounding is the most desired method of CPNC preparation, the extent of exfoliation it provides depends not only on the thermodynamic interactions but also on the thermal stability of the organoclay. To separate these effects, Chang and Park [2001a,b] prepared a series of CPNC with poly(ethylene terephthalate-co-ethylene naphthalate) (PETN) as the matrix via the solution intercalation method in *N,N*-dimethyl acetamide (DMAc). Thus, MMT was intercalated with: (1) hexadecyl amine (HDA), (2) trimethyl dodecyl amine (3MDDA), (3) dimethyl hydrogenated tallow 2-ethyl hexyl ammonium (2MHTL8 or Cloisite® 25A) and (4) methyl tallow dihydroxy ethyl (MT2EtOH or Cloisite® 30A). The interlayer spacing of these organoclays was, respectively, d_{001} = 2.60, 1.68, 1.96 and 1.79 nm. After the solution intercalation with PETN the spacing in CPNC containing 4 wt% of organoclay increased to: d_{001} = 3.10, 2.96, > 8.8 and 2.93 nm, respectively. Thus, exfoliation was obtained only for 2MHTL8. Specimens containing 0 to 6 wt% of clay were prepared for thermogravimetric analysis. The analysis showed only small effects of the type and amount of organoclay on thermal degradation – the onset degradation temperature of PETN (404 °C) was increased up to 419 °C by adding 6 wt% of MMT-2MHTL8; the residue at 600 °C was a sum of that of PETN and the amount of clay.

By contrast with the results of the thermogravimetric analysis, the tensile tests on cast films gave interesting, if not unexpected results. The initial modulus for PETN and its four series of CPNCs is shown in **Figure 166**, and the ultimate tensile strength in **Figure 167**. Clearly, the best overall performance is that of CPNC containing 4 wt% of clay in the form of intercalated MMT-HDA: modulus increased by a factor of 2.8, strength by a factor of 2.6. The system containing exfoliated Cloisite® 25A (C25A) equalled the performance but at 6 wt% clay loading. The data for Cloisite® 30A (C30A) are surprising, as the presence of hydroxyl groups should have induced good interactions between organoclay and the matrix. Possibly, the groups caused hydrolytic decomposition of the matrix, hence the poor performance may be due to a reduction of the matrix *MW*. The effects on the elongation at break (and probably on the impact strength) are similar for the four series – it decreases with clay loading from ε_b = 6 to about 2%.

The goal of the second publication [Chang and Park, 2001b] was to clarify the intercalation between PETN chains and MMT-HDA and to further improve the thermal stability and tensile properties. The study confirmed the results in the preceding publication. The transition temperatures (T_g and T_m) remained constant in the full range of organoclay loading. The films prepared by solvent

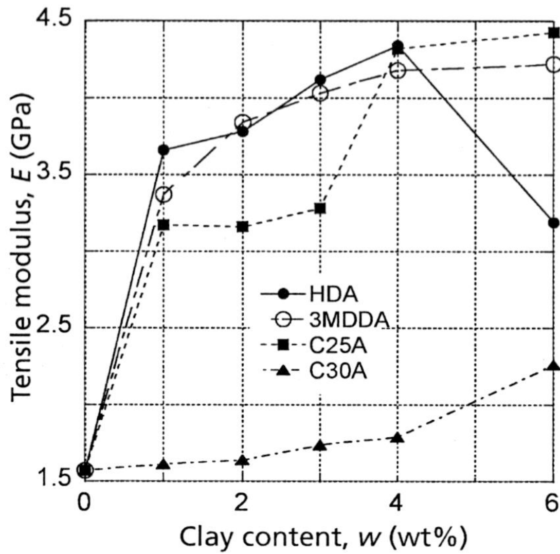

***Figure 166** The initial tensile modulus for CPNC with PETN as the matrix versus clay content (see text). Data [Chang and Park, 2001a].*

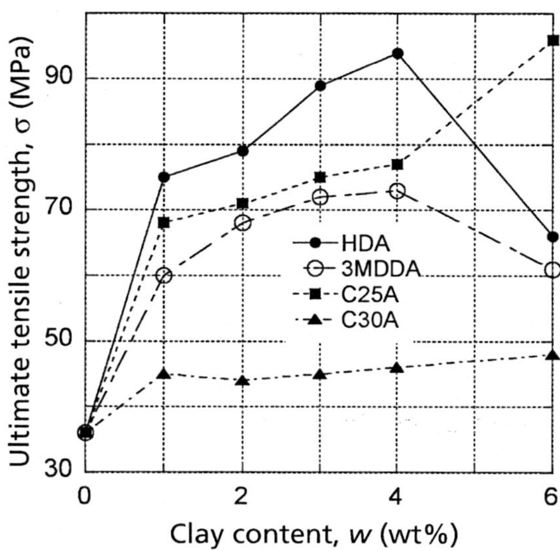

***Figure 167** The ultimate tensile strength for CPNC with PETN as the matrix versus clay content (see text). Data [Chang and Park, 2001a].*

casting were off-white, but translucent, not affected by MMT content from 0 to 4 wt%, but slightly decreasing at 6 wt%. PETN/organoclay hybrids were prepared by solution blending with PETN and HDA-MMT. The authors assigned responsibility for the significant improvement of the mechanical properties to the strong interfacial interaction between the matrix and organoclay, but without specifying the mechanism. Miscibility of PEST with the hydrocarbon chain of HDA is unlikely, thus interactions may be postulated between the PETN macromolecules and –OH groups on the clay surface or with the ammonium HD-NH_3^+ ion. Identification of the mechanism awaits.

In a recent paper Davis *et al.* [2002] offered a dramatic illustration of the thermal instability of ammonium intercalants. The authors intercalated Na-MMT (CEC = 0.92 meq/g from SCP) with either *N*,*N*-dimethyl-*N*,*N*-dioctadecyl ammonium bromide (2M2ODA) or 1,2-dimethyl-3-*N*-hexadecyl imidazolium tetrafluoroborate (2MHDI). The precipitated organoclay was filtered, washed 15 times and then Soxhlet extracted with EtOH for 10 h. Intercalation increased the Na-MMT interlayer spacing of d_{001} = 1.19 nm to 2.5 and 1.82 nm, respectively. The vacuum-dried powder was melt compounded with PET (T_m = 254 °C) in a corotating mini-TSE at 285 °C, under a N_2 blanket, at a screw speed of 200 or 300 rpm and residence time of 2, 5, or 7 min. Nanocomposites produced with 2M2ODA were black, brittle, and tar-like, while those with 2MHDI showed high levels of dispersion and delamination. The difference originates in the different intercalant decomposition temperature, *viz.* T_d = 250 °C for the former and 350 °C for the latter! Because of the evident degradation the PET/MMT-2M2ODA were not characterised. The compositions containing 2MHDI were all uniformly dispersed. The highest degree of dispersion was obtained under the mildest compounding conditions, *viz.* the slowest screw speed and the shortest residence time – TEM indicated the presence of individual platelets with short stacks with $d_{001} \cong 3$ nm, rarely having more than 3 parallel clay platelets.

Connell *et al.* [2004] described the preparation of exfoliated CPNC with high molecular weight polymer (primarily PEST or PA) as the matrix. The process involves: (i) melt compounding organoclay with an oligomer, and (ii) incorporating a high molecular weight matrix polymer, and (iii) to increase the *MW* to the desired level, solid-state polymerisation or reactive coupling may be involved. The product may be formed into single- or multilayered film, sheet, pipes, tubes, profiles, mouldings, preforms, and injection blow moulded containers (preferably bottles). The preferred organoclay was Na-MMT pre-intercalated with octadecyl methyl bis(polyoxyethylene) ammonium chloride (Ethomeen 18/25 – with 15 ethylene oxide units). The analysis gave the interlayer spacing of d_{001} = 3.38 nm, and the inorganic content of 52.1 wt%. The first stage compounding with 68.6 wt% of oligo(ethylene terephthalate) (OET; M_n = 382 g/mol) virtually left the gallery height unchanged (d_{001} = 3.4 nm), but the number of tactoids was reduced. The second stage melt compounding with 90 wt% of PET, followed by solid-state polymerisation (at 218 °C for 8 hours) resulted in exfoliated CPNC with a high molecular weight PET matrix. Examples of fully exfoliated CPNC prepared with poly(*m*-xylyladipoyl diamine) (MXD6) are also given.

To close this part on PEST-based CPNC a rare report on the effect of clay addition on weathering should be mentioned. Thus, PBT and PBT-based PNC injection moulded specimens were exposed to artificial weathering for up to 36 weeks (6048 h) [Goldman *et al.*, 1998]. Using XRD, the degree of crystallinity and crystal size were measured. In neat PBT samples exposed for < 400 h the crystallinity and crystal size increased, reaching a plateau, then after 1200 h, started to decrease. Incorporation of clay was found to have a profound effect. For example, the depth of UV radiation penetration in PBT was 50 μm, while in CPNC 25 μm. In consequence, the service lifetime of the PBT-based CPNC was significantly longer than that of the neat resin.

4.1.7 Polycarbonate (PC)

Polycarbonates being polyesters of carboxylic acid show similar behaviour to PEST. Thus, the methods of PNC formation described for PET can be adopted for PC. Moreover, in the 1980s, General Electric developed a polymerisation method involving the formation of a *cyclomer* (cyclic oligomer with up to 20 structural units) as an intermediary [Salem, 1991; Rosenquist and Miller, 1987, 1988]. The process is particularly suitable for the manufacture of high-fill GF composites with a high molecular weight PC matrix. The polymerisation occurs in two steps: (i) preparation of a cyclomer, (ii) catalytic ring-opening reaction at 200-250 °C, leading to PC with the degree of polymerisation $DP \leq 1000$. The cyclomer is a powder that melts at $T_m \cong 150$ °C and at higher temperatures turns into a low viscosity liquid. Since the cyclomer easily penetrates GF bundles, it has been expected that it may easier intercalate clay particles more easily (than PC) and afterwards it can be polymerised into PC.

In consequence, PC-based CPNC was prepared starting with MMT intercalated with 2M2TA (B34 from Rheox) and a cyclomer [Huang *et al.*, 2000]. Two methods of mixing were used: (1) in CH_2Cl_2, and (2) in the melt. Interlayer spacing of d_{001} = 2.47-3.62 nm was determined by XRD. Next, ring-opening polymerisation of the cyclomer was initiated in an internal mixer with Bu_4NBPh_4 as initiator. The reaction was carried out for 1 h at 180 °C and for 15 min at 240 °C. Exfoliated clay in PC having M_w = 40 kg/mol was obtained. By contrast, blending B34 with PC resulted in only an intercalated system. Since in the systems containing cyclomer exfoliation was obtained with or without a reaction, the authors concluded that molecular architecture (cyclic versus linear) and kinetics (low viscosity and shear) play the principal roles.

Severe *et al.* [2000] investigated the effects of clay on the thermal characteristics of CPNC with PC as matrix. The authors observed that alkyl-ammonium intercalants are not suitable for PC because of immiscibility, they are not thermally stable at the PC processing temperatures, $T \geq 240$ °C, and the presence of amine may cause chain scission of PC molecules through trans-reaction. In the study MMT pre-intercalated with C_{16}- and C_{18}-tributyl phosphonium ion and synthetic clay was used. Tethering groups with amino or epoxy functionality were added during the preparation of the synthetic clays. The synthetic clays were made by the hydrolysis of an aluminium salt ($AlCl_3$) with an alkyl alkoxy silane, $C_{18}Si(OCH_2CH_3)_3$, to produce a clay with layered structure. The clays were melt compounded with PC in a TSE at T = 290 °C, then the extrudates were compression moulded into sheets. The work focused on the effect of clay content and type on the T_g and thermal stability of PC. DSC results indicated that synthetic clays did not affect the T_g of PC, whereas addition of MMT pre-intercalated with phosphonium slightly reduced it. Furthermore, the onset temperature of thermal degradation significantly decreased in these nanocomposites. Phosphonium-intercalated MMT provided significantly better thermal stability for PC. TGA indicated that the presence of synthetic clays caused a significant reduction of PC thermal stability, whereas that of MMT intercalated with phosphonium improved it.

The patents on the preparation of CPNC with PEST from Kaneka [Suzuki and Oohara, 2001] (see Section 4.1.6) also specifically name PC as a suitable matrix and give several examples of the preparation and performance of these CPNCs. The authors modified clays (MMT or FM) with organosilanes, Y_nSiX_{4-n} (n = 0 to 3, Y is a C_{1-25} hydrocarbon, and X is a hydrolysable or a hydroxyl group), *viz.* R_i-amino-propyl trimethoxy silanes, where: R_1 is γ-(2-aminoethyl)-,

R_2 is γ-glycidoxy-, and R_3 is γ-(polyoxyethylene)-. MMT treated with these silanes were labelled as (*a*), (*b*), and (*c*), respectively. The fourth organoclay (*d*) was FM treated with R_1-amino-propyl trimethoxy silane similar to MMT (*a*). The PC with silane-treated clay nanocomposites were prepared by melt compounding in a TSE at T = 260 to 280 °C and 100 rpm. To prevent degradation 5 phr of phosphorus-containing stabiliser was added. The CPNC containing 6 wt% of organoclay (*a*), (*b*), (*c*), and (*d*) were prepared. The interlayer spacing for these samples was determined as: d_{001} = 6.8, 6.5, 4.8, and > 10 nm, respectively. The flexural tests demonstrated that the ratio of CPNC flexural strength to that of the neat PC, σ_f(CPNC)/σ_f(PC) = 1.38, 1.41, 1.38, and 1.46, respectively. Similarly, the ratio of flexural moduli, E_f(CPNC)/E_f(PC) = 1.66, 1.78, 1.38, and 1.80, respectively. It is interesting to note the reversal – for PET the spacing for CPNC with organoclay (*a*), and (*d*) was d_{001} > 10, and d_{001} = 8.0 nm, respectively, with significantly better mechanical performance and heat resistance for the former.

A melt compounding method has been used to prepare PC-based CPNC [Mitsunaga *et al.*, 2003; Yoon *et al.*, 2003]. The former authors used fluorohectorite pre-intercalated with 2MODA, compatibilised with PC by means of a styrene-maleic anhydride copolymer (SMA). Yoon *et al.* used MMT pre-intercalated with a series of mostly experimental ammonium ions. Both groups prepared the CPNC by compounding PC with organoclay in a CORI TSE at 240 to 260 °C. As discussed in Section 3.2.1, most ammonium intercalants have poor thermal stability, evidenced by discoloration, reduction of PC molecular weight, and a new XRD peak at d_{001} = 1.3 nm. In spite of the lack of exfoliation, at 5 wt% clay loading the relative modulus of non-compatibilised CPNC increased by *ca.* 50 to 60% (depending on intercalant) and that of compatibilised CPNC by *ca.* 80%.

4.1.8 Liquid Crystal Polymers (LCP)

In principle, nanocomposites are systems containing dispersed nanometre scale particles. In the preceding parts solid nanoparticles were incorporated into a matrix polymer. This book focuses on the production of nanocomposites containing well-dispersed clay platelets, as described, e.g., for PET. However, in the case of liquid crystal polymers (LCP) the system is multiphase to start with – one may produce genuine nanocomposite by controlling the LCP crystal structure on the nanoscale. In short, LCP and LCP blends may inherently be formed into nanoscale reinforced systems – the inherent 'nanocomposites'. Evidently, LCP may also be produced as a clay-containing nanocomposite, the CPNC. The performance of such a system is expected to depend on the interplay between the inherent nanocomposite and CPNC structures. In particular the orientation of these two types of solid particles is important – examples will follow.

Unidirectional, self-reinforced sheets or fibres of PET/LCP blends (LCP being based on PET and *p*-hydroxy-benzoic acid, HBA) were prepared using a SSE with a static mixer, followed by a die [Song and Isayev, 1998]. Various operating conditions, such as die geometry, die temperature, LCP content, extension ratio and screw speed during extrusion were used to optimise the processing conditions required for improving the mechanical properties of the blends. The main factors affecting the tensile strength and Young's modulus were the die temperature, LCP content and the extension ratio. These findings correlate well with the degree of microfibrillation of the LCP phase in the thermoplastic matrix, as evidenced by SEM. The mechanical properties of bars injection moulded at $T < T_m$ of LCP were investigated to find the optimum conditions for nanocomposite formation.

Another type of nanostructured system with LCP as a matrix was prepared by sequential polymerisation of two polymers [Gin *et al.*, 1998]. The key to this technology is the well-known ability of specific types of molecules to self-assemble into well-defined structures, *viz.* spherical dispersions of uniform size, lamellae or tubes that can form hexagonal structures. In the case of the patent from Gin *et al.*, the first monomer provides an organic template, which after polymerisation forms the ordered matrix for the PNC. Polymerisation, in the presence of an optional crosslinking agent, preserves the liquid crystalline (LC) order. The next step is polymerisation of the second monomer within the tubular channels of the matrix. Thus, the PNC has a matrix in the form of a lyotropic LC phase with a hexagonally-packed array of tubular channels, which provide structural integrity and control the PNC performance. For example, the tubular channels may contain a photoluminescent or electroactive conjugated polymer, semiconducting particles, metal salts, etc.

The method used to prepare these LCPs comprises three steps:

1. Combining inverse hexagonal-forming monomers, an aqueous or polar organic solvent, and tubular channel filler precursor to form a phase-separated mixture, in which monomer-1 forms an inverse hexagonal phase around the aqueous or polar organic solution.
2. Polymerising monomer-1 to form the PNC matrix with hexagonally-packed tubular channels.
3. Reacting the tubular channel precursor (monomer-2).

As an example of matrix-forming monomer-1, compounds having the chemical formula: HG-Y-$(X\text{-}PG)_{1\text{-}4}$ were identified as able to form an inverse hexagonal phase. In the formula, HG is a hydrophilic head group; Y is a bond or a template (e.g., aromatic ring) for the attachment of lipid tail groups; X is a lipid tail group having from 8 to 24 C-atoms in a linear or branched chain; and PG is a polymerisable group, *viz.* acrylate, acrylonitrile, ethylene, styrene, vinyl, diene, pyrrolidine, etc. After (or even during) the polymerisation of monomer-1, the tubular channel monomer-2 (or a polymerisable precursor comprising several components) can be converted. A variety of precursors can be used, *viz.* sol-gel silica prepared from a solution of tetraethyl-orthosilicate (TEOS), aluminophosphates, aluminosilicates, or zirconia-, chromium- or titanium-silicates. Other channel fillers are magnetic particles, semiconductors, metal or metal salt particles, conjugated functional polymers or the *water-soluble polymers*. A radical initiator (1-3 wt%) may also be added to the precursor.

The solvent (15-25 wt%) may be water or a polar organic solvent (e.g., DMF, DMSO, THF). The useful crosslinkers (10-20 wt%) include: divinylbenzene (DVB), *N,N*-bis-acrylamide, ethylene glycol dimethacrylate, etc. Polymerisation of monomer-1 results in formation of a regular, hexagonal array of tubular channels, of diameter $d \approx$ 3-6 nm. The diameter can be controlled by the structure of monomer-1 and composition. Polymerisation of monomer-2 (the precursor) completes the preparation of PNC. For example, a channel precursor, such as TEOS, can be polymerised by photoacid catalysis. The following example illustrates the invention. PNC with LCP as the matrix was prepared from Na-*p*-styryl-octadecenoate (monomer-1) and a solution of TEOS (monomer-2). A photoinitiator and a crosslinking agent were co-dissolved in the LC-phase to produce a crosslinked network. A water-soluble photoacid generator (2-hydroxy-2-methylpropiophenone) was used to initiate silica condensation. The hexagonal architecture was confirmed by XRD. The primary reflection yielded d_{100} =

3.54 nm, which corresponds to an inter-channel distance of 4.09 nm. The unit cell dimensions of the inverse hexagonal phase, and thus the diameters of the hydrophilic channels changed slightly when a hydrophilic TEOS solution replaced water.

Chang *et al.* [2002] provide an example of CPNC with LCP as the matrix. The selected LCP was of the thermotropic polyester type, with an alkoxy side-group:

$$\left[\overset{O}{\overset{\|}{C}} - \bigcirc - \overset{O}{\overset{\|}{C}}O - \bigcirc - O \right]_n \qquad \text{Br} \qquad CH_3CH_2O$$

The nanocomposites were prepared by dry-blending LCP and Cloisite® 25A (C25A; MMT-2MHTL-8) powders, and then compounding at T = 190 °C, i.e., within the nematic region of the polymer, for 30 min in an internal mixer. Compositions containing 0 to 6 wt% organoclay were prepared. The CPNCs were processed for fibres, and spun from a capillary rheometer at 190 °C. ^{1}H- and ^{13}C-NMR spectroscopy did not detect transesterification under these processing conditions. XRD of C25A established the interlayer spacing as d_{001} = 1.814 nm. For neat LCP and all the CPNC specimens a peak at 2θ = 4.69° was observed for the LCP spacing of d = 2.198 nm. Since no obvious clay peaks could be detected, the systems were exfoliated, which was confirmed by TEM. Addition of clay did increase the transition temperatures, *viz.* T_g by 6 °C, T_m by 7 °C and the decomposition temperature from 330 for neat LCP to 352 to 353 °C for CPNC with 2 to 6wt% C25A, respectively.

As shown in **Figure 168**, the ultimate strength and initial modulus of the CPNC fibres increased with organoclay content all the way to 6 wt% loading. The initial modulus increased linearly, doubling the neat polymer value at 6 wt%

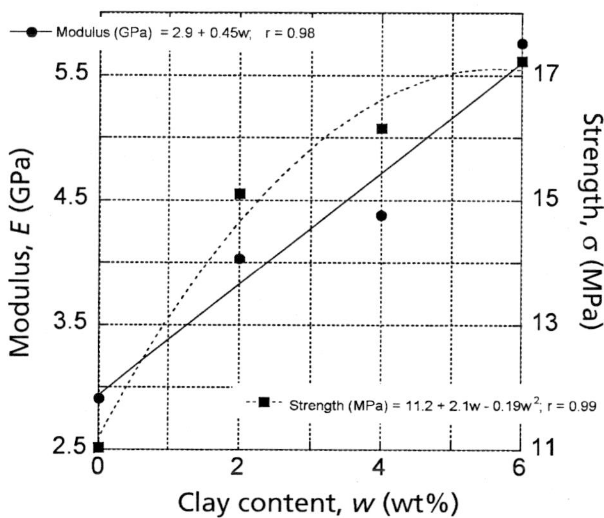

Figure 168 *Tensile modulus and ultimate tensile strength of LCP/C25A CPNC. The tests were conducted on oriented fibres. Data [Chang et al., 2002].*

loading. The efficiency of organoclay for the ultimate strength decreased with the organoclay concentration, suggesting progressively weaker interfacial bonding between the LCP matrix and the nanofiller. The data does not indicate any major effect of C25A on LCP crystallinity. Considering the molecular structure of the components, the only type of interaction between C15A and the LCP segments would be expected to involve hydrogen bonding to ≡Si-OH groups. In summary, the overall performance indicates that an organoclay loading of 2 to 4 wt% offers the best choice.

4.1.9 Fluoropolymers

Fluoropolymers are highly immiscible, thus they pose the ultimate challenge for CPNC preparation. The only solvents for fluoropolymers are fluorinated compounds, e.g., polyfluoro-aliphatic and perfluoro-aromatic. Nevertheless, several methods for the preparation of nanocomposites based on these polymers have been developed.

A Du Pont patent describes an improved coating composition composed of a fluoropolymer and a dispersed inorganic phase [Michalczyk *et al.*, 1998]. The PNCs were designed for coating articles made of glass, ceramic, plastics, elastomers, wood, or metal. The coating would be durable, anti-reflective, abrasion and chemical resistant, anti-soiling, anti-staining having low surface energy. In multilayer coatings, the PNC of this invention is to be used as a primer. The nanocomposites of this invention are not of the CPNC type – they do not use clay.

The method for the preparation of fluoropolymer-based PNCs requires:

1. Fluorinated solvent able to dissolve the two principal components: the fluoropolymer and a crosslinkable inorganic compound.
2. An inorganic oxide precursor is to be converted into a crosslinked inorganic oxide, dispersed in the fluoropolymer matrix into particles with diameter smaller than 75 nm.

The preferred inorganic oxides are: silica, titania, alumina, and zirconia. Compounds of Al, B, Ge, Si, Sn, Ti and Zr can be useful as precursors, provided that they are soluble in a fluorinated solvent, thus, they must contain fluorinated groups, e.g., $Si(CF_3C(O)CHC(O)CF_3)_2(OCH_2R_f)_2$, $Ti(OCH_2R_f)_4$, $Al(OCH_2R_f)_3$ or $Al(CF_3C(O)CHC(O)CF_3)(OCH_2)R_f)$, where R_f = C_{18} perfluoro-alkyl; $-[CF_2CF(CF_3)O]_rCF_2CF_2CF_3$. A transition metal catalyst such as Pt, or a free radical initiator may be used. The reactions can be conducted at T = 80-100 °C, P = 1 atm, under inert gas for t = 4 to 24 h. Solvent may be required to dissolve the reactants, but not for the reaction. The product can be isolated by devolatilisation. The fluoropolymer can be PTFE or its copolymers with other fluoro-monomers. The preferred are amorphous Teflon-AF and Teflon-FEP, which are soluble in perfluoro-(butyl-THF) (FC-75), hexafluorobenzene (HFB), perfluoro-(methyl-cyclohexane), etc. The fluoropolymer and inorganic oxide precursor are dissolved in a fluorinated solvent. When polysilicate is used as a precursor, a gelling agent is optional. For example, $C_8F_{17}C_6H_4Si(OCH_3)_3$ can be converted into a polysilicate by the reaction of formic acid in THF and then added to the dissolved fluoropolymer to form the nanocomposite. The inorganic oxide is formed within the fluoropolymer by allowing the inorganic oxide precursor to crosslink. The reactions with air-sensitive materials are carried out in a dry box or under nitrogen. The PNC may be recovered by drying, or the solution directly used for coating.

The following example was given to illustrate the adhesion promotion. Two solutions in HFB were prepared, one of Teflon-AF the other of tetra(trifluoroethoxy)silane [FES; $Si(OCH_2CF_3)_4$]. At room temperature the latter was added to the former while stirring. Next, a third solution of trifluoro-acetic acid (TFA) in HFB was dripped into it while stirring. The combined clear solution was used to dip coat a glass slide. The film (containing 14 wt% silica) was air dried at 200 °C. The contact angle of the dry film was 60° (after extraction in boiling water it was 53°), compared with 45° for PTFE. In another example, $C_8F_{17}C_6H_4Si(OCH_3)_3$ was converted into a polysilicate by reaction with formic acid in THF, then dried and re-dissolved in a solution of TEFLON FEP in perfluorophenanthrene (PP-11). The solution was used to flow coat a glass microscope slide. After drying, the coating showed excellent adhesion to the substrate. Numerous other examples illustrate other properties of the new coating materials, *viz.* adhesiveness, durability, thermal stability, high and random dispersion of the inorganic phase (particle diameter d = 10 to 12 nm), dimensional stability upon heating, etc.

In Xerox Corp. laboratories CPNC with fluoroelastomer as the matrix was developed [Badesha *et al.*, 1998]. The nanofiller was a mica-type silicate, e.g., MMT, bentonite, hectorite, vermiculite and saponite. The organoclay used in the examples from SCP (SCPX-984), was pre-intercalated with 5-10 wt% of dimethyl dioctadecyl ammonium (2M2ODA). The clay plates were about 1 nm thick and had the aspect ratio p = 50 to 1000. The fluoroelastomer was selected from between copolymers and terpolymers of vinylidene fluoride, hexafluoropropylene, and tetrafluoroethylene, known commercially as Viton, Fluorel, Aflas, Tecnoflon, etc. These are curable with, e.g., a bisphenol and organophosphonium salt. Other additives, *viz.* colouring agents, processing aids, conductive fillers, initiators and accelerators may also be added. The CPNCs developed could be used in printing machines (e.g., as a fuser, a donor, a pressure member, a layer of an intermediate toner image transfer member). Alternatively, they may be moulded into seals or O-rings, showing greater thermal stability than silicone nanocomposites. Finally, they may also be applied as coating (by spraying or dipping in a solution). The organoclay may be dispersed in a fluoroelastomer by melt compounding (prior to curing) for *ca.* 5 to 30 min. As is evident from the examples, the milling temperature is critical for the performance of CPNC. While mixing the organophilic clay with fluoroelastomer, the macromolecules penetrate the expanded interlayer space, causing each platelet to be surrounded by polymer chains, hence exfoliating. For example, MMT-2M2ODA was milled with Fluorel-FC in a two-roll Farrell rubber mill with a tight nip, at 27-38 °C for 15 min. XRD indicated that no intercalation had taken place. When the milling was carried out at 50 °C, d_{001} = 3.3 nm interlayer spacing indicated intercalation. Only milling at 66 °C produced exfoliated CPNC.

Ellsworth from Raychem Corp. also prepared clay-containing CPNC with fluoropolymer as the matrix. However, reading the patents one gets the impression that the prime task of the work has been the development of new, thermally stable intercalants. The new process comprises three steps [Ellsworth, 1998, 1999]:

1. Select 100 weight parts of melt processable fluoroplastic, e.g., poly(ethylene-co-tetrafluoroethylene) (ETFE), perfluorinated ethylene-propylene copolymer (FEP), or tetrafluoroethylene-perfluoropropyl-vinyl-ether copolymer (PFA).

2. Select 10-80 phr of an organoclay, intercalated with organophosphonium cations, $R_1P^+(R_2)_3$ where R_1 is a C_8-C_{24} alkyl or arylalkyl group and each R_2, which may be the same or different, is an aryl, arylalkyl, or a C_1-C_6 alkyl group. The interlayer spacing of the pre-intercalated organoclay should be $d_{001} \geq 3.5$ nm.
3. Combine parts (1) and (2) by melt compounding, to form the CPNC.

In the past it has not been feasible to produce CPNCs with a semicrystalline thermoplastic having high T_m (e.g., a fluoroplastic) or an amorphous polymer having a high T_g. The main obstacle was the ammonium intercalant of organoclays, which is stable at $T < 250$ °C. By contrast, the newly developed organoclays that are intercalated with organophosphonium cations are thermally stable up to 370 °C. To prepare the new-type organoclay, a clay must be reacted with an organophosphonium (OP-B) cation, which nearly completely replaces Na^+. The OP-Bs are arranged probably in a head-to-tail orientation, with the tails slanted. As a result of the intercalation, modified layered clay has the interlayer distance $d_{001} \geq 2.0$ nm (depending on the size of the organic groups R_1 and R_2), amenable to exfoliation during melt blending with a fluoropolymer.

The polymers of interest have a melt processing temperature of at least T = 250-270 °C. They include fluoroplastics, aliphatic polyketones, PEST, PPS, PES, PEK, PEI, polyimides (Aurum), PC, etc. Fluoroplastics, such as ETFE, are of main interest. The OP-B-modified clays can also be used to make nanocomposites with polymers having lower melting temperatures, such as aliphatic PA, but since for this application standard quaternary ammonium salts can be used, no special advantage would be gained. Conventional additives (e.g., antioxidants, UV stabilisers; flame retardants, acid scavengers, crosslinkers and pigments) can also be added to the mixture.

For example, Na-hectorite (10 g) was dispersed in aqueous alcohol solution at 90 °C with stirring. To this suspension a solution of tributyl-hexadecyl-phosphonium bromide (3BHDP) in isopropanol was added and the reaction carried on for 8 h. The organoclay was filtered out, washed and oven dried at 120 °C for 24 h. The dry powder was milled and screened through a 40 μm sieve. In an internal mixer a fluoropolymer (ETFE, Tefzel 280 from Du Pont, T_m = 268 °C) was melted while mixing at T = 275-350 °C at screw speed 10-40 rpm. The organoclay was added to the molten polymer, and then the mixing speed was increased to 50-100 rpm. After 5-15 min of mixing the compound was removed. Plaques were obtained by hot pressing at 275-300 °C and 2-7 MPa for 5-15 min, followed by cold pressing at 15-30 °C, 2-7 MPa pressure for 5-15 min. The properties of ETFE, FEP, and PFA, without and with 10 wt% organoclay are presented in **Table 91**. Evidently, the modulus has been improved but at a cost of the elongation at break and (to a lesser degree) of strength.

The patent is quite broad as far as the polymeric matrix is concerned. Thus, examples are also given for the preparation of CPNC with other than fluorinated polymers, e.g., with PEI, PEEK or PEK. The mixing conditions are not identified, but compounding and forming these polymers usually requires T = 350 to 420 °C. Comparative data on mechanical properties of the resins and their CPNCs containing either 2 or 5 wt% organoclay are listed in **Table 92**. As the table shows, these nanocomposites have significantly increased modulus while retaining useful levels of tensile strength and elongation at break. However, judging by the 20% increased modulus at 5 wt% loading, clay in these CPNC did not exfoliate.

Table 91 Properties of fluorpolymers and based on these CPNCs. Data [Ellsworth, 1998, 1999]

Polymer	Parameter	Polymer value	PNC value
ETFE	Young's modulus, E (MPa)	490±7	883±21
ETFE	Elongation at break, ε_b (%)	486±4	250±40
ETFE	Ultimate tensile strength, σ_b (MPa)	54±4	26±6
FEP	Young's modulus, E (MPa)	359±55	483±14
FEP	Elongation at break, ε_b (%)	440±40	400±20
FEP	Ultimate tensile strength, σ_b (MPa)	29±3	23±1
PFA	Young's modulus, E (MPa)	428±34	600±14
PFA	Elongation at break, ε_b (%)	400±40	300±27
PFa	Ultimate tensile strength, σ_b (MPa)	30±2	19±1

Table 92 Properties of PEI and PEK with organoclay. Data [Ellsworth, 1998, 1999]

Polymer/property	Organoclay (wt%)		
PEI (GEC)	0	2	5
Young's modulus, E (MPa)	2069±207	2172±103	2483±69
Elongation at break, ε_b (%)	20±5	8±1	5±2
Ultimate tensile strength, σ_b (MPa)	93±4.5	110±4.1	96±3.4
PEEK (ICI)	0	2	5
Young's modulus, E (MPa)	2207±138	2552±138	2690±138
Elongation at break, ε_b (%)	17±1	7±2	8±2
Ultimate tensile strength, σ_b (MPa)	87±0.7	94±1.4	97±0.7

Priya and Jog [2003] prepared PVDC/MMT-2M2HTA nanocomposites by melt compounding dried PVDF (M_w = 100 kg/mol, M_w/M_n = 2.5) with Cloisite® 20A (C20A; MMT-2M2HTA) in an internal mixer at 200 °C and 60 rpm for 5 min. XRD indicated intercalation; the interlayer spacing of C20A, d_{001} = 2.47 nm, increased to d_{001} = 2.7, 3.3, 3.1, 2.8 nm for CPNC with 1.5, 3, 5, and 7% wt% of C20A. DSC indicated that incorporation of clay increased T_m, and T_c by *ca.* 7 and 12 °C, respectively, but the total crystallinity was reduced from 60 to about 43%. As was observed for CPNC with PA-6 as the matrix, upon addition of clay the crystalline α-form of neat PVDF changed into β-form, while the crystallisation rate increased with clay loading. DMA in tensile mode indicated an increased storage modulus in the temperature range from -100 to 150 °C, e.g., E'(CPNC)/E'(PVDF) = 1.4 and 1.8 for 1.5 and 7 wt% of C20A at 20 °C.

4.1.10 CPNC with High Temperature Polymers

One of the main obstacles to the preparation of clay containing nanocomposites with high temperature polymers is the thermal stability of the ammonium intercalants, which have been universally used for the modification of layered clays. Several methods have been used to change the situation:

(a) Development of phosphonium intercalated clays, *viz.* [Ellsworth, 1998, 1999].

(b) Use a solution method, e.g., as for the preparation of CPNC based on PSF [Sur *et al.*, 2001], or on PAI [Ranade *et al.*, 2002].

(c) Commonly used exfoliation organoclays in a monomer, macromer or cyclomer, followed by polymerisation.

(d) Melt compounding with clay pre-intercalated with primary ammonium ions.

(e) Melt compounding into low melting-point polymer, which is miscible with the high temperature resin.

(f) Use of inorganic intercalants.

(g) Other methods.

Evidently, melt compounding would be the most attractive. However, the data available are not always consistent. For example, PEI-based CPNC were prepared by melt compounding by Huang *et al.* [2001b]. The authors achieved exfoliation by mixing PEI with up to 20 wt% (!) of MMT-HDA in an internal mixer at 370 °C. By contrast, Morgan *et al.* [2001], after finding that melt compounding of PEI with Na-MMT does not result in nanocomposites, used a two-step method, by first *in situ* tethering PEI to MMT via ω-amino dodecanoic acid (ADA), which resulted in a partial exfoliation and short, intercalated stacks (d_{001} = 1.36 nm) randomly dispersed in the matrix. However, during the second step – processing under standard conditions – the ammonium linkages were destroyed, causing the CPNC to lose the enhanced properties.

Polybenzoxazoles (PBO) with T_g of about 315 °C have good thermal and hydrolytic stability. From the point of view of chemistry and performance they resemble polyimides (polybenzimidazole, PBI), but are easier to process. When used for coating in the microelectronic industry, they have significantly higher thermal expansion coefficient than the substrate. To reduce the mismatch, organoclays have been incorporated during reaction in a solvent [Phiriyawirut *et al.*, 2001; Hsu *et al.*, 2002; Takeichi *et al.*, 2002]. Na-MMT intercalated with a variety of ammonium compounds was used. It seems that the best results were obtained with DDA. Films from these CPNC were prepared. Upon evaporation of a solvent d_{001} decreased significantly. The effect of melt processing is unknown.

During the last 10 years or so, many research papers have been published on CPNC with a diversity of polyimides. The early work involved solution intercalation of organoclays with polyamic acid, followed by film casting and a final polycondensation step by heating at $T \geq 300$ °C [Yano *et al.*, 1993; Yano *et al.*, 1997; Tyan *et al.*, 1999a,b; Kim *et al.*, 2001c; Chang *et al.*, 2001c; Chang and Park, 2001c; Liang *et al.*, 2003; Delozier *et al.*, 2003, etc.].

In 1997, starting with 2 wt% of DDA-intercalated clay, Yano *et al.* achieved exfoliation of MMT or synthetic clay (both had CEC = 1.19 meq/g), partial exfoliation for hectorite (CEC = 0.55 meq/g) and intercalation for saponite (CEC = 1.0 meq/g). Good correlation was reported between the relative permeability and the aspect ratio (p ranged from 46 for the hectorite to 1230 for the synthetic clay!), but, surprisingly, independent of the finesse of dispersion.

Two years later Tyan *et al.* pre-intercalated MMT with *p*-phenylene diamine, and then in solution reacted the compound with polyamic acid, achieving exfoliation for up to 7 wt% organoclay. Both XRD and TEM confirmed the stability of the clay dispersion in pyromellitic dianhydride-4,4′-oxydianiline. The CPNC had high modulus and heat stability. The thermal, dynamic and mechanical properties of these nanocomposites were reported.

Kim *et al.* [2001c] dispersed Cloisite® 30B (MMT-MT2EtOH) in polyamic acid solution, obtaining exfoliation for up to 6 wt% organoclay (and intercalation for CPNC with 9 and 14 wt% organoclay). The same year Chang and his colleagues followed a similar route incorporating into polyamic acid from 1 to 8 wt% organoclay (MMT pre-intercalated with either DDA or with HDA). Good dispersion was obtained – at 8 wt% loading, after curing the matrix to PI at 300 °C a small peak at d_{001} = 1.35 nm reappeared. This characteristic spacing of MMT was cited as evidence of thermal decomposition of the organoclay. Better performance was obtained for MMT-HAD than for MMT-DDA, e.g., at 2 wt% organoclay the tensile modulus increased, respectively, by 87 and 45%, the ultimate strength by 26 and 15%, and the oxygen permeability was reduced by a factor of 20 and 17, etc.

Two methods for the preparation of CPNC with PI as the matrix were compared [Gu and Chang, 2001; Gu *et al.*, 2001]. MMT pre-intercalated with cetyl pyridinium chloride (CPC) was used. Method 1 was the same as used by Yano *et al.*, and other authors, namely the addition of MMT-CPC organoclay to a solution of polyamic acid. Method 2 involved addition of a dispersed suspension of MMT-CPC organoclay to a solution of oxydianiline, followed by addition of dianhydride and polymerisation into polyamic acid. The resulting CPNCs with 1 to 10 wt% organoclay were only intercalated, with the interlayer spacing systematically larger for method 2 than for 1. The largest values for the tensile modulus and strength were obtained at 3 wt% loading (by 42 and 30%, respectively).

Hsueh and Chen [2003] prepared CPNC of layered double hydroxides and polyimide (LDH/PI) following the method proposed by Yano *et al.* By contrast with clays, LDH is a layered material that consists of positively charged layers and interlayer exchangeable anions. The composition is $[M(II)_{1-X}M(III)_X(OH)_2]^{X+}[A_{X/n}{}^{n-}\cdot mH_2O]^{X-}$, where M(II) and M(III) are divalent and trivalent cations, respectively, and A^{n-} is an exchangeable anion. LDHs have been used as catalysts, ion exchangers, adsorbents, ceramic precursors, and organic-inorganic nanocomposites Organo-LDH was prepared by ion exchange with 4-amino benzoic acid. A dispersion of organo-LDH was mixed with a solution of pyromellitic anhydride and then cured to generate intercalated nanocomposites. Alternatively, the organo-LDH was mixed with a solution of oxydianiline, then pyromellitic anhydride was added and the reaction carried out. The product cured at temperatures up to 350 °C was exfoliated, with only small peaks detectable at high concentrations (7 and 10 wt%) of organo-LDH, indicating the presence of intercalated stacks. The nanofilled systems showed excellent performance. The maximum tensile strength and elongation at break were obtained for organo-LDH loadings of 5 and 4 wt%, respectively. The initial tensile modulus and thermal stability increased with nanofiller content, while the coefficients of thermal expansion decreased.

Liang *et al.* [2003] synthesised two novel thermally stable, rigid-rod aromatic amino-phthalimides:

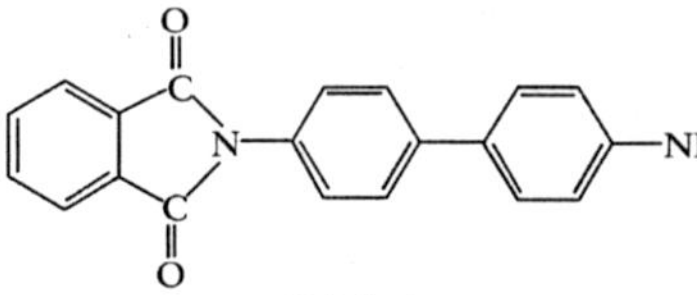

OM-1 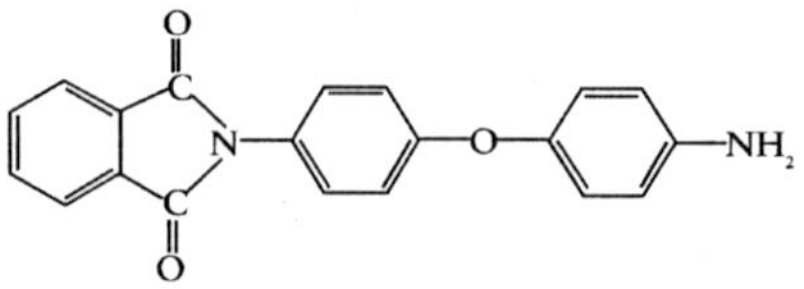 OM-2

N-[4-(4′-amino-phenyl)]phenyl phthalimide (OM-1) and *N*-[4-(4′-amino-phenoxy)]phenyl phthalimide (OM-2), to be used for the manufacture of CPNC with PI as the matrix. MMT was intercalated in these amino-phthalimides as well as with HDA. As a result, the interlayer spacing of MMT, d_{001} = 1.2 nm, increased to 2.90, 2.34, and 2.06 for OM-1, OM-2 and HDA, respectively. The thermal stability of these organoclays was 344.8, 320.3 and 252.1 °C, respectively. The three organoclays were used to prepare CPNC by *in situ* polymerisation of PI at $T \leq 280$ °C. As a result of the reaction, the d_{001} of MMT-HDA decreased, whereas that of the amino phthalimides increased towards exfoliation (at 3 wt% loading). The difference in the degree of dispersion was reflected in the performance. For example, the relative (with respect to neat PI) tensile modulus for CPNC with 3 wt% of MMT-OM-1, MMT-OM-2 and MMT-HDA was 3.71, 3.57, 3.16, respectively; the relative tensile strength was 1.61, 1.49, and 1.45, respectively; and the relative elongation at break was 1.16, 1.44 and 1.04, respectively. For MMT content not exceeding 3 wt%, the CPNC showed higher strength and toughness, the thermal stability was improved, T_g slightly increased, the coefficient of thermal expansion (CTE) markedly decreased, and the solvent uptake rate was significantly reduced.

Delozier *et al.* [2003] also used thermally stable intercalant for MMT, but instead of a mono-amine type (as OM-2) the choice was a dihydrochloride of 1,3-bis(3-aminophenoxy)benzene (APB):

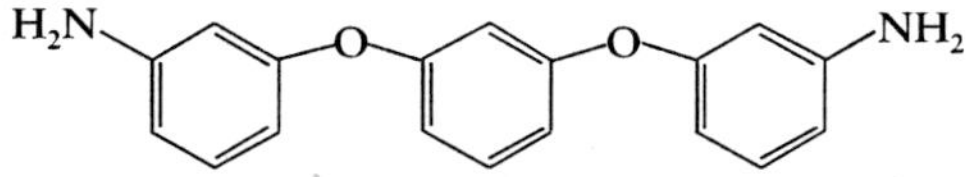

This APB was also used as a co-monomer in the synthesis of PI. To reduce the platelet-platelet interactions the authors converted Na^+-MMT into Li^+-MMT and kept the latter salt for 24 h in an oven at temperatures of 120, 130, 140, and 150 °C which reduced the anionic character of the clay's surface by incorporating Li^+ ions into the octahedral sheets of MMT. Thus, the MMT CEC was reduced from 1.11 to 0.71, 0.70, 0.63 and 0.49, respectively. These clays were ion-exchanged with APB for use in the preparation of CPNC with PI as the matrix. The thermal stability was excellent, *viz.* heated at a rate of 2.5 °C/min to 300 °C they lost 0.7 to 0.8 wt%, to be compared with 12.7 wt% loss of an alkyl-ammonium intercalated MMT (SCPX-2003). The interlayer spacing for MMT with CEC ≥ 0.63 was d_{001} = 1.46 nm, unchanged after the curing cycle, which resulted in the same peak position for the organoclay. The spacings for clays with lower CEC-values were still smaller, *viz.* 1.37 and 1.31 nm.

The organoclays were used for the preparation of CPNC PI films by polycondensation. The films containing 3 to 8 wt% of organoclay were heated to 300 °C to cause imidisation. The best performance was obtained using organoclay with CEC = 0.70 meq/g. The degree of clay dispersion is difficult to judge, as the XRD peaks are broad and located at high diffraction angles ($2\theta \cong 7°$). With densely packed clay platelets in the stacks the TEM images were not conclusive. The tensile tests of CPNC with the best organoclay showed that the relative modulus increased with clay content from 1.00 to 1.2, 1.21, and 1.33 for clay contents of 0, 3, 5, and 8 wt%. At the same time the tensile strength decreased by a factor of 1.22 and the elongation at break by a factor of 2.21. It would be interesting to check whether this poor performance could be improved by using an aromatic intercalant with a single ammonium group.

As the cited documents indicate, preparation of CPNC with PBO or PI as matrix is not too difficult as the reactive exfoliation takes place in a solution, before film casting and the final curing step. The technology is well described in the open literature.

4.1.11 Electroconductive CPNC

Electroconductive polymers have been used in batteries, displays, optics, EMI shielding, LED, sensors, and the aeronautical industry. It was reported that in 1990 a stretched, low density (ρ = 300 kg/m^3), I_2-doped polyacetylene (PAc), achieved the same conductivity as that of copper. Since, polymers are more flexible than metals, easier to process, and are corrosion resistant, there has been great interest in development of these materials. Unfortunately, most conductive polymers are sensitive to moisture, to oxidation and are quite brittle.

To ensure that the electrically conductive material meets all the performance criteria, it has been necessary to disperse the electrically conductive polymers in a matrix polymer which is inexpensive, flexible, and a matrix polymer which is resistant to environmental factors, i.e., to produce electrically conductive polymer blends. These blends have been studied since the 1970s. For example, PACE was polymerised into a polyethylene film impregnated with a catalyst. Similarly, using electrochemical methods, polypyrrole (PPY) was polymerised in a resin matrix, providing electrically conducting material with improved mechanical properties over those of neat PPY [Utracki, 1998].

The application of nanocomposite technology to polymer electroconductivity is more recent. There are two approaches for producing electroconductive CPNC: either by using a conductive polymeric matrix, or a conductive nanofiller. The former approach, especially using polyaniline (PANI) as the matrix has been favoured. The polymer has good chemical stability and respectable conductivity when doped or protonated. However, PANI is insoluble and it cannot be melt processed as it decomposes at temperatures below a softening or melting point.

Liao and Lin [2000a,b] patented a method for producing a conductive CPNC by dispersing clay in an aqueous solution of aniline, an acid (e.g., methyl-sulfonic acid), and an oxidising agent (e.g., ammonium persulfate), and then conducting oxidative polymerisation. The preferred clay was MMT, FM, saponite, beidellite, nontronite, hectorite, stevensite, vermiculite, halloysite, or sericite, having CEC = 0.5 to 2.0 meq/g. Prior to use the clay may be either acid-treated or pre-intercalated with PEG (MW = 0.1 to 50 kg/mol). The product contained 0.05 to 80 wt% of clay with interlayer spacing greater than 5.0 nm. The invention is illustrated by examples. Thus, MMT was pre-intercalated with PEG (MW = 3 kg/mol), which increased the interlayer spacing to d_{001} = 1.7 nm. The modified MMT was washed and freeze-dried, and then added to an aqueous solution of aniline, and methyl sulfonic acid. At 0 °C an aqueous solution of ammonium persulfate was slowly introduced to cause polymerisation (*ca.* 6.5 h with stirring at 600 rpm). The resulting CPNC of MMT/PANI was filtered out, washed, and dried. The XRD showed an absence of diffraction peaks, indicating exfoliation. The electrical conductivity of neat PANI, 0.269 S/cm, was found slightly reduced to 0.157 S/cm. Alternatively, FM was treated at 90 °C for 2 h with dilute sulfuric acid. The acid-treated FM was added to an aqueous solution of dodecyl-benzene sulfonic acid, aniline, and ammonium persulfate. The reaction proceed at 20-22 °C for 4 h. The resulting nanocomposite was found to be intercalated (d_{001} = 3.62 nm), and its electrical conductivity was a respectable 0.239 S/cm.

Kim *et al.* [2002] attempted to disperse Na-MMT directly into a PANI matrix during emulsion polymerisation. Dodecyl benzene sulfonic acid (DBSA) was used for both dopant and emulsifier. XRD showed that in reaction products the initial interlayer spacing of Na-MMT (d_{001} = 0.97 nm) was slightly expanded to d_{001} = 1.52 nm. The electrical conductivity of the PANI/DBSA complex showed a sharp maximum at a 1.25 molar ratio of DBSA. In the presence of clay, the conductivity of the complex PANI/DBSA/MMT increased with the dopant level up to 1.5 S/cm (at dopant molar ratio = 2). According to DSC and TGA results, the thermal stability of CPNC was improved.

Bora *et al.* [2000] used another approach to manufacture electrically conductive CPNC – modifying the clay to make it conductive. The publication describes the preparation of thermally stable organoclay by ion exchange between Na-MMT and $[Ni\text{-}(ligand)_2]^{2+}\ 2Cl^-$. For example, organoclays: $[Ni\{di(2\text{-}aminoethyl)amine\}_2]$-MMT and $[Ni(2,2':6',2''\text{-}ter\text{-}pyridine)_2]$-MMT were prepared with good thermal stability. Since this approach is general and several electroconductive complexes are known, this approach offers a simple way to produce stable and electroconductive CPNC.

4.2 Thermoset CPNC

Preparation of CPNC with a thermoset matrix is simpler for two reasons: (1) the monomers or oligomers show low viscosity, and (2) they are polar. Thus, there are many publications and patents, especially for epoxy-based nanocomposites.

4.2.1 Epoxy Resins

Wang and Pinnavaia [1994] observed exfoliation of an onium-modified MMT-ADA in Epon 828 epoxy resin by mixing the two components at 75 °C for ½ h, and then heating to T = 200-300 °C. XRD and TEM measurements of the composite suggested exfoliation in CPNC with 5 wt% organoclay. TEM micrographs showed 'accordion' like expanded MMT stacks, about 800 nm wide and several micrometres long. High temperature curing produced intractable powders instead of a solid epoxy. The reaction kinetics were found to depend on the heating rate and the intercalating cation. Lan and Pinnavaia [1994] prepared a series of CPNCs with MMT pre-intercalated with *n*-alkyl (primary) onium salts (n = 8, 12, and 18) in Epon 828, cured with polyether-amine (PEA; Jeffamine D2000). The degree of intercalation/exfoliation depended on the *n*-value for the intercalant and the curing conditions (as well as, probably, clay content). Thus, at 10 wt% MMT-ODA loading, the clay was fully exfoliated after curing, whereas in CPNC with MMT-OA only intercalation was obtained. The reason for using PEA instead of the more common *m*-phenylene-diamine was a desire to obtain rubbery (not glassy) CPNC at room temperature. The tensile test results for these nanocomposites were excellent, *viz.* more than a 10-fold increase of tensile strength and modulus at 15 wt% MMT-ODA

At about the same time [Messersmith and Giannelis, 1994; Giannelis and Messersmith 1996] epoxy CPNC were prepared using smectite clay pre-intercalated with an alkylammonium. The latter had functional groups able of reacting with an epoxy. The specified organoclay was MMT intercalated with methyl-tallow-bis (2-hydroxyethyl) ammonium (MT2EtOH), e.g., Cloisite® 30B (C30B). The organoclay was dispersed in a mixture of epoxy resin with the diglycidyl ether of bisphenol-A (DGEBA, e.g., DER 332 or Epon), and a curing agent. The authors postulated that at the first stage DGEBA reacts with the intercalant –OH groups, increasing the interlayer spacing to about 1.7 nm. Mixing at 90 °C further increased it (d_{001} = 3.5 nm), and then increasing T gradually increased d_{001} to about 3.8 nm. Thus, before curing only intercalation of the organoclay was observed – to achieve exfoliation in the final product a suitable curing agent must be used. Many tested agents resulted in CPNC with d_{001} = 3.0 to 4.0 nm. Addition of a bifunctional primary or secondary amine (e.g., methylene dianiline, MDA) caused immediate clouding, probably due to bridging two clay platelets thus preventing further increase of the interlayer spacing. However, the authors identified three exfoliating curing agents, *viz.* nadic methyl anhydride (NMA), benzyl-dimethyl amine (BDMA), and boron trifluoride monoethyl amine (BTFA).

In general, a curing agent might participate in three reactions: crosslinking the epoxy, reacting with organoclay, or catalysing the reaction between the organoclay and epoxy. The latter one makes it possible to cure at lower temperatures (*viz.* T = 100-200 °C) than before. The use of intercalants with functional groups also leads to chemical bonding between clay platelets and the crosslinked epoxy network. The curing resulted in exfoliated clay platelets ($d_{001} \geq 8.8$ nm) randomly dispersed within the crosslinked epoxy matrix. TEM showed individual silicate layers, *ca.* 1 nm thick, with short stacks of *ca.* 10 platelets ($d_{001} \approx 10$ nm).

It is noteworthy that the curing temperature (*ca.* 100 °C) corresponds to the temperature at which exfoliation has occurred. It seems that exfoliation exposes the hydroxyl groups of the alkyl ammonium chains in the interlayer to DGEBA. Interestingly, in the absence of organoclay full curing of the epoxy/NMA mixture did not take place, regardless of the heating. During the dynamic curing of this formulation two distinct exotherms are observed: a weak one at 180 °C, followed by a strong one at 247 °C. A possible sequence might involve a reaction between the organoclay -OH groups with NMA to form a monoester, which in turn reacts with the epoxide to form a network.

The viscoelastic properties of the crosslinked epoxy/BDMA composite containing 4 vol% of C30B were studied in DMA. The CPNC showed a broadened T_g = 122 °C, i.e., 4 °C higher than that of neat epoxy, the dynamic storage modulus of the CPNC in the glassy region was higher by 57% and in the rubbery region by a factor of 4.55 (at 4 wt% clay). This is a considerable enhancement of properties, considering that epoxy composites with classical filler particles exhibit only a small (< 10%) increase of E′.

In 1996, Pinnavaia and Lan applied for a series of patents for the manufacture of CPNC with epoxy as the matrix [Pinnavaia and Lan, 1998a,b,c; Pinnavaia and Lan, 2000a,b]. The work was also described in several publications [Lan *et al.*, 1995; Pinnavaia *et al.*, 1996; Shi *et al.*, 1996]. The professed goal of these patents was to produce elastic epoxy/clay CPNC with intercalated or exfoliated clay, to be used for seals and other thin layer applications, for example, as decorative or protective coatings and encapsulation, polymer concrete, trowel coatings and wood consolidation, composites for propeller and impeller blades, boats, filament-wound tanks and piping, and in the dampening of vibrating surfaces [Pinnavaia and Lan, 1998a,b]. The resulting CPNC showed superior tensile strength and/or solvent resistance as compared to cured epoxy resins without clay, or with non-intercalated clay. The nanocomposites were produced by dispersing organoclay in an epoxy + diamine matrix, reacting these and curing. The clay may be mineral, *viz.* MMT, hectorite, saponite, nontronite or beidellite, or synthetic, *viz.* FH, laponite, taeniolite or tetrasilicic mica. The intercalant is an alkyl (with 3 to 22 C-atoms) ammonium cation, capable of expanding the interlamellar gallery to a height of 0.7 to 30 nm. The epoxy is DGEBA or DGEBF. The preferred curing agent is a polyoxypropylene diamine (x = 4 to 40) sold as the Jeffamine D-series. CPNC interlayer spacing and mechanical properties were found to be related to the intercalant alkyl chain length; in a series with C4, C8, C10, C12, C16, and C18 the interlamellar gallery height increases from 1.0 to 1.51, 1,76, 2.02, 2.53 and 2.78 nm, respectively and the tensile strength from 8.1, to 9.0, 12.2, 13.8, 13.9 and 14.5 MPa. The system with C18 (ODA) was fully exfoliated. As evident from the data displayed in **Figure 169**, there is a close

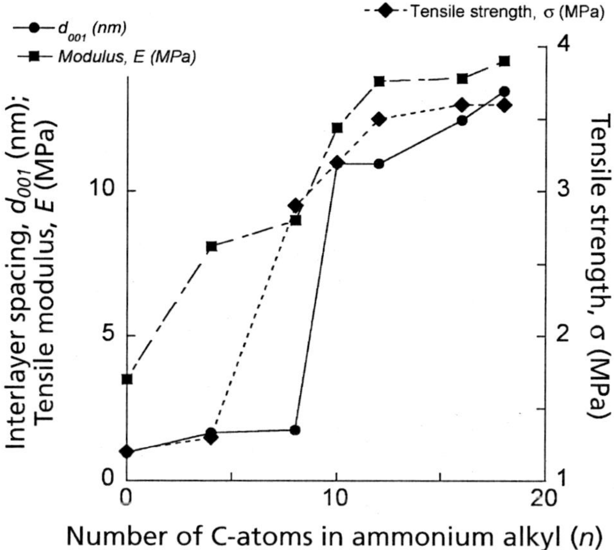

Figure 169 Interlayer spacing, tensile strength and modulus for Epon-828/Jeffamine D2000 (27.5/72.5) with 10 wt% of organoclay versus number of C-atoms in the intercalant alkyl chain: $CH_3(CH_2)_{n-1}NH_3^+Cl^-$ (n = 4, 8, 10, 12, 16 & 18). The values for n = 0 indicate absence of MMT. Data [Pinnavaia and Lan, 1998a,b].

correlation between the interlayer spacing and mechanical properties. Significantly, the tensile strength increases in parallel with the modulus.

In the following patent [Pinnavaia and Lan, 1998c] there is a significant change in the method of CPNC manufacture: (1) the clay is acidified into $MMT^- H^+$, (2) the use of ammonium intercalant is eliminated, and (3) the monomer(s) or oligomer(s) contains at least one basic group for reaction with the protons of the clay. The authors argued that the expensive alkyl-ammonium ions in the interlamellar galleries might prevent interactions between the epoxy matrix and the clay surfaces. Furthermore, the alkylammonium ions are toxic and require special handling procedures. Another problem that the patent aims to eliminate is the difficulty in obtaining total exfoliation in the CPNC.

The claims are broad, pertinent to thermoset as well as to thermoplastic CPNC. The nanofiller may be selected from a variety of layered inorganic compositions (e.g., silicates, phosphates, arsenates, titanates, vanadates, etc.), but primarily a mineral or synthetic smectite, vermiculite, mica or hydromica, etc. The polymerising component might be an amine, an amide, an imide, an anhydride, an epoxide, an isocyanate, an alcohol, a hydroxide, a pyridinyl, a pyrrolyl or a vinyl. Several thermoset polymers have been mentioned, *viz.* epoxy, polyurethane, polyurea, alkyd, polysiloxane, polyester and polyimide – preferably amine or amide. However, the principal reactants remained: DGEBA or DGEBF, Novolac, diglycidyl polyalkylene-ether, and epoxy-propoxy-propyl terminated polydimethylsiloxane. The preferred crosslinking and curing agents are: a polyoxypropylene diamine, a polyoxypropylene triamine, a polyamide, and an amino-propyl terminated PDMS. The thermoplastic polymer may be: PA, PEST, TPU, PSF, POM, PEG, PCL, PLA, PI, PO, etc.

The process is illustrated by numerous examples. The principal clay was Na-MMT converted to H^+-MMT and showing d_{001} = 1 to 1.5 nm (depending on the degree of hydration). Next, the protonated clay was dispersed in water with a small amount of ethanol, the curing agent (PEA, Jeffamine D2000) was added and the suspension was stirred vigorously at RT for 6 h, and then centrifuged. The dried powder had d_{001} = 4.6 nm and it contained 45 wt% of PEA. Alternatively, the H^+-MMT powder was placed in a blender running at high speed, and Jeffamine was slowly added. The resulting powder had d_{001} = 3.6 nm, indicating significant intercalation. The third method of combining these components was similar to the first one, but instead of centrifugation the H_2O + EtOH was boiled off, and then the residue was placed under vacuum (at 100 °C) obtaining thick gel. The interlayer spacings in the 'powder' and 'gel' compounds were comparable. The final step in the preparation of CPNC was mixing the concentrate with Epon 828 epoxy and Jeffamine D2000 curing agent at room temperature for 15 min, followed by curing at 75 °C for 3 h and at 125 °C. The CPNC with 5 wt% MMT was exfoliated. The tensile properties of these products with 0 to 15 wt% MMT were measured (see **Figure 170**). For comparison, the data for CPNC prepared from the same curing agent and epoxy resin, but using MMT-ODA in the place of H^+-MMT were included for comparison. The tensile modulus and strength increased linearly with MMT-content. The results show that the presence of ODA in the clay gallery decreased the effectiveness of the clay reinforcement. However, to achieve the optimum performance the relative rates of intercalation, chain formation and crosslinking must be controlled [Wang and Pinnavaia, 1998a,b].

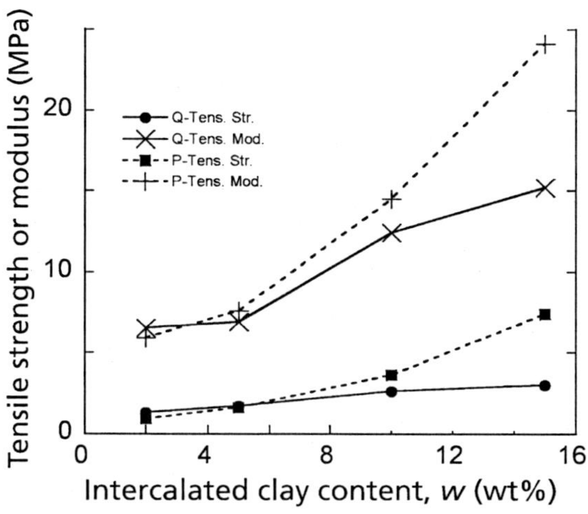

Figure 170 *Tensile strength and modulus for Epon-828/Jeffamine D2000 (27.5/72.5) with 2-15 wt% of intercalated MMT. Two ammonium salts were used, primary:* $CH_3(CH_2)_{17}NH_3^+Cl^-$ *(P) and quaternary:* $CH_3(CH_2)_{17}N(CH_3)_3^+Cl^-$ *(Q). Data [Pinnavaia and Lan, 1998c].*

While the epoxy matrix in the CPNC discussed above was amine-cured elastomeric, a couple of patents from AMCOL focused on rigid systems cured with anhydride [Lan, 2000; Lan and Westphal, 2001]. The work had two goals: (1) the development of new organoclays, and (2) using 0.05 to 60 wt% of these, producing rigid anhydride-cured epoxy-type CPNC with improved thermal (e.g., T_g increased by at least 30 °C), mechanical and/or barrier properties. The 'new' organoclay was a standard layered clay, intercalated with an onium, *viz.* primary, secondary, tertiary or quaternary, but the latter was favoured. Furthermore, as in several earlier patents, at least one alkyl radical, C_n, must have $n > 6$. The new aspect of this invention is the requirement that the matrix should not covalently bond either to the clay or to the intercalant. However, the epoxy matrix is to be a part of the intercalating composition in the amount from *ca.* 10 to 100 wt% (on dry clay). The composition may also include a curing accelerator for co-intercalation into the interlamellar galleries.

The professed logic of the process is that onium salt provides the initial expansion of the interlayer spacing, which is further expanded by diffusion of anhydride-curable epoxy resin. The intercalation expands the interlamellar galleries by about Δd_{001} = 2.5 to 3.5 nm. Thus, the onium ion and anhydride-curable epoxy resin are co-intercalants, which make the clay compatible with the epoxy matrix polymer to form the CPNC. The new organoclays may be compounded with all commercial epoxy resins (e.g., bisphenol-A-derived, epoxy cresol Novolacs, epoxy phenol Novolac, etc.) to produce CPNC.

The nanocomposite production requires melting the epoxy resin in an extruder, adding the organoclay, and dispersing it throughout the melt. The process may be conducted directly to the final concentration of organoclay, or it may go through masterbatch compositions, which later must be diluted with more anhydride-curable epoxy polymer. The resulting composition is cured with one or more anhydride curing agents. Examples of suitable epoxy and phenoxy resins, such as aliphatic-, cycloaliphatic-, or aromatic-, are shown in **Figure 171**. In the upper part (**Figure 171(a)**) the link 'A' is a divalent alkyl group, C_nH_{2n}, with n = 1 to 4; R is hydrogen or an alkyl; X is hydrogen, a halogen atom, an alkyl or alkoxy group. The preferred resins are of the oligomeric DGEBA type (see **Figure 171(a′)**, with the degree of polymerisation, n' = 2 to 30. Also useful are epoxy phenol (or cresol) Novolacs (see **Figure 171(b)**). Polyfunctional epoxy compounds, *viz.* tetraglycidyl ether of tetra-kis(4-hydroxyphenyl)ethane, *N,N,N′,N′*-tetraglycidyl-4,4′-diaminophenylmethane, triglycidyl *p*-aminophenol resins, or a triglycidyl isocyanurate may also be used. These epoxies are curable with commercial anhydride curing agents (e.g., anhydrides: nadic methyl, trimellitic, phthalic, etc.) following the appropriate cure schedule.

The clays of interest are smectites, especially MMT, nontronite, beidellite, volkonskoite, hectorite, saponite, etc. For example, MMT was intercalated with non-functionalised quaternary ammonium ions, e.g., 3MODA, 2MODA, 2MBODA, 2M2HTA, MOD2EtOH. The intercalated MMT is then subjected to grinding or milling in a hammer mill, jet mill, or air-classifier mill. Next, the organoclay (5 to 35 phr) may be mixed with the anhydride-curable epoxy resin first, and then the anhydride curing agent and accelerator may be added. To exfoliate the mixture may require shearing (at T = 120 °C and at a rate of 100 to 10,000 s^{-1}), heating, or applying low pressure. The shearing can be provided by mechanical means (e.g., during extrusion, injection moulding, or in internal mixers), thermal shock, pressure alteration, or by ultrasonics. The resulting CPNC

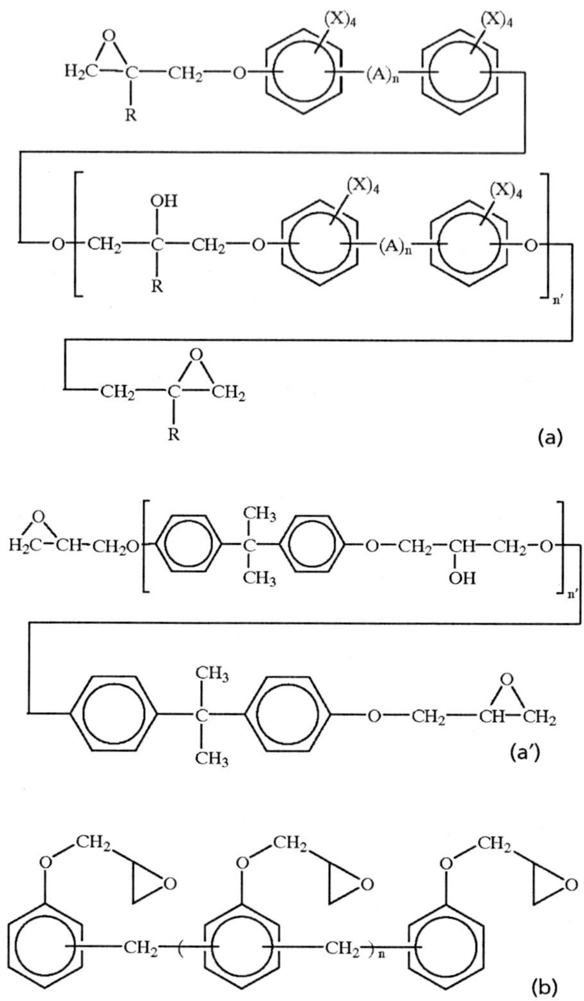

Figure 171 Examples of anhydride-curable epoxy resins suitable for preparation of CPNC with ammonium intercalated MMT [Lan and Westphal, 2001]. See text. Reprinted with permission from AMCOL.

is cured under standard conditions. The cured CPNCs are more rigid below or above the T_g. The latter transition temperature is significantly higher than that for unfilled epoxy. Furthermore, the CPNC have improved solvent and chemical resistance over a wide range of T.

As an example 10 wt% of organoclay was mixed with epoxy (DER 331 and a curing agent) at 75 °C. The clay-epoxy dispersions were stable at both 75 °C and RT. They had slight settle-down at extended storage time that could be eliminated by gentle mixing before use. The interlayer spacing of the MMT-3MODA and MMT-MOD2EtOH organoclays before mixing with epoxy are: d_{001} = 2.4 and 1.9 nm, and after mixing with epoxy 3.7 and 3.4 nm, respectively. The CPNC was cured at 110 °C for 1 hour, followed by an additional 4 hours at 160 °C. The product was light brown, semi-transparent, without visible particles at magnification x200. DMA temperature scans were conducted at a frequency of 1 Hz and at the scan rate of 2 °C/min. Incorporation of 10 phr organoclay increased the T_g from 120 to 140 °C. The storage modulus below T_g increased from E' = 3.2 (for neat epoxy) to 3.5 MPa. At 120 °C the difference in stiffness was more dramatic, *viz.* E' = 0.02 to 2.7 MPa, but at this temperature the neat epoxy was in the rubbery state while the CPNC was in the glassy state.

Epoxy-clay nanocomposites were also prepared by swelling Na-MMT (intercalated with primary *n*-ODA - d_{001} increased from 1.2 to 2.1 nm) in a DGEBA (Epon-828) followed by polymerisation with different di-amine curing agents [Kornmann *et al.*, 1999, 2002; Kornmann, 2001]. The agents were: bis(*p*-amino-cyclohexyl methane) or BACM, 3,3′-di-methy-4,4′-di-amino-dicyclohexyl methane (DDDHM), and Jeffamine D-230. Before adding a catalyst the ingredients were mixed at 75 °C for up to 24 h. The exfoliation very much depended on the clay CEC, curing agent and mixing. Thus, better exfoliation was obtained with clay having CEC = 0.9 than with 1.4 meq/g. Similarly, DDDHM gave larger spacing than BACM. Mixing for 6 or 12 h produced intercalated CPNC, whereas mixing for 24 h engendered exfoliation. The extent of intercalation/exfoliation depended on the relative reaction rate within the clay interlayers versus outside – when the reaction rate outside the interlayer is faster exfoliation is not to be expected. Evidently, the diffusion rate of the reactive species into the interlayers is the controlling factor.

Zilg *et al.* [1999a] investigated the basic correlations between polymer morphology, clay superstructures, T_g, stiffness and toughness of the epoxy-based CPNC as functions of the clay type and content. The CPNCs were based upon hexahydrophthalic anhydride-cured DGEBA and layered silicates such as FM, purified sodium bentonite and synthetic hectorite. The clays were intercalated with various mono- and di-functional alkyl ammonium ions. Enhanced toughness was associated with the formation of dispersed anisotropic laminated nanoparticles consisting of intercalated silicates. CPNCs were observed by TEM and AFM. The authors confirmed earlier reports from Pinnavaia's laboratory that d_{001} increases with the onium salt alkyl chain length, and that primary amines increase the interlayer spacing more than the quaternary ones. During the curing, the former reacts more easily with epoxy than the latter. With 10 wt% of intercalant in organoclay not all stacks were properly intercalated, but exfoliation was observed by TEM, e.g., for hectorite intercalated with 2MBODA. The most intriguing is the comparison of TEM and AFM observations. The spacings were quite different, *viz.* 6.8 and 11 nm, respectively. The offered explanation is that the individual clay platelets are flexible and bend under the force of the AFM test tip!

The Nichols and Chou patent [1999] has been mentioned several times in this book. It is unique in stressing the benefits of phosphonium intercalants as well as its wide applicability, which stretches from the engineering and speciality polymers to polyolefins, rubbers and thermosets. The claims describe PNCs comprising a polymer with layered material intercalated with an organic intercalant and an ionic or non-ionic inorganic intercalant.

- The polymer matrix may be a thermoset (phenolic; epoxide or epoxy; polyester; polyurethane; urea; melamine, furan, or vinyl ester), a thermoplastic or a rubber.
- The layered material is a phyllosilicate, an illite mineral, a layered double hydroxide or mixed metal hydroxide, $ReCl_3$ and FeOCl, TiS_2, MoS_2, MoS_3, $Ni(CN)_2$, $H_2Si_2O_5$, V_5O_{13}, etc.
- The inorganic intercalant is obtained by hydrolysing a metallic alcoholate, *viz.* $Si(OR)_4$, $Al(OR)_3$, $Ge(OR)_4$, $Si(OC_2H_5)_4$, $Si(OCH_3)_4$, etc.
- The organic intercalant may be a water-soluble polymer, a reactive organosilane, an onium salt, an amphoteric surface-active agent or a chlorine compound. After intercalation the organic intercalant is calcined or otherwise removed from the layered inorganic filler.

Clay is dispersed in alcohol and intercalated with an inorganic intercalant, which is hydrolysed at $T > 70$ °C (an organic intercalant may be added to react with the hydrolysed clay surfaces). Following intercalation, the complex can be centrifuged and dried at $T = 50\text{-}80$ °C. The interlayer spacing increased from about 0.4 to 60 nm. Next, 4 to 40 wt% of the intercalated clay (the preferred $d_{001} = 1.2$ to 3.0 nm) may be dispersed in monomer(s), and subsequently polymerised to form a CPNC.

In 1999 Polansky *et al.* described CPNCs based on an epoxy vinyl ester (Derakane that contains 30-50% of styrene) and/or unsaturated polyester. The aim of the work was to provide materials for the production of barrier films, foams, and extruded thermoset articles for diverse applications, such as infrastructures of roads and bridges, marine, transportation, electronic and business equipment, and packaging. The PNC might be reinforced with traditional fibres or mats. The claims specify that the layered inorganic material (e.g., MMT) must be intercalated with onium ions, *viz.* ammonium, phosphonium or sulfonium. Thus, for example, MMT intercalated with MT2EtOH was dispersed (2-6 wt%) by shear and sonication in a 1:1 mixture of unsaturated polyester and styrene (St). After curing with Co-naphthenate the interlayer spacing increased from $d_{001} = 1.9$ to 3.8 nm, hence intercalation. In another example the intercalated clay was dispersed in DGEBA and methacrylic acid was added (to form an epoxy vinyl ester), then St was mixed in and the CPNC was cured. During the polyester formation the interlayer spacing increased from $d_{001} = 1.5$ to 4.4 nm, but incorporating St and then curing resulted in full exfoliation. It is not clear where the difference in the degree of intercalation/exfoliation comes from. The second experiment was conducted at $T = 65\text{-}120$ °C for at least 2 h and the ingredients' viscosity at these temperatures is expected to be low. However, there is no information on the duration, temperature and ingredient viscosity for the first experiment. It is possible that all three factors contributed to the observed difference in behaviour.

During the last few years several interesting studies on CPNC with epoxy matrix have been published. For example, Lü *et al.* [2001] studied intercalation and exfoliation of organoclays (MMT-ODA or MMT-3MODA) in DGEBA epoxy. Two curing agents were used *p,p′*-di-amino-diphenyl-methane (DDM) and methyl-tetrahydrophthalic anhydride (MTHPA). The ingredients were combined either in melt mixing at 70-80 °C or solution mixing in chloroform. The organoclays were easily intercalated by epoxy oligomer to form a stable epoxy/clay hybrid. When either of the two organoclays was mixed with DGEBA the interlayer spacing increased from $d_{001} = 2.4$ to about 3.7 nm. When the mixtures were cured with

DDM in the temperature range of 80 to 200 °C, the XRD spectra for the MMT-ODA systems were featureless, indicating exfoliation, while those containing MMT-3MODA remained intercalated. Curing with MTHPA engendered exfoliation in either system.

To determine the mechanism responsible for exfoliation, gelling time and exfoliating time were measured during isothermal curing. It was found that for exfoliation to occur the rate of platelet separation should not be slower than the rate of gelation. There is a noteworthy catalytic effect of the primary ammonium ion in the MMT-ODA organoclay, already observed by Lan and Pinnavaia [1994]. Since the T_m of DDM is 89 °C it is solid during the mixing and possibly unable to diffuse into the interlamellar galleries, hence without the catalytic effect of the onium ion in MMT-3MODA the clay cannot be exfoliated. However, since MTHPA is a highly polar, low viscosity liquid it can easily diffuse into the interlamellar galleries, causing a rapid polymerisation reaction. From the FTIR analysis it has been shown that more MTHPA resided within the interlamellar galleries than in the DGEBA matrix. The authors postulated that the higher concentration of the curing agent in the galleries speeds the curing reaction there, causing the clay exfoliation. In short, it is the relative rate of polymerisation within the clay stack and outside it that determines whether the system is intercalated or exfoliated.

Butzloff *et al.* [2001] also reported the catalytic effects of MMT-ODA on DGEBA curing. The authors made a similar connection between the reaction kinetics and the degree of exfoliation. In addition, they noted that at organoclay loadings exceeding 2.5 wt% a mixture of intercalated and exfoliated dispersions was observed.

Chen and Curliss [2001] published extensive studies on the use of commercial organoclays with DGEBF and diethyl toluene diamine as a curing agent. Exfoliation was reported for MMT intercalated with primary ammonium salts, *viz.* MMT-ODA (I.30E from Nanocor) and MMT-HDA. The former organoclay was exfoliated ($d_{001} \geq 10$ nm) at concentrations up to 6 wt% (transparency ended at 3 wt%), whereas the latter was exfoliated ($d_{001} \geq 14$ nm) and transparent all the way to 7.2 wt%. The relative values of the coefficient of thermal expansion (CTE, α), tensile modulus, strength and elongation at break are shown in **Figure 172**. In spite of exfoliation the effects on the properties (at $T < T_g$) are small and mainly negative. The values of T_g as measured by DMA were also relatively unchanged, *viz.* $T_g = 154$ °C for neat epoxy and 156 °C for epoxy with 7 wt% MMT-ODA. The authors speculated that poor interaction between the matrix and dispersed clay platelets is to blame. Feng *et al.* published similar results in 2002.

Becker *et al.* [2002, 2003] also studied MMT-ODA in DGEBA and two other epoxy resins, *viz.* triglycidyl *p*-amino phenol (TGAP) and tetraglycidyl diamino diphenyl methane (TGDDM); diethyl toluene diamine (DETDA) was the curing agent. Good exfoliation was obtained. It was found that as the organoclay concentration increased, T_g decreased by 15 to 20 °C for 10 wt% loading. This decrease is most likely related to the increasing amount of ODA in the system and not to a reduced cure – DSC showed that the latter did decrease with organoclay loading. All three epoxy resins, bi-, tri- and tetra-functional, have shown the same relative improvement in stiffness (by *ca.* 20% for 10 wt% organoclay) as well as in toughness. In a later publication better MMT platelet dispersion was observed for DGEBA cured at higher temperatures. For example

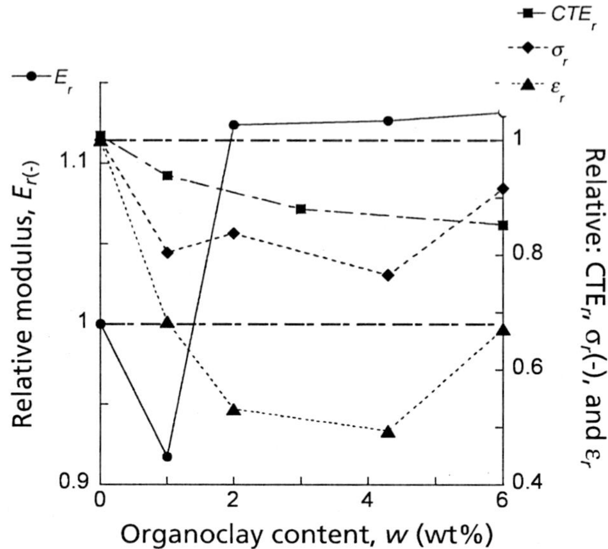

***Figure 172** Relative tensile modulus, strength, elongation at break and coefficient of thermal expansion (CTE) for MMT-ODA exfoliated in cured DGEBF. Data [Chen and Curliss, 2001].*

TGDDM cured at $T \leq 140$ °C was only intercalated, but it exfoliated when cured at 160 °C, but DGEBA was exfoliated at $T = 100$ as well as at 160 °C. However, the gains of mechanical performance were quite modest. Positron annihilation lifetime spectroscopy (PALS) showed small and erratic variation of the free volume fraction, expressed as the product of intensity and lifetime, $f \propto I_3 t^3_3$.

In a series of articles from the Karger-Kocsis laboratory CPNC with a matrix composed of unsaturated ester and epoxy (ratio 1:1) have been described [Karger-Kocsis *et al.*, 2003]. Organoclays (either Cloisite® 30B or Somasif®-ODA, -HEDI or -AGE) were used in the range of 5 to 15 wt%. After curing at 150 °C the system was phase-separated in the form of an interpenetrating polymer network (IPN) with organoclay being enrobed by the epoxy. Incorporation of organoclay reduced the T_g and modulus of the matrix resin, but the apparent fracture energy ($G_{c, app}$) doubled at about 5 wt% organoclay loading. The fracture surface of the specimens without and with organoclay showed a dramatic difference.

4.2.2 Unsaturated Polyester Resin

The earliest patent on a thermoplastic polyester/clay system dates from 1974 [Burns, 1974]. The author mechanically dispersed *ca.* 20 wt% clay in a solution of polyol in polyester. The resulting paste was added to the formulated liquid polyester in such a quantity that the clay concentration was kept below 3 wt%. The aim was to make the polyester formulation thixotropic for hand layout applications with glass fibre mat. Considering the significant effect of a small amount of clay on viscosity, significant intercalation was probably achieved.

Kornmann *et al.* [1998] investigated the possibility of producing CPNC based on MMT and unsaturated polyester (UP). Thus, Na-MMT was intercalated at

50 °C in MeOH with two different silane coupling agents, *viz.* a cationic styryl amine and a methacrylate-type, then dried. However, since the former agent did not react with Na-MMT, all the experiments were carried out with the latter. The treated MMT was dispersed in the resin (containing 42 wt% of styrene) and Co-octanoate, mixing it for 4 h at 60 °C, then 1 wt% of peroxide was added. The mixture was cured at room temperature for 3 h then post cured at 70 °C for another 3 h. XRD of Na-MMT gave d_{001} = 0.95 nm, that of PNC containing 5 or 10 wt% clay showed no peaks, hence exfoliation was achieved. The effect of MMT content on the mechanical properties was investigated. The data are presented in **Figure 173**. They seem to suggest that the optimum concentration of clay in UP is about 3.5 vol%. Notably, at 5 wt% organoclay loading the modulus increase (over the neat resin) was 32%, while the stress at break decreased by 31%.

Suh *et al.* [2000] studied the properties and formation mechanism of the UP/MMT nanocomposite. The CPNC was prepared by either simultaneous mixing of UP, St and organoclay for 3 h at 60 °C, or by sequential mixing. In the first step of this process a mixture of the UP and organoclay was prepared, and then St was added. The structure and properties of CPNC prepared using the two methods were compared. The results helped to understand the mechanism of UP/MMT nanocomposite formation, which in turn resulted in increased crosslinking density and better dispersion. In principle, UP is soluble in St. During the free radical (e.g., initiated by benzoyl peroxide, BPO) polymerisation (curing) St acts as a curing agent by bridging adjacent UP molecules. Two organoclays were used, the first was MMT-ODA the other MMT-2M2HT (Cloisite®20A). All UP/St with 5 wt% organoclay mixtures contained 0.01 wt% hydroquinone as an inhibitor to prevent reaction during mixing. The curing was carried out at 80 °C for 3 h, and then for 4 h at 120 °C. The XRD analysis determined the interlayer spacing of MMT, MMT-ODA and MMT-2M2HT as: d_{001} = 1.19, 1.84 and 2.4 nm, respectively. In the CPNC prepared by simultaneous mixing the interlayer spacing

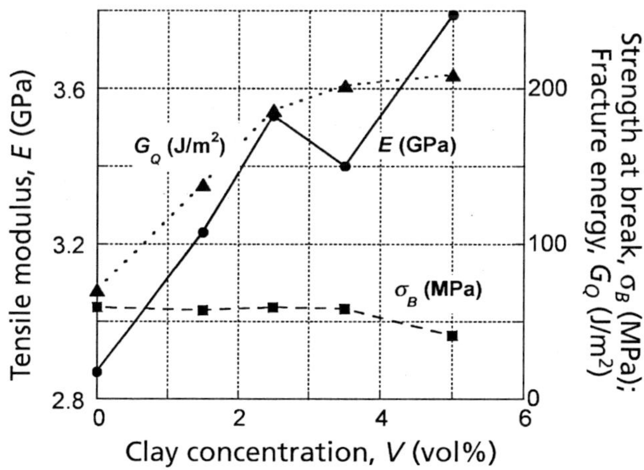

***Figure 173** Tensile modulus (E), strength at break (σ_B) and the fracture energy (G_Q) versus clay content (vol%) in unsaturated polyester-based CPNC. Data [Kornmann et al., 1998].*

increased from d_{001} = 1.84 to 3.153 nm. By contrast, the sequential method (with MMT-2M2HT) yielded CPNC with exfoliated structures.

DMA showed that the T_g of the cured CPNC (UP/MMT-ODA) was lower by more than 10 °C compared with that of the cured neat UP. Since the main factor affecting the T_g is the crosslink density, it may be concluded that the simultaneously mixed UP/MMT-ODA nanocomposite has low crosslinking density. Considering the different diffusion coefficients for UP and St it is probable that PS concentrates in the interlamellar galleries and its polymerisation within the clay stacks does not produce sufficient matrix crosslinking. To test this proposition the sequential method was used. DMTA tests showed that the same cured composition prepared by either simultaneous or sequential mixing, shows significant differences in behaviour, *viz.* the T_g of the latter is higher by *ca.* 50 °C, and the storage modulus, G′, by *ca.* 50%. Thus, to maximise the performance of CPNC one must carefully considered the mechanisms of exfoliation and the manufacturing process. Furthermore, in these multi-intercalating nanocomposite compositions, the use of compatibilisers should be considered as well.

Crosslinked polyester CPNCs were prepared by dispersing up to 10 wt% of MMT-MT2EtOH (Cloisite® 30B, C30B) in pre-promoted polyester resin (i.e., containing an accelerator for decomposition of peroxide) and subsequently crosslinking using peroxide [Bharadwaj *et al.*, 2002]. The XRD indicated that CPNC interlayer spacing was d_{001} > 2.6 nm, while TEM demonstrated the presence of randomly dispersed intercalated aggregates. As the organoclay content increased the thermal degradability increased as well, while the dynamic loss and storage moduli, and the tensile modulus decreased. A possible source of these unexpected results is a reduction of the degree of crosslinking of the polyester resin in the presence of clay. In particular, the CPNC with 2.5 wt% C30B displayed a more significant reduction of properties than the other compositions. This was attributed to the higher degree of exfoliation, which resulted in a greater decrease in the degree of crosslinking. The O_2 permeability data were fitted to the theoretical dependence (see Equation 23 in Section 2.3.10) with the aspect ratio p = 28 to 100.

Rare information on the use of MMT in polyester formulations as a filler was published by Abu-Jdayil *et al.* [2002]. The clay was ground and sieved through a 0.5 mm screen. Cured composites, containing up to 42 wt% clay, were tested in compressive mode. The properties depended on the ratio of styrene to polyester, with the best overall performance offered for about 1:3 ratio, at which the compressive strength and hardness increased by a factor of about two, as compared to the basic resin.

Nanocomposites were prepared by dispersing up to 10 wt% of Cloisite® 30B in styrene and functionalised plant oil derivative (e.g., acrylated epoxidised soybean oil, maleated acrylated epoxidised soybean oil, or soybean oil pentaerythritol maleate) [Lu *et al.*, 2004]. According to XRD and TEM a high level of clay dispersion was obtained, especially at the low clay loading of 3 wt%. The flexural modulus versus organoclay content went through a shallow maximum (30%) at *ca.* 8 wt%. There was no significant effect on flexural strength, T_g, or thermal stability.

4.2.3 Polyurethanes

For several decades polyurethanes (PU) have been mixed with nanosize particles, *viz.* carbon black (CB), metal, salts and silica particles. Three examples illustrate the approach.

Metal particles. PU was dissolved in dimethyl acetamide (DMAC) and a solution of metal ions (iron, cobalt, nickel and copper) was added [Chen *et al.*, 1996]. The metal ions were reduced by sodium borohydride under mild conditions into amorphous fine powders. Polar polymer and low metal concentration favoured smaller particles. The polymer chains prevented excessive aggregation of the metal atoms and had a protective effect on the fine metal powders. An energy disperse X-ray spectrometer connected with TEM proved that the dispersed particles were metal aggregates with sizes ranging from 10 to 150 nm.

Silica. A series of PU nanocomposites was prepared with 0-50 wt% of silica, having a particle size of about 12 nm [Havni and Petrovic, 1998]. SEM showed regular distribution and small spacing between neighbouring particles at all concentrations. For all samples XRD showed a single broad maximum at about 6° and a shoulder at 20°. The T_g of the PU increased with the silica concentration. A parallel series of PU filled with micron size silica had higher density, hardness and modulus than the nanosilica filled systems. However, the tensile strength and elongation at break were dramatically better for the systems with nanosilica. Composites with nanosilica were clear and transparent while those with micron size silica were opaque. A sol-gel process has also been used to generate nanosized silica particles within a PU matrix [Goda and Frank, 1997]. At lower concentration the silica penetrated the hard segment domains of PU, at higher it disrupted its ordered structure.

Cadmium sulfite particles (CdS). Nano-CdS particles were prepared in reverse micellisation [Hirai *et al.*, 1999]. These were dispersed into PU via surface modification with 4-hydroxythiophenol or 2-mercaptoethanol, followed by polyaddition of ethylene glycol with toluenedyl-2,4-diisocyanate. The resulting CdS-PU powder could be dissolved in DMF to make nano-CdS/PU transparent films that showed quantum-size effects. The films were used for photocatalytic generation of hydrogen.

Nanoclays. Several broad patents cited in this book state that the technology is applicable also to a PU matrix, e.g., [Usuki *et al.*, 1989; Nichols and Chou, 1999; Beal *et al.*, 1996; Serrano *et al.*, 1998; Wang and Pinnavaia, 1998].

By contrast with epoxy-based CPNC, well studied and patented by several teams of researchers, that of PU is not as well described. One of the earlier exceptions is a document by Kosiński *et al.* [1997] reporting the preparation of PU with exfoliated clay for the manufacture of fibres or films. The process started by dispersing clay in a PU solution of dimethyl acetamide (DMAC), agitating the suspension until sufficient exfoliation takes place, and forming fibres or films by spinning or casting, respectively. The solution may contain dyes and/or other additives.

The clay mentioned in this patent is a smectite, e.g., MMT, saponite, beidellite, nontronite, hectorite, stevensite, bentonite or swellable mica (CEC = 0.5 to 2 meq/g). The swellable synthetic micas such as Somasif, or tetrasilicic mica (containing a Li^+ or Na^+ ion in the interlayers), taeniolite, etc., may also be used. The clays are to be pre-intercalated with an organic onium ion to generate short stacks, and contain not more than five clay platelets, with the interlayer spacing, $d_{001} \geq 1.4$ nm. The onium ions may be selected from between a primary, secondary, tertiary or quaternary ammonium (or pyridinium, imidazolinium, phosphonium, sulfonium) ion. The PU of interest includes elastomers, segmented resins, PU-urea, etc. However, the patent focuses on segmented PU used in the manufacture of Spandex. This PU is composed of 'soft' and 'hard' segments, the former being polyether-based, e.g., derived from a poly(tetramethylene ether)glycol (PO4G, *MW* = 1750 to 2250), while the latter are derived from MDI and a mixture of 1,2-diaminoethane (EDA) and 2-methylpentamethylene diamine (MPMD), e.g., PO4G(1800):MDI:EDA/MPMD(90/10).

The concentration of the pre-intercalated clay in PU should be within 0.1 to 12 wt% on the solid mass, resulting in a significant reduction of tackiness in PU fibres and films as well as in increased dyeability. Dispersion of organoclay in a PU solution should increase d_{001} (in dry film) to more than 3.5 nm, and cause at least a partial exfoliation. To achieve this, the dispersion is agitated in a high shear mixer, a static mixer, mills, gear pumps, etc. Following agitation, the clay is dispersed into individual platelets or their doublets. Conventional agents (e.g., dyes, antioxidants, thermal stabilisers, UV stabilisers, pigments, lubricating agents, anti-tack agents, and additives to enhance resistance to chlorine degradation) can be added as long as these do not engender negative effects in the PU or the organoclay. The CPNC dispersion may then be formed into fibres or film.

A dye may be introduced to a suspension containing PU and organoclay, or it may constitute the clay's intercalant or co-intercalant, ionically associated with the clay. Thus, for example, a dispersion of 3 wt% Na-MMT in deionised water was heated to 70 °C, and then combined with an aqueous solution of dehydroabietyl ammonium and Nile Blue dye. After vigorous stirring the flocculated clay dispersion was filtered, washed and dried at 110 °C under vacuum. The organoclay-dye pigment had d_{001} = 2.2 nm and it contained about 5.2 wt% of Nile Blue dye attached to the clay. Numerous examples well illustrate the advantage of the invented process. For example, solution cast and dried films retained good elastomeric properties while the tensile, yield modulus and modulus at 10% elongation improved significantly. Addition of clay reduced the tack and its increase over time, extending the shelf stability. The CPNC films showed much greater dyeability in comparison to PU formulations without clay, *viz.* 25 times deeper blue dye, and 2-3 deeper yellow absorptions. The dye-fastness was also improved. Since addition of organoclay may cause yellowing, small amounts of pre-dyed clay added to the composition largely offsets this effect.

Kuwabara *et al.* [1998] used clay to thicken a polyol stock solution during the preparation of rigid PU foams. In this process the foaming agent is CO_2 generated in a reaction between water and isocyanate. However, this kind of foam has poor dimensional stability in comparison to that prepared with a Freon – the CO_2 diffuses through the cell walls causing the foam to shrink. When the water content in the polyol was reduced the resulting foam had high density and reduced flame retardancy. Hence, a flame retarding agent (e.g., Sb_2O_3) and a filler had to be added. Since the filler tends to sediment, in turn a thickener had to be used. This caused the viscosity of the polyol composition to increase. To bypass this problem, the temperature had to be increased, which provided only a few hours of workable suspension. CPNC technology was explored to solve this series of problems.

The thickening effect of solvent-swollen clay added to polyol was found to be proportional to concentration. The high viscosity of the thixotropic system stabilised a dispersion of the flame retardant, but on shearing the suspension viscosity dramatically decreased, making it easy for transfer and foaming. The resulting hard PU foam showed stable low density and good flame retardancy.

The composition of the stock solution comprised 10-90 wt% polyol, 0.1-40 wt% of H_2O, 0.1-10 wt% of a flame retardant, and 0.1-20% of a solvent-swollen clay. The stock solution had steady shear viscosity η = 1 to 10 Pas reduced to 0.001-0.5 Pas when sheared. Polyols of phthalic acid ester, ethylenediamine, tolylene-diamine, sugar amine, and polyether type, as well as mannitol or sorbitol could be used. The flame retardants were, e.g., solids: Sb_2O_3, $Al(OH)_3$, CB, or melamine; as well as liquids: DOP and tris-β-chloropropyl phosphate or triethyl phosphate (TEP). The indicated clays are the usual, *viz.* MMT, halloysite, mica, beidellite, etc. These clays can be swollen in, e.g., diethylene glycol and dipropylene glycol, water, or TEP. To obtain good quality hard PU foam, catalysts, surfactants, etc., may also be added.

The foam was produced in a conventional process, e.g., by reacting the stock solution with a polyisocyanate in a 1:1 volume ratio. Thus, stock solutions were prepared using an organophilic synthetic (LiMgNa)-silicate smectite (Lucentite, an oleophilic smectite from Coop Chem. Co.), swollen in a solution containing a polyol mixture (e.g., ethylene diamine, a sugar amine, tolylene diamine type diethylene glycol); water (2 wt%); amine, dibutyl-tin alkyl maleate catalysts (5 wt% total); TEP and Sb_2O_3. After the initial dispersion at high shear rates the suspension was allowed to stand for 15 days at RT, no precipitation of Sb_2O_3 was observed. The compositions containing TEP and 10-15 wt% of clay displayed high thixotropy, offering an easy transfer and foamability. Good dimensional stability of foam and fire resistance were obtained. The industrially produced hard PU foam was uniform and had improved properties, particularly in flame retardancy. Apparently the authors did not measure the clay interlayer spacing, thus it is not know what exfoliation was achieved. However, on the basis of the earlier cited work on the use of clay as a thickener it is probable that the system was exfoliated.

Zilg *et al.* [1999b], prepared PU-based CPNC by intercalation of synthetic FM (Somasif ME 100; aspect ratio of p = 10-20!) with methyl-dodecyl bis(2-hydroxyethyl) ammonium chloride (MDD2EtOH), then dispersing the organoclay in trihydroxy terminated oligopropyleno-oxide (at 100 °C in a high shear mixer) with *N,N*-dimethyl benzyl amine and (after cooling to room temperature) reacting with diisocyanato-diphenylmethane. After degassing, the mixture was placed in a mould and cured at 80 °C. Mouldings showed multiple XRD peaks with *d*-spacings of 2.7, 4.4 and 8.8 nm – evidently uniform exfoliation was not obtained. However, exfoliated platelets were present alongside short stacks. The CPNCs were reinforced by the platelets bonded to the PU matrix. Upon incorporation of 2.5 to 10 wt% of the organoclay the tensile modulus decreased (in comparison to neat PU) by about 20%, but the strength and elongation at break increased, by 60-240% and by 130-400%, respectively.

One of the general patents granted to Pinnavaia and Lan [2000a] is applicable to PU, but the examples given are for epoxy systems. The method of CPNC preparation is original since to start with it involves transformation of Na-MMT into its acidic form: H^+-MMT. Three methods: acidification with HCl for 6 h, ion exchange and pyrolysis of NH_4-MMT, all yielded material with identical properties and the interlayer spacing reduced from 1.32 to 1.05 nm. The second step is intercalation of the acidified clay with a basic compound. For the epoxy this role is well played by amine curing agents. For example, Jeffamine increased d_{001} to a respectable spacing of 4.5 nm. In other words, the method of CPNC formation hinges on the presence of an ingredient that has a Lewis base character.

While amines are preferred, oxygenated compounds can also be used to intercalate the protonated clay. For example, glycerol can be intercalated to proton-exchanged clay. The intercalation of the protonated clay can be conducted in a solvent (if inert toward the reactions) or in a solvent-free process. Particularly suitable solvents are water or water-ethanol, water-acetone and, similar polar co-solvent systems. Upon removal of the solvent, intercalated particulate concentrates are obtained. In the solvent-free process, a high shear blender is usually required to conduct the intercalation reaction.

The authors noted that many polymerising components contain basic groups. The invention is generalised and applied to thermoset systems, such as epoxy, polyurethane, polyurea, polysiloxane and alkyds, where polymer curing involves coupling or crosslinking reactions. The same principle applies to thermoplastic polymers. The thermoplastic polymer CPNC can be prepared from protonated clay intercalated by a monomeric reagent. Alternatively, the concentrate may be combined with molten polymer to form the CPNC. The thermoplastics benefiting from this technology include PU, PA, PEST, POM, PPE, PO, PSF, PCL, PI, PLA, etc.

Polyurethanes, PU, are prepared by reaction of polyol and isocyanate:

$$\text{HO-R-OH} + \text{OCN-R}'\text{-NCO} \rightarrow \text{-(-CO-NH-R}'\text{-NH-CO-O-R-O-)}_n\text{-}$$

The polyol component R usually contains a basic centre such as a secondary or tertiary amine. Therefore, they may react with the protons in the gallery of the proton-exchanged clay to intercalate it, forming a concentrate. These are diluted with appropriate amounts of polyols and isocyanate to form reaction-cured PU-based PNC.

Wang and Pinnavaia [1998b] prepared thermoset PU-based CPNC by curing a PU network via a more traditional route, namely using MMT intercalated with DDA or ODA from Nanocor. These organoclays were solvated by several polyols commonly used in PU synthesis, *viz.* ethylene glycol, polyethylene glycol, polypropylene glycol, and glycerol propoxylates. The solvation occurred at RT, but it may be accelerated at higher temperatures, *viz.* T = 50 °C for 12 h. The mixture of polyol with ≤ 20 wt% organoclay is easily pourable. The chain length of the onium ion determines the interlayer spacing. As shown in **Table 35**, the observed d_{001} values agree with the calculated ones (d_{calc}). The clay platelets are 0.96 nm thick. Initially, in the non-solvated clay the alkyls of onium ion are oriented parallel to the platelets (judging from d_{001} = 2.22 and 2.30 nm), and then they reorient to optimise solvation by the polyol. The following procedure was used for the preparation of CPNC: (1) polyol was mixed with organoclay and degassed under vacuum at 100 °C for 6 h; (2) the mixture was combined with methylene diphenyl diisocyanate prepolymer (MDDP; MW = 1050; functionality = 2.0) and the suspension was stirred at 70 °C for 15 min before degassing at 95 °C in a vacuum oven. The mixtures remained pourable at ≤ 10 wt% of organoclay. The bubble-free mixture was poured into a mould for curing at 95 °C for 10 h under N_2. The alkylammonium exchange ions of the organoclay were considered to react with isocyanate and were counted as contributors to the stoichiometry for polymerisation, by reducing the amount of polyol in proportion to the cation exchange capacity of the clay. No catalyst was used.

In CPNC cured at 95 °C the d_{001} spacing increased with time from 3.74 to 5.08 nm, thus the clay platelets were intercalated. However, considering the relatively large interlayer spacing, matrix reinforcement has been expected. In **Figure 51** the tensile modulus, strength, and strain-at-break versus organoclay concentration are plotted versus the MMT-ODA content for PU prepared from

V230-238 and MDDP. Clearly, incorporation of nanoclay simultaneously improved the tensile strength, stiffness and toughness in comparison with neat PU. Even at high organoclay content, where the clay forms tactoids, it does strengthen, stiffen and toughen the matrix. At a loading of 10 wt% the tensile strength increased by about 110%, the modulus by 120%, and the elongation at break by 150%. This notable reinforcement was accomplished without a loss of high optical transparency.

Chen *et al.* [1999b] prepared PU-based CPNC by using polycaprolactone-based nanocomposite as an intermediary. Thus, first Na-MMT was treated in an aqueous solution of 12-amino-lauric acid (ADA), washed, dried, ground, and screened through a 325-mesh sieve. Next, the organoclay was dispersed in ε-caprolactone, which was subsequently polymerised at 170 °C in 3 h under stirring. PU was synthesised in two steps: (1) mixing 4,4′-diphenyl-methane diisocyanate (MDI), polycaprolactone diol and DMF for 2 h at 70 °C then reacting it with 1,4-butanediol for 30 min. (2) The PCL-based PNC was added and the whole mixture reacted for 30 min. The prepolymer solution was spread, dried under vacuum and cured at 80 °C in 10 h. XRD showed that in the CPNC the clay was exfoliated. In the system obtained segmented PU formed the matrix, which was chemically bonded to the randomly dispersed PCL-based CPNC. SEC (or GPC of old) indicated that the M_n of PCL was only 3,620 g/mol, and that for PU M_n decreased with PCL-based CPNC content, from M_n = 52 to 13 kg/mol. As a consequence, the mechanical properties versus clay content plot show the presence of local maxima (see **Figure 52**). For example, the tensile strength increases linearly with the clay content, but the elongation at break for the PU-based CPNC goes through a sharp maximum at 0.74 wt% clay. In the lap tests it is the modulus that increases linearly with clay content while the strength goes through a local maximum at 1.3 wt% of clay.

In the following publication the same authors explored another method of PU-based CPNC preparation [Chen *et al.*, 2000]. First, Na-MMT (CEC = 0.76 meq/g) was complexed (3 h stirring in aqueous suspension at 60 °C) with either ADA or benzidine (BZD, H_2N-ϕ-ϕ-NH_2), forming the complexes: MMT-ADA and MMT-BZD. These were washed, dried under vacuum at 80 °C for 12 h, then ground and screened. Next, MDI and polytetramethylene glycol (PTMEG) at a molar ratio of 2:1 were dissolved in DMF, and heated to 90 °C for 2.5 h to form a prepolymer. Then, 1,4-butanediol was added with rapid mixing at 90 °C for 10 min. Different amounts of organoclay (1, 3, 5 wt%) dispersed in DMF were added and the reaction completed after 3 h mixing at RT. The final concentration of PU in DMF was 30 wt%. The organoclays containing MMT-ADA and MMT-BZD showed similar interlayer spacing, d_{001} = 1.7 and 1.54 nm, respectively. The difference was attributed to the intercalant size. The PU-based CPNCs prepared with 1, 3 and 5 wt% of MMT-ADA, as well as with 1 and 3 wt% MMT-BZD were reported to be exfoliated. This conclusion was partially supported by TEM micrographs – randomly placed, about 1 nm thick layers of organoclay were observed. However, CPNC with 5 wt% MMT-BZD had a broad peak indicating intercalated stacks with d_{001} = 2.47 nm.

The molecular weights of PU in neat resin and in the CPNCs were quite similar: $M_n \cong 11$ kg/mol, for all the seven samples: $-58 \le T_g \le -59$ °C, and the water absorption varied randomly from 1.3 to 1.7 wt%. During the TGA analysis the CPNC with MMT-ADA started to degrade faster than neat PU. This may be due to the presence of extra ADA, unattached either to clay or to PU. However, at higher temperatures

(T > 350 °C) the PU/clay showed better thermal resistance than PU. The stability of the CPNC with MMT-BZD was better than that with ADA. The mechanical properties of the new PU-based CPNC are shown in **Figure 54**. In spite of full exfoliation, addition of MMT-ADA only slightly improved the tensile strength and elongation at break over that of neat PU. By contrast, the partially exfoliated MMT-BZD enhanced the tensile strength by a factor of 2.0 and the elongation at break by a factor of 2.8. The authors suggested that the large difference in efficiency of these two organoclays is related to the interactions between the intercalating agent, clay and the PU matrix. ADA has one terminal -NH_2 group that transformed into HOOC-$(CH_2)_{11}$-NH_3^+ ammonium ion, formed a complex with MMT^- and still was able to react with -NCO to form urea. BZD has two terminal -NH_2 groups that can participate in these interactions/reactions. Therefore, ADA can be part of linear chains while BZD of crosslinked ones. Schematics of these interactions/reactions are reproduced in **Figure 55**. Thus, the authors stipulate that it is not the presence of clay platelets as such, but rather the increased crosslink density that is responsible for the enhancement of the mechanical properties in **Figure 54**. However, if this supposition is true then R_2 in the schematic must be rather long and elastic – the CPNC is not becoming more brittle as the organoclay content increases. The other possible source of the observed disparity of structure/properties relation between CPNC with ADA and BZD is the thermal instability of the former.

Hu *et al.* [2001] prepared CPNC with PU as the matrix by dispersing MMT-3MHDA in polyether (MW = 1 to 3 kg/mol). Next TDI dissolved in DMF was added into the suspension, and then diglycol + glycerin were added and vigorously mixed at 80 °C. Finally, the mixture was poured into a mould and cured at 100 °C in 24 h. XRD was used to determine the interlayer spacing: upon pre-intercalation d_{001} increased from 1.26 nm (in MMT) to 1.96 nm. Intercalation with polyether increased d_{001} to 3.15 nm and curing a bit further – the latter spacing varied with the organoclay content from 4.85 to 3.99 nm at 10 to 25 wt% loading, respectively. The behaviour of these intercalates is unknown.

In 2002 Chang and An reported on the role of organoclay in CPNC/PU on the thermomechanical properties, morphology, and gas permeability. Three organoclays were used, *viz.* MMT-HDA, MMT-3MDDA, and MMT-2MHTL8 (Cloisite® 25A, C25A). These were dispersed in DMAC, and then a PU solution was added. After vigorous stirring at room temperature for 1 h the suspension was cast into films 10-15 μm thick. All cast films were optically transparent. In all fractured CPNCs SEM showed voids and deformed regions that probably originated in agglomeration of clay particles (the surface of cast PU was smooth). As shown in **Table 93**, TEM of these specimens indicated intercalation.

TGA indicated that the initial degradation temperature (at a 2% weight loss) increased with organoclay loading, but the effect was relatively small, with a maximum of 18 °C for 4 wt% MMT-3MDDA or 6 wt% C25A. The tensile tests indicated that the ultimate strength increased with organoclay content only up to 3 wt% loading, *viz.* from 45 MPa for neat PU to 59, 57 and 76 MPa for MMT-HDA, MMT-3MDDA, and C25A, respectively. A similar increase of the initial modulus was observed, e.g., at 3 wt% C25A loading the modulus increased from 7.24 GPa for neat PU to 13.55 GPa, i.e., by a factor of 1.87. The elongation at break for CPNC films with MMT-HDA and MMT-3MDDA was lower than that of neat PU, whereas that for CPNC with C25A was slightly higher.

One may try to rationalise this behaviour on the basis of the interlayer spacing data listed in **Table 93**, but since all three intercalants are alkyl-type, there is

Table 93 Interlayer thickness (d_{001}) and O_2 permeability (P) for PU-type CPNC. Data [Chang and An, 2002]

Conc. (wt%) Organoclay	d_{0001} (nm)			P (L/m²/day)		
	MMT-HDA	MMT-3MDDA	C25A	MMT-HDA	MMT-3MDDA	C25A
0	–	–	–	7.214	7.214	7.214
1	Exfoliated	2.712	Exfoliated	–	–	–
2	Exfoliated	2.712	3.260	5.447	6.953	5.817
3	3.122	2.712	3.260	5.074	6.381	5.622
4	3.122	2.712	3.260	3.587	6.136	4.725
6	3.122	2.712	3.260	–	–	–
100	2.596	1.685	1.963	–	–	–

little difference in the interactions between the PU matrix and the organoclays. The difference might be expected because of the ability of polar groups of the PU macromolecule to interact with either primary or quaternary ammonium ions. Indeed, CPNC with MMT-HDA seems to be better exfoliated than the two others. Some justification for this argument can also be found in the O_2 permeability data listed in the table. However, since the fractured CPNC specimens showed the presence of voids, the basis for any statement on mechanical properties or permeability are on shaky ground. It is remarkable that in spite of the puzzling porosity, incorporation of 3 wt% of organoclay resulted in such a significant improvement of tensile and gas-barrier properties.

CPNC with a PU matrix was prepared using Na-MMT [Yao *et al.*, 2002]. Thus, different amounts of the clay were mixed with modified polyether polyol (MPP) at 50 °C for 72 h. The suspension was blended with a known amount of modified 4,4′-diphenylmethylate diisocyanate (M-MDI) for 30 s at 20 °C. The mixture was cured at 78 °C for 7 days, yielding an elastic film. As a result of intercalation by MPP, the interlayer spacing of Na-MMT, d_{001} = 1.1 nm, increased to 1.6 nm at 21.5 wt% clay loading.

In spite of the small interlayer spacing, the tensile properties increased with the clay loading (see **Figure 174**). Thus, at 10 wt% of clay the tensile strength and strain-at-break increased by 10 and 12%, respectively. By contrast, the storage modulus below T_g increased by more than 350%. Addition of clay strongly reduced the heat capacity and slightly increased thermal conductivity.

Kim *et al.* [2003] developed clay-reinforced emulsions of PU following a multi-stage process:

1. Dispersing 1/2 of the organoclay (Cloisite® 25A; C25A = MMT-2MHTL8, with $d_{001} \approx 2$ nm) in a soft segment polyol, either polytetramethylene glycol (PTMG, MW = 1 kg/mol) or polybutylene adipate (PBA, MW = 1 kg/mol). Agitation for 5 h at 70 °C exfoliated the organoclay.

2. The exfoliated suspension of C25A in polyol was reacted with isophorone diisocyanate (IPDI) and dimethylol propionic acid (DMPA) in the presence of dibutyltin dilaurate (DBT catalyst) for 1 h at 80 °C, and then with 1,4-butanediol (BD chain extender) for 1 h at 80 °C. This NCO-terminated prepolymer (*MW* = 5 kg/mol), was then mixed at 80 °C for 3 h with the remaining organoclay.
3. After cooling to RT the resulting NCO-terminated hard segment prepolymer/organoclay mixture, triethylamine (TEA) was fed into the reactor and agitated for 1 h to neutralise DMPA in PU macromolecules.
4. Next, using a pump at constant flow rate, water at 30 °C was fed to the mixture for 10 min. An aqueous solution of the crosslinking agent, triethylene tetramine (TETA), was incorporated and the reaction continued at 50 °C for 1 h. The resulting PU emulsion contained 30 wt% solids, with an average particle size (for PU with PTMG) that increased with organoclay content from *ca.* 39 nm in neat PU to *ca.* 108 nm at 5 wt% organoclay; the increase in PU with PBA was smaller, *viz.* from about 60 to 76 nm, respectively.
5. Emulsion cast films were dried at 60 °C for 24 h, and then at 30 °C under vacuum for 2 days. TEM indicated good dispersion of clay platelets.

The tensile properties of the CPNC with a PU matrix based on PTMG are shown in **Figure 175**. These were measured within the temperature range between the T_gs of the soft and hard segment domains. As a result secant modulus, tensile strength, elongation at break, and Shore A hardness, are presented in the figure in relative (to the neat PU) terms. As expected, the modulus and hardness increase with organoclay content. At 5 wt% loading these parameters increased by a factor of *ca.* 3.6 and 1.15, respectively, while the elongation at break decreased by *ca.* 35%. The tensile strength versus composition dependence goes through a

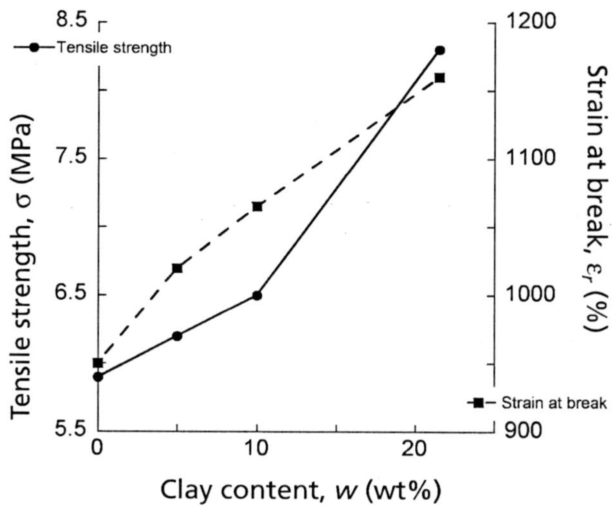

Figure 174 *Tensile properties of PU CPNC with Na-MMT. Data [Yao et al., 2002].*

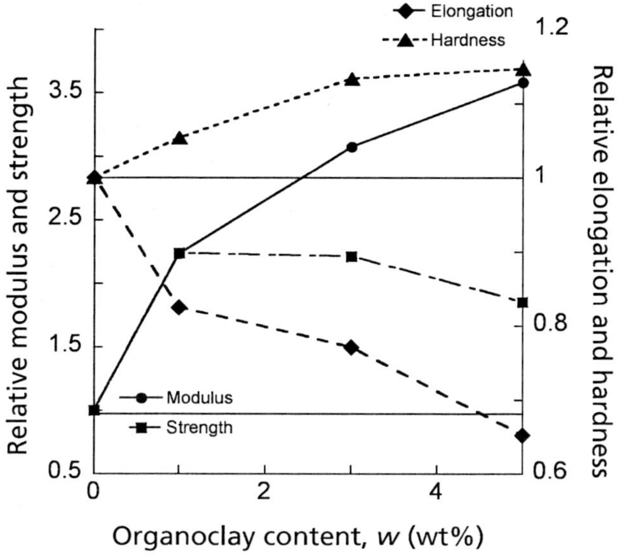

Figure 175 Relative (to neat PU) performance characteristics for CPNCs of PTMG-type with C25A – see text. Data [Kim et al., 2003].

local maximum at about 2 wt% organoclay loading – the expected increase at this concentration would be by a factor of *ca.* 2.4. During the CPNC preparation the authors attempted to disperse the organoclay in both soft and hard segment domains, but it is possible that the second portion of the organoclay incorporated during the second step of the process migrated to the soft segment domains. The local maximum on the strength versus concentration curve may indicate that above *ca.* 2 wt% of organoclay the platelets within the polyol domains are overcrowded which reduces the clay-matrix interactions.

The thermal stability of these CPNCs was measured by TGA. For systems with PTMG containing either 3 or 5 wt% organoclay the thermogram of neat PU was shifted up-scale by *ca.* 30 °C for CPNC. Similarly, the presence of clay reduced the rate of water absorption by the PU matrix. The greatest effect seems to be at low loading, *viz.* 1 to 3 wt% organoclay, where coincidently the light transmittance is virtually unaffected, indicating good dispersion of clay platelets.

4.2.4 Other CPNC with Thermoset Matrix

Besides epoxy, unsaturated polyester and polyurethane resins discussed in Sections 4.2.1 and 4.2.2, another category of thermoset polymers, polyimides, are considered in Section 4.1.10. Many more polymer types belong to the thermosetting category, such as melamine-formaldehyde, phenol-formaldehyde, urea-formaldehyde, allyl polyesters, bismaleimides, some silicones, hard rubbers, etc. However, for the latter resins CPNC technology is at an early stage.

4.3 Elastomeric CPNC

The first patent on 'elastomer reinforced with a modified clay' was deposited in 1947 [Carter *et al.*, 1950]. The bentonite was to be pre-intercalated with an onium ion, *viz.* ammonium, pyridinium, phosphonium, arsonium, stibonium, oxonium, sulfonium, selenonium, stannonium, or iodonium, but in particular with an ammonium ion, *viz.* triethanol ammonium $(HO\text{-}C_2H_4)_3NH^+ Cl^-$, or *n*-butyl ammonium chloride. Onium ions with alkyl, aromatic or heterocyclic radicals; primary, secondary, tertiary or quaternary type, etc., are also named in the patent. The process comprises two steps: (1) preparation of organoclay suspension, and (2) combining the suspension with elastomer latex to produce reinforced latex films. Alternatively, the reinforced latex may be precipitated, washed and formed by standard methods, with standard additives to reinforce, crosslink, stabilise or to facilitate the processing.

Numerous examples are given with bentonite or hectorite intercalated using a variety of onium ions: DDA, triphenyl dodecyl phosphonium (3PDDP), aniline, rosin-amine, melamine, lauryl pyridinium, etc. In some cases mixed organoclays were used. The amount of organoclay per 100 weight parts of elastomer ranged from 24 to 43. The properties of interest were tensile and hardness. For example, the performance of a CPNC (with 36 phr organoclay) and that of neat resin were measured. The improvements were significant, *viz.* the tensile strength increased by a factor of 3.7, elongation by 1.5, modulus by 2.0 and Shore hardness by a factor of 1.2.

One of the general patents from Toyota described the preparation of CPNC with a matrix resin selected from between a thermoplastic polymer, a thermosetting resin or a rubber [Usuki *et al.*, 1989]. Judging by the examples and claims, it seems that the latter polymer was the main objective. In particular, rubbers reinforced with an organoclay and carbon black (CB) were developed for use in the transport and building industries, as well as for thickening of paints or greases. Owing to poor bond strength between the inorganic particles and the organic matrix the earlier CPNCs were brittle. Furthermore, the amount of clay that could be incorporated was limited. Thus, the main goal of this new patent was to produce CPNC with superior mechanical characteristics, heat resistance, as well as water and chemical resistance. The CPNC comprised an elastomer and clay uniformly dispersed in it with CEC = 0.5 to 2.0 meq/g. The customary processing additives could also be added.

The process followed three steps: (1) intercalation of clay to obtain d_{001} > 3 nm, (2) dispersing the organoclay with a monomer and/or oligomer in a mortar or vibration mill, and (3) either polymerisation of the monomer and/or oligomer incorporated in the second step, or kneading the mixture with rubber. The latter polymer could be a natural, synthetic, or a thermoplastic elastomer, e.g., 1,2-polybutadiene, SBS, SIS, or their blends, but preferably a diene rubber co-vulcanised with a liquid rubber. It is desirable that the onium intercalant used in the first step has a functional group, which tethers the macromolecules, formed in step three, to the clay surface. The polymerisation may be carried out in bulk, in suspension or in solution using, *viz.* water, CS_2, CCl_4, glycerin, toluene, aniline, benzene, chloroform, *N,N*-DMF, DMSO, etc. The polymerisation may be radical, cationic, anionic, coordination, or polycondensation. It is desirable that the onium salt becomes a constituting unit of a polymer. Thus, polymerisation takes place in the interlayer space, expanding the interlayer distance and causing the clay to be bound to the matrix and to uniformly disperse in it.

To illustrate the invention several examples were provided. For example, liquid polybutadiene (PBD; MW = 3.4 kg/mol; acrylonitrile content: AN = 16.5%) was dispersed in DMSO and water. The mixture was acidified with HCl and Na-MMT was added. The reaction product was filtered and dried under vacuum. The MMT-PBD complex was characterised by pulse NMR. Two spin-spin relaxation times (T_2) were observed, the short T_{2S} = 10 μs for the glassy state and the long T_{2L} > 1 ms for the rubber. The latter component originates in the part of the system where the molecular mobility is restricted by bonding between MMT and PBD - about 20% of the rubber molecules was restrained by the MMT surface. XRD indicated that the layers of MMT were exfoliated and uniformly dispersed in the PBD matrix, with d_{001} > 8.8 nm.

Next, the complex was hammer-milled at liquid N_2 temperature into particles d < 3 mm. The powder was mixed with NBR (33% AN), sulfur (vulcaniser), dibenzothiazyl disulfide (vulcanisation accelerator), ZnO (vulcanisation auxiliary), and stearic acid. To obtain the final composition, the CPNC mixture (containing 5-10 phr MMT) was kneaded in a roll mill at 50 °C. The specimens were vulcanised at 160 °C. At about 25% clay loading the performance characteristics of CPNC were clearly superior to those obtained for a mechanical mixture of MMT with the same rubber formulation, *viz.* storage modulus at 25 and 100 °C increased by a factor of 1.96 and 2.04, respectively, the stress at 100% deformation increased by a factor of 1.75, swelling in benzene reduced by a factor of 1.19. The glass transition temperature remained unchanged at –13 °C, but the Mooney viscosity at 100 °C increased by a factor of 2.14.

This Toyota technology was pursued in the following series of patents more specifically addressing preparation of the elastomeric CPNCs, but with wide implications for other matrices as well. The first of these [Usuki *et al.*, 1999, 2000a, 2000b] provides a method for dispersing clay in a low polarity polymer. The method is similar to that used for the preparation of CPNC with a PO matrix, *viz.* PP (see Section 4.1.21). Thus, the new CPNC comprises:

a) Layered mineral or synthetic clay, CEC = 0.5 to 2.0 meq/g; e.g., MMT, FH, ...

b) Onium intercalant, *viz.* ammonium, sulfonium or phosphonium intercalant having at least six C-atoms, e.g., ODA, 2M2OA, 3OA, 2M2ODA, 1-hexenyl ammonium, 9-dodecenyl ammonium, etc.

c) An organic guest molecule (MW = 0.1 to 100 kg/mol) having a polar (hydroxyl, halogen, carboxyl, anhydrous carboxylic acid, thiol, epoxy or amino) group that bonds to clay and expands the interlamellar gallery, and possibly an unsaturated group to co-vulcanise with the matrix polymer.

d) Matrix polymer, miscible with the guest organic molecule.

The polymer may be a PO (e.g., PE, PP, PS, PIB, PU, SBR, acrylic resin) or one selected from between the rubbers: natural, isoprene, chloroprene, styrene, nitrile, ethylene-propylene, ethylene-propylene-diene, butadiene, styrene-butadiene, butyl, epichlorohydrin, acrylic, urethane, fluoro-, and silicone rubber. The CPNCs are prepared by kneading and crosslinking the components. The most important aspect of the invention is the use of 2 types of clay intercalants, the first being a traditional onium ion and the second that bonds to the clay surface via hydrogen bonding and chemically or physically bonding to the matrix polymer. Thus, the organic guest molecule is a compatibiliser between the two immiscible phases: the organoclay and the matrix polymer.

The following example illustrates the invention. Na-MMT (1 wt%; CEC = 1.2 meq/g) was dispersed in water at 80 °C, and then the equivalent amount of 2M2ODA was added. The precipitate was washed and dried. The organoclay, containing 54.2 wt% MMT, had d_{001} = 3.28 nm. Next, the guest molecule (Polytail-H, a PBD with –OH end groups, MW = 3 kg/mol) was dissolved in toluene and the organoclay was added. The suspension was mixed for 6 hours. After solvent evaporation the complex showed d_{001} = 3.87 nm, indicating that Polytail H diffused into organoclay galleries. The interlayer spacing was found to increase linearly with the weight ratio, R, of Polytail-H-to-organoclay content: $d_{001} \cong 3.28 + 0.524R$. For $R \geq 10$ full exfoliation was obtained. In another example, the interlamellar gallery was expanded by addition of either Polytail-H or its homologues with either –COOH, epoxy or –Cl groups. The following rating was obtained: –COOH > -OH > epoxy ≅ –Cl. Other 'guest molecules' were also used, notably PP-MA (Yumex1010; MW = 30 kg/mol; T_m = 145 °C), which in later documents became the main compatibiliser for CPNC with PP as the matrix.

To illustrate another aspect of the invention MMT-2M2ODA was dispersed in toluene with stearic acid and liquid butyl rubber then mixed for 6 hours. After evaporation of toluene d_{001} = 5.48 nm was determined, showing that adding stearic acid and liquid butyl rubber can expand the interlamellar gallery height. Next, the clay complex was kneaded with butyl rubber, CB, ZnO, sulfur and vulcanisation accelerator. The CPNC was vulcanised for 40 minutes at 150 °C. The product showed superior mechanical and barrier properties.

The advantage of the Toyota two-step process may be contrasted with the results of a single-step process [Akelah *et al.*, 1995]. The authors used amine-terminated butadiene-acrylonitrile rubber (18%AN, M_w = 7.34, M_n = 1.09 kg/mol) to intercalate Na-MMT in a slightly acidified aqueous dioxane solution. The reaction with stirring continued overnight, and then the product was filtered and dried. The amount of the grafted rubber was *ca.* 60 wt%. The clay formed clusters with an average size of 60 nm and an interlayer spacing of about 1.52 nm.

Exxon patented the use of organoclay to reduce air permeability in tyre liners and inner tubes [Kresge and Lohse, 1995]. The tyre inner-tube composition was to be made of two parts solid rubber (MW > 10 kg/mol) and a complex comprising a reactive (cationic) rubber (MW > 0.45 kg/mol) and 1 to 45 phr of a layered silicate uniformly dispersed with d_{001} > 1.2 nm. The solid rubber was either natural or synthetic (e.g., PBD), a thermoplastic elastomer or a blend, reinforced with CB. The reactive rubber may be selected from between: PBD, CPI, EPR, EPDM, butadiene copolymer with St, AN, isoprene, etc. The amine-terminated rubber (e.g., NBR) was preferred. The reactive rubber should have good miscibility with the solid rubber or able to co-crosslink with the solid rubber, characterised by T_g

≤ -25 °C. Suitable layered silicates comprise smectite clays (i.e., MMT), vermiculite and halloysite.

By controlling the size, spacing, and orientation of the clay platelets a rubber composition could be obtained with reduced air permeability and without a significant decrease in the flexibility, important in the use of rubber in tyres. For good performance, the clay platelets should be between 0.7 to 1.2 nm thick, with the interlayer spacing between them $d_{001} > 1.2$ nm, and they should be aligned perpendicular to the flux direction. The orientation might be imposed by extruding, stretching or shearing the material before crosslinking. The following example illustrates the process. First, MMT was purified by centrifugation of its aqueous slurry, and then ion exchanged. Next, the acidified clay was contacted in a Waring blender with a toluene solution of an amine-terminated NBR (Hycar 1300; $M_n \cong 1.3$ kg/mol). After removal of solvents, the light brown, transparent material contained 25 wt% of intercalated clay with $d_{001} \cong 1.4$ nm and $T_g \cong -37$ °C, thus 4 °C higher than neat rubber. The oxygen permeability through the clear, flexible CPNC compression moulded film was about 22 times smaller than that through neat resin. Furthermore, it was reported that deposition of a 0.38 mm (15 mil) layer of this film to the inner surface of a passenger car tyre reduced the tyre loss of inflation pressure from 1.5 to less than 0.2 psi/month.

Exxon also found it useful to incorporate nanoclay into an elastomeric material containing asphalt, designed for use as automotive undercoating, noise absorbing and carpet backing compositions [Eidt *et al.*, 1997]. The CPNC comprises: (1) an elastomer (e.g., SBR, M_n = 30 to 600 kg/mol) with T_g < 25 °C, (2) 0.5 to 15 wt% of onium ion modified clay (e.g., MMT, saponite, beidellite, hectorite, vermiculite, etc.), and (3) 0.1 to 15 wt% of sulfonated and neutralised asphalt. As an example, the elastomeric composition containing 9 wt% of sulfonated asphalt and 10 wt% of Bentone-34 (MMT-2M2TA, d_{001} = 2.47 nm, from Rheox), showed significant improvement of performance, e.g., modulus increased by a factor of 2.7.

Another invention from Exxon considers the preparation of CPNC with improved mechanical properties and reduced air permeability by the method of emulsion polymerisation of a thermoplastic elastomer in the presence of organoclay [Elspass *et al.*, 1999]. Since the method has been discussed in Section 4.1.4.1, there is no need to repeat.

Elspass and Peiffer [2000] developed elastomeric CPNC with low air permeability for tyre inner liners. The material comprised two non-ionic, miscible elastomers and dispersed organoclay. The novelty of this patent is the use of melt compounding for the preparation of the CPNC with 1.0 to 25 wt% of a customary organoclay, 45 to 98.5 wt% of a melt processable first polymer (natural rubber, CPI, PBD, SBR, copolymers of isobutylene and isoprene, particularly an isobutylene bromo isoprene copolymer) having *MW* > 50 kg/mole, and 0.5 to 30 wt% of a second polymer of lower *MW*, e.g., M_n = 0.5 to 50 kg/mol. The preferred second polymers are polymers based on the same monomers as those of the first one. The CPNC may be used for pneumatic tyres with liners, for inner tubes, bladders and for other substantially air impermeable membranes. To achieve the desirable dispersion, compounding is carried out in an internal mixer, an extruder, or a calendering unit. However, full exfoliation is not desired – the document specifies that mixing results in short stacks of clay platelets, with an average thickness of 5.0 to 140 nm. The preference is to have about 80 vol% of clay in stacks thicker than about 10 nm.

The following example illustrates the process. MMT pre-intercalated with a dialkyl-ammonium and a low-*MW* PBD (M_n = 5 kg/mol) were manually mixed together, then kept at 70 °C for 36 h without agitation. The paste was then mixed with *cis*-1,4-oligobutadiene (M_n = 45 g/mol) in a Brabender at 100 rpm for 10 minutes at 130 °C, and then a crosslinking package consisting of ZnO, stearic acid, sulfur and an accelerator was added and the composition mixed for an additional 3 min. The average thickness of the clay stack was 11 nm with 99.7 vol% of the clay forming stacks thicker than 10 nm. Compression moulded films were prepared at 165 °C from the CPNC as well as from the base resin. These were tested for oxygen transmission and uniaxial tensile properties. The results are presented in **Figure 176.** Incorporation of 9 wt% of clay increased the modulus by a factor of 6.4 and the tensile strength by a factor of 1.9, but at a cost of reduced tensile strain by a factor of 2.7. The principal goal of this invention was the reduction of permeability – as the data show, the oxygen permeability was reduced by a modest factor of 3.73. One has to wonder whether the professed desire to prevent exfoliation of the organoclay has something to do with the achieved low barrier properties. Having stacks instead of individual platelets makes the orientation to the film surface so much more difficult.

CPNC with fluoroelastomers, to be used in electrostatographic printing machines, were of interest to Xerox Corp. [Badesha *et al.*, 1998]. The developed CPNC comprised two components: 5 to 20 wt% of organoclay and a fluoroelastomer. The clay, with the aspect ratio p = 50 to 1000, was either MMT, bentonite, or hectorite, while the fluoroelastomer was a copolymer of vinylidene-fluoride, hexafluoropropylene, and tetrafluoroethylene. The CPNC was prepared by shear-mixing the ingredients for 10 to 30 min at T = 49 to 93 °C, which

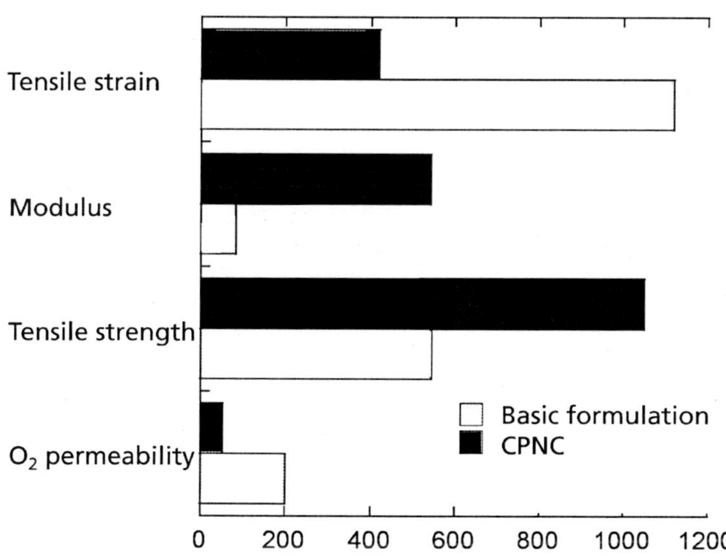

***Figure 176** Oxygen permeability and uniaxial tensile properties of PBD film and its homologue containing 9 wt% clay. Data [Elspass and Peiffer, 2000].*

resulted in clay intercalation or exfoliation. The intercalants mentioned in the documents are long chain alkyl ammonium salts or amino acids such as 2M2ODA, or ADA. The organoclay used in the examples was SCPX-984, i.e., MMT-2M2ODA.

The motivation for developing these CPNCs was the need to enhance the thermal conductivity, stiffness, thermal stability and swell resistance. The fluoroelastomers of interest are known commercially, *viz.* Viton, Fluorel, Aflas or Tecnoflon. To the formulation the customary additives, e.g., crosslinking or colouring agents, processing aids, conductive fillers (thermal and electrical), polymerisation initiators and accelerators could be added. For example, 10 pph of SCPX-984 was mixed with 100 pph of Fluorel FC 2145 at 27, 49 or 66 °C for 15 minutes in a two-roll rubber mill with a tight nip. XRD of the resulting three specimens indicated that the first one had the same interlayer spacing as the organoclay, d_{001} = 2.88 nm. For the second, d_{001} = 3.3 nm was determined, thus fluoroelastomer managed to diffuse into the interlamellar galleries. The third specimen prepared at the highest temperature of T = 66 °C, showed no diffraction peaks hence was judged to be fully exfoliated.

These examples demonstrate that to achieve exfoliation one must supply sufficient chain mobility to the intercalating polymer segments. For Fluorel FC 2145 the critical temperature needed to provide sufficient thermal mobility seems to be somewhere between 49 and 66 °C. The fluoroelastomer is poly(vinylidene fluoride-*co*-hexafluoropropylene) or PVDF-HFP, whose T_g depends on composition, e.g., for 11 to 41 mol% HFP the glass transition occurs at: T_g = –32.6 to –5.3 °C [Bonardelli *et al.*, 1986]. However, if the emulsion polymerisation results in a certain degree of block formation, then the HFP block may have a significantly higher T_g than the random copolymer. It is of note that the extrapolated T_g of polyhexafluoropropylene is T_g = 152 to 167 °C. The presence of short blocks may act as physical crosslinks and free molecular motion is only possible once these last glassy segments disintegrate – in the discussed case at $T \geq T_g + 60$.

During the last few years several articles have been published on the preparation of CPNC with an elastomeric matrix. It is worth mentioning that the 'general approach' to the preparation of CPNC by Ishida *et al.* [2000] provides several examples of successful exfoliation of 10 wt% of MMT-ADA + 2 wt% of Epon 825 in elastomeric matrices, *viz.* PB, PIB, CPI, and CR, by simple 30 min mixing at 75 °C, followed by overnight heating under vacuum at 140 °C.

Wang *et al.* [2000] compared two methods (latex and solution) for CPNC preparation with three elastomeric matrices. The latex method required dispersing clay in water, then adding latex, coagulating the system, washing and drying. The solution method required dispersing an organoclay in toluene, adding a toluene solution of an elastomer, followed by solvent evaporation. Significantly, the former, less expensive method resulted in intercalated CPNC (d_{001} = 1.46 nm) with better performance parameters, *viz.* tear strength, hardness, tensile strength and elongation at break.

To prepare CPNC a rubber was melt compounded in an internal mixer at $T \leq 60$ °C with either Na-MMT or an organoclay from SCP [Vu *et al.*, 2001]. To improve the chances for exfoliation before use the organoclay was pre-swollen in toluene. The compounded formulation also comprised other standard ingredients: sulfur, stearic acid, ZnO, Ca-stearate, and vulcanisation accelerators, either mercaptobenzothiazole (MBT) or *N*-cyclohexyl-2-benzothiazole sulfonamide (CBS). The elastomer was either a synthetic *cis*-1,4-polyisoprene

(NR) or epoxidised natural rubber having 25 or 50 mol% epoxide, ENR25 and ENR50, respectively. The organoclays were: MMT-MT2EtOH, MMT-2M2HT and MMT-2MHTL8, or Cloisite® -30B, -15A and -25A, respectively. XRD indicated intercalation of 10 phr of Na-MMT by ENR25 and ENR50, *viz.* d_{001} = 3.17 and 3.27 nm, respectively. At the same concentration level, exfoliation was obtained for C30B in ENR50 – for all other systems XRD indicated the presence of intercalated short stacks. The mechanical properties of these CPNCs were found to depend on the extent of the dispersion of clay platelets and clay loading.

Chang *et al.* [2002] and Usuki *et al.* [2002] published studies on similar CPNCs of MMT-ODA dispersed in EPDM. The former publication presented results for three EPDM systems (each at 5, 10 and 20 phr clay content): (1) with Na-MMT, (2) with MMT-ODA, and (3) with MMT-ODA and low molecular weight (liquid) EPDM. None of these systems showed exfoliation, but increased interlayer spacing was observed for systems (2) and (3), with stacks 50-80 nm thick. As expected, the mechanical (modulus, strength and strain at break) and gas barrier properties depended on the degree of dispersion and clay content. Thus, the worst performance was for system (1) and the best (but still quite modest) for system (3).

Usuki *et al.* adopted a different strategy. EPDM was melt compounded with 7 phr of MMT-ODA in a TSE (D = 30 mm; L/D = 45.5) at 200 °C. To the extrudate standard vulcanisation ingredients were added (with the exception of accelerator) and the formulation was compounded using mixing rolls, then vulcanised in a hot press at 160 °C. Five vulcanisation accelerators were used – it was found that the degree of dispersion very much depends on their selection. While standard MBT or CBS resulted in intercalation, the use of tetramethyl thiuram sulfide (TS) or zinc-dimethyl dithio carbamate (PZ) resulted in a high degree of exfoliation. The performance of these two systems was also superior – in comparison to EPDM the CPNC formulation with PZ showed significant improvements: the tensile strength by a factor of 2.0; elongation at break by 1.9; modulus by 1.9 and N_2 permeability reduced by a factor of 1.4. The explanation offered by the authors is based on the decomposition of TS or PZ into organoradicals that graft themselves to the unsaturations in EPDM, making it more polar, thus able to interact with the clay surface.

Excellent performance was reported for vulcanised PDMS/fluorohectorite (FH) systems [LeBaron and Pinnavaia, 2001]. However, to achieve good performance PDMS molecules had to be hydroxyl-terminated and the FH pre-intercalated with 3MHDA. Mixing these two ingredients for 12 h resulted in intercalation – the interlamellar gallery height was found to depend on the molecular weight of PDMS (MW = 0.4 to 4.2 kg/mol). Exfoliation was achieved by curing the system by means of tetraethyl ortho silicate (TEOS) and Sn-(diethyl hexanoate) catalyst. At 10 wt% organoclay loading the mechanical properties significantly improved: tensile strength by a factor of 5.8, modulus by 1.9 and elongation at break by a factor of 4.7. At 5 wt% the cyclohexane uptake went down by 58% while at 8 wt% the oxygen permeability was reduced by 25%. This modest reduction was related to random orientation of the FH platelets and their low aspect ratio.

Schön *et al.* [2002] prepared CPNC with SBR as the matrix and three commercial organoclays: from Südchemie FH with phenol(4-nonyl, 2 methyl(bi-ethyl) amine) (FHO) and MMT with 2M2ODA, and from Nanocor I.42E (MMT-2M2ODA). The degree of clay dispersion and performance increased in this order (from FHO to I.42E).

Natural rubber (NR) was melt compounded with several fillers, sulfur cured and its performance evaluated [Varghese and Karger-Kocsis, 2004)]. Five fillers were used: precipitated non-layered silica (used as reference), two pristine clays (synthetic sodium fluorohectorite (FH), and natural sodium bentonite (MMT)), as well as two organoclays [MMT-ODA and MMT-MT2EtOH]. Compounding was done on a two-roll mill at room temperature. The specimens were cured at 150 °C. It was observed that addition of organoclay reduced scorch time and accelerated curing. The degree of filler dispersion was determined using XRD and TEM. Silica, being amorphous did not show any diffraction peak. The two clays, FH and MMT, showed one strong peak with d_{001} = 0.94 and 1.24 nm, respectively. In the case of MMT-ODA and MMT-MT2EtOH one peak ($d_{001} \cong$ 1.3 nm) and a shoulder ($d_{001} \cong 3.0$ nm) were obtained. After compounding, the spacing in NR/organoclay specimens was found to be smaller than that observed for neat organoclays. This effect the authors explained by the formation of zinc coordination complexes with amines extracted from the interlamellar galleries. The DMTA, tensile properties, and hardness of specimens containing 10 phr of filler, were determined. In comparison to silica-filled specimens, incorporation of clay or organoclay increased the performance parameters in the following order: MMT, FH, MMT-ODA and MMT-MT2EtOH. A partial intercalation of the organoclays by NR was probably the source of high tear strength.

It is of note that the highest value of the elongation at break and tensile strength was obtained for specimens containing neat FH. For several applications, where black rubber has to be covered or painted over, the possibility of replacing CB by organoclay offers a simple solution [Arroyo *et al.*, 2003; López-Manchado *et al.*, 2003]. Performance of NR with 10 and 40 CB was compared with that containing 10 phr of either Na-MMT or MMT-ODA. The formulation was mixed on a two-roll mill, and then vulcanised at 150 °C. Incorporation of Na-MMT into NR did not change the spacing – d_{001} = 1.26 nm remained after curing. However, the interlayer spacing of MMT-ODA was transformed from d_{001} = 1.76 nm into a high degree of dispersion. Both CB and organoclay accelerated the vulcanisation, but the effects were quite different. The time for 97% cure, t_{97}, at 150 °C for NR, NR + 10 phr CB, NR + 40 phr CB and NR + 10 phr organoclay was, respectively: 19.97, 16.64, 14.66 and 4.09 min. The crosslinking density was determined by swelling in toluene and *n*-heptane. The relative (to neat NR) number of crosslinks for the four formulations above were determined as: 1.0, 1.16, 1.34 and 1.74, respectively. Thus, the presence of organoclay drastically increased the rate of cure and increased the degree of crosslinking.

The mechanical performance was expressed in terms of the modulus at different strains (100, 300 and 500%), the strength and elongation at break. The data are displayed in **Figure 177**. NR with 10 phr organoclay shows the best performance. For example, in comparison to vulcanised NR tensile strength increased by 353%. The abrasion resistance of the NR+10 phr organoclay is slightly lower than that of NR or NR with 10 phr of CB, but better than that with 40% CB. The organoclay improved the strength of the NR without reduction of the elongation at break – both NR and NR + 10 phr organoclay showed elongation > 700%. The same properties were also measured for NR with 10 phr of Na-MMT. The filler showed rather negative effects, *viz.* it reduced the 100% modulus, strength, elongation at break, and hardness. Poor interactions between the mineral and the matrix had to be expected.

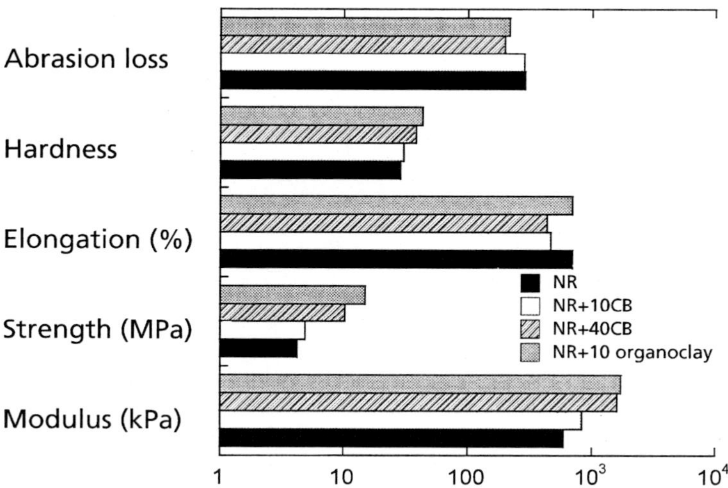

Figure 177 Mechanical properties of natural rubber (NR), NR with 10 phr carbon black (CB), NR with 40 phr of CB, and NR with 10 phr of MMT-ODA. Data [Arroyo et al., 2003].

Part 5

Performance

Clay-Containing Polymeric Nanocomposites

5 Performance

As discussed in the text, addition of clay to a polymer reduces the free volume content (see Section 3.1). The melt flow of PNCs shows increased viscosity, significant augmentation of yield stress and possible exfoliation during the testing. The flow behaviour can be understood by considering the concentration and the aspect ratio of the dispersed clay platelets. The rheological tests offer information regarding the structure and interaction in PNC, hence they constitute an important aid for optimisation of CPNC formulation. Most properties of CPNCs have been discussed in the preceding parts along with the methods used to prepare the systems. Furthermore, the fundamentals of the enhancement of mechanical properties were presented in Section 3.5. In the following parts, a few examples of the mechanical behaviour, flammability and permeability of CPNCs will be discussed.

5.1 Mechanical Properties

It has been reported that the impact strength of crystalline polymers is reduced by incorporation of nanoparticles. For example, the addition of 2.2 wt% of organoclay to PA-6 reduced the impact strength by a factor of 4.3. However, the problem is also related to the crystallinity. By proper selection of process variables, the same impact strength of PNC as that of the matrix resin may be obtained [Graff, 1999]. For example, modification of PA-6 behaviour by addition of MMT is illustrated in **Table 65** and **Table 94**.

The mechanical behaviour of some CPNCs based on PP is also encouraging (e.g., see **Table 79**). The tensile modulus increased nearly 8-fold with 1 wt% of the nanofiller. The tensile strength also increased, but at least a 10% loading was required to achieve a significant result, and at that level the Izod impact strength was reduced. However, even at a 30% loading, the Izod impact strength was 28 J/m, comparable to that of PP (33 J/m). The addition of chain-grafted filler particles retarded the rate of crack growth. At 10% loading, the time to failure increased 4.5 times. The new PO-based PNCs are used as moulding resins. They replace PE or PP when decreased crack growth rates are desired.

5.2 Flame Retardancy of CPNC

The flammability of polymers is of growing concern and consumption of fire retardants is growing. For example, in the USA between 1975 and 1990 it grew from 100 to 320 kton/y [Boryniec and Przygocki, 1999]. This market-driven tendency reinforces the commercial attractiveness of nanocomposite technology – the unexpected improvement of the matrix flame resistance by the presence of dispersed clay platelets in most resins is becoming one of the major advantages of this new technology. A recent review of polymer stabilisation methods concluded that CPNCs have significant advantages over traditional flame retardant systems [Beyer 2001].

Table 94 Mechanical properties of PA-6 and PNC based on it. Data from [Ube Industries, Ltd., 2002]

Property	ASTM	Units	PA-6	PNC
Tensile strength, σ	D-638	kg/cm^2	800	910
Tensile elongation, ε_b	D-638	%	100	75
Flexural strength, σ_f	D790	kg/cm^2	1100	1390
Flexural modulus, E_f	D-790	kg/cm^2	28,500	35,900
Impact strength, *NIRT*	D-256	kg cm/cm	6.5	5
HTD (18.56 kg/cm^2)	D-648	°C	75	140
HTD (4.6 kg/cm^2)	D-648	°C	180	197
H_2O permeability, P_{H2O}	JIS Z208	G/m^2 24 h	203	106
Density, ρ	D-792	kg/m^3	1140	1150

In 1965, Blumstein reported that the presence of MMT increased the thermal decomposition temperature of PMMA by 40-50 °C. Thirty years later Burnside and Giannelis published similar results for PDMS and polyimide [1995]. Since 1997 several publications from NIST described the dispersed organoclays as 'revolutionary new flame retardants' [Gilman *et al.*, 1997; Gilman *et al.*, 1998 a,b]. For example, the flammability of PA-6-based nanocomposites was compared with that of neat PA-6 and other flame retarded PAs. Cone calorimetry data showed that for CPNC comprising 5 wt% clay the peak heat release rate (PHRR) was reduced by 63%, thus better than caused by incorporation of 4 wt% of phosphorus as co-monomer triphenyl phosphine oxide (PHRR reduced by 58%). Significantly, several negative effects (e.g., 16-fold higher CO generation, 7-fold higher soot generation, costs, density, etc.) accompany the use of phosphorus.

The clay platelets enhance formation of char and reinforce it. Furthermore, by contrast with the customary flame-retardants that reduce resin performance, in CPNC the other physical properties are improved by the incorporation of clay.

To explain the reduction in the flammability of polymer nanocomposites Nyden and Gilman [1997] used molecular dynamics simulations (MD). The computations started with a Hamiltonian derived from the consistent valence force field (CVFF):

$$H = \sum_{i}^{3N}\left(p_i^2 / 2m_i\right) + \sum_{ij}^{n-bonds} V_b\left(r_{ij}\right) + \sum_{ijk}^{n-angles} V_a\left(\theta_{ijk}\right) + \sum_{ijkl}^{n-torsions} V_t\left(\phi_{ijkl}\right) + \sum_{ij}^{n-pairs} V_{nb}\left(r_{ij}\right) \quad (184)$$

In this relation the sums represent, respectively from left to right, the kinetic energy of atoms, the potential energy for bond stretching (V_b), bending (V_a), torsion (V_b), and the non-bond potential energy (V_{nb}), resulting from the interactions between pairs of atoms separated from each other by at least two atoms. The computations were carried out using a MD-REACT computer program developed earlier by Nyden and his colleagues. The program accounts for a series of reactions present during thermal degradation, *viz.* chain scission, new bond formation from free radical fragments, depolymerisation, hydrogen transfer, chain stripping, cyclisation, and crosslinking. For the simulation a nanocomposite model of PP with graphite was used. The system element consisted of four PP chains (DP = 48) and a graphite sheet made of 600 C-atoms and 80 H-atoms at the edges. The computations were carried out at 600 °C, considering that the $=CH-CH_3$ bond scission is followed by β-scission of the backbone, resulting in fragments with either vinyl end-groups or free radicals, which in turn unzip the monomer.

The rates of mass loss as a function of time for neat PP and for its macromolecules in between the graphite sheets were computed for graphite sheet separation varying from 2.5 to 5 nm. The effect of PP-macromolecule interactions with the graphite was separated from that of nanoconfinement by comparing the results obtained from simulations when the nonbonding interactions between the polymer and the graphite were either included or excluded. The results indicate that in either case the average rate of mass loss in the nanocomposite goes through a local minimum at a separation of about 3 nm (see **Figure 178**). Thus, there is significant stabilisation effect for this specific gallery height, Δd_{001} = 3.0 nm. It is noteworthy that the analysis indicates that for intercalated systems with Δd_{001} < 2.8 nm the interaction with graphite sheets accelerates the degradation – only for larger gallery heights are the beneficial effect of nanocomposite to be expected.

There are several problems with translating these results into the actual degradability of CPNC. The first one is related to the possibility of a catalytic effect of clay – such effects on polymerisation have been observed by numerous researchers. The

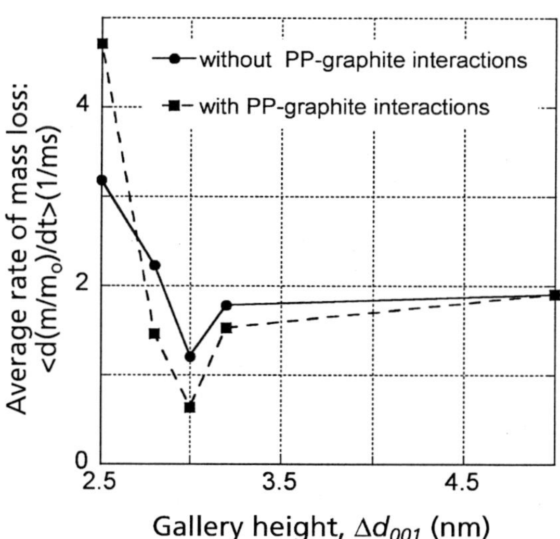

***Figure 178** Computed thermal decomposition rates at 600 °C for PP/graphite nanocomposites as a function of the distance between graphite sheets. The rates were computed assuming either a presence or an absence of non-bonded interactions between PP and the graphite. Data [Nyden and Gilman, 1997].*

second one originates from the required gallery height of Δd_{001} = 3.0 nm. Evidently, every CPNC manufacturer strives for exfoliation, whereas these computations seem to indicate that the effect is negligibly small. However, as the authors pointed out, during degradation of CPNC, the structure and the distance between the initially exfoliated clay platelets is reduced, possibly to the stabilisation level.

Gilman published an excellent review on these topics summarising decades of observations of the beneficial effects of clays on the thermal stability of diverse polymers (e.g., PMMA, PDMS, PEI, PA-6, PS, PP, etc.) [Gilman, 1999]. The author stressed the similarity of the combustion process for different polymers and different stabilisers – small differences could be accounted for by the presence of different intercalants modifying the reaction kinetics. The best results were observed for exfoliated systems, but intercalation also provides significant reduction of flammability. The microscopic analysis of CPNC char shows a sandwich structure with carbon-clay layers of similar thickness (*ca.* 1 nm). This structure acting as an insulator and mass transport barrier may be responsible for the reduction of flammability in CPNC.

The mechanism of flammability reduction in CPNC with PP + PP-MA or with PS has been studied using a cone calorimeter [Gilman *et al.*, 2000a]. Three organoclays were used: FH intercalated with tetradecyl ammonium (FH-TDA), and MMT with either ODA or 2M2ODA. Three PS CPNC specimens were prepared using solvent intercalation, static melt intercalation and a melt-mixing (in an internal mixer) intercalation method. Both the nanocomposites and the residual char were studied by XRD and TEM. The cone calorimeter data showed that all CPNCs (matrix: PA-6, PS, or PP-*g*-MA) with MMT show reduction of PHRR by 50 to 75%. While the clay should be well dispersed, it does not need to be fully exfoliated. However, FH in the intercalated PS/FH systems was ineffective for reducing PS flammability. The authors speculate that this may be caused by its geometry, but the chemical reactivity may also be a factor. The results demonstrated the existence of a common mechanism of flammability reduction. However, the type of clay, the degree of dispersion, and the method of CPNC manufacture influence the effectiveness. The flame retarding effect of clays originates in the char formation on the burning surface, which insulates the underlying material and slows the mass loss rate of decomposition products.

More extensive studies on CPNC flammability are available in the NIST report [Gilman *et al.*, 2000b]. In 1998 an industrial consortium was formed to study the flammability of CPNC. The initial goal was to compare the flammability properties of: (1) intercalated versus exfoliated nanocomposites, (2) tethered versus non-tethered nanocomposites, (3) nanocomposites with different clays, (4) nanocomposites crosslinked to different degrees, (5) nanocomposites with different melt viscosities, (6) nanocomposites with different silicate loading levels, and (7) nanocomposites containing PPE charring-resin.

The preliminary findings are summarised in **Table 95**. They confirm and reinforce the previous conclusions regarding the formation of clay-reinforced carbonaceous char during combustion of CPNCs. The char is responsible for the reduced mass loss rates and hence the significantly lower PHRR. The reductions of the peak HRR are significant, but the ignition time, t_{ign}, is not necessarily longer. Significantly, reduction of the peak HRR has also been observed for systems with matrix resin that normally does not produce char (PS, PP-*g*-MA, PA-6 and EVAc). This char is responsible for the reduced mass loss rates, and hence the lower flammability. The virtual absence of flammability depression by well-

Table 95 Summary of the consortium findings. Data [Gilman *et al.*, 2000b]

Result/critical experiment	EVAc	PA-6	PS	Epoxy	PP
Reduced peak HRR (%)	70	80	80	0-20	70
Shorter ignition time, t_{ign}	no	yes	no	no	no
High initial HRR or MLR	yes	no	yes	yes	yes
Carbonaceous char formed	yes	yes	yes	no	yes
Intercalated *vs.* delaminated	nm	equivalent	moderate	nm	nm
Effect of molecular weight	nm	nm	moderate	nm	nm
Effect of clay loading (% at max)	5	10	5	tbd	5

Notes: MLR = mass loss rate; nm = not measured

dispersed MMT in an epoxy matrix, as well as the lack of effect when char-enhancing PPE was added to CPNC with PA-6 as matrix is puzzling. The study showed that intercalated CPNCs performed as well as exfoliated ones. The effect of tethering, if it exists, is negligible. Similarly, the effect of CPNC melt viscosity could not been detected. One of the more important findings was that the effectiveness of dispersed clay for the reduction of flammability reaches a maximum at about 5 wt% clay loading.

To clarify the role of clay on the flammability reduction, exfoliated PA-6 based CPNCs (2 and 5 wt% clay) were prepared [Kashiwagi *et al.*, 2004]. Formation of floccules of blackened residues during the flammability tests was video-recorded. The floccules migrate to the burning surface, forming a protective crust. Analysis of the residue showed that up to 80 wt% of their mass consisted of clay and the rest of thermally stable graphitic components. The initially well dispersed (in CPNC) platelets, formed stacks in the residue with $d_{001} \cong 1.3\text{-}1.4$ nm. Evidently, during the burning/gasifying process the polymer has been pyrolysed from the clay surface and the bare clay platelets, pushed by rising bubbles toward the surface, were concentrated to form aggregates of collapsed stacks. The aggregates formed individual floccules hardly protecting the burning sample. Consequently, as reported by Hu *et al.* [2003], an addition of traditional flame retardant, e.g., decabromodiphenyl oxide and/or antimony oxide, may be needed to pass the UL-94 test.

Rossetti [1999/2000] studied the flammability of three types of CPNC with PP, PU and epoxy as the matrix. Different organoclays were used – based either on FH (Somasif ME100) or MMT (Cloisite® series). The flammability reduction was observed for all systems provided that the organoclay level was high enough, *viz.* 10 wt%. The effect for PP and PU could be expected from the observations at NIST cited above. However, Rossetti reported a significant reduction of MLR for the epoxy (EP). The latter systems were prepared by dispersing modified-FH in DGEBA crosslinked with an anhydride. The series of organoclays was prepared using primary ammonium ions, *viz.* $C_nH_{2n+1}NH_3^+ Cl^-$, where n = 4, 8, 12 or 18. FH with butyl ammonium did not change the flammability of EP, but the three other organoclays reduced the MLR from 1.5 to about 0.9 to 1.0%/s, thus by about 40%. This reduction was accompanied by reduction of the maximum temperature (from 133 to 108 °C).

Vaia *et al.* [1999] examined the use of PA-6-based CPNCs as ablative materials. The authors reported that during ablation of CPNC with 2 wt% of clay, an inorganic char formed, reducing the mass loss by at least one order of magnitude (relative to PA-6). Significantly, the presence of exfoliated clay platelets did not alter the inherent thermal decomposition kinetics and the total generated heat. As proposed by Gilman, the effect was related to the insulating layer of char formed on the ablated surface and as such is to be expected to be general for CPNC with exfoliated clay platelets.

The ablative layer and the role played by O_2 in thermal degradation in CPNC with a PA matrix were examined [Dąbrowski *et al.*, 2000]. In accord with the previous observations the authors reported that during thermal degradation of CPNC a protective ablative layer is formed that slows down the fuel diffusion. A large difference between the degradation rates of PA-6 and PNC was reported, especially at higher temperature and lower conversion rates.

During the last few years Wilkie and his collaborators have published a series of articles on the preparation of CPNC and on their thermal stability, flammability and performance. Most of this work was discussed in Sections 3.2 and 4.1, *viz.* [Zhu *et al.*, 2001, 2002; Du *et al.*, 2002; Wang *et al.*, 2001; Wang and Wilkie 2002]. For example, Zhu *et al.* [2001] prepared three CPNCs with PS as the matrix starting with MMT pre-intercalated with synthesised onium ions as shown in **Figure 158**. The cone calorimetry measurements provide a host of characteristic parameters from which the peak heat release rate (PHRR) is probably the most important. In **Figure 179** PHRR is plotted versus organoclay content in PS nanocomposites. Of the three intercalants it is not the phosphonium, P16, but the non-tethering OH16 that provided the best suppression of HRR.

Recently a new and general method of CPNC preparation for a variety of polymers was described [Su and Wilkie, 2003a,b]. The two cited papers provide details on the manufacture and properties of CPNC with PS, HIPS, ABS, PMMA, PP or PE as the matrix. The preparation of CPNC follows several steps, for example: (1) Free radical copolymerisation of vinyl benzyl chloride with either styrene (COPS) or methyl methacrylate (MAPS) ($MW \cong 5 \pm 1$ kg/mol). (2) Conversion of the copolymer into ammonium salt by reacting it with *N,N*-dimethyl hexadecyl amine. (3) Ion-exchange of Na-MMT (in an aqueous solution of THF) with the macromolecular ammonium salt. Considering the high molecular weight of the intercalant, the organic content in the organoclay produced is *ca.* 80%. (4) While in principle CPNC may be prepared using these organoclays in the reactive or melt compounding process, the authors selected melt compounding in an internal mixer. Compositions containing 0, 2.5, 5, 15 and 25 wt% organoclay were prepared with the six polymers listed above.

Evidently, since these organoclays are miscible with some neat polymers but not with others the 12 types of CPNC produced must show different degrees of clay dispersion, phase morphology and performance. For example, COPS should be miscible with PS, and MAPS with PMMA. However, according to XRD these

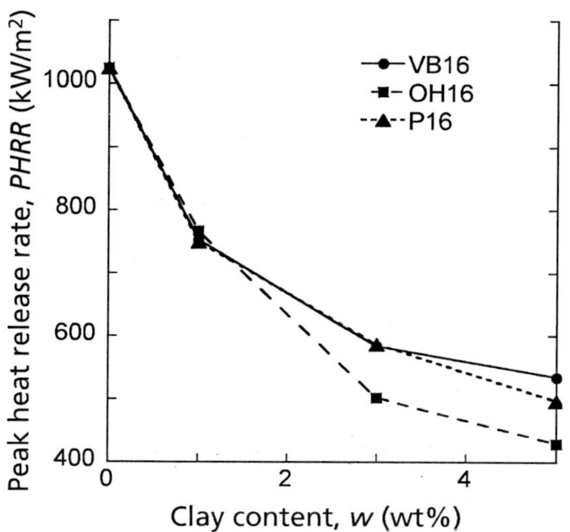

Figure 179 *Peak heat release rate for CPNC with PS as the matrix. Three organoclays were used (see* ***Figure 158*** *and text). Data [Zhu et al., 2001a].*

systems are only intercalated. By contrast, the two organoclays must be immiscible with PO, but XRD shows a surprising lack of diffraction peaks. The situation can be reconciled considering that clay resides in a minor, amorphous phase of the system, being largely unaffected by the matrix. TEM at low magnification seems to show lamellar morphology with submicron thickness. The weak XRD peaks observed for these organoclays in CPNC may be reduced in intensity not by exfoliation, but by dilution. These CPNCs, especially at an organoclay content $w \geq 5$ wt%, may be treated as non-compatibilised blends. Accordingly, one should not expect a significant improvement of mechanical properties, which indeed was observed [Su and Wilkie, 2003b]. However, the new organoclays showed better thermal stability than that of standard ones. TGA also shows an increase of the thermal decomposition temperature. The PHRR versus organoclay content for the CPNC with PP and PE as a matrix is shown in **Figure 180**. Considering the large scatter of data the overall suppression of the HRR can be approximated by a straight line: $<PHRR> = 2022 - 30w$ with respectable value for the correlation coefficient squared: $r^2 = 0.99$.

Other publications from the same laboratory describe the thermal stability and flame retardancy (by TGA and Cone Calorimetry, respectively) of CPNC with PE [Zhang and Wilkie, 2003] or PP as the matrix [Wang and Wilkie, 2003]. These intercalated nanocomposites have also been prepared by melt blending in an internal mixer. The PO-organoclay systems showed a mixed immiscible-intercalated structure that improved when PO-MA was present. The results from cone calorimetry suggest that a high degree of dispersion has occurred, since there is a significant PHRR reduction (by 30 to 40%).

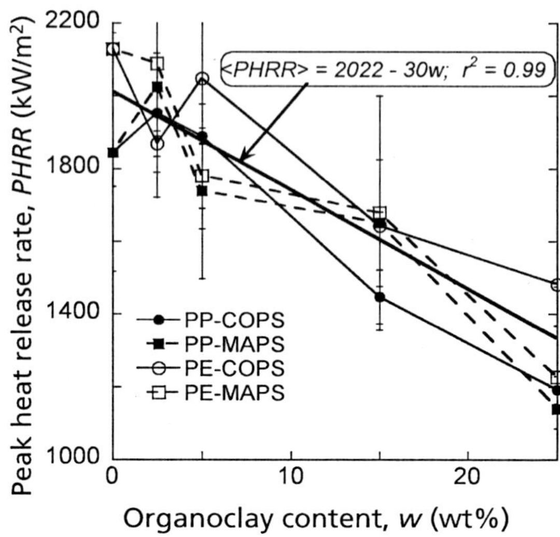

Figure 180 *Peak heat release rate for CPNC with either PP or PE as the matrix and two organoclays (see text). Error bars as well as a mean-average response are shown. Data [Su and Wilkie, 2003b].*

5.3 Permeability Control

Along with mechanical properties and flame retardancy the barrier properties constitute one of the principal advantages of CPNCs, especially for packaging applications. On the preceding pages the improvements in the gas barrier properties have been frequently mentioned, e.g., see **Table 3** listing commercially important CPNC. Simultaneous improvement of mechanical, thermal, flame resistance and barrier properties of CPNC with PA or PARA; PET, PC or LCP; PO, PS, PMMA, PVC, PI, PU, rubbers or thermoplastic elastomers, epoxy resins and water-soluble and/or biodegradable polymers (e.g., PVAl, PLA, PCL) have been discussed.

A simple theory for the barrier properties through a multiphase polymer has been presented in Section 2.3.10. The model is based on the concept of an increased pathway for the diffusion of penetrant molecules through a two-phase polymer film (also known as the tortuosity model). According to the model the barrier properties are determined by three factors: the inherent permeability through neat polymers, the aspect ratio (in the flux direction) of the dispersed phase, and its volume fraction. Accordingly, as far as CPNC is concerned, the barrier property depends on the degree of clay platelet dispersion, their aspect ratio (p), orthogonal orientation to the flux direction and concentration (ϕ).

Another model, developed 80 years ago by Fricke [1924] to describe the electrical conductivity of a suspension of homogeneous spheroids may be adopted to describe the barrier properties in CPNC. The derived equation expressed the conductivity, k, in terms of the volume fraction of the suspended particles, ϕ, and the eccentricity or the aspect ratio, p:

$$\frac{k/k_1 - 1}{k/k_1 + x} = \phi \frac{k_2/k_1 - 1}{k_2/k_1 + x} \tag{185}$$

where k, k_1 and k_2 are conductivities of the suspension, the matrix and the suspended particles. The parameter x is a function of two relative numbers: the relative conductivity k_2/k_1, and the aspect ratio:

$$x = -\frac{(k_2/k_1 - 1) + \beta k_2/k_1}{(k_2/k_1 - 1) + \beta} \tag{186}$$

where β is a shape factor, which depends on the ellipsoid shape and p. For spheres, $x = 2$ and the Clausius-Mosotti relation for the dielectric constant is recovered.

In CPNC it is usually assumed that the clay platelets are impermeable, thus setting $k/k_1 = P_r$ and $k_2/k_1 = 0$ the following dependence is recovered:

$$P_r = (1-\phi)/(1+\beta\phi) = (1-\phi)/(1+p\phi/2) \qquad \therefore \beta = p/2 \tag{187}$$

For oblate ellipsoids with aspect ratio p, the shape parameter is given as $p/2$ and the expression for the reduced permeability by the presence of well-aligned clay platelets (see Sections 2.4.2 or 3.1.7) is recovered. The advantage of Equations 185-187 is that they are more general, applicable to systems with semi-permeable particles or their aggregates of any shape.

Bharadwaj [2001] proposed an extension of this simple model, which includes correlation between the aspect ratio, p, the volume fraction of the dispersed phase, ϕ, the orientation, and the degree of clay dispersion. The model attempts to be general, considering dispersed solid particles of different shape, *viz.* spheres, plates, fibres, etc. However, in the present case of clay-containing nanocomposites only platelets will be discussed. The platelets are characterised by three orthogonal dimensions: length, width and thickness (w), but as it turns out, the important factor is the aspect ratio: $p = L/w$, where L is the diameter of an average clay platelet. As before the increased pathway of the penetrant molecules that results in the reduction of permeability is given by the tortuosity factor $\tau = 1 + \phi p/2$ and consequently the permeability ratio is written as:

$$P_r \equiv P/P_o = \frac{1-\phi}{1+\phi p/2} \tag{188}$$

where P is the permeability through CPNC and P_o is that through neat matrix resin. Thus, this simple consideration of platelets oriented orthogonally to the flux direction yields an expression equivalent to Equation 187. Now, the first modification of Equation 188 involves the introduction of an orientation factor S that corrects the orthogonal (to the flux direction) platelet aspect ratio considering the average platelet orientation across the membrane:

$$S = \langle 3\cos^2\theta - 1 \rangle / 2 \tag{189}$$

where θ is the angle between the average platelet orientation and the flux direction, thus for orientation perfectly parallel to the membrane surface $\theta = 0$ and $S = 1$, for platelets perfectly aligned with the flux direction $\theta = \pi/2$ and $S = -1/2$, and $\theta = \pi/4$ and $S = 1/4$ for random orientation. In consequence, Equation 188 is modified to read:

$$P_r = \frac{1-\phi}{1+F(S)\phi p/2} = \frac{1-\phi}{1+\phi p(S+1/2)/3} \tag{190}$$

For perfectly planar platelets orientated perpendicular to flux ($\theta = 0$ and $S = 1$) Equation 190 reverts to Equation 188. For the perfect orientation of platelets perpendicular to the membrane surface ($\theta = \pi/2$ and $S = -1/2$) Equation 190 predicts that $P_r = 1 - \phi$. Finally, for random orientation Equation 190 predicts that the effective aspect ratio is $p_{effective} = p/2$.

Another problem associated with predicting the barrier properties of CPNC is a need to include some aspect of the degree of dispersion. Bharadwaj solves this problem by suggesting that the aspect ratio should be modified by assuming that one must consider, not individual clay platelets, but rather their short stacks. Furthermore he assumed that L remains unchanged but the stack thickness depends on the number of platelets and the interlamellar gallery height. For example, if for a perfectly exfoliated system $p = L/w$, for stacks made of two platelets $p = L/3w$, for stacks made of three platelets $p = L/5w$, etc. Thus, the relative permeability is expected to strongly depend on the extent of dispersion.

Evidently this approach is arbitrary since rarely does the interlayer spacing in short stacks equal the platelet thickness, i.e., $d_{001} = 1.12$ nm. Furthermore, the above approach excludes the possibility of lateral displacement of clay platelets in respect to each (constant L). Finally, the unspoken assumption here is that CPNC is made of strictly two homogeneous phases: permeable matrix and impermeable clay platelets or their stacks. The aspect of solidification of organic molecules on the clay surface, associated with reduction of segmental mobility as well the free volume and resulting disproportionate increase of the barrier properties has not been addressed. However, the new model constitutes a significant step in the right direction and offers an insight into the mechanisms controlling permeability through CPNC sheets and films.

For analysis of the permeability data Equation 188 may be rearranged in linear form with an average, apparent aspect ratio readily calculable from the slope:

$$(1/P_r) - 1 = (1 + \langle p \rangle / 2)\frac{1}{(1/\phi) - 1} \tag{191}$$

Examples are shown in **Figure 181** and **Figure 182**. The value of the apparent aspect ratio can be used to judge the degree of clay exfoliation and/or orientation. The figures show not only different values of $\langle p \rangle$, but also different tendencies: in **Figure 181** the clay content goes up to *ca.* 18 vol%. The linearity is observed only up to about 5 vol%, at higher clay loading the efficiency of the permeability reduction significantly decreases, but a second order polynomial well describes the behaviour. In **Figure 182** the last point may be an indication of worsening intercalation. The interlayer spacing in both these systems was $d_{001} = 1.7$ to 2.6 nm and without any additional information the interpretation of the barrier properties must stop with $\langle p \rangle$.

As discussed in Section 3.1.7, the high-energy surface of nanoclays causes adsorption of macromolecules on the surface, which results in reduced segmental mobility and lower free volume. The magnitude of this effect very much depends on the system chemistry and morphology. For example, bare clay surfaces readily adsorb polar macromolecules. Thus, the volume fraction of the nanofiller particles

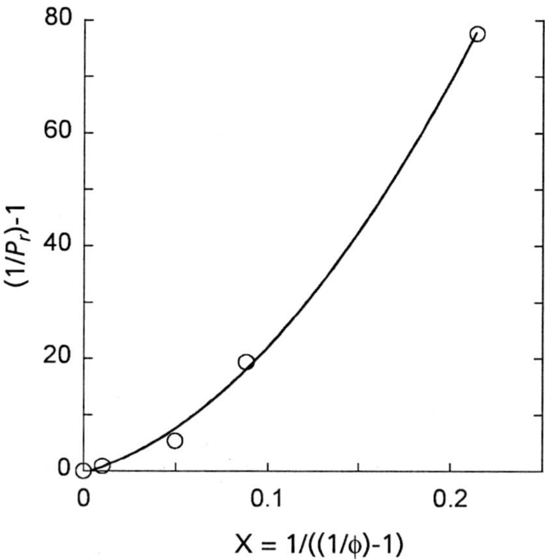

***Figure 181** Relative permeability of dichloromethane through PCL plotted as Y = $1/P_r$ versus X = $\phi/(1 - \phi)$. Data [Gorrasi et al., 2002]. The line is the least square fit to the relation: $Y = 1 + (1 + \langle p \rangle/2)X + bX^2$, with the average aspect ratio: $\langle p \rangle$ = 197 and the parameter b = 1240.*

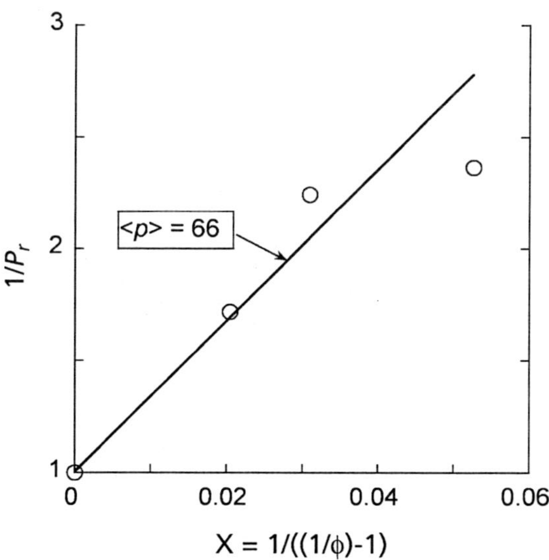

***Figure 182** Relative permeability of O_2 through PLA plotted according to Equation 191. Data [Chang et al., 2003]. The average aspect ratio is $\langle p \rangle$ = 66.*

greatly increases (for PA-6 by a factor of *ca.* 30!), which at low filler concentration reduces the relative permeability by a factor of 1-29ϕ. Evidently, the smaller the size of filler particle the greater is the effect. Another discrepancy in the permeability results was pointed out in Section 4.1.2.4. The data from Gorrasi *et al.* [2003] indicated quite different apparent aspect ratios for clay platelets in the same film, *viz.* p = 826, and 107 for diffusing vapours of *n*-pentane, and dichloromethane, respectively. It seems that the diffused layer of intercalant that coats the clay platelets may impose a barrier to diffusion by some penetrants. This is particularly evident at higher organoclay concentrations, where the oriented domains of pre-intercalated clay overlap.

Merkel *et al.* [2002] reported on the reverse problem – enhanced permeability of some gases through a membrane with nanoparticles. In the study poly(4-methyl-2-pentene) (PMP) was filled with fumed silica spherical particles of different diameters. The authors reported $P_r > 1$, for methane diffusing through PMP filled with particles of diameter $d < 500$ nm. Going back to Equation 187 one notes that as far as the particle geometry is concern, the relative permeability depends only on the aspect ratio – once p = constant, P_r only depends on concentration. However, in the case of PMP/silica P_r decreased hyperbolically from $P_r \cong 2$ for $d \cong 10$ nm to $P_r \cong 1.1$ for $d \cong 100$ nm (at a constant loading of $\phi = 0.13$). Furthermore, for small silica particles the P_r increases with filler content, reaching the value of 3.4 at $\phi \cong 0.26$. The explanation given is based on TEM. High immiscibility between filler and matrix results in clustering – evidently, the smaller the particle, the higher the surface energy hence the larger the clustering effect and the lower the percolation threshold concentration for the diffusion paths. Since permeability depends on the solubility (S of the penetrant in the membrane polymer) and the diffusivity (D) through it, the permeability enhancement is most likely related to the latter – the presence of solid particles was found to disrupt polymer chain packing hence increasing D, e.g., see [Gorrasi *et al.*, 2003].

Porosity in CPNC poses an obvious and serious problem to the barrier properties. The porosity may originate from the voids between organoclay platelets [Krook *et al.*, 2002]. These authors prepared CPNCs of a biodegradable polyester-amide with 5 and 13 wt% MMT-ODA by melt mixing at either 135 or 175 °C. The interlayer distance, d_{001} = 3.2 to 3.6 nm was observed. Nevertheless, the density indicated the presence of voids, constituting *ca.* 36 vol% of the total volume of incorporated organoclay. Compression moulding only slightly reduced the void content. Attempts to improve barrier properties by uniaxial stretching resulted in opposite effects – more voids forming and higher permeability. According to TEM the voids were located inside the clay stacks. Processing at lower temperature generated more voids. However, depending on the processing conditions, at 13 wt% loading the modulus increased by a factor of 2.6 to 3.5, the strength by a factor of 1.3 to 2.1, at a cost of the elongation at break which reduced by a factor from 1.9 to 4.3. Thus, while at low deformation the voids do not seem to affect the stiffness, they play a role at higher deformation. Krook *et al.* labelled the origins of this phenomenon as 'shear-induced', but the initial presence of low molecular weight volatiles could have contributed to the final effect.

Clay suitable for use in CPNC must be purified with great care (see Section 2.2.3). The non-colloidal mineral content should be kept below 0.5 wt%. These foreign materials (e.g., quartz or salts formed during pre-intercalation) generate voids in biaxially stretched CPNC films dramatically increasing the penetrant flux. As discussed in Section 4.1.1, two patent applications from AMCOL point out the influence of particulate impurities in organoclay on the barrier properties [Lan *et al.*, 2002a,b]. The first of these specifies that organoclays must have less than 2 wt% of quartz. The motivation for the second application is similar, but here the presence of inorganic salts formed during ion exchange is the culprit. The presence of these contaminants causes the formation of voids. Their presence, especially in biaxially stretched films, is responsible for a dramatic increase of fluid permeability. Furthermore, the presence of these inorganic contaminants is responsible for the reduced film transparency.

Control of vapour diffusion through a polymeric wall is of particular concern to engineers designing automobile fuel tanks – Ellis proposed to use CPNCs for this purpose [2003a]. The fuel tank is composed of three principal layers, the one in the middle being responsible for the barrier properties. With gasoline, satisfactory performance can be achieved using EVAl alone, but its effectiveness is reduced in the presence of methanol in the fuel. The patent application proposes to melt compound EVAl (32 mol% ethylene) with 40 to 90 wt% of a commercial CPNC containing 4 wt% of exfoliated clay in a PA-6 matrix. A plot of the permeability coefficient of MeOH versus the composition of the middle layer indicates a reduction of permeability for CPNC content exceeding about 50%. At higher EVAl content immiscibility seems to affect the organoclay dispersion. However, as the related publication indicates, the process is not without some challenges [Ellis, 2003b]. For example, blends containing 75 wt% of EVAl showed higher permeability than the neat blend components (EVAl and CPNC with PA-6 as the matrix) as well as the blends containing 25 or 50 wt% of EVAl. XRD and TEM indicated that while in blends containing ≤ 50 wt% EVAl the organoclay was exfoliated; this was not the case for high EVAl content – a strong peak of $d_{001} \cong 2.2$ nm was observed. There are three possible sources of this 'reversed exfoliation': (1) thermodynamic immiscibility of the three component system: EVAl/PA-6/clay; (2) thermal de-intercalation of organoclay at the processing temperature of 240 °C; and (3) segregation of clay platelets during crystallisation. Obviously, a combination of any of these three should also be considered. It is interesting that this reverse exfoliation process does not lead to total collapse of the interlayer spacing (e.g., as that resulting from Hofmann elimination – see Section 3.2). The interlamellar gallery height of *ca.* 1.3 nm is large enough to accommodate a double layer of polyamide chains. Furthermore, the permeability of this blend corresponds to what could be expected from a mixture of two polymers without clay: EVAl and PA-6. Thus, from the three possible sources the first, thermodynamic phase separation, seems to be the most likely culprit.

Part 6

Closing Remarks

Clay-Containing Polymeric Nanocomposites

6 Closing Remarks

The technology of polymer nanocomposites is in an early stage of development. Several complementary strategies have developed, of which the incorporation of layered inorganics is but one. To the latter category belong clay-containing polymeric nanocomposites, or CPNC – the principal topic of this book. However, even in this quarter-century old domain there are many unanswered questions.

CPNCs are designed for structural applications; hence for large-volume production where improved performance, low cost and high reproducibility are the principal criteria. With this in mind let us first summarise the current status of knowledge about these materials and then outline what should be done in the near future.

6.1 Summary

1. The most frequently used nanofiller is a purified, ion-exchanged sodium montmorillonite, Na-MMT, an organoclay. The optimum cation exchange capacity is CEC = 90-120 meq/100 g. The aspect ratio of exfoliated platelets ranges from p = 10 to 1000. The higher the aspect ratio the better the CPNC barrier properties and rigidity, but the more difficult is their intercalation and processing (controlled orientation without attrition). It is important to recognise the need to preserve the high aspect ratio, especially for the barrier properties and mechanical performance.
2. The 'maximum packing volume fraction', ϕ_m, is a hyperbolic function of the aspect ratio (see Equation 1). From the rheological point of view it represents the concentration above which the yield stress and shear viscosity start to increase rapidly. This is particularly important for the shear processing of CPNC, e.g., extrusion or injection moulding. By contrast, the processes dominated by the extensional flow (e.g., blow moulding, wire-coating, calendering) are less affected by ϕ_m.
3. Intercalation is a process during which the intercalant molecules diffuse into the galleries between individual clay platelets. These, *ca.* 1 nm thick layers are elastic, but there is a limit to the amount of bending they can undergo. Thus, efficient strategies of intercalation involve reduction of the stack size and step-wise increase of the penetrant size, starting with water, then (usually) an onium cation. The process depends on the balance of forces: an attractive one that causes the intercalant molecules to bond to the clay surface and a repulsive one that resists the penetration due to geometric constraints, *viz.* the initial gallery size, the size and shape of the intercalant molecule, the matrix viscosity, reduction of entropy, etc.
4. Intercalation by exchange of Na^+ for onium cations is the most common. There is a great diversity of these intercalants, mostly based on ammonium or phosphonium cations. Quaternary ammonium salts are mainly used, with

the highest *pK* – the ion bonding strength. However, primary and secondary ammonium cations seem to be more thermally stable and they perform better in specific applications, e.g., where the ammonium group may react and/or catalyse polymerisation as in epoxy systems.

5. Besides the anionic sites on the clay surface, there are hydroxyl groups and cations, which have been (rarely) used to improve the clay/polymer dispersion. For example, hydroxyl groups can participate in reactions with known 'sizing agents', e.g., organosilanes, organotitanates or organozirconates. This method of (usually secondary) intercalation is quite attractive in the melt compounding preparation of PO-based PNC. An additional advantage of this approach is the possibility of scavenging with these agents for free amine groups, thus improving the thermal stability of the system. The cations are mainly located on the platelet edges, thus are not efficient for intercalation, but may help during the exfoliating step.
6. There is little information about the effects of T and P on the interlayer spacings during melt compounding.
7. The theories as well as experiments indicate the importance of direct bonding between the clay platelets and the matrix. Such bonding can be accomplished by end-tethering functionalised macromolecules that either are identical with the main polymer or miscible with it.
8. The main disadvantage of the quaternary ammonium intercalants with alkyl chains is their lack of thermal stability. Their degradation starts at *ca.* 165 °C and is quite severe at $T > 220$ °C. Even when CPNC is prepared at low temperature (e.g., by polymerisation or solution methods) the processing of industrially interesting polymers is usually carried out above 220 °C. The following solutions have been offered:
    - Some improvement of stability can be achieved by eliminating excess ammonium from organoclay. Extraction of organoclays, as well as neutralisation-complexing during compounding of CPNC has been used. In the latter case, chemical reactions with 'sizing' agents (*viz.* organosilanes, organotitanates, and organozirconates) or epoxy compounds were described.
    - Use of the more stable phosphonium intercalant.
    - Use of organometallic intercalants that form inorganic pillared structures.
    - Use of a bifunctional monomer that can bind to the clay surface and subsequently be incorporated into the matrix polymer chain.
    - Use of β-carbon branched alkyls large, thermally stable aromatic molecules that play multiple roles, e.g., that of an intercalant and a die, polymer stabiliser, bio-active agent, etc.
9. Amines and their salts are toxins with severity rating (on a scale 0 to 5) from 3 to 5, and the lethal dose values, LD_{50}, ranging from 5 to 5,000 mg/kg. Thus, one must be cautious regarding these intercalants as well as the products that may be generated during processing via the Hofmann decomposition process. Alternatively, strategies should be devised to eliminate the toxic amines or ammonium ions by means of chemical scavengers, e.g., suitable epoxy compounds.
10. In CPNCs the clay should be exfoliated, i.e., the distance between individual platelets should be larger than about 8.8 nm, with only a few short stacks

(tactoids) remaining. Total randomisation of clay lamellae is only possible below ϕ_m, above it the clay platelets are oriented parallel to each other, with the mean distance between them inversely proportional to concentration.

- During the preparation of CPNCs by polymerisation the growing macromolecules should preferably bond to clay platelets. Thus, their growth within the interlamellar galleries leads to exfoliation and prevents the exfoliated platelets from re-assembling under the flow condition. The presence of inorganic multivalent ions, e.g., Ca^{2+}, or Fe^{3+}, may interfere with exfoliation.
- During the preparation of CPNC by melt compounding the use of a functionalised secondary intercalant (preferably an end-functionalised compatibiliser) may be necessary. In general, the main polymer must be miscible with the organic part of this intercalant. The process is diffusion controlled, hence sufficient time must be provided to obtain expansion of the interlayer spacing from about 3-4 nm to > 10 nm. Low melt viscosity and the use of ultrasonics (*ca.* 50 kHz) is helpful.

11. In systems with thin, dispersed platelets with the aspect ratio p = 100-1000, the performance is dominated by the surface effects. In semicrystalline polymers the presence of nanoparticles changes the microstructure, usually increasing the crystallisation rate (which may lead to less thermodynamically stable crystalline forms), total crystallinity and brittleness.
12. The recognised advantages of CPNCs (in comparison to the matrix polymer) are: increased modulus, strength and elongation at break, greatly reduced permeability to gases, vapours and liquids, as well as increased flame resistance, good transparency, etc. The disadvantages are related mainly to flow (enhanced viscosity and yield stress) and cost.
13. Only CPNCs with aliphatic polyamides are fully commercial. CPNCs with aromatic PA, PET and PO are close to commercialisation. It is of note that CPNC may be produced on standard processing lines following standard procedures. In principle, CPNCs with enhanced performance can be produced for any polymeric matrix: thermoplastic, thermoset or elastomer.
14. The technology is market-driven, thus a reliable supply of CPNC with proven, reproducible performance advantages at reasonable cost is required.
15. The usual route for market penetration of new materials is by substitution – this approach seems to be working well for CPNCs with PA-6 or PA-66 as the matrix.

As the summary indicates, there are several levels of problems that need to be solved if this technology is to fulfill the expectations for large commercial success. These may be grouped around three topics: composition, method of preparation, characterisation and testing.

6.2 The Future

6.2.1 Composition

- Because of tradition (oil and grease applications) the layered nanofillers readily available for use in CPNCs have been based on the mineral montmorillonite (a component of bentonite). The advantage is the general availability and low cost of the raw material. The disadvantages are: the difficulty of ensuring adequate (for CPNC) purification, the variability of composition recognised from the beginning (chemical composition as well as type and concentration of crystallographic defects) and the limited range of aspect ratios. More attention should be given to alternatives: the synthetic and semi-synthetic (based on natural talc) layered nanofillers. Little is know regarding their chemical and physical characteristics, as well as the available and potential range of structural characteristics. Their disadvantage is the limited sourcing, uncertainty about their properties and (slightly) higher cost. However, in principle their composition and structure can be controlled, thus the reproducibility and repeatability of performance should be better. Furthermore, the synthetic route may inherently modify the platelet reactivity and reduce the need for extensive intercalation with onium ions.

- There is a need to develop better intercalants – better as far as thermo-oxidative stability, expansion of the interlamellar galleries and reduced toxicity is concerned. Currently, ammonium intercalants dominate. Intercalation with phosphonium somehow improves the thermal stability, but the effect on mechanical performance is disappointing. The relatively unexplored molecules are organometallic or rather metallo-organic.
- For hydrophobic polymers (e.g., PO or PS) the use of a compatibiliser may be required. Currently two approaches have been used: the treatment of organoclay with epoxy compound, or the use of maleated macromolecules. For both, the exact chemistry is unknown – is the functional group reacting directly with a clay functionality (anion, cation or –OH), or with some functionality of the intercalant? Are these potential reactions pressure, temperature and catalyst dependent? Different chemistry would lead to a widely different phase structure in CPNC and performance. Systematic studies on well-characterised model systems are badly needed. If experimental verification is difficult molecular modelling may offer the simplest route.
- The fourth element on the composition list is the relative concentration of all the ingredients. Modelling showed that when a functionalised compatibiliser is to be used, the density of grafting clay with intercalant should be reduced, but where is the optimum? What is the role of the standard commercial additives on the phase equilibrium in CPNC? These diverse additives, may be used in large quantities (exceeding 5 wt%). This is bound to affect the delicate balance of interactions within the CPNC system.

6.2.2 Method of Preparation

The two principal routes for CPNC preparation are chemical and mechanical. The former involves dispersing organoclay in a monomer, cyclomer or oligomer and cause a suitable polymerisation reaction to take place. The latter is based on the principles of blending/compounding. The availability of the chemical route is limited to resin manufacturers. Furthermore, the degree of dispersion achieved is stable for highly polar polymers able to develop strong interactions with the clay surface, *viz.* polyamides, or water-soluble polymers. When hydrophobic polymer is formed during reactive exfoliation, the degree of dispersion may change during processing and forming. There are several advantages of the mechanical dispersion route, the main one being the wider accessibility of this technology to compounders and manufacturers of finished products, the other the generation of a more thermodynamically stable dispersion (under the condition that the intercalant is thermally stable!).

- Preliminary studies indicate the beneficial effect of extensional flow mixing on the degree of clay dispersion and performance. Studies on the principal process variables: type and intensity of stresses, residence time, throughput, pressure and temperature are needed.
- The effect of process variables on clay platelet attrition should be investigated. Efforts should be made to develop procedures providing exfoliation with limited damage to the lamellae.
- During compounding some clay platelets bend while others break. The extent of attrition seems to vary with the clay as well as with the type of intercalant. However, bending as well as breaking platelets reduces their efficiency. It is important that during the forming stage platelets are stretched (flattened out) as well as being randomly dispersed in the matrix or well oriented in the desired direction. Owing to the sluggishness of large diameter platelets the process is not simple and the best methods for orientation should be studied.

6.2.3 Characterisation and Testing

- The present methods of characterisation of clay dispersion are laborious, costly and inadequate. There is a need to develop rapid, reliable test procedures.
- There is no method to characterise the aspect ratio – the ones used (flow and barrier properties) are long and prone to errors (mainly due to orientation). Due to the thinness of clay platelets TEM is not suitable. New ideas and new approaches are needed.
- There have been several reports of bubbles being formed in the vicinity of clay platelets. These strongly affect the performance, particularly permeability and mechanical properties. Elimination of these nanobubbles is difficult. A procedure should be developed to eliminate this problem, either by suitable prevention or suitable specimen forming and testing procedures.
- The knowledge of the thermodynamics and rheology of nanocomposites is very limited. CPNCs were found to engender a wide range of responses, from high non-linear viscoelastic behaviour with very sluggish orientation effects to a response similar to polymer composites with low filler loading. Can one use the non-linear viscoelastic characterisation of CPNCs to quantify the thermodynamic interactions? Can this be related to tensile stress in the solid specimen (clay would be expected to reside in the amorphous regions – glassy or rubbery)?
- There is little doubt that molecular modelling using either a lattice model or numerical simulation by a MC or MD procedure can be of great help to understanding the CPNC systems, differentiate the degree of importance for numerous variables and suggest routes for optimisation. Not enough of this activity is being carried out around the globe.

ars longa, vita brevis

Part 7

Appendices

Clay-Containing Polymeric Nanocomposites

7 Appendices

7.1 General and Chemical Abbreviations

(*n*, *m*)	Topological characterisation of a nanotube by the chiral vector
A	Amorphous polymer
A.-G.; AG	Aktiengesellschaft
AA	Acrylamide
AAc	Acrylic acid (monomer)
AEA	*N*,*N* Dimethyl aminoethyl acrylate
AFM	Atomic force microscopy / atomic force microscope
AIBN	*N*,*N*´-Azo-bis(iso-butyro nitrile)
AMPS	2-Acryl-amido-2-methyl-1-propane sulfonic acid
AN	Acrylonitrile
APES	Aliphatic polyester
APP	Ammonium polyphosphate
APS	α-Amino-propyl triethoxysilane
APTS	*p*-Amino-phenyl trimethoxysilane
aq	Aqueous
ASTM	American Society for Testing and Materials
ATPS	Amine-terminated polystyrene
ATRP	Atom transfer radical polymerisation
BACM	*bis*-(*p*-Amino-cyclohexyl methane)
BD	1,4-Butanediol (chain extender)
BDMA	Benzyl-dimethyl amine
BHET	*bis*-Hydroxy-ethyl terephthalate
BMI	*bis*-Maleimide
BN-NT	Boron nitride nanotube
BOPP	Biaxially oriented polypropylene film
BPA	Bisphenol-A
BPO	Benzoyl peroxide
BTDA	Benzophenone tetra-carboxylic di-anhydride
BTFA	Boron tri-fluoride mono-methyl amine
BZD	Benzidine or 1,1´-biphenyl-4,4´ di-amine: H_2N-ϕ-ϕ-NH_2
CB	Carbon black
CBS	*N*-Cyclohexyl-2-benzo-thiazole sulfonamide (vulcanisation accelerator)
CD	Coefficient of determination
C-D	Convergent-divergent flow geometry
CEC	Cation exchange capacity
CED	Cohesive energy density
CGC	Constrained geometry catalyst

cgs	Old units: centimetre - gram - second
CHO	Cyclohexanone
CHX	Cyclohexane
CMC	Critical micelle concentration
CNC	Ceramic matrix nanocomposite
CNT	Carbon nanotube
COPET	PET copolymer comprising also other di-hydroxyl compounds than di-ethyl glycol
CORI	Co-rotating intermeshing TSE
CP	Chemical shift (in NMR measurements)
$Cp^{*}Ti(OMe)_3$	Penta-methyl cyclopentadienyl trimethyl titanium
CPC	Cloud point curve; also cetyl pyridinium chloride
CPNC	Clay-containing polymeric nanocomposite
CRNI	Counter-rotating non-intermeshing TSE
CRT	Constant residence time
CST	Critical solution temperature
CTBN	Carboxyl terminated nitrile rubber
CTC	Constant thermal conduction
CTE	Coefficient of thermal expansion
CTM	Cavity transfer mixer
CUT	Continuous use temperature
CVD	Chemical vapour deposition
CVFF	Consistent valence force field
DAAM	Di-acetone acrylamide: CH_2=$CHCONHC(CH_3)_2CH_2COCH_3$
DAM	Dry as moulded
DBSA	Dodecyl benzene sulfonic acid
DBT	Di-butyl tin dilaurate (catalyst)
DCA	Di-chloro acetic acid
DCB	Di-chloro-benzene
DCP	Di-cumyl peroxide
DD	Degree of dispersion
DDA	Dodecylamine
DDDHM	3,3′-Di-methy-4,4′-diamino-dicyclo hexyl methane
DDM	*p,p′*-Di-amino-di-phenyl-methane
DDP	1-Dodecyl-2-pyrrolidone
DEFM	Dynamic extensional flow mixer
DETDA	Di-ethyl toluene diamine
DFT	Density functional theory
DGEBA	Di-glycidyl ether of bisphenol-A
DGEBF	Di-glycidyl ether of bisphenol-F
DIBAH	Di-isobutyl aluminium hydride
DIN	Deutsches Institut für Normung
dl-; DL-	Racemic
DLC	Diamond-like carbon
DMA	Dynamic mechanical analysis
DMAc	*N,N*-Dimethyl acetamide
DMAEM	*N,N*-Dimethyl amino-ethyl methacrylate
DMC	Dough moulding compound
DMF	*N,N*-Dimethylformamide
DMPA	Di-methylol propionic acid

DMSO	Di-methyl sulfoxide (solvent)
DMT	Di-methyl ester of terephthalic acid (TA)
DMX	Dynamic melt mixer
DOP	Di-(2-ethylhexyl) phthalate
DP	Degree of polymerisation
DR	Draw ratio
DSC	Differential scanning calorimetry
DSIMS	Dynamic secondary ion mass spectrometry
DTA	Differential thermal analysis
DTUL	Deflection temperature under load
DuPont	E.I. Du Pont de Nemours & Co.
DVB	Divinylbenzene
D-W	Dee and Walsh eos
DWNT	Double-wall nanotube
ed.	Edited, edition, editor
EDA	1,2-Diaminoethane
EDX	Energy dispersive X-ray
EE	Edge-to-edge clay platelet interactions
EFM	Extensional flow mixer
EME	Elastic melts extruder
emf	Electromotive force
EMI	Electromagnetic interference
E-MMT	Epoxidised MMT
emu	Electron magnetic unit
EOA	Oligo(oxy ethylene amine)
eos	Equation of state
EP	Engineering polymer
EPB	Engineering polymer blends
EPR	Electron paramagnetic resonance
Eq	Equation
ER	Electro-rheological fluid
ESCA	Electron-spectroscopy for chemical analysis
ESCR	Environmental stress crack resistance
ESD	Electrostatic dissipation
ESR	Electron spin resonance
esu	Electrostatic unit
Et_3Al	Tri-ethyl aluminium
EtAc	Ethyl acetate
EtOH	Ethanol
FA	Formic acid
FCM	Farrel continuous mixer
FCP	Mechanism of fatigue crack initiation and propagation
FE	Face-to-edge clay platelet interactions
FED	Field emission device
FEM	Field emission microscopy
FENE	Finite extendable nonlinear elastic model
FF	Face-to-face clay platelet interactions
FH	Fluorohectorite
FM	Fluoromica (synthetic clay, e.g., Somasif ME-100)
FOV	Flory-Orwoll-Vrij eos

FP	Freezing point
FTIR	Fourier-transform infrared spectroscopy
FTR	Fourier-transform rheology
GC	Gas chromatography
GF	Glass fibre, sometimes glass fibre reinforced plastic
GIC	Graphite intercalated compounds
GLC	Gas-liquid chromatography
GMA	Glycidyl methacrylate (monomer)
GMT	Glass mat reinforced plastics
GP	Gear pump
GPC	Gel permeation chromatography (now: size exclusion chromatography, SEC)
GPO3	Glycerol propoxylate
HALS	Hindered amines (antioxidants)
HAO	Higher alpha-olefins
HBA	*p*-Hydroxy-benzoic acid
HCP	Hairy clay platelet model
HDT	Heat deflection temperature
H-H	Hartmann and Haque
H-MMT	Protonated montmorillonite
HPB	Homologous polymer blend
HPLC	High-pressure liquid chromatography
HPMC	Hydroxy propyl-methyl cellulose
HRC	Rockwell hardness (C scale)
HREM	High-resolution electron microscopy
HRR	Heat release rate
HRTEM	High-resolution transmission electron microscopy
HT	Hectorite (one of the smectite clays)
HTBR	Hydroxy-terminated polybutadiene
HV	Vickers hardness number
HVEM	High voltage electron microscope (-py)
I.V.	Intrinsic viscosity (or limiting viscosity number [η], dL/g) – a measure of MW
ICRR	Intermeshing counter-rotating TSE
ICT	International Critical Table
ID	Inside diameter
IGC	Inverse gas chromatograph
Im	Imaginary part of a complex function
IPN	Interpenetrating polymer network
IR	Infrared spectroscopy
ISO	International Standards Organization
IU	International Unit
IUPAC	International Union of Pure and Applied Chemistry
JSR	Japan Synthetic Rubber Co., Ltd.
JSW	Japan (Nippon) Steel Works
K	Kaolinite
L/D	Extruder screw length-to-diameter ratio
LA	Lactic acid
LALLS	Low angle laser light scattering
LAOS	Large amplitude oscillatory shear

LC	Liquid crystal
LCB	Long chain branching
LCD	Liquid crystal display
LCP	Liquid crystal polymer
LCST	Lower critical solution temperature
LD_{50}	Lethal dose with 50% mortality; usually listed in mg per kg body mass
LDH	Layered double hydroxides
LED	Light-emitting diode
LHS	Left hand side
LLDPE	Linear low density polyethylene
LMW	Low molecular weight
LSc	Light scattering
m-	meta-
MAH	Maleic anhydride
MAO	Methyl aluminoxane
MAS-NMR	Magic angle spinning nuclear magnetic resonance
MBT	Mercapto-benzo-thiazole (vulcanisation accelerator)
MC	Monte-Carlo computational method
MD	Molecular dynamics (also 'machine direction')
MDDP	Methylene diphenyl diisocyanate prepolymer
MDI	Methylene-bis-(4-phenylisocyanate)
MEK	Methyl ethyl ketone (solvent)
MEMS	Micro-Electro-Mechanical Systems = integration of mechanical elements, sensors, actuators, and electronics on a common silicon substrate
MeOH	Methanol
meq	Milliequivalent
MeTHPA	Methyl-tetra-hydro phthalic anhydride (curing agent)
MFI	Melt flow index; now MFR
MFR	Melt flow rate
MI	Melt index
MMA	Methyl methacrylate (monomer)
MMDA	4,4´-Di-amino-3,3´-dimethyl-diphenyl methane
MMT	Montmorillonite
MNC	Metallic matrix nanocomposite
MPDA	*m*-Phenylene diamine
MPMD	2-Methylpentamethylene diamine
MPP	Modified polyether polyol
MTHPA	Methyl-tetra-hydro phthalic anhydride
MW	Molecular weight
MWD	MW distribution
MWNT	Multi-wall nanotube
Na-MMT	Sodium montmorillonite
NC	Nanocomposite
n-C_{10}	*n*-Decane
n-C_6	*n*-Hexane
n-C_7	*n*-Heptane
NCL	National Chemical Laboratories (Pune, India)
NDB	Negatively deviating blends

NG	Nucleation-and-growth
NI	Notched Izod impact strength
NIRT	Notched Izod impact strength at room temperature
NIST	National Institute of Standards and Technology
NMA	Nadic methyl anhydride
NMF	*N*-Methyl formamide
NMP	*N*-Methyl-2-pyrrolidone
NMR	Nuclear magnetic resonance
NRCC	National Research Council Canada
NRET	Non-radiative energy transfer
NTP	Normal temperature, and pressure (25 °C, and 101.3 kPa or 1 atm)
NVC	*N*-Vinylcarbazole
OCC	Cyclic carbonate oligomer (LMW cyclomer of PC)
OD	Outside diameter
ODA	Octadecylamine
ODAn	4,4′-Oxy-di-aniline
ODPA	3,3′-Oxydiphthalic anhydride
OET	Oligo(ethylene terephthalate)
OM	Optical microscopy
OMP	Optical microscopy in polarised light
ORMLAS	Organically modified layered alumino-silicates
o-	ortho-
OWM	Optical waveguide microscopy
OZ	Methyl-vinyl oxazoline
p-	*para-*
PAA	*N,N*-Dimethyl aminopropyl acrylamide
PAB	Polymer alloys and blends
PALS	Positron annihilation lifetime spectroscopy
PamA	Polyamic acid
PANC	Polyamide-6/MMT nanocomposite
PDB	Positively deviating blends
PEA	2-Phenylethylamine
PEBAX	Poly(ether-*b*-amide)
pH	Negative logarithm of the effective hydrogen ion concentration
phr	Concentration in parts per hundred of resin
PHRR	Peak heat release rate
PICS	Pulse induced critical scattering
pK	Negative (decimal) logarithm of an equilibrium constant K
PLL	Persistent lamellar length
PMDA	Pyromellitic di-anhydride
PMIM	Phase measurement interference microscopy
PMR	Proton magnetic resonance
PNC	Polymeric matrix nanocomposite
PNDB	Positively and negatively deviating blends (sigmoidal)
POA	Oligo(oxy propyl amine)
POSS	Polyhedral oligomeric silsesquioxanes
PPA	Poly(phosphoric acid)
ppb	Parts per billion (10^9)
ppm	Parts per million (10^6)

PS-VO	Styrene-methyl vinyl oxazoline copolymer
PTA	Phospho-tungstic acid
PTES	Phenyl-tri-ethoxy silane
PTM	Photon tunnelling microscopy
PTS	Phenyl tri-methoxysilane
PZ	Zinc-di-methyl di-thio carbamate (rubber vulcanisation accelerator)
QA	Quality assurance
QC	Quality control
QDS	Quality data statistics
QMC	Quick moulding change
RAM	Random access memory
Re	Real part of a complex function
Ref.	Reference
REX	Reactive extrusion
r-f, RF	Radio frequency
RH	Relative humidity (in %)
RHS	Right hand side
RIM	Reaction injection moulding
RLM	Reactive liquid polymer
RMS	Rheometrics Mechanical Spectrometer
rms	Root-mean square
rpm	Rotations per minute
RRIM	Reinforced reaction injection moulding
RT	Room temperature
RTD	Residence time distribution
RTM	Resin transfer moulding
RTP	Reinforced thermoplastic
RTPO	Reactor-blended thermoplastic olefinic elastomer
RTS	Reinforced thermoset
RTV	Room temperature vulcanisation (of silicone rubber)
RVE	Representative volume element (in CPNC modelling)
S. A.	Société Anonyme
SANS	Small angle neutron scattering
SAW	Self-avoiding random walk
SAXS	Small angle X-ray scattering
SCB	Short chain branching
SCF	Self-consistent field
SCP	Southern Clay Products
SD	Spinodal decomposition
SEC	Size exclusion chromatography
SEI	Secondary electron image
SEM	Scanning electron microscopy
SFA	Surface force analyser
SH	Strain hardening
SI	Le Système International d'Unités (International System of Units)
SIN	Simultaneous interpenetrating network or semi-interpenetrating network
SiNT	Silica nanotubes
SIPN	Sequential interpenetrating polymer network

SIS	Solvent-induced shift
S-L	Sanchez-Lacombe eos
SLS	Sodium lauryl sulfate, $CH_3(CH_2)_{10}CH_2OSO_3Na$
SM	Static (motionless) mixer
SMC	Sheet moulding compound
SMM	Synthetic mica-montmorillonite clay
S-N	Imposed stress versus number of cycles to failure diagram
SNOM	Scanning near-field optical microscopy
SPM	Scanning probe microscopy (includes STM, AFM, etc.) *or* Surface plasmon microscopy
SPN	Di-ethyl-methyl-oligo(oxypropylene)-ammonium chloride, $(C_2H_5)_2(CH_3)N^+(O\text{-}Pr)_{25}$ Cl
S-S	Simha-Somcynsky eos or theory
SSE	Single-screw extruder
St	Styrene
STEM	Scanning transmission electron microscopy
STM	Scanning tunnelling microscopy
STN	Methyl-tri-octil ammonium chloride, $CH_3(C_8H_{17})_3N^+ Cl^-$
STS	Scanning tunnelling spectroscopy
SWNT	Single wall nanotube
TA	Terephthalic acid
TCE	Tetra-chloro ethane
TD	Transverse direction
TDI	Toluene di-isocyanate
TEA	Tri-ethylamine
TEM	Transmission electron microscopy
TEOS	Tetra-ethyl ortho-silicate
TEP	Tri-ethyl phosphate
TETA	Tri-ethylene tetramine
TFA	Tri-fluoro-acetic acid
TFF	Thermal fatigue failure mechanism (S-N test)
TGA	Thermogravimetric analysis
TGAP	Tri-glycidyl *p*-amino phenol
TGDDM	Tetra-glycidyl di-amino di-phenyl methane
TGIC	Tri-glycidyl isocyanurate
THF	Tetrahydrofuran
TLC	Thin layer chromatography
TMA	Thermo-mechanical analyser
TMR	Twente mixing ring
TP	Thermoplastic resin
TPP	Tetra-phenyl phosphonium
TREF	Temperature rising elution fractionation
TS	Tetra-methyl thiuram sulfide (rubber vulcanisation accelerator)
TS	Thermoset resin
TSC	Thermally stimulated current
TSE	Twin screw extruder
TTGG	*Trans-trans-gauche-gauche* chain conformation
TTTT	Planar all-trans 'zigzag' conformation
TW	Glass temperature width (°C)
UCST	Upper critical solution temperature

UP	Unsaturated polyester
US	Ultrasonics, ultrasonication
UV	Ultraviolet
VA	Vermiculite
VAc	Vinyl acetate
VCM	Vinyl chloride (monomer); also VC
VDAC	Vinyl-benzyl-dimethyl dodecyl ammonium chloride (2MVBDDA)
VE	Vinyl ester
VOC	Volatile organic compound
VONT	Vanadium oxide nanotubes (V_2O_5)
W&P	Werner and Pfleiderer
WAXS	Wide angle X-ray scattering
XDMS	Poly(*p*-xylenylene *di*-methyl sulfonium bromide)
XPS	X-ray photoelectron spectroscopy
XRD	X-ray diffraction
Z-N	Ziegler-Natta
ZSK	Zwei Schnecken Kneter (two-screw kneader or extruder)
ε-CL	ε-Caprolactam

7.2 International Abbreviations for Polymers

ABA	Acrylonitrile-butadiene-acrylate copolymer
ABM	Acrylonitrile-butadiene-methyl acrylate copolymer
ABMA	Acrylonitrile-butadiene-methacrylic acid copolymer
ABS	Thermoplastic terpolymer, an acrylonitrile-butadiene-styrene copolymer
ACM	Acrylic elastomer, e.g., alkyl acrylate-2-chloroethyl vinyl ether copolymer
AEM	Elastomeric ethyl (or other) acrylate-ethylene copolymer
AES	Terpolymer from acrylonitrile, ethylene-propylene elastomer, and styrene (ACS)
ASA	Thermoplastic copolymer from acrylonitrile, styrene, and acrylates
ATPS	Amine-terminated polystyrene
BOPLA	Bi-axially oriented PLA
BR	Butadiene rubber
CA	Cellulose acetate
CAB	Cellulose acetate-butyrate
CB	Cellulose butyrate (also carbon black reinforcing pigment)
CFM	Polychlorotrifluoroethylene; also CEM, CTFEP, PCTFE
CHR	Elastomeric copolymer from epichlorohydrin, and ethylene oxide
CMC	Carboxy methyl cellulose (or critical micelle concentration)
COP	Cyclo-olefin polymers or copolymers
COPET	PET copolymer comprising also other di-hydroxyl compounds than di-ethyl glycol
COPO	Poly(ethylene-co-propylene-co-carbon monoxide), a linear, alternating terpolymer
CP	Cellulose propionate, or chlorinated polyethylene (also CPE)
CPE	Chlorinated polyethylene
CPP	Chlorinated polypropylene
CPVC	Chlorinated polyvinylchloride
CR	Chloroprene, or Neoprene rubber
CSR	Chlorosulfonated polyethylene
CTBN	Carboxy-terminated nitrile rubber
CTFE	Poly(chlorotrifluoroethylene)
EAA	Ethylene-acrylic acid copolymer
EBA	Ethylene-butyl acrylate copolymer
EC	Ethyl cellulose
ECTFE	Poly(ethylene-co-chloro tri-fluoro ethylene)
EEA	Elastomeric copolymer from ethylene, and ethyl acrylate
EGMA	Ethylene-glycidyl methacrylate copolymer
EMA	Copolymer from ethylene, and maleic anhydride or ethylene-methyl acrylate
EMAc	Copolymer from ethylene, and methacrylic acid
EMM	Copolymer from ethylene, and methyl methacrylate
EO-ECH	Copolymer of ethylene oxide, and epichlorohydrin (also ECO, CHR)
EP	Epoxy resin
EPR, EPDM	Elastomeric copolymer of ethylene, propylene, and a non-conjugated diene

ETFE	Copolymer from ethylene and tetrafluoroethylene
EVAc	Copolymer from ethylene and vinyl acetate
EVAl	Copolymer of ethylene and vinyl alcohol (also EVOH)
FEP	Fluorinated EPR; tetrafluoroethylene/hexafluoropropylene rubber
FPM	Vinylidene fluoride/hexafluoropropylene elastomer
HDPE	High density polyethylene (*ca.* 960 kg/m^3)
HEC	Hydroxyethyl cellulose
HIPS	High impact polystyrene
HPC	Hydroxypropyl cellulose
IR	Polyisoprene
LCP	Liquid crystal polymer
LDPE	Low density polyethylene (*ca.* 918 kg/m^3)
LLDPE	Linear low density polyethylene
MABS	Copolymer from methyl methacrylate, acrylonitrile, butadiene and styrene
MAN	Copolymer from methyl methacrylate and acrylonitrile
MAS	Copolymer from methyl methacrylate, acrylonitrile and styrene
MBA	Copolymer from methyl methacrylate, butadiene and acrylonitrile
MBS	Copolymer from methyl methacrylate, butadiene and styrene
NBR	Elastomeric copolymer from butadiene and acrylonitrile; nitrile rubber
NCR	Elastomeric copolymer from acrylonitrile and chloroprene
NR	Natural rubber
P3FE	Poly(trifluoroethylene)
P4VP	Poly-*N*-vinylpyrrolidone or poly(4-vinyl pyrrolidone)
PA	Polyamide; PA; the abbreviation is followed by a number, a letter or their combination. A single number refers to the PA from an α,ω-amino acid or lactam (e.g., PA-6). In a combination of two numbers the first number indicates the number of methylene groups of aliphatic diamine, the second that in an aliphatic dicarboxylic acid, *viz*. PA-66. An '*I*' stands for isophthalic acid, a '*T*' for terephthalic acid. For example, co-polyamide from caprolactam, hexamethylenediamine condensed with isophthalic, and terephthalic acids is abbreviated as: PA-6IT6, or that from caprolactam, *m*-xylylene diamine and adipic acid as: PA-mXD6, etc.
PA-6	Poly-ε-caprolactam
PA-66	Poly(hexamethylene diamine-adipic acid), polyhexamethylene-adipamide
PA-6IT6	Poly(caprolactam-co-hexamethylene diamine-isophthalic, and terephthalic acids)
PAA	Polyacrylic acid
PAAE	Polyaryl-amide-polyether
PAAM	Polyacrylamide
PABM	Polyamino bismaleimide
PAC	Polyacrylonitrile fibre (also PAN); polyacrylate
PACE	Polyacetylene
PAE	Polyarylether
PAEI	Polyacrylic ester imide
PAEK	Polyaryletherketone

PAES	Polyarylethersulfone
PAI	Polyamide-imide
PAMS	Poly-α-methylstyrene
PA-mXD	Poly(*m*-xylylene adipamide)
PA-mXD6	Poly(*m*-xylylene-diamine, and adipic acid-*co*-caprolactam)
PAN	Polyacrylonitrile
PANI	Polyaniline
PAPI	Polymethylene-polyphenylene isocyanate; also PMPPI
PAr	Polyarylate, $[-\phi-C(CH_3)_2-\phi-CO_2-\phi-CO_2-]_n$, an amorphous polyester
PARA	Polyaryl amide (aromatic, usually amorphous polyamide)
PArSi	Poly(aryloxysiloxane), e.g., poly(dimethylsiloxybiphenyleneoxide)
PAS	Poly(acetoxystyrene)
PAS, PASU	Polyarylsulfone $[-\phi-SO_2-\ \phi-O-]_{0.875}[-\phi-O-]_{0.125}$
PB	Poly-1-butene, polybutylene; elastic polydiene fibre
PBA	Polybutylacrylate [also poly(1,4-benzamide)]
PBAN	Poly(butadiene-*co*-acrylonitrile)
PBCD	Poly(butylene cyclohexane dicarboxylate)
PBD	Poly(butadiene)
PBE	Poly(1-butene-*co*-ethylene)
PBI	Polybenzimidazole
PBMA	Poly-*n*-butyl methacrylate
PBMI	Polybismaleimide
PBN	Poly(butylene-2,6-naphthalene dicarboxylate)
PBNDC	Poly(butylene-2,5-naphthalene-dicarboxylate)
PBO	Poly(*p*-phenylene benzobisoxazole), (also PBZ)
PBOa	Polybenzoxazine
PBS	Copolymer from butadiene, and styrene; see also GR-S, SBR
PB-SMA	Styrene-maleic anhydride-grafted polybutadiene
PBT	Polybutylene terephthalate
PC	Polycarbonate of bisphenol-A
PCD	Polycarbon-di-imide
PCDT	Poly(1,4-cyclohexylene dimethylene terephthalate)
PCHMA	Polycyclohexyl methacrylate
PCI	Poly(1,4-cyclohexylenedimethylene isophthalate)
PCL	Poly(ε-caprolactone)
PCME	Poly(2,2-dichloromethyltrimethylene ether)
PCO	Polycycloolefin
PCTFE	Polychlorotrifluoroethylene; see also CEM, CFM, CTFE
PCTG	Poly(cyclohexane terephthalate-glycol); with $\leq$ 34 mol% ethylene glycol and $\geq$ 66 mol% cyclohexylene dimethanol
PDMS	Polydimethylsiloxane
PDPS	Polydiphenylsiloxane
PE	Polyethylene
PEA	Polyetheramide
PEAc	Polyethylacrylate
PEBAX	Thermoplastic elastomer, polyether-block-amide
PEC	Polyestercarbonate or chlorinated polyethylene; usually CPE
PECO	Polyethylene carbonate,
PEEI	Polyesteretherimide
PEEK	Polyetheretherketone

PEG	Polyethylene glycol (polyoxyethylene)
PEI	Polyetherimide
PEIm	Polyetherimine
PEK	Polyetherketone
PEKEKK	Poly(ether-ketone-ether-ketone-ketone)
PE-MA	Maleated polyethylene (also PE-*g*-MA)
PEMA	Poly(ethyl methacrylate)
PEN	Poly(ethylene 2,6-naphthalene dicarboxylate) or Polyethylene naphthalate
PEOX	Poly(2-ethyl-2-oxazoline)
PEPA	Polyether-polyamide copolymer
PES	Polyethersulfone $[-\phi-SO_2-\phi-O-]_n$
PEsA	Polyesteramide
PEST	Thermoplastic polyesters, e.g., PBT, PET, (also TPES)
PET	Polyethylene terephthalate
PETG	Polyethylene terephthalate glycol; copolymer with 66 mol% ethylene glycol and 34 mol% cyclohexylene dimethanol
PEtI	Polyethyleneimine
PETN	Poly(ethylene terephthalate-co-ethylene naphthalate)
PF	Phenol-formaldehyde resin
PFA	Polyfluoro-alkoxy alkane; copolymer of tetrafluoroethylene and perfluorinated propyl-vinyl-ether
PFEP	Copolymer from tetrafluoroethylene and hexafluoropropylene; also FEP
PH	Phenolics
PHAE	Polyhydroxy amino ether – a thermoplastic reaction product of an amine with diglycidyl ether and epoxy
PHB	Poly(*p*-hydroxybenzoic acid)
PHEMA	Poly-2-hydroxyethyl methacrylate
PHMT	Polyhexamethylene terephthalate
PhPS	Poly(*p*-phenyl styrene)
PHZ	Polyphosphazene
PI	Polyimide
PIB	Polyisobutene
PISU	Polyimidesulfone
PKS	Polyketone-sulfide $[-\phi-S-\phi-CO-]_n$
PLA	Polylactic acid
PMA	Polymethylacrylate
PMAA	Polymethacrylic acid
PMAN	Polymethyl acrylonitrile
PMMA	Polymethylmethacrylate
PMP	Poly-4-methyl-1-pentene; also TPX
PMPhS	Polymethylphenylsiloxane
PMPPI	Polymethylenepolyphenylene isocyanate; also PAPI
PmPV	Poly(*m*-phenylene vinylene-co-2,5-dioctoxy-*p*-phenylenevinylene)
PMQ	Silicone rubbers with methyl and phenyl substituents
PMS	Poly-α-methylstyrene
PMSQ	Poly(methyl silsesquioxane) (ladder-like POSS)
PO	Polyolefin, but also: Elastomeric polypropylene oxide and Phenoxy resin

POSS	Polyhedral oligomeric silsesquioxanes
PP	Polypropylene (the common isotactic; syndiotactic must be marked as *s*PP)
PPA	Polyphthalamide (also polypropylene adipate)
PPAc	Polypropyl acrylate
PPBA	Polyparabanic acid
PPD-T	Poly(*p*-phenylene terephthalamide), *Kevlar*™
PPE	Poly(2,6-dimethyl 1,4-phenylene ether)
PPG	Polypropylene glycol
PPhA	Polyphthalamide
PPI	Polymeric polyisocyanate
PPMA	Poly(phenyl methacrylate)
PP-MA	Maleated polypropylene (also PP-MAH, PP-*g*-MA)
P*p*MS	Poly-*p*-methylstyrene
PP-OH	Hydroxyl modified poly(propylene)
PPP	Poly-*p*-phenylene
PPR	Polypyrrole (also PPY)
PPrA	Poly(*n*-propyl acrylate)
PPS	Polyphenylsulfide
PPSK, PKS	Polyphenylene-ketone-sulfide $[-\phi\text{-S-}\phi\text{-CO-}]_n$
PPSS	Polyphenylene sulfide-sulfone, polythioethersulfone
PPSU	Polyphenylene sulfone or polysulfone; also abbreviated as PSU, PSF, PSO
PPT	Poly(propylene terephthalate) (also PTT)
PPTA	Poly(1,4-phenylene terephthalamide)
PPV	Poly(*p*-phenylenevinylene)
PPX	Poly(*p*-xylylene)
PPY	Polypyrrole (also PPR)
PS	Polystyrene (*atactic*); *isotactic*-PS (*i*PS) or *syndiotactic*-PS (*s*Ps)
PS3Br	Poly(3-bromostyrene)
PSF	Polyphenylene sulfone; Polysulfone, (also PSU, PSO)
PS-GMA	Styrene-glycidyl methacrylate copolymer
PSIR	Polystyrene-*b*-polyisoprene
PSU	Polysulfone $[-\phi\text{-}SO_2\text{-}\phi\text{-O-}\phi\text{-C}(CH_3)_2\text{-}\phi\text{-O-}]_n$, (also PSF, PSO)
PS-VPh	Poly(styrene-*b*-vinyl phenol) block copolymer
PTA	Polytetramethylene adipamide
PTFE	Polytetrafluoroethylene (also TFE)
PTHF	Polytetrahydrofuran (also known as polybutylene glycol, PBG)
PTMA	Polytetramethylene adipate
PTMC	Poly(trimethylene carbonate)
PTMEG	Poly(tetramethylene ether glycol)
PTMG	Polytetramethylene glycol
PTMT	Poly(tetramethylene terephthalate) = polybutylene terephthalate (PBT)
PTO	Polytransoctanylene
PTT	Poly(trimethylene terephthalate)
PU	Polyurethane elastomer
PVAc	Polyvinyl acetate
PVAl	Polyvinyl alcohol
PVBu	Polyvinyl butyrate

PVC	Polyvinyl chloride
PVCH	Poly(vinyl-cyclohexane)
PVDC	Polyvinylidene chloride; also PVC_2
PVDF	Polyvinylidene fluoride; also PVF_2
PVDF-HFP	Poly(vinylidene fluoride-co-hexafluoropropylene)
PVF	Polyvinyl fluoride
PVK	Poly-*N*-vinylcarbazole (also PNVC)
PVME	Polyvinyl methyl ether
PVMO	Polyvinyl methyl-oxazolidone
PVMQ	Silicone rubber with methyl, phenyl and vinyl substituents
PVO	Polyvinyl oxazolidone (PVO)
PVP	Poly-2-viny pyridine or poly(4-vinyl pyridine)
PVPh	Poly(4-vinylphenol), poly(*p*-hydroxy styrene)
SAA	Styrene-acrylic acid copolymer
SAMA	Styrene-acrylonitrile-methacrylic acid copolymer
SAN	Thermoplastic copolymer from styrene and acrylonitrile
SBR	Styrene-butadiene elastomer
SBS	Symmetric (styrene-butadiene-styrene) block copolymer
SCR	Elastomeric copolymer from styrene and chloroprene
SEBS	Styrene-ethylene/butylene-styrene tri-block polymer
SEP	Styrene-ethylene-propylene block copolymer
SIR	Elastomeric copolymer from styrene and isoprene
SIS	Styrene-isoprene-styrene tri-block polymer
SMA	Copolymer from styrene and maleic anhydride
SMI	Copolymer from styrene and maleimide
SMMA	Styrene-methyl methacrylate copolymer
*s*PP	Syndiotactic polypropylene
*s*PS	Syndiotactic polystyrene
SVPh	Styrene-*p*-vinyl phenol copolymer
TMPC	Tetramethyl bisphenol-A polycarbonate (or MPC, TMBPA-PC)
TOR	*t*-Polyoctenamer rubber
TPO	Thermoplastic olefinic elastomer
TPU	Thermoplastic urethanes
TPV	Thermoplastic vulcanisate
TPX	Poly(4-methyl-1-pentene); see also PMP
UHMWPE	Ultrahigh molecular weight polyethylene (over 3 Mg/mol)
ULDPE	Ultra low density polyethylene (*ca.* 900 to 915 kg/m^3)
UP	Unsaturated polyester

7.3 Abbreviations for Organic Cations Used as Clay Intercalants

(see also Tables 17 and 69)

[Ni-(ligand)$_2$]1	[Ni{di(2-aminoethyl)amine}$_2$]
[Ni-(ligand)$_2$]2	[Ni(2,2′:6′,2′′-*ter*-pyridine)$_2$]
2M2BhA	di-methyl di-behenyl ammonium chloride; $(CH_3)_2$-$(C_{22}H_{47})_2$-N^+ Cl^-
2M2DDA	di-methyl di-lauryl (or dodecyl) ammonium chloride
2M2HTA	di-methyl di-hydrogenated tallow ammonium chloride (Arquad 2HT)
2M2OA	di-methyl-di-octyl ammonium chloride
2M2ODA	di-methyl-di-octadecyl ammonium bromide, or chloride
2M2SA	di-methyl di-stearyl (or octadecyl) ammonium chloride (2M2ODA)
2M2TA	di-methyl di-tallow ammonium chloride
2M2TDA	di-methyl di-tetradecyl ammonium chloride
2MBHDA	dimethyl benzyl-*n*-hexadecyl ammonium chloride
2MBHTA	di-methyl benzyl hydrogenated tallow ammonium chloride
2MBODA	di-methyl benzyl octadecyl- (or stearyl) ammonium chloride
2MBSA	di-methyl benzyl stearyl-(or octadecyl) ammonium chloride (2MBODA)
2MEtOHHDA	di-methyl ethoxy hexadecyl ammonium chloride
2MHDI	1,2-dimethyl-3-hexadecyl imidazolium
2MHDODA	di-methyl hexadecyl-octadecyl ammonium chloride
2MHTL8	di-methyl hydrogenated tallow 2-ethylhexyl ammonium methyl sulfate
2MODA	di-methyl octadecyl ammonium chloride
2MVBDDA	di-methyl vinyl-benzyl dodecyl ammonium chloride
2MVBHDA	dimethyl-*n*-hexadecyl-(4-vinylbenzyl) ammonium chloride
3BHDP	tri-butyl-hexadecyl-phosphonium bromide
3MCA	tri-methyl cetyl (or hexadecyl) ammonium chloride (3MHDA)
3MDDA	tri-methyl dodecyl ammonium bromide
3MHDA	tri-methyl hexadecyl ammonium bromide
3MHTA	tri-methyl hydrogenated tallow ammonium chloride
3MODA	tri-methyl octadecyl ammonium chloride
3MSA	tri-methyl stearyl –(or octadecyl) ammonium chloride (3MODA)
3MVBA	tri-methyl vinyl-benzyl ammonium chloride
3OA	tri-octyl ammonium chloride
3PDDP	tri-phenyl dodecyl phosphonium bromide
3PDPP	tri-phenyl di(methoxycarbonyl)phenoxy decyl phosphonium bromide
4MA	tetra-methyl ammonium chloride
AADA	ω-aminoacid-diacetylene
AC12A	ω,ω-dodecyl amino acid (also ADA)
ADA	ω-amino dodecyl (or lauric) acid, or 12-amino dodecyl acid (AC12A)
AHA	ω-amino-hexanoic acid
AHDA	ω-amino-hexadecanoic acid
Allyl-16	dimethyl-*n*-hexadecyl allyl ammonium chloride
AODA	ω-amino octadecyl acid
APB	*di*-hydrochloride of 1,3-bis(3-aminophenoxy)benzene
APTS	γ-aminopropyl-triethoxy silane

ATDA	ω-amino-tetradecanoic acid
ATPS	amine-terminated polystyrene
AUA	amino-undecanoic acid
B3HTA	benzyl tri-hydrogenated tallow ammonium chloride
BDMA	benzyl-dimethyl amine
Bz-16	dimethyl benzyl-*n*-hexadecyl ammonium chloride (also 2MBHDA)
BZD	benzidine or 1,1´-biphenyl-4,4´ di-amine: H_2N-ϕ-ϕ-NH_2
CPC or CA	cetyl (or hexadecyl) ammonium chloride (HDA)
DADA	di-amine-diacetylene
DDA	dodecyl ammonium chloride
DDP	1-dodecyl-2 pyrrolidone
DIP	di-methyl isophthalate tri-phenyl-phosphonium
HA	hexyl ammonium chloride
HDA	hexadecyl ammonium chloride
HDP	hexadecyl (or cetyl) pyridinium (CPC)
M	melamine
M2EPPOH	methyl di-ethyl poly propylene glycol ammonium chloride
M2HTA	methyl di[hydrogenated tallow] ammonium chloride
M3HTA	methyl tri-hydrogenated tallow ammonium chloride
M3OA	methyl tri-octyl ammonium chloride
MADA	mono-amine-diacetylene
MB2HTA	methyl benzyl di-hydrogenated tallow ammonium chloride
MC2EtOH	methyl coco-alkyl di-2-hydroxy ethyl ammonium chloride
MDD2EtOH	methyl dodecyl di-2-hydroxy ethyl ammonium chloride
MHA	2-methacryloyl-oxyethylhexadecyldimethyl-ammonium bromide
MOD2EtOH	methyl octadecyl di-2-hydroxy ethyl ammonium chloride
MR2EtOH	methyl-rapeseed-di-(hydroxy-ethyl) ammonium chloride
MS2EtOH	methyl stearyl (or hexadecyl) bis[2-hydroxyethyl] ammonium (MHD2EtOH)
MSA	methyl stearyl (or hexadecyl) ammonium chloride
MT2EtOH	methyl tallow di-2-hydroxy ethyl ammonium chloride (Ethoquad T12)
OA	octyl ammonium chloride
ODA	octadecyl (or stearyl) ammonium chloride
OH-16	dimethyl-*n*-hexadecyl-(4-hydroxymethylbenzyl) ammonium
OM-2	*N*-[4-(4´-amino-phenoxy)]phenyl phthalimide
OM-l	*N*-[4-(4´-amino-phenyl)]phenyl phthalimide
P18	stearyl-tributyl phosphonium bromide
POE-DPA	polyoxyethylene decyl propyl ammonium chloride
QD1	2-(*N*-methyl-N,N-diethyl ammonium iodide) ethyl acrylate
QD4	2-(*N*-butyl-*N,N*-diethyl ammonium bromide) ethyl acrylate
S2EtOH	stearyl (or octadecyl) di-2-hydroxy-ethyl amine (OD2EtOH)
SA	stearyl (or octadecyl) ammonium chloride (ODA)
SPN	di-ethyl-methyl-oligo(oxypropylene)-ammonium chloride, $(C_2H_5)_2(CH_3)N^+(O\text{-}Pr)_{25}$ Cl
STN	methyl-tri-octil ammonium chloride, $CH_3(C_8H_{17})_3N^+ Cl^-$ (M3OA)
TADA	tallow di-ammonium chloride
TDA	tetradecyl ammonium chloride
TPP	tetra phenyl phosphonium bromide

VB-16	dimethyl –(*p*-vinyl benzyl) *n*-hexadecyl ammonium chloride (2MVBHDA)
VDAC	di-methyl vinyl-benzyl dodecyl ammonium chloride (2MVBDDA)
VM	vinyl benzyl trimethyl ammonium chloride (3MVBA)

7.4 Notations

7.4.1 Notation Roman Letters

$3c$, $c\times 3$	Number of external degrees of freedom in S-S eos (Equations 86-90)
a	Radius of a particle
$a(x)$	Local concentration
$\vec{a}_1, \vec{a}_2$	Unit vectors
a_i	Equation parameters
a_T	Temperature shift factor
A	Hamaker constant for attraction between two infinitely thick slabs
A	Surface or contact area (*viz.* Equations 6, 67-68)
A_2	Second virial coefficient
$b = V_o$	van der Waals occupied volume (Equation 81)
B	Thermodynamic interaction parameter; $B = \chi_{12}RT/V_{1u}$
B, B_o	Isothermal bulk modulus and its value extrapolated to $T = 0$, $P = 0$ (Equation 84)
$B_{1/2}$	Peak width at half-height
c, c_i, C	Concentration
c_i^o	Universal constants in WLF equation
c_s^i	Chemical shift (in NMR)
CED	Cohesive energy density (Equation 91)
C_p	Heat capacity at constant pressure
$\vec{C}_n$	Chiral vector of graphene sheet
C_v	Heat capacity at constant volume
d, d_i	Diameter, diameter of i-th generation of particles in polydispersed suspensions
d_f	Diameter of a filler or fibre particle
$d_{v/s}$	Volume-to-surface average particle diameter
d_{00n}	The n-order interlayer spacing as defined by the simplified Bragg's Equation 2. For the reflection order $n = 1$, the principal spacing is: $d_{001} = nd_{00n}$ (e.g., $d_{001} = 2\ d_{002}$)
D	Diffusion coefficient
D	Extruder screw diameter
$D(t, \sigma_E)$	Tensile creep compliance
D/a^2	Effective diffusion rate (Equation 16)
D_c, D_e	Capillary and extrudate diameter, respectively
DD	Degree of dispersion
D_M	Mutual-diffusion coefficient
D_r	Rotational Brownian diffusion coefficient
D_s	Self-diffusion coefficient
e	Electron; electronic charge
E	Tensile, or Young's, modulus
$E(t)$	Tensile relaxation modulus
$E(t, \varepsilon)$	Tensile stress relaxation modulus
E_a	Activation energy
E_i	Interaction energy
E_{ij}	Exchange energy of i-j contact
E_n	Nucleating energy (Equation 140)

E_R	Relative tensile modulus = E_{CPNC}/E_{matrix}
E_t	Total energy (Equation 33)
f	Free volume fraction
$f(x)$	Function of a parameter x
F	Flexural modulus
F	Helmholtz free energy ($F = E - TS$)
g	Acceleration due to gravity; g = 9.80621 m/s^2 (see level, lat. 45°)
g'	Initial slope of the storage shear modulus: $g' \equiv (d \log G'/d \log \omega)_{\omega < 10}$
g''	Initial slope of the loss shear modulus: $g'' \equiv (d \log G''/d \log \omega)_{\omega < 10}$
G	Crystallisation growth rate (Equation 141)
G	Effective mobility parameter (Equation 73)
G	Gibb's free energy ($G = E - TS + PV = H - TS$)
G	Shear modulus (modulus of rigidity)
$G''(\omega)$	Shear loss modulus
$G(t, \gamma)$, $G(t)$	Shear stress relaxation modulus
$G'(\omega)$	Shear storage modulus
$G^*(\omega)$	Complex shear modulus
$G_i(z)$	Configurational probability in SCF model (Equation 53)
G_Q	Fracture energy
G'_y, G''_y	Yield values for G' and G''
h	Separation distance between plates, e.g., grapheme sheets
$h(P, T)$	Hole fraction in the S-S theory (Equation 87)
H	Enthalpy; also Hamiltonian (Equation 184)
$H(\lambda)$	Relaxation spectrum (Equation 126)
$H_G(s)$	Gross frequency relaxation spectrum
I	Scattering intensity (in XRD, PALS, etc.)
I_s	Intensity of segregation (Equation 17)
J	Flux (Equation 78)
$J''(\omega)$	Shear loss compliance
$J'(\omega)$	Shear storage compliance
J_c	Fracture energy parameter (with two parts, elastic and plastic: $J_c = J_{el} + J_{pl}$)
k	Rate constant (Equation 27)
k, k_i	Electrical conductivity (Equation 185)
k_B	Boltzmann s universal constant
k_H	Huggins constant for solution viscosity
K	Bulk modulus
K, K', K^*	Equation constant (Equations 4-5, 12)
K_{Ic}	Stress intensity factor in mode I
l	Length or distance
l_f	Length of a filler or fibre particle
L	Length (e.g., of extruder screw, CN-tube, fibre, etc.)
m	Number of platelet layers in a stack (Equation 8)
m	Strain ratio in asymmetric extension
M	Maximum value of elastic modulus
M, M_w, M_n	Molecular weight, its weight and number averages
M_e	Entanglement molecular weight (Equation 129)
M_s	Statistical segment molecular weight in S-S eos
M_v	Viscosity-average molecular weight
MWD	Molecular weight distribution = M_w/M_n

n	'Power law' exponent between melt viscosity and the deformation rate
n	Avrami's exponent in the crystallisation rate equation (Equations 144-150)
n	Number of C-atoms in a paraffin group, C_nH_{2n+1} (Equation 103)
n_i^D	Index of refraction (for 20 °C and sodium light)
n_c	Number of divisions engendered by each static mixer element
n_e	Number of elements in a static mixer
N	Extruder screw speed in number of rotations per minute (Equation 29)
N, N_i, $\overline{N}$	Number of molecules, statistical or Kuhn segments (Equations 32-34)
N_1, N_2	First and second normal stress difference, respectively
N_e	Number of polymer segments between entanglements
N_s	Number of striations
p	Anisometric particle aspect ratio (in this book $p \geq 1$ for fibres and disks)
p	Extent of reaction (Equation 26)
P	Pressure
P. P_O	Permeability and oxygen permeability, respectively
P_e	Entrance-exit pressure drop in capillary flow
P_i	Non-functionalised polymer chain length (Equations 54-58)
P_R	Relative permeability = P_{CPNC}/P_{matrix}
q	Wavevector
qz	Number of interchain contacts in a lattice of coordination number z
Q	Heat input or a throughput of an extruder
Q	Mass of substance diffused into interlamellar gallery (Equation 16)
$\left\langle r_\Theta^2 \right\rangle^{1/2}$	Unperturbed radius of gyration
$<r^2>$	Mean square end-to-end distance
r^2	Correlation coefficient squared, precision of data
R	Radius
R	Universal gas constant; 8.31432 J/mol deg
R', R''	Rate of G′ or G″ changes from non-rheological sources (Equation 119)
$R_g = \left\langle r_g^2 \right\rangle^{1/2}$	Radius of gyration
$R_n(\omega)$	Relative magnitude of the odd harmonic peaks (n = 3, 5, 7, ...) divided by the first: $R_n(\omega) = I(n\omega)/I_1(\omega)$ (Equation 117)
R_T	Trouton's ratio of the extensional to shear viscosity (at zero deformation rate)
s	Number of statistical segments per macromolecule (Equations 86-90)
s_i	Specific surface area of i-th particle
S	Clay platelet orientation factor (Equation 189)
S	Entropy
SH	Strain hardening (Equation 136)
$S_{organoclay}$	Surface area of organoclay (Equation 182; in m^2/g)

S_p	Degree of dispersion per 1 μm² of its surface (Equation 21)
S_v	Entropy per unit volume
t	Thickness of a clay platelet or their stack
t	Thickness of a stack of clay platelets (Equations 7-8)
t	Time
$t_{1/2}$	Crystallisation half-time (Equation 145)
$\bar{t}$	Average residence time (in a TSE – see Equations 19, 20)
$\bar{t}_p$	Period of rotation for anisometric particles (Equation 105)
T	Temperature
T_c	Crystallisation temperature
T_g	Glass transition temperature
T_m	Melting point
T_x	Temperature for the mass loss in TGA test by (usually) x = 10 or 50 wt%
u	Exponent characterising polydispersity of interacting domains (Equation 130)
U	Total energy of the system
v	Valence
v^*	Repulsive volume parameter in S-S eos
v_i	Molar volume (Equations 46-47)
V	Volume (specific)
V_f	Free volume
V_L, V_S	Ultrasonic velocity; transverse and shear
w	Concentration in weight percent (wt%) or weight fraction
w_i	Weight fraction of specimen *i*
x	Variable
x_i	Mole fraction
X	Crystalline content (wt%)
$y = 1 - h$	Occupied volume fraction in S-S eos
z	Lattice of coordination number (Equation 53)
z	Variable representing direction perpendicular to the clay platelet ($h_o \leq z \leq h$)
Z	Partition function

7.4.2 Notation – Greek Letters

Greek alphabet (lower case and capital letter)

α, Α	alpha	ν, Ν	nu
β, Β	beta	ξ, Ξ	xi
γ, Γ	gamma	π, Π	pi
δ, Δ	delta	ρ, Ρ	rho
ε, Ε	epsilon	σ, Σ	sigma
ζ, Ζ	zeta	τ, Τ	tau
η, Η	eta	ϕ, (φ), Φ	phi
θ, (ϑ), Θ	theta	χ, Χ	chi
κ, Κ	kappa	ψ, Ψ	psi
λ, Λ	lambda	ω, Ω	omega
μ, Μ	mu		

α	Reduced interlamellar gallery height: $\alpha \equiv h / Na$ (Equation 68)
α_i	Thermal expansion coefficient: $\alpha_i \equiv (\partial \ln V / \partial T)_{P=const}$
α_m	the *m*-th positive root of the Bessel zeroth-order function (Equations 16, 77)
α(t)	Weight fraction of crystalline part after crystallisation for time *t* (Equation 144)
β	Shape factor (Equations 186, 187)
β_i	Compressibility coefficient: $\beta_i \equiv (\partial \ln V / \partial P)_{T=const}$
γ	Shear strain
γ_f	Strain fraction
γ_R	Recoverable shear strain
$\dot{\gamma}$	Shear rate
$\dot{\gamma}_c$, $\dot{\gamma}_y$	Critical value of $\dot{\gamma}$ for onset of dilatancy or yield
Γ	Bead friction (Equation 52)
δ	Solubility parameter
δ_e, δ_v	Adjustable parameters (close to unity) in Equation 101
δ_d, δ_p, δ_h	Hansen's solubility parameters: dispersive, polar and hydrogen bonding, respectively (Equation 50)
Δ	Increment
ΔC_p	Heat capacity gradient
ΔE_η, ΔH_η	Activation energy or enthalpy of flow
ΔG_m	Gibbs free energy of mixing
ΔH_m	Heat of mixing
ΔS_m	Entropy of mixing
ΔT_p	Degree of supercooling: $\Delta T_p = T_m - T_c$ (Equation 140)
Δl	Thickness of the interphase
ε, ε_{11}	Tensile or Hencky strain
$\dot{\varepsilon}$	Strain rate in elongation
ε_b	Maximum Hencky strain at break
ε_{ij}	Pairwise interaction energy between substance i and j (Equations 48, 98-103)
ε^*	Attractive segmental energy of interaction in S-S eos
$\zeta(t)$	Time-dependent fraction of melt-intercalated organoclay (Equations 77, 80)
η, η_σ	Shear viscosity; its value at constant stress
$\eta(z - z')$	Short range interaction function in Equation 53
η_o	Limiting viscosity at zero shear rate, i.e., at the upper Newtonian plateau
η_∞	Limiting viscosity at infinite shear rate, i.e., at the lower Newtonian plateau
η_s	Viscosity of solvent or of continuous medium
η_r	Relative viscosity (η/η_s)
η_{sp}	Specific viscosity ($\eta_r - 1$)
$[\eta]$	Intrinsic viscosity
η_m	Viscosity of a matrix liquid
η_{app}	Apparent viscosity
$\eta^*(\omega)$	Complex viscosity
$\eta'(\omega)$	Dynamic viscosity
$\eta''(\omega)$	Out-of-phase component of complex viscosity
$\eta_E^+(t, \dot{\varepsilon})$	Tensile stress growth coefficient

$\eta_E^-(t,\dot{\varepsilon})$	Tensile stress decay coefficient
η_E	Elongational or tensile viscosity
θ	Scattering angle in, e.g., XRD experiment
θ	Total amount of adsorbed intercalant: $\theta = \rho N_i$
κ	Compressibility (see β)
λ, Λ	Wavelength, e.g., of the X-ray beam
λ	Thermal conductivity
λ	Longest relaxation time (Equations 137, 138)
μ, μ_T^*	Mass absorption coefficient in XRD analysis (Equations 4, 76)
μ_i	Chemical potential of substance i
ν_m	Poisson ratio
ν_{ij}	Interfacial tension coefficient between liquids i and j (Equations 49, 50, 179)
ν, ν_o	Dynamic interfacial tension coefficient and its equilibrium value
ν	Radial frequency in rheological tests (Hz)
Ξ	Components weight ratio, e.g., MMT/PP-MA
ρ	Density
σ	Standard deviation
σ	Characteristic, distance-reducing parameter in Lennard-Jones potential
σ	Tensile strength (Equations 168)
σ^e	Specific surface energy of a growing crystal (Equations 69, 140)
σ^2	Variance (Equation 17)
σ_{ij}^a	*ij* component of the stress tensor
σ_y	Yield shear stress
σ_c	Critical shear stress for droplet break-up
σ_m	Critical shear stress for melt fracture
σ_E	Elastic energy (Equation 111)
$\sigma^+(t,\dot{\gamma})$	Shear stress growth function
$\sigma^-(t,\dot{\gamma})$	Shear stress decay function
$\sigma(t, \gamma)$	Shear stress relaxation function
ρ	Density
ρ	Grafting density of an intercalant onto clay in SCF model
τ	Relaxation time
τ	Thickness of the interphase (Equations 165, 166)
τ_y	Characteristic time of the yield cluster (Equation 130)
τ_3	Ortho-positronium lifetime in PALS experiments
$\tau\zeta$	Lifetime of the density fluctuation
Φ	Electric potential
Φ_i	Farris volume fraction of component i in the mixture
ϕ, ϕ_m	Volume fraction and its maximum value, respectively
χ_{ij}	Thermodynamic interaction coefficient between species *i* and *j*
χ_s	Surface adsorption energy in Equation 72
$\psi(t)$	Retardation function
ψ	First normal stress coefficient approximated by the expression: $\Psi \equiv G'_{corr} / \omega^2 = \Psi_o\left[1 + \left(G'' / G_\Psi\right)^2\right]^{-m_2}$
$\Psi_1^+(t,\dot{\gamma})$	First normal stress growth coefficient
Ω	Vorticity
ω	Radial frequency in dynamic testing (rad/s)

7.4.3 Subscripts

c	critical, crystallisation, or composite
d	diffusion
E	uniaxial extension
f	filler or fibre in composites
g	glass
i	counting subscript, inversion or dispersed phase
m	mixing, melt, matrix
n	number average
o	Initial (or reference) value
p	polymer matrix
s	suspension
o	initial (or reference) value
R	reference or relative
t	total
u	monomer unit
w	weight average
y	yield
z	Z-average

7.4.4 Superscripts

E	excess value, elongation
L	lattice gas model
+	stress growth function
-	decay function
~ (tilde)	Reduced variable
*	complex or reducing variable
s	surface

7.4.5 Mathematical Symbols

<>, ^	Average
Π	Product
π	3.141592654
e	2.302585093
Σ	Sum
→	Vector

7.5 Dictionary

Ablation The decomposition of a material caused by heat.

Accelerated ageing Ageing by artificial means to obtain an indication of how a material will behave under normal conditions over a long period. The conditions are intensified to reduce the time required to obtain deteriorating effects, similar to those resulting from normal service conditions.

Accelerated weathering Duplicating weather conditions or accelerating the normal weathering by means of a device.

Acrylics The name given to plastics produced by the polymerisation of acrylic acid derivatives, primarily methyl methacrylate, resulting in amorphous thermoplastic. In technological jargon either a polymethylmethacrylate, PMMA, or polyacrylonitrile fibre with at least 85-wt% of PAN.

Additive A material added to a polymer during the final synthesis stages or in subsequent processing to improve or alter some characteristic of the polymer. Additives as a class of materials are not intended to increase strength properties. Examples: pigments, lubricants, anti-static agents, flame retardants, and plasticisers.

Adhesion State in which two surfaces are held together at an interface by mechanical or chemical forces, by interlocking action, or both.

Advanced composites Composite materials that are reinforced with continuous fibres having a modulus higher than that of glass fibres. The term includes polymeric matrix, metal matrix, and ceramic matrix composites, as well as carbon-carbon composites.

Ageing The change of a material over time under defined natural or synthetic environmental conditions, leading to improvement or deterioration of properties. Also, changes caused by exposure to physical, and chemical factors (*viz.* light, temperature, chemicals, weather), leading to irreversible deterioration. See also ***Accelerated ageing, Artificial ageing, Chemical ageing, Physical ageing.***

Alkyds Thermosetting resins from polyhydric alcohols, and polybasic acid or anhydrides, prepared by esterification of a polyfunctional alcohol (e.g., glycerin) with phthalic anhydride in combination with fatty acids or rosin acids (molecular weight about 2000 to 5000). These resins are frequently modified by incorporation of, e.g., nitrocellulose (NC), or phenolics. Alkyds are used mainly as lacquers.

Alloy, polymer alloy A material made by blending together polymers or copolymers under selected conditions, e.g., SAN blended with NBR. A mixture of at least two chemically different polymers to form a material with controlled, and stable dispersion, as well as strong interphase adhesion in the solid state.

Ambient temperature Temperature of the medium surrounding an object. The term has been mainly used to denote room temperature (RT).

Amorphous polymer A non-crystalline polymeric material that has no definite order or crystallinity. A polymer in which the macromolecular chain has a random conformation in solid (glassy or rubbery) state. It is of note that as an amorphous polymer may show a short-range order, a crystalline polymer may be quenched to the amorphous state (*viz.* polyethylene terephthalate (PET)).

Anionic polymerisation Chain polymerisation in which the active centre is an anion, usually a carbanion. The method is mostly used to polymerise vinyl monomers carrying electron-withdrawing substituents (e.g., -CN, -COOR, -COR, -aryl). The polymerisation is frequently initiated by *n*-butyl lithium.

Anisometry The difference in the magnitude of the dimensions of a particle that depend on the direction. Thus, a sphere is isometric – it has a minimum of anisometry. In physics it is customary to define anisometry in terms of the aspect ratio, p. Representing a particle by an equi-biaxial ellipsoid of rotation, p is calculated by dividing the diameter in the direction of the main axis of rotation, by the diameter perpendicular to it, *viz.* $p = a_1/a_2$, where $a_2 = a_3$. Thus, for fibre-like particles, p is the length-to-diameter ratio, i.e., $p \geq 1$, while for platelets, p is the thickness-to-diameter, thus $p' \leq 1$. However, in most publications on CPNC the aspect ratio is defined as the diameter of a clay platelet divided by thickness, hence $p = 1/p' > 1$. Macromolecules show high anisometry with a typical value for the chain length-to-diameter being, $p = 1{,}000$.

Anisotropy The material properties being dependent on the direction. Most multiphase polymeric systems show some degree of anisotropy. The mechanical performance in the machine direction can be as much as a hundred times higher than that in the transverse direction. In homopolymers, the anisotropy is a reflection of the molecular orientation in either a glassy or a semicrystalline state (see ***Birefringence***).

Annealing To heat a moulded plastic article to a predetermined temperature, and slowly cool it to relieve stresses. Annealing of moulded or machined parts may be done dry, as in an oven, or wet, as in a heated tank of mineral oil. To relieve the stresses without introducing a major change of the molecular structure in the formed article the annealing is frequently carried out at a temperature being a few degrees *below* the glass transition temperature, T_g. The treatment is also used to increase polymer crystallinity. The process requires keeping the polymer at a temperature $T < T_d$ (where T_d is the thermal degradation temperature). The best results are usually obtained when $(T_g + T_m)/2 \leq T < T_m$, where T_m is the melting temperature.

Antioxidant A substance that, when added in small quantities to the resin during mixing, prevents its oxidative degradation, and contributes to the maintenance of its properties.

Aromatic polymer A polymer containing aromatic ring structures, *viz.* polyamides, polyesters, polyethers, polysulfides, polysulfones, polysiloxanes, etc.

Aromatic Chemicals that have at least one unsaturated ring derived from benzene in their chemical structure. The description is very general, and covers a wide range of chemicals. Many chemicals classified as aromatics have a very different smell or no smell at all.

Artificial ageing The exposure of a plastic to conditions that accelerate the effects of time, such as heating, exposure to cold, flexing, application of electric field, immersion in water, exposure to chemicals, and solvents, ultraviolet, light stability, resistance to fatigue, etc. The accelerated testing of plastic specimens to determine their changes in properties is carried out over a short time. The tests indicate what may be expected of a material under service conditions over extended periods. Typical investigations include those for dimensional stability, mechanical fatigue, chemical resistance, stress cracking resistance, dielectric strength, and so forth, under the conditions that reflect the conditions under which the article will be used. (See also ***Ageing***.)

Aspect ratio, p (see ***Anisometry***) The relative comparison of one dimension of an object to another. For complex objects like a particle of clay, *p* is a relative number approximating the ratio of the longest dimension to the shortest. The ratio determines how much stress can be transferred to the fibres or platelets before being transferred back into the polymer matrix, thus it is key to the effectiveness of a reinforcing particle – the higher aspect ratio, $p < 500$, the larger the reinforcing effect. Particles with $p > 500$ behave as reinforced with infinitely large fibre or flake.

Atactic polymer A polymer in which at least one chain atom in a mer can exhibit stereoisomerism (e.g., -CH_2C^*HX-), but has no preference for one particular configuration, e.g., atactic PVC, PS or PP.

Average molecular weight Summation over the distribution of molecular weights of a polymeric substance, e.g., with respect to the number, M_n or weight, M_w. Depending on the method of determination, M_n, M_w, or higher average molecular weight is obtained.

Batch A quantity of materials formed during the same process or in one continuous process, and having identical characteristics throughout. Also a *Lot*.

Bimodal distribution A probability distribution in which the differential distribution function has two maxima.

Binder The resin or a cementing constituent that holds the other components together. The agent applied to mats or imodal to bond the fibres before moulding.

Binodal The line on the temperature versus composition phase diagram, which separates the metastable region from the two-phase regions. Hence, it represents the limits of stability in a two-phase system, *viz.* a polymer solution or polymer blend.

Binodal distribution A discrete probability distribution based on two possible outcomes, which may be labelled success (with probability *p*) or failure (with probability $q = 1 - p$); the probability function expresses the number of *X* successes in *n* independent trials.

Biopolymer Polymer produced by biosynthesis in nature, e.g., polysaccharides, nucleic acids, proteins, cellulose, lignin, and natural rubber.

Birefringence (Dichroism) The difference between the index of refraction in two directions, measured with polarised light. The birefringence originates in the molecular orientation in either a glassy or crystalline phase. Positive birefringence occurs when the principal optic axis lies along the chain; negative, when it is perpendicular.

Blending Preparation of polymer blends or alloys, usually involving mixing of two polymeric liquids with an appropriate compatibiliser.

Blends see ***Polymer Blends***

Block copolymers A copolymer synthesised from two or more monomers in such a way that monomers of the same kind are arranged in homopolymeric blocks.

Bonds Forces between atoms that hold them in relative proximity to each other, resulting in molecules. Primary bonds, formed by sharing two electrons of two atoms, are the strongest. Secondary bonds are between atoms of different molecules or remote sections of the same molecule. They are due to attractions by polarity, induced polarity, or temporary polarity caused by vibration or spinning. These bonds are weaker than the primary bonds.

Branched chains Side chains attached to the main original chain.

Branched polymer A polymer in which the molecules consist of a linear main chain to which there are randomly attached secondary chain branches, *viz.* low density polyethylene. The fraction of repeat units in a polymer that statistically contain one branch is defined as the *branching density:* $\lambda = \alpha b/n$, where α is the branching coefficient (dependent on the functionality of the branch point), b is the number of branch points, and n is the number of repeat units.

Brittle failure Failure resulting from inability of material to absorb energy, leading to an instant fracture upon mechanical loading.

Brittle or brittleness temperature Temperature at which a polymer exhibits brittle failure under impact conditions – the lowest temperature at which the material withstands a given condition without failure.

Brittle point The highest temperature at which a material fractures in a prescribed impact test procedure.

Bulk or ***apparent density*** Average density of material in a loose or powdered form of plastic (granular, nodular, etc.) expressed as a ratio of weight to volume.

Bulk polymerisation Polymerisation where only monomer, and initiator (or catalyst) are involved. Owing to the heat of polymerisation, and the difficulty of safe dissipation of generated heat the conversion rarely exceeds 50%.

Calendering Passing of sheet material (e.g., PVC) between sets of pressure rollers to produce a smooth finish sheet of desired thickness.

Calorimeter An instrument capable of making absolute measurements of energy absorbed in a material by measuring changes of temperature.

Capillary rheometer Viscometer for measuring the flow properties of molten polymer, made of a capillary tube of specified geometry, a means for forcing the melt through the capillary, a means for maintaining the desired temperature, and a means for measuring differential pressures and flow rates. The stress field in capillary flow is heterogeneous, making this type of measurement unsuitable for the characterisation of multiphase/multicomponent polymeric systems.

Carbon black A black pigment or filler produced by the incomplete burning of natural gas or oil. It is widely used in the rubber industry, and for wire/cable applications. Since it possesses ultraviolet protective properties, it is also used in formulations intended for outside weathering applications.

Carbon fibre Fibres produced by the pyrolysis of organic precursor fibres, such as rayon, polyacrylonitrile (PAN), and pitch, in an inert environment. The term is often used interchangeably with the term graphite; however, carbon fibres, and graphite fibres differ. The differences lie in the temperature at which the fibres are made, and heat-treated, as well as in the amount of elemental carbon produced. Carbon fibres typically are carbonised at around 1315 °C, and contain 94 ± 1% carbon, while graphite fibres are graphitised at 1900 to 2480 °C, and contain ≥ 99% carbon.

Carbon nanotubes (CNT) Hollow tubular structures with a wall thickness of 0.07 nm, and an interlayer spacing of 0.34 nm, are either single- or multi-walled (SWNT or MWNT, respectively). While a SWNT consists of only a single cylinder, a MWNT consists of 2 to 30 concentric tubes. A SWNT has an average diameter of 1.2-1.4 nm, and density ρ = 1.33 to 1.40 g/ml. The theoretical, and experimental values of tensile modulus are E = 1 to 1.5 Tpa and tensile strength 11 to 63 GPa. The electrical resistivity is about 10^{-4} Ω-cm, thermal conductivity *ca.* 2 kW/m/K, thermal stability in vacuum up to 2800 °C, and current density up to 10^{13} A/m^2.

Carreau equation between viscosity, η, and the deformation rate, ω, for monodispersed polymers: $\eta_o / \eta = \left[1 + (\tau\dot{\gamma})^2\right]^{(1-n)/2}$, where η_o is the zero-shear viscosity, τ, is the principal relaxation time, and n is the power-law exponent. For polydispersed systems, the above generalised equation is known as the Carreau-Yasuda equation with two empirical exponents, m_1 and m_2: $\eta_o / \eta = \left[1 + (\tau\dot{\gamma})^{m_1}\right]^{m_2}$ with $n = 1 - m_1 m_2$.

Catalyst A substrate that changes the rate of a chemical reaction without changing its composition or becoming a part of the product molecular structure.

Cation A positively charged ion.

Cationic polymerisation Chain polymerisation in which the active centre is a cation, usually carbonium ion, $-C^+$. The method is used to polymerise vinyl monomers carrying electron-releasing substituents (e.g., alkyl or alkoxy groups). The polymerisation is initiated by an initiator, and co-initiator, *viz.* $BH_3 + H_2O$.

Ceiling temperature Temperature at which the free energy of polymerisation is zero. Thus, above this temperature no further polymerisation takes place.

Cellular plastic A plastic with greatly decreased density because of the presence of numerous cells or bubbles dispersed throughout its mass. See also *Foams*.

Chain polymerisation An addition polymerisation (radical, anionic or cationic) in which a monomer is converted to polymer in a chain reaction. Here initiator activates the monomer to which other monomers are added:

$$I^* + M \rightarrow IM^*$$

$$IM^* + M \rightarrow IMM^*$$

preserving the active status of the terminal mer.

Charpy impact test A destructive test measuring impact resistance, consisting of placing the specimen in a horizontal position between two supports, then striking the specimen with a pendulum striker swung from a fixed height. The magnitude of the blow is increased until the specimen breaks.

Chemical ageing The long-term deleterious effects on a material under defined natural or artificial environmental conditions (*viz.* light, temperature, humidity), leading to irreversible deterioration of properties. A process of exposing plastics to natural or artificial factors for prolonged time. See also: ***Ageing, Accelerated Ageing, Artificial Ageing, Physical Ageing.***

Chemical resistance Ability of a material to retain utility, and appearance following contact with chemical agents.

Chromatography The separation, especially of closely related compounds, caused by allowing a solution or mixture to seep through an absorbent, such that each compound becomes adsorbed in a separate layer.

Clay-chemical complex An association of MMT with an intercalant ionically bonded to the surface. The association is supposed to make the clay compatible with the matrix.

Clay-containing polymeric nanocomposite (CPNC) A polymer or copolymer having dispersed *Exfoliated* individual platelets obtained from an *Intercalated layered material.*

Coefficient of expansion or thermal expansion The fractional change in a specified dimension or volume of a material for a unit change in temperature. Values for plastics range from 10 to 200 parts per million per one °C.

Coefficient of friction A measure of the resistance to sliding of one surface in contact with another. Its value is calculated for a known set of conditions, *viz.* pressure, surface, speed, temperature, and material, to develop a number, either static or dynamic, of the resistance of the material to slide.

Coefficient of regression, or correlation, r Measure of the degree of relationship between a model obtained by regression (curve fitting), and the independent variables.

Co-intercalation Intercalation by means of a charged spacing agent, and another compound, *viz.* a monomer, oligomer, or polymer.

Colloid A system in which at least one component exists in state of fine dispersion with particle diameter d = 1 to 1000 nm. Three types: colloidal dispersions, lyophilic colloids, and colloidal associations are distinguished.

Colorimeter Instrument for matching colours with results approximately the same as those of visual inspection but more consistent.

Commodity resin The term associated with high-volume low-price resins having low-to-medium physical properties, used for less critical applications. The principal five resin types are: PE, PP, styrenics, acrylics and vinyls.

Compatibility Ill-defined term pertaining to the ability of one material to coexist with another without undesirable effects – to be avoided.

Compatibilisation Modification of interfacial properties of an immiscible system (polymer blend, or composite) leading to stable dispersion of a desired size, and shape of the dispersed phase, and good adhesion between phases in the solid state.

Complex modulus The ratio of stress to strain in which each is a vector that may be represented by a complex number. It may be measured in tension or flexure, E^*; compression, K^*; or shear, G^*.

Compliance Tensile compliance is the reciprocal of Young's modulus, E. Shear compliance is the reciprocal of shear modulus, G. The term is also used in the evaluation of stiffness, and deflection.

Composites A solid material that consists of a combination of two or more types retaining their separate identity. In polymer technology the term is reserved for those systems in which additions of solid particles result in a reinforcing effect. Composites are divided into: reinforced filled systems (particle size: $d \leq 50$ nm), short fibre composites, long fibre composites, and continuous fibre composites.

Compound A mixture of polymer(s) with the additives necessary for the finished product. In reinforced plastics, and composites, the intimate admixture of a polymer with fillers, softeners, plasticisers, reinforcements, catalysts, pigments, dyes, etc.

Compounding Preparation of a *compound*.

Concentrate A composition that needs to be diluted by a matrix polymer before the forming stage. In CPNC an intercalate formed by intercalation of a multi-charged spacing/coupling agent, and a co-intercalant polymer, having a clay concentration greater than needed to improve one or more properties of the matrix polymer, so that the concentrate can be mixed with additional matrix polymer to form a nanocomposite composition or a commercial article.

Condensation polymer A polymer obtained in a step-growth polymerisation, often accompanied by elimination of small molecules (e.g., water). Polyesters, polyamides, phenol-, melamine-, and urea-formaldehyde resins are typical.

Conditioning The subjection of a material to standard environmental, and/or stress history before testing, so that it will respond in a uniform way to subsequent testing or processing. The term is frequently used to refer to the treatment given before testing, e.g., $T = 23 \pm 2$ °C, and relative humidity $RH = 50 \pm 5\%$.

Configuration Chemical constitution, a spatial arrangement of atoms in a molecule that cannot be changed without breaking chemical bonds.

Conformation Arrangement of chain elements in space (scale 1 to 30 nm) with a specific reference to bond rotation. For polymers with the carbon-carbon main chain two conformations are important: *trans* or *t,* and *gauche,* or *g*:

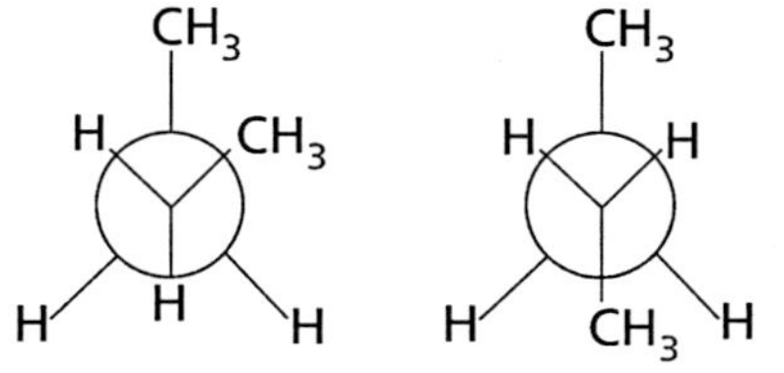

The helical conformations are defined by the long sequences of these two, *viz. tgtgtgtgtgt*, or *tggtggtggtggt*, etc.

Conjugated polymer A polymer with sequence of conjugated double bonds, such as polyacetylene, polyphenylene, dehydrochlorinated PVC, etc.

Continuous use temperature (CUT) Maximum temperature at which material may be subjected to continuous use without fear of premature thermal degradation.

Conventional composite In CPNC technology a composite where the clay acts as conventional filler, and is not nano-dispersed.

Co-polycondensation Polycondensation of different monomers having either a different constitution or different functional groups following different reaction mechanisms.

Copolymer A polymer obtained from polymerisation of two or more monomers, where the repeating structural units of both are present within each molecule. In most cases the term refers to a polymer containing two monomer types, e.g., styrene, and butadiene, SBR. When a copolymer contains three or four different mer species terms: *terpolymer*, *tetrapolymer*, or *multipolymer* may be used. Seven types of copolymers are recognised: *statistical, random, alternating, periodic, graft, block*, and *core-shell.* The copolymers may be prepared by reactive blending, with properties between those of polymers that could be obtained by separately polymerising the composing monomers.

Copolymerisation parameter Ratio of the velocity constants in copolymerisation.

Copolymerisation Polymerisation with more than one species of monomer, which can react with one another, forming a copolymer.

Coupling agent A material used to improve the interfacial properties between two phases. The term most frequently refers to the material used to improve adhesion between polymer matrix, and filler or reinforcing particles.

Covalent bond A bond where one or pairs of electrons are equally shared between two atoms producing a stable electron configuration, and a stable molecule. Covalent are the strongest of the molecular bonds.

Crack A fracture, a separation of material, visible on opposite surfaces of the part, and extending through the thickness.

Crazing Formation of fine cracks on the surface of a polymeric material. The cracks may extend in a network on or under the surface or through the whole thickness. These result from stresses either caused by moulding shrinkage, machining, flexing, impact shocks, temperature changes, action of solvents or ultraviolet radiation.

Creep modulus (apparent modulus) Ratio of initial applied stress to creep strain.

Creep rupture strength Stress required to cause fracture in a creep test.

Creep strength Maximum stress causing specific creep in specific time.

Creep The change in dimension of a plastic under load over a period of time (excluding the initial instantaneous elastic deformation). Owing to viscoelastic nature, a plastic subjected to a load for a period of time tends to deform more than it would from the same load released immediately after application. The degree of this deformation depends on the load duration. Creep is the permanent deformation resulting from prolonged application of stress below the elastic limit. Data obtained in a creep test is presented as creep versus time, with stress, and temperature constant. The slope of the curve is the creep rate, and the end point of the curve is the time for rupture.

Crosslinked polymer A polymer in which the linear macromolecules are joined by a covalent bond or a short sequence of chemical bonds either during the polymerisation or in a post-polymerisation crosslinking reaction (chemical or irradiation crosslinking, curing or vulcanisation). The crosslinked materials are insoluble, and they do not flow when heated.

Crosslinking Chemical reaction between polymeric molecules to form covalently bonded 3D macromolecules. The reaction progresses from a linear chain to branched elastomeric macromolecules, than to hard, and brittle resin, as in thermosetting resins. Crosslinking (by irradiation or chemical means) leads to formation of a single, infusible macromolecule.

Crystalline polymer A material having an internal structure in which the atoms are arranged in an orderly three-dimensional configuration. More accurately *a semicrystalline polymer,* since only a portion (5 to 70%) of polymer is in crystalline form.

Crystallinity A regular arrangement of atoms in a solid state, or the presence of solid crystals with a definite geometry. Such structures are characterised by uniformity, and compactness. The crystalline regions are submicroscopic volumes in which there is a degree of regularity sufficient to obtain X-ray diffraction patterns. High crystallinity causes a polymer to be less transparent, or opaque.

Curing Crosslinking or vulcanising a polymer to improve such properties as modulus, strength, thermal stability, etc.

Cyclopolymerisation Polymerisation leading to ring structures, with usually low molecular weight, and low viscosity. The resulting *prepolymers* or *cyclomers* can be used at a higher temperature in the subsequent catalysed reaction to generate a high molecular weight, linear polymers, *viz.* polycarbonates, polyesters, etc. The *cyclomer* technology facilitates preparation of polymer alloys, composites or nanocomposites.

Deflection temperature under load (DTUL) The temperature at which a simple beam has deflected a given amount under load (formerly called heat distortion temperature).

Deformation Any change of form or shape in a body, in particular a linear change of dimension of a body in a given direction produced by the action of external forces.

Degradation A deleterious change in the chemical structure, physical properties, and/or appearance of a plastic, usually caused by exposure to heat, *viz.* thermal, hydrolytic, oxidative, photo, bio, radiation, etc.

Degree of freedom The number of degrees of freedom in statistical analysis is the number of independent elements used in the computation of that statistic.

Degree of polymerisation The number of mers in a macromolecule, i.e., the number of repeat units in the chain of a molecule, *DP*. In a condensation polymer a repeating unit is composed of a monomer group from each reactive species.

Density The weight per unit volume of a substance, expressed in kg/m^3.

Depropagation or ***unzipping*** A degradation reaction in which the consecutive mers are gradually removed from one macromolecular chain end to another. Few polymers undergo this reverse kinetics process, *viz.* PMMA, POM, PTFE, etc.

Devolatilisation The removal of volatile components during processing.

Diblock copolymer A block copolymer made of two blocks, one having a chain of AAAAA mers, and the other of BBBBB to form AAAAAABBBBBBB polymer. The two block copolymers are used as compatibilisers in the poly-A + poly-B mixtures.

Dichroic ratio The ratio of absorbencies of polarised radiation, usually the infrared region. The dichroic ratio is used as a measure of molecular orientation in oriented polymers. The dichroic ratio may provide information on the orientation in the glassy, and at the same time in the crystalline phase.

Dichroism The dependence of absorbency of polarised radiation on the direction of polarisation. For polymers the magnitude of dichroism, expressed as *dichroic ratio*, depends on the orientation of the radiation absorbing groups, thus the macromolecules. In consequence, infrared dichroism is a powerful method to measure the molecular orientation.

Dielectric constant or ***permittitivity*** Normally the relative dielectric constant, in practice, a ratio of the capacitance of a given configuration of electrodes with a material as dielectric to the capacitance of the same electrodes' configuration with a vacuum or air as dielectric. A relative measure of non-conductance. A low dielectric constant is desired for plastic components used to insulate, and isolate electrical components from each other. High dielectric constant materials are desirable for use as the insulator portion of capacitors, so that the electrical energy can be stored in as small a volume of material as possible.

Dielectric strength The maximum electrical voltage a material can sustain before it is broken down, or 'arced through'. Also an electric voltage gradient at which an insulating material is broken down or 'arced through'.

Dielectrometry An electrical technique to measure changes in loss factor, and capacitance during cure of the resin. Also called ***Dielectric spectroscopy***.

Differential scanning calorimetry (DSC) Thermal analysis technique that measures the quantity of energy absorbed or evolved (given off by a specimen as its temperature is changed). Also measurements of the energy absorbed (endotherm) or produced (exotherm) while undergoing glass transition, melting, crystallising, curing, evaporating of solvents, and other processes involving an energy change.

Differential shrinkage Non uniform material shrinkage in a part caused by non-uniform distribution of stresses, thus orientation.

Differential thermal analysis (DTA) An analysis method in which a specimen and a control are simultaneously heated, and the difference in their temperatures is monitored. The difference provides information on the relative heat capacities, presence of solvents, changes in structure, and chemical reactions. See also ***DSC***.

Diffusion The movement of a material, such as a gas or liquid, in the body of a polymer. If the gas or liquid is adsorbed on one side of a test piece, and given off on the other, the phenomenon is called permeability. Diffusion, and permeability are controlled by chemical not physical mechanisms.

Dimensional stability Ability to retain the precise shape to which it was moulded, cast, or otherwise fabricated.

Discolouration Either a change from an initial colour possessed by a plastic, or lack of uniformity in colour over the whole area of an object, caused either by overheating, light exposure, irradiation, or chemical attack.

Dispersion Finely divided particles of one material suspended in another.

Dispersive mixing A mixing process in which agglomerates are reduced in size by fracture due to stresses generated during mixing, and/or drops of the dispersed phase are deformed and broken.

Dissipation factor Ratio of the conductance of a capacitor in which the material is dielectric to its substance, or the ratio of its parallel reactivity to its parallel resistance. Most plastics have a low dissipation factor, a desirable property because it minimises the waste of electrical energy as heat.

Distribution function A differential or integral description of population. For polymer molecular weight a mathematical description of the polydispersity.

Distributive mixing A mixing process in which the dispersed phase domains are uniformly distributed – a reduction of non-uniformity.

Domain A morphological term used in non-crystalline systems, such as block copolymers or polymer blends, in which the chemically different parts generate amorphous phases.

Double strand polymer Rigid rod 'ladder polymer', consisting of two parallel chains of polymer regularly joined by covalent bonding, *viz.* pyrrones, polyquinoxalines, polyphenylsilsesquioxane, etc.

Drop impact test Impact resistance test in which a predetermined weight is allowed to fall freely onto the specimen from varying heights.

Dry as moulded (DAM) Term used to describe a part immediately after it is removed from a mould, and allowed to cool down. Parts and test bars in this DAM state are felt to be their weakest in some properties as they have not had time to condition or relieve the moulded-in stresses.

Dry blend A compound containing all necessary ingredients mixed in a way that produces a dry, free flowing, particulate material (commonly used for PVC formulations).

Ductility The amount of plastic strain that a material can withstand without fracturing, the extent to which a solid material can be drawn into a thinner cross-section without breaking. Also, the ability of material to deform plastically before fracturing.

Dyes Intensely coloured synthetic or natural chemicals that are soluble in most common solvents, and can be dissolved in a resin to impart colour. Dyes are characterised by good transparency, high-tinctural strength, and low specific gravity.

Dynamic mechanical measurement A technique in which either the modulus, and/or damping of a substance under oscillatory load or displacement is measured as a function of temperature, frequency, time, or their combination.

Elastic deformation A deformation in which a substance completely returns to its original dimensions when the load is removed.

Elastic limit The greatest stress a material is capable of sustaining without permanent strain remaining after the complete release of the stress. A material is said to have passed its elastic limit when the load is sufficient to initiate plastic, or non-recoverable, deformation.

Elastic recovery The fraction of a given deformation that behaves elastically. A perfectly elastic material has an elastic recovery of one, while a perfectly plastic material has an elastic recovery of zero.

Elasticity That property of plastic materials because of which they tend to recover their original size, and shape after removal of a force causing deformation. If the strain is proportional to the applied stress, the material is said to exhibit Hookean or ideal elasticity.

Elastomers A customary name for substances showing the plastic-elastic behaviour, characteristic for vulcanised rubber-like synthetic or natural polymers, *viz.* rubbers, weakly crosslinked polyether, and polyester urethanes, etc.

Electrical strength [dielectric strength] The highest electric stress that an insulating material can withstand for a specified time without the occurrence of electrical breakdown. The property of an insulating material that enables it to withstand electric stress.

Elongation at break Elongation recorded at the moment of rupture of a specimen, often expressed as a fraction or percentage of the original length.

Elongation Deformation caused by stretching, or fractional increase in length of a material in tension, expressed as a percentage difference between the original length, and the length at the moment of the break – higher elongation indicates higher ductility.

Embrittlement Reduction of ductility caused by physical or chemical changes.

Emulsion polymerisation Free radical polymerisation of an emulsion, consisting of an aqueous phase containing an initiator, and an emulsified oil phase containing the monomer.

Emulsion A stable dispersion of one liquid in another created in the presence of an emulsifier (that has affinity with both phases). The emulsifier engenders formation of the third phase, the *Interphase*.

Endothermic Action or reaction that absorbs heat.

Endurance or fatigue lifetime Maximum fluctuating stress a material can endure for an infinite number of cycles – determined from the *S-N* diagram.

Engineering polymer alloy A processable engineering material containing two or more compatibilised polymers, capable of being formed to precise, and stable dimensions, exhibiting high performance at a continuous use temperature above 100 °C, and having tensile strength exceeding 40 MPa.

Engineering polymer blend A polymer blend either containing engineering polymer(s) or having properties of an engineering polymer.

Engineering polymer A processable polymer capable of being formed to precise, and stable dimensions, exhibiting high performance at high temperature, and high tensile strength. Its *CUT* > 100 °C, and the tensile strength > 40 MPa. Five types of polymers belong to this category: PA, PC, PEST (*viz.* PBT, PET, PAr), POM, and PPE.

Environmental stress cracking, ESC The susceptibility of a thermoplastic resin to crack or craze when in the presence of surface active agents or other environments, e.g., under the influence of certain chemicals or ageing, weather, and stress.

Epoxy plastic A thermoset polymer containing one or more epoxide groups, and curable by reaction with amines, alcohols, phenols, carboxylic acids, acid anhydrides, and mercaptans. It has been primarily used as a matrix resin in composites, and adhesives.

Epoxy resin An oligomer containing two or more epoxide groups per molecule crosslinked with a hardener, usually diglycidyl ether of bisphenol-A or pentaerythritol: RH + CH(O)CH- ⇒ -CIH-C(OH)H-. The OH group may further react with the hardener, which leads to highly crosslinked thermoset polymer.

Ethylene plastics Plastics based on polymers of ethylene or copolymers of ethylene with other monomer, the ethylene being in greatest amount by mass.

Exfoliate A surface treated nanoclay, which possesses a sufficiently large interlayer spacing to fully delaminate in a matrix.

Exfoliated layered material Individual platelets (of an *Intercalated layered material* in the form of tactoids comprising 2-10 platelets) dispersed in a carrier material or a matrix polymer with the distance between them > 8.8 nm. The platelets can be oriented, forming *Short stacks* or *Tactoids* or they can be randomly dispersed in a medium.

Exfoliation A process of converting *Intercalate* into *Exfoliate*. During exfoliation platelets from the top, and bottom of the stack peel off – compatibilisation is essential.

Exothermic Pertaining to an action or reaction that gives off heat.

Extensibility (or Extendibility) The ability of a material to extend or elongate upon application of sufficient force, expressed as a percentage of the original length.

Extensometer Instrument for measuring changes in linear dimensions (*viz.* strain gauge).

Extrudate swell (previously *Die swell*) Ratio of the outer parison diameter to the outer diameter of the die. The swell is influenced by polymer nature, die construction, land length, extrusion speed, additives (*viz.* external lubricants), and temperature.

Fatigue ductility The ability of a material to plastically deform before fracturing in a constant-strain amplitude, low-cycle fatigue test. See ***S-N Diagram.***

Fatigue failure The failure or rupture of a plastic under repeated cyclic stress, at a point below the normal static breaking strength. See ***S-N Diagram.***

Fatigue limit The stress below which a material can be stressed cyclically for an infinite number of times without failure. See ***S-N Diagram.***

Fatigue strength Magnitude of fluctuating stress required to cause failure in a fatigue test specimen after specified number of cyclic loading – determined from the *S-N* diagram. Also the maximum cyclic stress a material can withstand for a given number of cycles before failure. The residual strength after being subjected to fatigue. See ***S-N Diagram.***

Fatigue Permanent structural changes that occur in a material subjected to fluctuating stress and strain that cause decay of mechanical properties. See ***S-N Diagram.***

Fibre Often the term is used synonymously with filament having a finite length, $L \geq 100d$, where the diameter is d = typically 100-130 μm. In most cases it is prepared by drawing from a molten bath, spinning, or depositing on a substrate. Fibres can be continuous, long (10 to 50 mm), or short (about 3 mm). In the plastics industry, almost synonymous with thin strands of glass used to reinforce both thermoplastic, and thermosetting materials.

Fibreglass reinforcement Major material used to reinforce plastic, available as mat, roving, fabric, and so forth. It is incorporated into both thermosets, and thermoplastics.

Fibreglass Filaments made by drawing molten glass. Continuous filaments have indefinite length. Staple fibre mat is made of glass fibres with length of *ca.* ≤ 430 mm, depending on the forming or spinning process used.

Fibre-reinforced plastic (FRP) A general term for a plastic that is reinforced with cloth, mats, strands, or any other fibre form.

Filament The smallest unit of a fibrous material. The basic units formed during drawing and spinning, which are gathered into strands of fibre for use as reinforcements. Filaments usually are of great length, and small diameter, d < 25 μm.

Filler A relatively inert substance added to plastics to improve their physical, mechanical, thermal, electrical, or other properties, or to lower cost or density. A compound or substance added to a polymer during the initial synthesis process or in subsequent processing to decrease the volume of resin needed to produce a given product. Fillers are generally much lower in cost than the resins they are used in, thus reducing resin cost per part. Fillers or extenders are generally not used with engineering resins.

Flakes Resin residue formed on the inside of pipes during material transfer, created by the friction of the pellets against the surface of the transfer piping. With time, they build up, flake off, and can cause feed problems at the throat of the extruder.

Flame resistance Ability of material to extinguish flame once the source of heat is removed.

Flame retardants Chemicals used to reduce the tendency of a polymer to burn.

Flame retarded A resin modified by flame-inhibiting additives so that exposure to a flame will not burn or will self-extinguish. Some resins will not burn; others can be modified to meet specifications, while others may not be able to be modified.

Flame treatment A type of surface treatment that oxidises a plastic surface for better reception of paint, inks, and adhesives. See also ***Surface treatment.***

Flammability Measure of the extent to which a material will support combustion.

Flex life The time of heat-ageing that an insulating material can withstand before failure when bent around a specific radius (used to evaluate thermal endurance).

Flexural modulus The ratio within the elastic limit, of the applied stress to specimen's strain during flexural deformation mode testing – a measure of relative stiffness.

Flexural strength Ability of a material to flex without permanent distortion or breaking.

Flock Short fibres of cotton, wood, glass, etc., used as an inexpensive filler.

Flocking A decorating, and/or sound-deadening technique where fibres of different materials are attached to the surface of a plastic part.

Flow curve A log-log plot of isothermal/isobaric viscosity versus deformation rate.

Flow length The actual distance a material will flow during injection moulding. Special mould and test conditions are used to characterise a new compound.

Flow marks Wavy surface appearance of an object moulded from thermoplastic resins, caused by improper flow of the resin into the mould. Also see ***Splay marks.***

Flow or weld line A mark on a moulded piece made by the meeting of two flow fronts during moulding. Also called weld mark.

Flow A qualitative description of the fluidity of a plastic material during processing. A quantitative value of fluidity is expressed by the *flow curve, melt index,* or *melt flow rate (MFR).*

Fluoropolymers The family of fluorinated polymers, *viz.* PTFE, PCTFE, PVDF, characterised by good thermal, and chemical resistance, non-adhesiveness, low dissipation factor, and low dielectric constant.

Foamed or cellular plastics Plastic with numerous cells disposed throughout its mass. Cells are formed by a blowing agent or by the reaction of the constituents. Resins in sponge form may be flexible or rigid; the cells may be open or closed.

Forming The term usually applied to a process in which the shape of plastic pieces such as sheets, rods, or tubes is changed to a desired form.

Fourier transform An analytical method used in advanced forms of spectroscopic or rheological analysis such as IR, NMR, or dynamic flow.

Fractionation A method for separating, and isolating fractions, each with the most uniform molecular weight, and thus of low polydispersity. The process also serves for ascertaining the distribution function.

Fracture strength The normal stress at the beginning of fracture. Calculated from the load at the beginning of fracture during a tension test, and the original cross-sectional area of the specimen.

Fracture Rupture of the surface without complete separation of the laminate, and as complete separation of a body because of external or internal forces.

Free radical polymerisation Polymerisation in which the active centres of reaction are radicals. The polymerisation can be initiated by thermally activated or redox initiator, irradiation or through thermal activation of monomer.

Freeze drying A method of removing volatiles from solidified material at low temperatures.

Gallery Space between parallel layers of clay platelets. The gallery spacing changes depending on what substance occupies the space.

Gaussian (or normal) distribution A symmetrical, bell-shaped distribution function: $y = \left[1/\sigma(2\pi)^{1/2}\right]\exp\left\{-\left[(x-\bar{x})/\sigma\right]^2/2\right\}$, where x is a variable, and σ is the standard deviation.

Gel or Trommsdorf effect Auto-acceleration at the end of chain growth polymerisation. With increasing size of the macroradicals their mobility decreases, and terminations are less frequent. However, the diffusion of the monomer remains unhindered, and the polymerisation proceeds exothermally, resulting in auto-acceleration.

Gel permeation chromatography (GPC); see ***Size exclusion chromatography (SEC)***

Gel point or time The stage at which a liquid begins to exhibit pseudoelastic properties.

Gel solutions Concentrated solutions, intermediate between gel and sol.

Gel That crosslinked part of polymer in liquid state, which having its network character may swell but not dissolve.

Glass transition temperature (T_g) Temperature range in which amorphous solid changes from being vitreous to being viscous. For organic polymers (-160 $\leq T_g$ °C $\leq$ 400) for the transition to occur 20 to 50 main chain atoms must be able to move in a cooperative manner. The transition is kinetic in nature, depending on the rate or frequency.

Glass A material that solidifies from the molten state without crystallisation, a supercooled liquid whose shear viscosity is $\eta \geq 10^{12.5}$ Pas, a liquid whose rigidity is great enough to be put to use. A typical glassy material is hard and brittle (tensile modulus $E \approx 70$ GPa, tensile strength $\sigma \approx 0.5$ GPa). Typical polymeric glasses are: atactic PS, PMMA, PC, etc.

Gloss Brightness or luster of a plastic resulting from a smooth surface. See ***Specular gloss.***

Grades Polymers that belong to the same chemical family, produced by the same manufacturer with performance affected by differences in molecular weight, additives, or other structural features, e.g., flame resistant, glass fibre reinforced, conductive, etc.

Graft copolymers A copolymer whose macromolecules consist of two or more macromolecular parts of different composition, covalently joined in such a way that one of the parts forms the main chain (polymer A) and the other(s) the side chains (polymer-B). Graft copolymers are frequently used as compatibilisers.

Graphite fibre A fibre from either pitch or polyacrylonitrile (PAN), precursor by oxidation, carbonisation, and graphitisation processes (see also ***Carbon fibre***).

Graphite A crystalline allotropic form of carbon.

Hardeners Polyfunctional substances able to cause crosslinking in thermosets.

Hardening Chemical crosslinking of material into a viscous liquid or hard solid.

Hardness The resistance to compression and surface indentation, usually measured by the depth of penetration of a blunt point under a given load using a particular instrument according to a prescribed procedure. Among the most important methods of testing this property are *Barcol, Brinnel, Knoop, Mohs, Rockwell,* or *Shore hardness.*

Haze The cloudy or turbid aspect of appearance of an otherwise transparent specimen caused by light scattered from within the specimen or from its surface.

Heat resistance The property or ability of plastics and elastomers to resist the deteriorating effects of elevated temperatures.

Heat stability Resistance of a plastic to chemical deterioration caused by exposure to heat during processing.

Heat stabiliser An ingredient added to a polymer to improve its processing or end use resistance to elevated temperatures. The term is used in different contexts depending on the benefit to be derived from the additive. For processing, it retards changes of colour. For end-use, it protects the surface of the part exposed to elevated temperatures.

Heat-distortion point or ***temperature (HDT)*** Temperature at which a standard test bar deflects by an arbitrary value under a stated load. In ASTM D 648, it is defined as a total deflection of 250 μm in a rectangular bar supported at both ends under a load of 0.5 or 1.8 MPa. The temperature is increased at a rate of 2 °C/min.

Helix A helical conformation of a polymeric chain in which all main chain atoms can be placed on a cylindrical surface in such a way that all elements on that surface are cut at constant angle, or in other words, that the conformation is exactly repeated at constant intervals. For example, 3_1 helix in PP has three repeating units per one helix turn.

Heterogeneous Materials consisting of identifiable dissimilar constituents separated by internal boundaries. It is noteworthy that not all non-homogeneous materials are necessarily heterogeneous.

H-MMT Protonated montmorillonite.

Homologous polymer blend A mixture of two or more homologous polymers, usually narrow molecular weight distribution fractions of the same polydispersed polymer.

Homologous polymers Polymers identical in structure and composition, differing only in *MW*. A polydispersed polymer is a miscible blend of homologous polymers.

Homopolymer The product of polymerisation of a single monomer, i.e., a polymer containing one type of repeating units, for example: $-[CH_2CHX-CH_2CHX]_n-$.

Hooke's solid An ideal elastic material where stress is directly proportional to strain.

Hoop stress The circumferential stress in a material of cylindrical form subjected to internal or external pressure.

Hydrolysis Chemical decomposition of a substance involving the addition of water.

Hydrophilic Materials characterised by strong dipole moments capable of absorbing water, virtually immiscible with an organophile.

Hydrophobic Material capable of repelling water.

Hygroscopic Material capable of adsorbing and retaining atmospheric moisture from air. Plastics such as PA, PEST, or ABS, are hygroscopic, and must be dried before moulding.

Hysteresis The failure of a property that has been changed to return to its original value when the cause of the change was removed. The area of the resulting elliptical hysteresis loop is equal to the heat generated in the system.

Immiscible polymer blend Blend whose free energy of mixing is positive: $\Delta G_m > 0$.

Imogolite A weathered pumice, also synthetic paracrystalline phyllosilicate of composition $Al_2O_3 \times SiO_2 \times nH_2O$, and molecular weight per crystalline unit cell of 4754. Imogolite occurs as tubes of several micrometres in length, having inside diameters of 1.0 and outside diameters of 2.0 – 2.52 nm; occasionally the tubes may branch out. The nanotubes have good rheological, adsorptive, and surface properties caused by their unique structure, and functional groups on the surface.

Impact strength The ability of a material to withstand shock loading, expressed as an amount of energy required to fracture a specimen subjected to impact. The relative susceptibility of plastic articles to fracture under stress applied at high speeds.

Impact test Often associated with the Gardner (ball or falling dart) test, with a known weight falling at a known distance, and hitting a part, thereby subjecting it to an instantaneous high load. ASTM impact tests for material properties are the Izod, Charpy, and Tensile Impact tests. The test can also be a pendulum-type.

Imperm® High barrier nanocomposite from Nanocor, based on Nylon MXD6*, with enhanced barrier to gases, water vapour, and hydrocarbon fuels. Used as interlayer in PET containers.

Infrared spectroscopy or spectrometry A technique used to observe the wavelengths in the electromagnetic spectrum lying beyond the red, from about 750 nm to a few mm.

Infrared (IR or ***FTIR)*** Pertaining to that part of the electromagnetic spectrum between the visible light range, and the radar range. Radiant heat is in this range, and infrared heaters are frequently used in the thermoforming and curing of plastics, and composites. Infrared analysis is used for identification of polymer constituents. The powerful Fourier transform infrared spectroscopy (FTIR), uses a method of splitting a beam into two waves, and the spectral information is obtained from the phase difference between the two waves, recombining them in the Michelson interferometer. The interferogram is obtained by digitising the detector signal and transforming it, by means of the Fourier transform, from the time domain into a conventional IR spectrum.

Inhibitor A substance that reacts with the active polymerisation site to form either a totally non-reactive product or reducing the system reactivity. In radical polymerisation the radical scavengers, *viz.* diphenyl-picryl-hydrazyl or ***Initiator*** Either an additive mixed in a material to cause a chemical or physical reaction in the melt, or a substance able to engender reaction of a monomer, radical or ionic.

Inorganic polymer A polymer with high proportions of non-carbon atoms. In polymer science the inorganic materials containing organic groups are considered inorganic polymers; e.g., polysiloxanes (silicones), phosphonitrilic elastomers, polycarboranes, organometallic polymers, polymetaphosphates, polyphosphazenes, sulfurnitrides, etc.

Inorganic A mineral compound not composed of carbon atoms.

Intercalant Material capable of diffusing into a clay gallery, and binding to the surface, forming an *Intercalate.* Often the intercalant is an onium salt, for example, $C_{18}H_{37}$-$NH_4^+Cl^-$, that bonds ionically with the platelet anion. Intercalation may involve organic or inorganic salts, monomers, polymers, etc.

Intercalated material *Layered material* with organic or inorganic molecules inserted between *Platelets,* thus increasing the interlayer spacing between them to at least 1.5 nm. As a result of single-step intercalation with an onium compound the interlayer spacing usually expands to d_{001} = 1.5 to 3.0 nm. In co-intercalation the spacing is expected to expand by an additional *ca.* 2 nm (the optimum interlayer spacing of an intercalate is 3.5 to 4.5 nm).

Intercalating carrier A carrier comprising water with or without an organic solvent used to form an *Intercalating composition* with onium compound, capable of achieving *Intercalation* of the *Layered material.* During co-intercalation with polymerisable monomers or oligomers, or with a macromolecular species a suitable solvent, frequently non-aqueous, must be used.

Intercalating composition A composition comprising an onium compound, and/or co-intercalating monomer, oligomers or polymer (*Intercalant*), an appropriate *Intercalating carrier* and a *Layered material.*

Intercalation forming an *Intercalate.*

Interface The boundary or surface between two different, physically distinguishable media. With fibres, the contact area between the fibres, and sizing or finish. In laminates, the contact area between the reinforcements, and the laminating resin. In CPNC the *Interphase* adjacent to the platelet surface is distinct from the matrix as it is composed of ionic species, hydroxyl groups, intercalant molecules, and (hopefully) some matrix polymers.

Interlamellar gallery thickness Interlayer *d*-spacing less the mineral layer thickness.

Interlayer spacing (d-spacing, d_{001} or basal spacing) The thickness of the repeating layers as seen by XRD. In MMT d_{001} is the mineral layer thickness (0.96 nm) plus the galley thickness.

Interpenetrating polymer network (IPN) Historically, any material containing two or more polymers, each in network form, without induced crosslinks between the individual polymers, usually produced by polymerising, and/or crosslinking at least one component in the immediate presence of the other, thus thermoset in character. Currently, the term *IPN* encompasses thermoplastic co-continuous polymer blends, as well as ionomers, block, and graft copolymers. The latter materials are known as *Thermoplastic IPN.*

Interphase The boundary region between two phases in polymer blends, the matrix polymer, and the dispersed phase, or (in the case of phase co-continuity) between two polymeric phases. In compatibilised blends, the ***interphase*** contains the compatibiliser as well as low molecular weight additives, and fractions.

Intrinsic viscosity The limiting value (at infinite dilution) of the ratio of the specific viscosity of the polymer solution to concentration.

Ion exchange resins Crosslinked polymers that form salts with ions from aqueous solutions.

Ion-dipole interaction Chemical bond formed between an ion, and a molecule with a dipole moment having a partial localised charge. For organoclays the complex has a definite mass ratio of organic to clay component.

Ionic polymers or ***Ionomers*** Polymers of linear or network structure with ionic groups which by addition of the appropriate counter ions can be ionically crosslinked. A copolymer of ethylene and acrylic acid is used as a compatibiliser in polyamide blends. Converted to ethylene-zinc acrylate copolymer, *Surlyn*™, is used as packaging film. Other ionic polymers are applied as polyelectrolytes, ion exchange resin, etc. Ionomers show transparency, tenacity, resilience, and increased resistance to oils, greases and solvents.

Isomeric polymers Polymers that are basically homogeneous but in which, by secondary alterations or by a small number of different kinds of linking of the primary molecules (e.g., branching), variations are introduced.

Isotactic polymer A polymer in which the constitutional repeating units have the same stereochemical configuration, e.g., isotactic PP.

Izod impact strength Determination of the resistance of a plastic material to a shock loading, in which a notched specimen bar is held at one end and broken by striking. The absorbed energy is measured.

Izod impact test A type of pendulum impact test in which a notched sample bar is held at one end, and broken by a blow. This is a test for shock loading.

Ladder polymer A rigid rod polymer, consisting of two parallel macromolecular chains regularly joined by covalent bonding, forming a sequence of interconnecting rings, e.g., pyrrone, polyquinoxalines, polyphenylsilsesquioxane, poly(bisbenzimid-azobenzo-phenanthroline).

Lamella The basic morphological unit of a crystalline polymer, usually ribbon-like or plate-like in shape. Generally (if ribbon-like), about 10-50 nm thick, 100 nm wide, and 1000 nm long. Clay platelets, or lamellae, are about 1 nm thick, usually rectangular, e.g., 100x500 nm hence having an aspect ratio of $p= 250$.

Lamellar thickness A characteristic morphological parameter, usually estimated from XRD studies or electron microscopy, usually 1 to 50 nm.

Laminar flow Flow of thermoplastic resins accompanied by solidification of the layer in contact with the mould surface that acts as an insulating tube through which material flows. This type of flow is essential for duplication of the mould surface.

Latex, latices An aqueous dispersion of polymeric particles, a polymer emulsion, a product of emulsion polymerisation used in paints, adhesives, coatings, etc.

Layered double hydroxides (LDH) Materials with the idealised structure: $[M^{II}_xM^{III}_{1-x}(OH)_2]_{intra}[A^{m-}_{x/m}\cdot nH_2O]_{inter}$, where M^{II} and M^{III} are metal cations, A is the anion, and intra and inter denote the intra-layer domain and the inter-layer space, respectively. The structure consists of edge-sharing $M(OH)_6$ octahedra. Partial M^{II} to M^{III} substitution induces a positive charge for the layers, balanced with the presence of the interlayer anions. LDH are often prepared via co-precipitation using M^{II} and M^{III} salts at constant *pH*, mostly basic conditions. Their charge density = 0.25 to 0.40 nm^2/charge, CEC = 0.6 to 4.8 meq/g.

Layered material Synthetic or mineral crystalline inorganic compound, such as smectite clay (e.g., see Montmorillonite), formed of adjacent layers with a thickness, for each layer being in the range of 0.3 to 1 nm.

Light-resistance The ability of a plastic material to resist fading after exposure to sunlight or ultraviolet light. Light stability is the measure of this resistance.

Li-MMT Lithium montmorillonite,

Liquid crystal polymer (LCP) A thermoplastic polyamide, and/or polyester that contains primarily benzene rings in its backbone, is melt processable, and can be highly oriented during processing.

Living polymer An ionic or radical polymer in which, in the absence of a monomer, the active centres of polymerisation are preserved.

London dispersion forces Weak intermolecular forces based on dipole-dipole interactions.

Long chain branching Branches of comparable length as the main polymer chain as in low density polyethylene, polyvinyl chloride, etc.

Loss modulus A quantitative measure of energy dissipation, defined as the ratio of stress 90° out of phase with oscillating strain to the magnitude of strain. The loss modulus may be measured in tension, E'', compression, K'', or shear, G''.

Lower critical solution temperature (LCST) The lowest temperature of immiscibility, where ***Lubricants*** Processing aids to assist material flow during extrusion or injection moulding. Internal and external lubricants are recognised. Internally lubricated resins use oils, ***Teflon***™, MoS_2, or other materials to give the moulded part a lower coefficient of friction. The external lubricant can be a solid, such as sodium or zinc stearate, a fluoropolymer or silicone resin, or liquid.

Macromer An oligomeric or telomeric chain capable of polymerisation.

Macromolecule A large molecule in which neither the end groups nor the substitution of a group has any significant influence on the material properties.

Mass spectrometer An instrument capable of separating ionised molecules of different mass/charge ratios, and measuring the respective ion currents.

Masterbatches A resin that contains high loadings (40-70%) of pre-dispersed additive, *viz.* pigment, or nanoclay. Such a masterbatch may be diluted with additional resin to form the desired compound, e.g., CPNC with clay loadings of 1 to 5 wt%.

Matrix polymer A thermoplastic, thermosetting or elastomeric polymer in which the *Intercalate* or *Exfoliate* is dispersed to form a *CPNC*.

Mean or Average The sum of values divided by their number.

Mechanical properties The properties related to the relationships between stress and strain, such as compressive and tensile strengths and moduli, associated with elastic and inelastic reaction to the applied force.

Melt flow index (MFI), now melt flow rate (MFR) The amount of a thermoplastic forced through a 2.10 mm orifice when subjected to the prescribed force, e.g., 2.16 kg force during 10 min at the prescribed temperature using an extrusion plastometer. It is customary to refer to the flow rate of PE as MFI, but the term MFR should now be used

Melt fracture An elastic strain set up in a molten polymer as the polymer flows through the die. It may show up as irregularities on the surface of the plastic. Several stages (and different mechanisms responsible for these) are recognised, *viz.* sharkskin, pressure oscillation (or spurt), gross distortions, etc.

Melt strength The strength of a plastic while in the molten state.

Melting point Temperature at which a resin changes from a solid to a liquid.

Metallocenes Metallo-organic sandwich compounds in which two cyclopentadienyl (Cp), rings form a sandwich around a metallic ion of, e.g., Fe, Co, Ni, Cr, Ti, V, Zr, etc. They have been used to catalyse the coordination polymerisation of olefinic or vinyl monomers into highly regular macromolecules, *viz.* with narrow *MWD*, high regularity of co-monomer placement, and/or high tactic purity. For example, ethylene was catalysed with $R'_s(Cp)_2MeQ$ [Me is metal from Group 4b, 5b, or 6b (preferably Zr), R′ is a C_1-C_4 alkylene radical, a dialkyl germanium or silicone, Q is an alkylidene radical having from 1 to about 20 carbon atoms, s = 0-1, p = 0-2, m = 4-5], in combination with alumoxanes.

Microstructure The molecular structural features of a single polymer chain: tacticity, isomerism, chain branching, structural irregularities, etc.

Miscibility A system, homogeneous down to the molecular level, associated with the negative value of the free energy of mixing, $\Delta G_m \leq 0$, and $\partial^2 \Delta G_m / \partial f^2 > 0$. In CPNC only the thermodynamic miscibility ascertains stable dispersion of clay platelets in a homogeneous polymeric matrix.

Miscible polymer blend A polymer blend, homogeneous down to the molecular level, in which the domain size is comparable to the macromolecular dimension, associated with the negative value of the free energy, and heat of mixing, $\Delta G_m \approx \Delta H_m \leq 0$, and $\partial^2\Delta G_m/\partial f^2 > 0$. Operationally, it is a blend whose domain size is comparable to the dimension of the macromolecular statistical segment.

Mixing General term associated with the physical act of homogenisation (e.g., mixing of fractions). Mixing of liquids is usually called *blending* (e.g., preparation of polymer blends or alloys), while incorporation of solids into molten polymer is usually called *compounding* (e.g., preparation of a *compound*).

MMT Montmorillonite – a smectite clay formed by hydrothermal alteration of volcanic ash. These 2:1 phyllosilicates have a triple layer sandwich structure that consists of a central octahedral sheet dominated by alumina, bonded to two silica tetrahedral sheets by oxygen ions that belong to both sheets. The thickness of the sandwich is *ca.* 0.96 nm, including the weakly bonded counterion/water expandable layer. MMT shows a broad spectrum of composition and properties.

Modulus of elasticity The ratio of the stress to the strain in elastically deformed material.

Modulus of resilience The energy that can be absorbed per unit volume without creating a permanent distortion. Calculated by integrating the stress-strain curve from zero to the elastic limit, and dividing by the original volume of the specimen.

Moisture absorption The pick-up of water vapours from the atmosphere. It must be distinguished from *water absorption*, defined as water take-up during immersion.

Moisture adsorption Surface retention of moisture from the atmosphere.

Moisture vapour transmission rate (MVTR) The rate at which water vapour permeates through a plastic film or wall at a specified temperature, and relative humidity.

Molecular weight distribution (MWD) A statistical description of the sizes, and frequency of occurrence of different molecular chain lengths within a polydispersed polymer, i.e., the distribution of molecular sizes. Several analytical functions *f(M)* have been proposed, *viz.* general statistical expressions (e.g., log-normal distribution, q.v.) or based on the assumed kinetics of polymerisation (e.g., Schultz-Flory distribution). *MWD* is normally determined using size exclusion chromatography, SEC. Wide and skewed distributions result in significant variance in properties.

Molecular weight, or average molecular weight (MW) The sum of the atomic masses of the elements forming the molecule expressed in units of 1/12 the mass of ^{12}C atom, or a mass of one mole of the substance (kg/mol), indicating the relative size or typical chain length of the polymer molecule. Owing to polydispersity the molecular weight of a polymer is expressed as an average: $M_k = \sum_i N_i M_i^{k+1} / \sum_i N_i M_i^k$

For $k = 1$ the M_n, for $k = 2$, the M_w, for $k = 3$ the z-average M_z, for $k = 4$ .the M_{z+1}-average molecular weight is generated.

Molecularly homogeneous Polymers with molecules having the same chemical structure.

Molecule A group of atoms bonded together that forms the fundamental structural unit of an organic substance. The number of atoms can range from two to millions. A molecule is the smallest unit of a substance that still retains the properties of that substance.

Monomer (= single unit) A small molecule of an organic substance capable of entering into a polymerisation reaction with itself or with other similar molecules, *viz.* vinyls, dienes, α, ω-lactams, diamines, etc. Monomers are either gases or liquids. When bonded together in long chains they form polymers.

Montmorillonite (MMT) A common smectite, the most frequently used in CPNC.

Morphology The study of the physical form and structure of a material. The overall physical form of the physical structure of a material, e.g., domains, lamellae, spherulites, etc.

Multiblock copolymer A block copolymer with more than three blocks, e.g., $-[AB]_n-$.

Multi-charged agent An organic compound that includes at least two positively charged atoms, such as two or more protonated N^+-atoms (quaternary ammonium), P^+-atoms (phosphonium), S^+-atoms (sulfonium), O^+-atoms (oxonium) or their combination spaced by 3 to 6 *C*-atoms. When dissolved, an anion may dissociate leaving a multi-charged cation molecule having positively charged atoms on opposite ends of the molecule.

Na-MMT Sodium montmorillonite.

Nanocomposite technology The materials and processes required for dispersing nanoscale particles in a matrix.

Nanocomposites (NC) Materials that comprise a dispersion of nanometre-size particles in a matrix. The matrix may be single or multi-component. It may contain additional materials that add other functionalities to the system (e.g., reinforcement, conductivity, toughness, etc.). The matrix may be metallic, ceramic, or polymeric.

Nanomer® MMT-based organoclays (or masterbatches containing these) from Nanocor that are used for the preparation of CPNC.

Nanoparticles Particles having at least one dimension below 10 nm. They are classified as: lamellar, fibrillar, tubular, spherical, and others.

Necking The localised reduction in cross-section that may occur in a material under tensile loading during a tensile test. Necking is disregarded in calculation of the engineering stress, but is taken into account in determining the true stress.

Network polymer A crosslinked polymer forming infinite network, obtained in a step-growth polymerisation with multifunctional monomers.

Newtonian fluid An ideal fluid characterised by a constant ratio of the shear stress to the shear rate with zero normal stress difference (non elastic).

Non-destructive evaluation (NDE) An analysis to determine whether the material is acceptable for its function.

Non-destructive inspection (NDI) A process or procedure, such as ultrasonic or radiographic inspection, for determining the quality or characteristics of a material, part, or assembly, without permanently altering the subject or its properties.

Nonpolar Incapable of having a significant dielectric loss (PS and PE are nonpolar).

Non-rigid plastic A plastic that has a modulus of elasticity (either in flexure or in tension) of $\leq$ 69 MPa at 25 °C and 50% relative humidity.

Notch sensitive Material is notch sensitive if it will break when it has been scratched, notched, or cracked. Glasses are highly notch sensitive.

Notch sensitivity A measure of reduction in load-carrying ability caused by stress concentration in a specimen. Brittle plastics are more notch sensitive than ductile.

Nuclear magnetic resonance spectroscopy (NMR) When an organic molecule containing certain atoms (e.g., ^{13}C, H, D, F), is placed in a strong magnetic field, and irradiated with radio frequency, a transition between different nuclear spin orientational states take place, and energy is absorbed at specific frequencies. Different types of NMR tests have been developed, *viz.* high resolution NMR of polymer solutions, wide-line of solid state, magic-angle spinning, pulse-induced, etc.

Nucleating agent A foreign substance, often crystalline, usually added to a crystallisable polymer to increase its rate of solidification, and reduce the size of spherulites.

Nucleation Any additive that assists or starts crystallisation of a polymer. These initiators can reduce cycle time by speeding up crystalline formation, thereby causing the part to solidify faster so its ejection from the mould can occur sooner.

Nylon™ A generic term for polyamides (a trade mark of E.I. Du Pont de Nemours, from New York and London). To be avoided – use *Polyamide* (PA), instead.

Olefins Plastics produced from olefins, *viz.* polyethylene or polypropylene.

Oligomer Low molecular weight polymeric material with the degree of polymerisation, $10 < DP < 50$; from the Greek *oligos* = few, little.

Onium ion intercalation Formation of a clay-chemical complex using an intercalant containing an ammonium or phosphonium functional group. The groups modify a nanoclay surface by ionically bonding to it, converting the surface from hydrophilic to organophilic.

Opalescence The limited clarity of vision through a sheet of transparent plastic at any angle, caused by light scattering within or on the surface of the plastic.

Opaque Materials that will not transmit light; non-transparent.

Ordered polymer A polymer with monomers arranged in regular sequence, *viz.* alternating or block copolymers.

Organic Refers to substances whose composition is based on the element carbon.

*Organophilic*Chemical structure having affinity for (or strong attraction to) organic compounds such as hydrocarbons. An antonym to hydrophilic, i.e., miscible with water.

Organosol Fine PVC suspension in a volatile, organic liquid. At room temperature the resin is swollen, but not appreciably dissolved. At elevated temperatures the liquid evaporates, and the residue upon cooling forms homogeneous plastics. Plasticisers may be dissolved in the volatile liquid (see also *Plastisols*).

Orientation The alignment of the molecular structure in polymeric materials to produce anisometric material properties. Drawing or stretching during fabrication, especially at low temperatures, may accomplish orientation.

Oxidation Any chemical reaction involving oxygen to form new compound, e.g., an oxide; the deterioration of an adhesive film due to atmospheric exposure; the break down of a hot melt adhesive due to prolonged heating, and oxide formation. Degradation of a material through contact with air.

Oxygen index The minimum concentration of oxygen expressed as a volume percentage in a mixture of oxygen and nitrogen that will just support flaming combustion of a material initially at room temperature under the specified conditions.

Permeability The passage or diffusion of a gas, vapour, liquid, or solid through a barrier without affecting it.

pH $= -\log[H^+]$, a measure of the acidity or alkalinity of a substance. Acid solutions have pH < 7, at neutrality pH = 7, and in alkaline solutions pH > 7.

Phase A separate, but not necessarily separable, portion of a system.

Photodegradation Degradation caused by visible/ultraviolet irradiation which is the main cause of outdoor weathering.

Photoelasticity Experimental technique for the measurement of stresses and strains in material objects by means of mechanical birefringence.

Physical ageing The relaxation process that takes place in plastics after fabrication. Upon cooling a melt, the molecular mobility decreases, and when the relaxation time exceeds the experimental time scale, the melt becomes a glass in non-equilibrium thermodynamic state (density, enthalpy, etc.). Thus the value of the thermodynamic parameters continues to change toward an equilibrium state. The process may lead to the development of cracks, and crazes that initiate critical failure. See also: ***Ageing, Accelerated Ageing,*** etc.

Physical crosslink A physical bond that joins two or more chains together. This may arise from crystalline portions of a semicrystalline polymer, the glassy or crystalline portion of a block copolymer, or the ionic portion of an ionomer. The temperature affects the physical crosslink forces.

Pigment Imparts colour to plastic while remaining a dispersion of undissolved particles.

Pigmented A resin comprising pigments to produce a desired colour after moulding. The pigments can be either organic- or inorganic-based.

Plastic deformation Any portion of the total deformation of a body that occurs immediately when load is applied but that remains permanently when the load is removed. The deformation of a material under load that is not recoverable after the load is removed. Opposite of elastic deformation.

Plastic A synthetic or natural organic substance (excluding rubbers) formable or pliable during its formation. Thus, a thermoplastic polymer is shaped through plastic flow under the influence of deforming forces, while a thermoplastic polymer is chemically 'set' during formation.

Plasticating extruder An extruder fed with solid polymer that melts and plasticates it while conveying toward the die.

Plasticity A property of plastics that upon the application of a force allows the material to be deformed continuously, and permanently without rupture (opposite of 'elasticity').

Plasticisation Softening, enhancement of flexibility engendered by incorporation of low molecular weight liquid, a plasticiser.

Plasticise To make a material mouldable by application of heat, pressure or addition of a plasticiser.

Plasticiser A material (e.g., 2-ethylhexylphthalate (DOP)) that when incorporated in a thermoplastic reduces its T_g, to increase its flexibility.

Plastisols Mixture of PVC with a plasticiser that can be moulded, cast, or converted to continuous films by heating.

Platelets Individual layers or lamellae of a *Layered material,* e.g., MMT.

Poisson distribution A probability function derived by Simeon Poisson in 1837, for cases when the probability of a single event is very small, but their number approaches infinity: $P(x) = (X^x/x!)exp\{-X\}$, where X is the variance, and mean of variable x.

Poisson's ratio (v_m) The ratio of latitudinal to longitudinal strain, which describes the extent to which a material is distorted in a direction perpendicular to an applied stress: $v_m = \varepsilon_{\perp}/\varepsilon_{11}$. It is a material constant that relates the modulus of rigidity, G, to Young's modulus, E, in the relation: $E = 2G(v_m + 1)$.

Polyaddition Step-growth polymerisation from two types of bi- or multi-functional molecules, e.g., PU formation. This irreversible, rapid process proceeds by bond displacement without forming of low molecular weight byproducts.

Polyallomers Crystalline thermoplastic polymers made from two or more different monomers, usually ethylene and propylene.

Polycondensation A polymer synthesis from bi- or multi-functional monomers with liberation of a low-molecular weight, volatile byproduct.

Polyhedral Oligomeric SilSesquioxanes (POSS) The chemical formula of POSS is: $(RsiO_{1.5})_{\Sigma n}$, thus as the name indicates (sesqui = 'one and a half') it is intermediate between silicas $(SiO_2)_n$ and silicones $(R_2SiO)_n$. The POSS compound may be visualised as a semi-organic cage with Σn nodules. In most cases Σn = 8, but Σn = 6, 10 or 12 have also been prepared. The size of the cage varies from about 0.7 to 3 nm. The octa-silasesquioxanes (Σn = 8) have cube-shaped molecules that consist of a Si_8O_{12} core, and eight reactive sites that may be differently functionalised. From the point of view of polymer technology the mono-substituted species, $R'R_7Si_8O_{12}$ (where R′ ≠ R are substituents) are the most interesting – the group R′ may provide ability of the octa-silasesquioxane to enter polymerisation, copolymerisation or grafting reaction, whereas the other functionalities, R, make the system miscible.

Polymer alloy An immiscible polymer blend having a modified interphase, solid interface and morphology.

Polymer blend A mixture of at least two macromolecular materials: two or more polymers, polymer with copolymers, two or more copolymers, etc.

Polymer conversion Preparation of polymer derivatives during which the number of macromolecules or the degree of polymerisation is preserved.

Polymer Material composed of many (Greek *poly*) units (Greek *meros*). A high-molecular-weight organic compound, natural or synthetic, formed by a chemical reaction in which two or more small organic units join to form large units composed of repeating small units, the mers. Synthetic polymers are formed in addition or condensation polymerisation. Some polymers are elastomeric, others are thermoplastic or thermoset. The term was coined by Berzelius in 1832 to describe hydrocarbons of a general formula $(CH_2)_n$ with n = 1 to 4 (sic!).

Polymeric nanocomposite (PNC) A polymer or copolymer having dispersed in it nanosized particles, *viz.* platelets, fibres, spheroids, etc.

Polymerisation A chemical reaction in which the molecules of a monomer are linked together to form macromolecules whose molecular weight is a multiple of that of the original substance $nM \rightarrow [M]_n$, where n is degree of polymerisation, DP. It is said that the polymerisation leads to telomer if DP = 1 to 6, to oligomer if $10 < DP < 50$, and to polymer if $DP \geq 50$. When two or more monomers are involved, the process is called copolymerisation. Most polymerisation processes are classified as condensation (step) reactions or addition (chain) reactions.

Polymolecularity Practically all polymers are mixtures of impossible to separate homologues or fractions. Mathematically the polymolecularity is expressed by a molecular weight distribution, *MWD*, q.v.

Porosity A condition of trapped pockets of air, gas or vacuum within a solid material, expressed as a percentage of the total non-solid volume to the total volume.

Prepolymer A chemical intermediate with a molecular weight between that of the monomer or monomers, and the final polymer.

Propagation A series of reaction steps in a chain polymerisation in which the monomers are being added to the active polymerisation centre.

Quenching Method of rapidly cooling thermoplastic parts when they are removed from the mould, usually by submerging the parts in water.

Radical polymerisation Free radical polymerisation in which the active centres of reaction are radicals. The polymerisation can be initiated by thermally activated or redox initiator, irradiation, or through thermal activation of monomer.

Radio frequency (RF), preheating A method of preheating used to mould materials to facilitate the moulding operation, and/or reduce the moulding cycle. The frequencies commonly used are between 10 and 100 MHz/sec.

Random copolymer A copolymer in which the different monomers are randomly placed in the main chain. A perfectly random copolymer is produced by polymerisation of different mers having identical reactivity ratios, $r_A = r_B = r_C$.

Randomness A condition in which individual values are not predictable, although they may come from a definable distribution.

Reactive extrusion Execution of chemical reactions during extrusion of polymers, and/or polymerisable monomers. The reactants must be in a physical form suitable for extrusion processing. Reactions have been performed on molten polymers, on liquefied or dispersed monomers, or on polymers dissolved or suspended in or plasticised by a solvent.

Reaction injection moulding (RIM) A semi-continuous manufacturing process in which two liquid components are metered in the calculated ratio by high pressure piston pumps, mixed by impingement mixing, and injected into a mould cavity or cavities, where the reactants are polymerised or cured. The process has been used to polymerise polyamides, elastomeric polyurethanes, and polyurethane foams.

Reactivity ratio A ratio of two kinetic constants $r = k_{AA}/k_{AB}$ where k_{AA} represents the self-propagation, and k_{AB} the transfer from A* active centre to B* active centre caused by addition of monomer B to a growing copolymer chain.

Recycled plastics A plastic material prepared from previously used or processed plastic materials that have been cleaned and reground.

Redox initiator An initiator capable of generating free radicals at low temperature by oxidation-reduction reaction between two components, *viz.* $H_2O_2 + FeSO_4$.

Regrind Waste material from industrial operations that has been reclaimed by shredding or granulating. Regrind is usually incorporated, at a predetermined percentage, with virgin material.

Reinforcement A substance or material added to a polymer during the final synthesis stages or in subsequent processing to improve the strength properties of the polymer. Usually, a high strength material bonded into a matrix to improve its mechanical properties. Reinforcements are usually long fibres (glass, carbon or aramid), chopped fibres, whiskers, particulates (glass beads, mica, clay, and organic fibres), and so forth. The term is not synonymous with filler.

Relative viscosity Ratio of the kinematic viscosity of a polymer solution to the kinematic viscosity of the pure solvent.

Relaxation time The time required for a stress under a sustained constant strain to diminish by a stated fraction of its initial value.

Repeatability The variation obtained when one person measures the same quantity several times using the same instrument.

Reprocessed plastic A thermoplastic material, prepared from melt-processed scrap or reject parts, or from non-standard or non-uniform virgin material. The term scrap does not necessarily connote feedstock that is less desirable or usable than the virgin material from which it may have been generated. Reprocessed plastic may or may not be reformulated by the addition of fillers, plasticisers, stabilisers or pigments.

Reproducibility The variation in measured averages obtained when several persons measure a quantity using the same instrument, or when one-person measures using different instruments.

Residence time distribution The distribution of residence time provides information about how long different parts of the resin reside in the processing equipment. The spread of the residence times reflects, on the one hand, the uniformity of flow inside the processing unit, and on the other, the quality of the product, the degree of mixing, or the extent of a chemical reaction.

Residence time The time a resin spends in a given processing machine (an extruder, injection moulding unit, etc.), and is subjected to heat and stress.

Residual stress The stresses remaining in a plastic part as a result of thermal or mechanical treatment.

Resin An organic material, usually of high molecular weight, that tends to flow when subjected to stress. 'Resin' is often used as a general term for polymers or plastics, and denotes a class of material. It usually has a softening or melting range, and fractures conchoidally. Most resins are polymers. Also any of a class of solid or semisolid organic products of natural or synthetic origin, generally of high molecular weight with no definite melting point (also see ***Polymer***). In reinforced plastics, the material used to bind together the reinforcement material; the matrix.

Rheology The study of the deformation, and flow of materials, of the interrelations between the force, and its effects. The science considers deformation of all materials, from the elastic deformation of Hookian solids to the flow of Newtonian liquids.

Rigid plastics A plastic that has a modulus of elasticity either in flexure or in tension greater than 690 MPa at 23 °C and 50% relative humidity (RH).

Rockwell hardness A common method of testing materials for resistance to indentation in which a diamond or steel ball, under pressure, is used to pierce the test specimen.

Rubbers Crosslinked polymers having glass transition temperatures below room temperature that exhibit highly elastic deformation, and have high elongation.

Rupture A cleavage or break resulting from physical stress. Work of rupture is the integral under the stress-strain curve between the origin, and the point of rupture.

Salt and pepper blends Resin blends of different concentrate additives, in pellet form, mixed with virgin resin to make a different product. Usually associated with colour concentrate blends, that, when melted, and mixed by the injection moulding machine's screw, yield a uniform coloured melt for a part.

Scanning electron microscopy (SEM) A microscopy technique in which the surface of a specimen is scanned, point-by-point, with a finely focused electron beam. Detecting the secondary electrons emitted by the specimen's surface makes the image. Even tough SEM resolution can be as high as 4 nm, the main advantage of SEM over the other microscopy techniques is its very large depth of field.

Scanning probe microscopy (SPM) A microscopy technique in which the surface of a specimen is scanned, point-by-point, using a very sharp probe (d = 10 nm). Accurate piezo-electric devices are utilised to maintain the separation distance between the lowest atom on the probe tip, and the highest atom on the specimen constant, and in the range of 1-100 nm. In this range of tip-to-sample spacing, phenomena like tunnelling current (scanning tunnelling microscopy, STM) or interatomic repulsion/attraction (atomic force microscopy, AFM) can be used for determining specimen topography with resolution from a few microns, down to atomic level.

Scanning transmission electron microscopy (STEM) A microscopy technique in which an ultra thin specimen is scanned with a finely focused electron beam. Detecting the electrons transmitted through the specimen forms the image.

Semicrystalline Polymers that exhibit localised, partial crystallinity.

Sequential polymerisation Formation of an alternating or block copolymer through careful control of addition of different monomers at specific stages of the reaction.

Sequential-IPN (SIPN) An intimate combination of two polymers in network form. During preparation of SIPN the first polymer-A is swollen in a mixture of monomer-B, crosslinking agent, and initiator, then polymerised *in situ*.

Shear fracture Breaking caused by the action of equal and opposed forces, located in the same plane.

Shear heating The rise in temperature created by the compression, and longitudinal pressure on the sheared resin, e.g., in an extruder barrel by the screw's pumping action.

Shear joining An ultrasonic welding joint design where the welding action is parallel to each part surface.

Shear rate The overall velocity over the cross-section of a channel with which molten polymer layers are gliding along each other or along the wall in laminar flow, defined as a change of shear strain within one second.

Shear strain Deformation relative to the reference configuration of length, area, or volume. The tangent of the angular change, caused by a force between two lines originally perpendicular to each other through a point in a body is called *Angular strain*.

Shear strength The maximum shear stress that a material is capable of sustaining. The maximum load required to shear a specimen in such a manner that the resulting pieces are completely clear of each other. Shear strength (engineering) is calculated from the maximum load during a shear or torsion test, and is based on the original cross-sectional area of the specimen.

Shear stress Stress developed because of the action of the layers in the material attempting to glide against or separate in a parallel direction. In other words, the stress developed in a polymer melt when the layers in a cross-section are gliding along each other or along the wall of the channel (in laminar flow).

Shelf life Time span during which stored material maintains its original physical or functional properties.

Shore hardness A method of determining the hardness of a plastic material using a scleroscope or sclerometer. The device consists of a small conical hammer fitted with a diamond point, and acting in a glass tube. The hammer is made to strike the material under test, and the degree of rebound is noted on a graduated scale. Generally, the harder the material, the greater the rebound (ASTM D2240).

Short stack or Tactoid *Intercalated* clay platelets aligned parallel to each other.

Shrinkage In a plastic, the reduction in dimensions after cooling. The relative change in dimension from the length measured on the mould when it is cold to the length of the moulded object 24 h after it has taken out of the mould.

SI units *Système international* of metric units.

Silicones Chemicals derived from silica used in moulding as a release agent, and general lubricant. Also silicon-based plastic materials. Polyorganosiloxanes (e.g., polydimethylsiloxane, silicone rubber), structures (linear or network), and molecular weight, are used as high temperature oils, resins, or elastomers.

Simultaneous-IPN (SIN) IPNs prepared by mixing two monomers, their respective crosslinking agents, and initiators, then polymerising simultaneously by way of non-interfering modes.

Size exclusion chromatography (SEC) Current name for what has been known as ***Gel permeation chromatography (GPC)*** Liquid chromatography in which the polymers are separated by their ability or inability to penetrate the material in the separation columns. Column chromatography technique employing a series of columns containing closely packed rigid gel particles. The polymer to be analysed is introduced into the column, and then eluted with a solvent. The macromolecules diffuse through the gel at rates depending on their molecular size. As they emerge from the columns they are detected by differential refractometer, viscometer, FTIR device, etc. From the output of these detectors a *MWD* curve is obtained.

Skewness The degree to which a distribution is asymmetrical; *Negative* or *Positive skewness* is observed when the distribution peak is shifted to the upper or lower side, respectively.

S-N Diagram Plot of stress, *S*, versus number of cycles, *N*, required to cause failure of similar specimens in fatigue test. Data for each curve on the *S-N* diagram are obtained by determining fatigue life of a number of specimens subjected to various amounts of fluctuating stress. The stress axis may represent stress amplitude, maximum stress, or minimum stress. A log scale is usually used, especially for the *N*-axis.

Softening temperature Temperature at which amorphous polymer (or the amorphous part of crystalline polymer) passes from the hard glass to the soft elastic or liquid state.

Sol-gel hybrids Nanocomposites comprising nanoparticles prepared from precursors [e.g., metal alkoxides $M(OR)_n$] by hydrolysis, and condensation that leads to the formation of metal oxo-polymers. The sol-gel process is taking place under relatively mild, well-controlled conditions, hence inorganic, and organic components can be formed on the nanometre scale, in a wide range of composition. The hybrids usually have ≥ 10% of the dispersed phase. Two types of hybrids are recognised: (1) those where only weak interactions between the organic, and inorganic species exists (*viz.* van der Waals, hydrogen bonding or electrostatic interactions), and (2) where the inorganic, and organic components are chemically bonded by either covalent or ionic-covalent bonds.

Solid state polymerisation Polymerisation of crystalline monomer, e.g., vinyl, using high energy radiation, or polyester under high vacuum at $T < T_m$. Topochemical, topotactic, and canal polymers also belong to this group.

Solvent Any substance, but usually a liquid, that dissolves another substance.

Solvent-casting This process consists of mixing, and dissolving the ingredients in a suitable carrier that conveys the solution of 'dope' through a drier where the solvent is subsequently evaporated, the resulting film is removed from the substrate surfaces, and wound into rolls.

Spacing in CPNC Two measures are used: *Interlayer spacing,* [also known as *d-spacing*, d_{001} or *basal spacing*], and *Interlamellar gallery spacing.* The former comprises the latter plus the platelet thickness. For example for MMT: d_{001} = *Interlamellar spacing* + 0.96 (nm).

Specific gravity The ratio of the mass of a given volume of a substance to the mass of an equal volume of a reference substance, usually water, at specified temperature. Also, the dimensionless ratio of a substance density to that of a reference material.

Specific heat The quantity of heat required to raise the temperature of a unit mass of a substance by one degree (1 °C) at constant pressure or volume.

Specific volume Reciprocal of density.

Spectrometry A method based on designation of the wavelengths within a particular portion of a range of radiation or absorption, for example, ultraviolet (UV), emission, and absorption spectrometry.

Spectrophotometer An instrument that measures transmission or reflectance of visible light as a function of wavelength, permitting accurate analysis of colour or accurate comparison of luminous intensities of two sources of specific wavelengths.

Spectroscopy The study of spectra using an instrument for dispersing radiation for visual observation of emission or absorption.

Specular gloss The relative luminous reflectance factor of a specimen at the specular direction.

Sphaerocolloids The colloidal particle has a spherical shape, formed either by single macromolecule or an association of low molecular weight species.

Spinodal decomposition (SD) The phase separation that occurs when the single-phase system is abruptly brought into the spinodal region, by either a rapid change of temperature, pressure or flash evaporation of a solvent, *viz.* in polymer blends. Owing to spontaneous phase separation in the system (no nucleation!) the morphology generated is co-continuous.

Spinodal The line on the temperature versus composition phase diagram for a mixture of two components, which separates the single-phase region from the two-phase regions. Hence, with *viz.* in polymer solutions or polymer blends. See also ***Binodal.***

Stabiliser An ingredient used in the formulation of some plastics to assist in maintaining the physical and chemical properties of the compounded materials at their initial values throughout the processing, and service life of the material.

Standard deviation The Greek letter sigma (σ) is used to indicate the standard deviation of a population, defined as the square root of the variance:

$$\textit{Normal distribution}: \quad y = \left[1/\sigma(2\pi)^{1/2}\right]\exp\left\{-\left[(x-\bar{x})/\sigma\right]^2/2\right\}$$

$$\textit{Variance}: \quad \equiv \sigma^2 = \sum(x-\bar{x})/(N-1)$$

$$\textit{Standard deviation}: \quad \sigma \equiv (\textit{Variance})^{1/2} = \sqrt{\sum(x-\bar{x})/(N-1)}$$

Statistical chain (Kuhn) Hypothetical free rotating polymer chain units made of N_{Kuhn} statistical segments, each of length L_{Kuhn}, defined to reproduce the unperturbed chain length; *viz.* the square end-to-end distance may be expressed as:

$$\langle r_\Theta^2 \rangle = 2n(\sigma l)^2 = \langle r_{Kuhn}^2 \rangle \equiv N_{Kuhn} L_{Kuhn}^2$$

$$\Rightarrow L_{Kuhn} = \sigma l (2n / N_{Kuhn})^{1/2}$$

where σ represents the internal, short range interactions – a measure of chain stiffness.

Stereoregular polymers Polymers exhibiting tacticity, i.e., regularity in the stereochemical configuration of the repeating units: isotactic, syndiotactic, erythro and threo.

Storage modulus A quantitative measure of elastic properties, defined as the ratio of the stress, in-phase with strain, to the magnitude of the strain. The storage modulus may be measured in tension or flexure, E', compression, K', or shear, G'.

Strain The change in length per unit of original length, expressed as a fraction of the original length, $\lambda = (L - L_o)/L_o$, in percent, $\Delta l = 100\lambda$, or in extensional flow as $\varepsilon = ln(L/L_o)$. The dimensionless numbers that characterise the change of dimensions of a specimen during controlled deformation.

Strength of material Refers to the structural engineering analysis of a part.

Stress concentration factor (SCF) Ratio of the maximum stress in the region of a notch, or another stress raiser, to the nominal corresponding stress. SCF is a theoretical indication of the effect of stress concentration on mechanical behaviour. Since it does not take into account the stress relief due to plastic deformation, its value is usually larger than the empirical fatigue notch factor or strength-reducing ratio.

Stress concentration The magnification of the level of applied stress in the region of a notch, crack, void, inclusion, or other stress raisers. Sections or areas in a part where the moulded-in or physical forces are high or magnified.

Stress crack External or internal cracks in a plastic caused by imposed stresses.

Stress cracking A process of cracking under induced mechanical stress. Stress cracking generally initiates with microscopic surface cracks causes by chemical attack or other degrading influence such as ultraviolet radiation. Under mechanical stress, the microcracks propagate eventually producing a localised failure.

Stress optical sensitivity The ability of materials to exhibit double refraction of light when placed under stress.

Stress relaxation The gradual decrease in stress with time under a constant deformation (strain), and temperature. Stress relaxation is determined in creep tests. Data is often presented as stress versus time plots.

Stress The ratio of applied load to the original cross-sectional area of a test specimen, or force per unit area that resists a change in size or shape of a body.

Stress-induced crystallisation The production of crystals in a polymer by the action of stress, usually during elongation. It occurs in fibre spinning, or during rubber elongation, and is responsible for enhanced mechanical properties.

Stress-strain curve Simultaneous readings of load, and deformation, converted to stress (ordinate), and strain (abscissas) to obtain a stress-strain diagram for tensile, creep, or torsional loadings.

Styrenics Indicates a group of plastics materials that are polymers, either whole or partially polymerised from styrene monomer.

Substituted macromolecules Linear macromolecules with side chains consisting of definite, and usually homogeneous substituents. (In branched macromolecules the side chains consist of the same primary molecules as in the main chain, are of varied length, and irregularly arranged.)

Surfactant A compound that usually reduces interfacial tension between two liquids.

Suspension polymerisation Chain polymerisation of vinyl monomer dispersed in the form of large drops in aqueous medium. Monomer-soluble initiator starts the polymerisation, thus each drop can be treated as individual bulk polymerising volume.

Swell A dimensional increase caused by exposure to liquids, and/or vapours.

Swelling Swelling is the ability of a body to take up liquids. It depends on the size, and shape of the macromolecules. Linear or lightly branched polymers in a good solvent first swell then dissolve. The crosslinked polymers show limited swelling.

Syndiotactic polymer A stereoregular polymer in which at least one monomeric carbon atom exhibits stereochemical configurational isomerism, and in which the configurations alternate between the neighbouring units, *viz.* syndiotactic PVC, isotactic or syndiotactic PP.

Tacticity A regularity of configurational isomeric unit placement in the polymeric chain.

Tapered block copolymer Gradient block copolymer in which there is a gradual change of composition at the junction between the two blocks from pure AAAAAAAA type to pure BBBBBBBB type. The tapered block copolymers are reported to be more efficient compatibilisers in polymer blends than AB block copolymers.

Telechelic polymer A polymer with purposely-introduced chain-end groups of a specific type, e.g., ionic, hydroxyl, acidic, etc.

Telomer Low molecular weight radical polymerisation product obtained in a reaction in which extensive chain transfer to a solvent (or specifically introduced chain transfer agent) has occurred, so that the telomer contains fragments of these reactants as end groups, and has low molecular weight (DP = 1 to 6).

Telomerisation Mainly a free radical solution polymerisation with high transfer constant, leading to *telomers* containing built-in fragments of the chain transfer agent.

Tensile impact energy The energy required to break a plastic specimen in tension by a single swing of a calibrated pendulum.

Tensile impact test A test whereby the sample is clamped in a fixture attached to a swinging pendulum. The swinging pendulum strikes a stationary anvil causing the test sample to rupture. This is similar to the Izod test.

Tensile strength or stress The maximum tensile load per unit area of the original cross-section, within the gauge boundaries, sustained by the specimen during a tension test. Ultimate strength of a material subjected to tensile loading.

Tensile strength The pulling stress at any given point on the stress versus strain curve, usually just before the material tears or breaks. The area used in computing the engineering strength is the original, rather than the necked-down area.

Terpolymers A copolymer composed of three different repeat units or monomers, where the repeating structural units of all three are present within each molecule. The influence of all three types of monomer is evident in the property profile of the polymer. Common terpolymers include ABS and ASA.

Tetrapolymers Copolymers that contain four different monomers.

Thermal conductivity Ability of a material to conduct heat. The coefficient is expressed as the quantity of heat that passes through a unit cube of the substance in a given unit of time when the difference in temperature of the two faces is 1 °C.

Thermal degradation Degradation caused by exposure to an elevated temperature. In the absence of oxygen the term *pyrolysis*, while in its presence the term *thermo-oxidative degradation* is used.

Thermal expansion The linear rate at which a material expands or contracts due to a rise or fall in temperature. Each material has its own rate of expansion and contraction.

Thermal polymerisation Free radical polymerisation initiated either by thermal homolysis of an initiator or caused by the action of heat on the monomer itself.

Thermoelasticity Rubber-like elasticity exhibited by a rigid plastic, and resulting from an increase of temperature.

Thermogravimetric analysis (TGA) The study of the change in mass of a material, either in oxygen, air or an inert atmosphere. The test can be conducted under various conditions of time, temperature, and pressure.

Thermomechanical analysis (TMA) An analytical technique consisting of measuring the physical dimensions of a material or changes in its moduli as a function of temperature, and/or frequency.

Thermoplastics (TP) A class of plastic materials that is capable of being repeatedly softened by heating, and hardened by cooling, *viz.* ABS, PVC, PS, PE. Generally, a polymer that, upon heating softens, changing from a solid into elastic or liquid mouldable state without having undergone chemical changes.

Thermoplastic elastomer (TPE) An elastomer that upon heating turns into regularly behaving linear polymer. Polystyrene-polybutadiene block copolymers, polypropylene blends with ethylene-propylene-diene terpolymer provide examples.

Thermoplastic IPN IPN in which the individual polymers are thermoplastic. The polymers may contain physical crosslinks as in ionomers where ionic clusters join two or more chains together. Phase-separated systems, e.g., block and graft copolymers or thermoplastic polyurethanes, are frequently considered thermoplastic IPNs.

Thermoplasticity The ability of material to be deformed without breaking with a relatively fast flow, when (at a suitable temperature) this material is properly stressed.

Thermosets (TS) Materials that undergo a chemical reaction by the action of heat and pressure, catalysts, ultraviolet light, etc., leading to an infusible state. Examples are the amines (melamine and urea), unsaturated polyesters, alkyds, epoxies and phenolics. A common thermoset goes through three stages. *A-Stage* An early stage when the material is soluble in certain liquids, fusible, and will flow. *B-Stage* An intermediate stage at which the material softens when heated, and swells in contact with certain liquids, but does not dissolve or fuse. Moulding compound resins are in this stage. *C-Stage* The final stage is the TS reaction when the material is insoluble, infusible, and cured.

Thixotropy A time dependent decrease of apparent viscosity under shear stress, followed by a gradual recovery when the stress is removed. Its antonym is *rheopexy*.

Topochemical polymerisation Solid-state polymerisation of crystalline monomers without any intermediate loss of order. The *topotactic oligomers* have been produced, but the order is lost as the polymerisation progresses beyond a low *DP*.

Torsional A twisting motion. Torsional stress is created when one end of a part is twisted in one direction while the other is held rigid or twisted in the other direction.

Toughness A measure of the ability of a material to absorb energy. The work per unit volume of material that is required to rupture it. Toughness is proportional to the area under the load-elongation curve from the origin to the breaking point.

Transesterification An ester interchange reaction occurring in the presence of hydroxy compound (alcoholysis) or acid compound (acidolysis). Since esterification is reversible, the transesterification occurs between mixed esters in the presence of a (thermally activated) low concentration of volatile reaction byproducts. Ester-amide exchange can also be accomplished by a similar (catalysed) process.

Translucent The quality of material for transmitting light without being transparent.

Transmission electron microscopy (TEM) Electron microscopy in which an ultra-thin specimen is illuminated by an electron beam. Detecting the electrons transmitted through the specimen forms the image. The short wavelength of electrons allows high resolution (0.2 nm).

Transparent A material that can be easily seen through.

Triblock polymer A block copolymer consisting of three $A_nB_mA_n$ blocks.

Tunnelling Release of longitudinal portions of the substrate in incompletely bonded laminates, and deformation of these portions to form tunnel-like structures.

UCST (Upper critical solution temperature) The highest temperature of immiscibility, where ***Ultimate strength*** Strength at the break point in tensile test.

Ultrasonics Branch of acoustics dealing with periodic waves with frequencies above the audible range, i.e., greater than 16 kHz.

Ultrasonic testing Measurement of ultrasonic velocity, and absorption to determine such structure-related factors as T_g, crosslink density, branching, morphology, composition, etc. Also a non-destructive test to locate internal flaws or structural discontinuities by high-frequency reflection or attenuation ultrasonic beam.

Ultraviolet (UV) The region of the electromagnetic spectrum between the violet end of visible light and the X-ray region, including wavelengths from 10 to 390 nm. Because UV photons have more energy they may initiate chemical reactions, and degrade plastics, particularly aramids and polypropylenes.

Unimodal distribution A distribution with a single peak.

Unsaturated polyester A low molecular weight polyester with unsaturated, double bonds able to enter into crosslinking reaction with added unsaturated monomer (usually styrene) by the free radical mechanism, usually initiated by methyl ethyl ketone peroxide.

Unzipping or depropagation A degradation reaction in which the consecutive mers are gradually removed from one macromolecular chain end to another. Few polymers undergo such a reverse propagation reaction, *viz.* PMMA, POM, PTFE, etc.

UV stabiliser Any chemical compound that, when added to thermoplastic material, selectively absorbs ultraviolet rays.

Vanadium pentoxide (V_2O_5) A nanofiller showing a wide range of morphologies from fibres to ribbons, sheets and (rolled) nanotubes. These may be prepared by polycondensation of vanadic acid (HVO_3) in water. The individual flat fibres are *ca.* 10 nm wide, *ca.* 1.5 nm thick, and up to several microns long. A single fibre has a double layer structure, each consisting of two V_2O_5 sheets. Consequently, each fibre consists of only four layers of vanadium atoms, and represents a wire of molecular dimensions. Owing to the disassociation of surface V-OH groups (V-OH + $H_2O \leftrightarrow$V-O^- + H_3O^+) the fibre surface is negatively charged. The V_2O_5 fibres are explored for electronic applications.

Vicat softening point The temperature at which a flat-ended needle of 1 mm circular or square cross-section penetrates a thermoplastic specimen to a depth of 1 mm under a specified load using a uniform rate of temperature rise.

Vinyl chloride plastics Plastics based on PVC or copolymers of vinyl chloride with other monomers, the vinyl chloride being the major component.

Vinyl Usually polyvinyl chloride, PVC, but may be used to identify other polyvinyl plastics.

Virgin plastics Material not previously used or processed, and meeting manufacturer's specifications.

Viscoelasticity A property involving a combination of elastic and viscous behaviour. A material having this property is considered to combine the features of an elastic solid and Newtonian liquid.

Viscometer An instrument used for measuring the viscosity and flow properties of fluids.

Viscosity The property of resistance to flow exhibited within the body of a material, expressed in terms of the relationship between applied shearing stress and resulting rate of strain. A measurement of resistance of flow due to internal friction when one layer of fluid is caused to move in relationship to another layer.

Viscous deformation Any portion of the total deformation of a body that remains permanently when the load is removed, also referred to as non-elastic deformation.

Void A void or bubble occurring in the centre of a heavy thermoplastic part, usually caused by excessive shrinkage.

Volume resistivity The electrical resistance between opposite faces of 1 ml of insulating material. It is measured under prescribed conditions using a direct current potential after a specified time of electrification.

Vulcanisation Process of converting of raw rubber compounds into lightly crosslinked network elastomer. Vulcanisation of diene rubbers involves compounding it with sulfur or sulfur compounds, then heating at about 140 °C for sometimes several hours. The process can be sped up by addition of a catalyst, e.g., ZnO.

Warpage Dimensional distortion in a plastic object after moulding.

Water absorption The amount of water absorbed by a polymer when immersed in water for a stipulated time.

Weathering A term encompassing exposure of polymers to solar or ultraviolet light, temperature, oxygen, humidity, snow, wind, pollutants (e.g., O_3, NO_2, CO_2), cyclical changes of temperature, and moisture, etc. Outdoor degradation of material, exposed to adverse weather factors.

Weatherometer An instrument used for studying the accelerated effects of weather on plastics, using artificial light sources, and simulated weather conditions.

Weibull distribution function: $y = (\beta/\alpha)(x - y)^{b-1} \exp\{-(x - \gamma)^{\beta/\alpha}\}$ *for* $x \geq \gamma$; $\Rightarrow y = 0$ where x is a variable, and α, β and γ are the distribution parameters

Welding Joining thermoplastic pieces by one of several heat-softening processes. Butt fusion, spin welding, ultrasonic, and hot gas are examples of such methods.

WLF equation Williams-Landel-Ferry equation that relates the value of the shift factor, a_T, (associated with time-temperature superposition of viscoelastic data) required to bring log-modulus (or log-compliance) versus time or frequency curves measured at different temperatures onto a master curve (at a particular reference temperature, T_o, usually taken at 50 °C above T_g: $a_T = C_1 (T - T_o) / [C_2 + (T - T_o)]$, where the constants, C_1 and C_2, are approximately identical for all polymers, *viz.*: -8.86 and 101.6 K, respectively.

Yield point elongation In materials that exhibit a yield point, the difference between the elongation at the completion, and the start of discontinuous yielding.

Yield point The point at which permanent deformation of a stressed specimen begins to take place. Only materials that exhibit yielding have a yield point.

Yield strength The stress at the yield point – the stress at which a material exhibits a specified limiting deviation from the proportionality of stress to strain, or the lowest stress at which a material undergoes plastic deformation. When the material is elastic at lower stresses and viscoelastic at higher, the stress at the border of this change is the yield stress. Yield stress is often defined as the stress needed to produce a specified amount of plastic deformation, usually a 0.2% change in length. In tensile testing, the yield stress is that at which there is no increase in stress with a corresponding increase in strain – usually the first peak on the curve. It may also be define as a specific limiting deviation from the proportional stress-strain curve.

Young's modulus The ratio of normal stress to corresponding strain under tensile or compressive loading at stresses below the linearity limit of the material.

Ziegler-Natta polymerisation Chain polymerisation on a Ziegler-Natta catalyst, *Z-N*, based either on $TiCl_4$, VCl_5, or $CoCl_3$ mixed with either $Al(C_2H_5)_3$ or $Al(C_2H_5)_2Cl$, e.g., reacting AlR_3 (R is an alkyl group) with crystalline $TiCl_3$ in an inert solvent. For example, three catalytic systems led to polymerisation of HDPE, (1) molybdena-alumina, (2) hexavalent CrO_3 on silica, and (3) aluminium trialkyl (e.g., $AlEt_3$) with $TiCl_4$. The polymerisation occurs under relatively mild conditions. *Z-N* polymerisation is frequently used to obtain stereoregular polymers, *viz.* isotactic or syndiotactic PP.

Dictionary References

Alger, M. S. M., ***Polymer Science Dictionary,*** Elsevier Applied Science, London (1989).

Beall, G. W., S. Tsipursky, A. Sorokin, and A. Goldman, *US Pat.*, **5,552,469,** 03.09.1996a, to AMCOL International Corp.

Brandrup, J., and Immergut, E. H., ***Polymer Handbook,*** Intersci. Pub., New York (1966).

Batzer, H., and Lohse, F., ***Einführung in die makromoleculare Chemie,*** Hütingt & Wepf Verlag, Basel (1976).

Flory, P. J., ***Statistical Mechanics of Chain Molecules,*** Intersci. Pub., New York (1969).

Frado, J., Ed., ***Plastics Engineering Handbook***, 4-th., Reinhold Pub. Corp., New York (1976).

Harper, C. A., Ed., ***Handbook of Plastics and Elastomers***, McGraw-Hill, New York (1975).

Lan, T., J. W. Gilmer, J. Ch. Matayabas, Jr., and R. B. Barbee, *US Pat.*, **6,387,996,** 14.05.2002a, Appl. 01.12.1999, to AMCOL International Corp.

Pebly, H. E., ***Glossary of Terms in Composites, Vol. 1, Engineering Materials Handbook***, ASM International (1987).

Richardson, T. A., ***Industrial Plastics: Theory and Application***, South-Western (1983).

Standard Abbreviation of Terms Relating to Plastics, D1600, Annual Book of ASTM Standards, American Society for Testing and Materials.

Standard Definitions and Descriptions of Terms Relating to Conditioning, E41, Annual book of ASTM Standards, American Society for Testing and Materials.

Standard Definitions and Descriptions of Terms Relating to Dynamic Mechanical Measurements on plastics, D4092, Annual book of ASTM Standards.

Standard Definitions and Descriptions of Terms Relating to Methods of Mechanical Testing, E6, Annual book of ASTM Standards.

Standard Definitions and Descriptions of Terms Relating to Plastics, D883, Annual book of ASTM Standards, American Society for Testing and Materials.

Standard Definitions and Descriptions of Terms Relating to Reinforced Plastic Pultruded Products, D3918, Annual book of ASTM Standards.

Standard Definitions and Descriptions of Terms Relating to Resinography, E375-75 (reproved in 1986), Annual book of ASTM Standards.

Standard Guide for Identification of Plastic Materials, D4000, Annual book of ASTM Standards, American Society for Testing and Materials.

Standard Terminology Relating to Radiation Measurements and Dosimetry, E170, Annual book of ASTM Standards, American Society for Testing and Materials.

Thewlis, J., Glass, R. C., and Meetham, A. R., Eds., *Encyclopaedic Dictionary of Physics,* Pergamon Press, Oxford (1967).

Utracki, L. A., *Polymer Alloys and Blends – Thermodynamics and Rheology*, Hanser Pub., Munich (1989).

Utracki, L. A., *Encyclopaedic Dictionary of Commercial Polymer Blends*, ChemTec Pub., Toronto (1994).

Utracki, L. A., *Commercial Polymer Blends*, Chapman & Hall, London (1997).

Utracki, L. A., Ed., *Polymer Blends Handbook*, Kluwer Academic Pub., Dordrecht (2002).

Van Nostrand's Scientific Encyclopaedia, D. Van Nostrand Co., Inc., Princeton (1958).

Whittington, L. R., *Whittington's Dictionary of Plastics*, 2nd., Technomic, Lancaster, PA (1978).

7.6 Companies Active in Organoclay, and/or CPNC Technology

Akzo Nobel
Algonquin Automotive
AlliedSignal Inc. (now Honeywell)
AMCOL Intl. Corp (Nanocor, Inc)
Argonide
Basell Poliolefine S. p. A.
BASF AG
Bayer AG
Cabolex Inc.
CarboLex (CNT)
Carbon Nanotechnologies (CNT)
Ciba Specialty Chem./Vantico
Clariant Masterbatches
Claytech, Inc.
CO-OP Chem. Co. Ltd.
Corning, Inc.
Creanova
Crompton Corp.
Deal International (CNT)
Decoma International Inc.
Degussa
Dow Chemical Co.
DSM N. V.
Du Pont de Nemours & Co.
Eastman Chemical Co.
EMS-Chemie
Euronil S. p. A. (Nilit group)
Exxon Research & Engineering Co.
Foster Corp.
Gabriel-Chemie Group
GE Plastics
General Motors Co.
Gitto Global
Honeywell
Hybrid Plastics
Hyosung Chemicals Co.
Hyperion Catalysis
International Paper Co.
Kabelwerk Eupen
Kanegafuchi Chem. Ind.
Kaneka Corp.
Kao Corp.
Kunimine Ind. Co.
Kuraray
Laporte Ind., Ltd.
LG Chemicals
Magna International
Mark IV Automotive

Mitsubishi
Mitsui
Montell (now Basell)
Nanocor, Inc., 1998, AMCOL subsidiary
NanoLab (CNT)
Nanophase Technologies Corp
NanoProducts
Nippon Gohsei
Nippon Shokubai Co.
Nishihinbo Ind. Ltd.
NL Ind., Inc.
Nova Chemicals
Oxford Polymers
Procter & Gamble
Pacific Nanotechnology, Inc.
PolyOne (Nanocor partner)
Raychem Corp.
Reade Advanced Materials
Rheox, Inc.
Rhodia
Rohm & Haas Co.
RTP Company
Samsung General Chem. Co.
Sealed Air
Sekisui Chem. Co.
Showa Denko K. K.
SNC TEC
Solutia
Southern Clay Products
Süd Chemie AG (Nanofil)
Tetra Laval
Topy Co.
Toray Ind. Ltd.
Toshiba Silicone Co.
Toyota Jidosha K. K.
Triton Systems Inc.
Ube Industries, Ltd.
Unitika Ltd.
W. R. Grace & Co.
Xerox Corp.
Yantai Haili Ind.
Zhejiang Fenghong Clay Chemicals Co., Ltd.

Part 8

References

8 References

Abu-Jdayil, B., K. Al-Malah and R. Sawalha, "*Study on bentonite-unsaturated polyester composite materials*", *J. Reinf. Plast. Compos.*, **21**, 1597-1607 (2002).

Adur, A. M., R. A. Volpe and K. S. Shih, "*Clay-filled polymer barrier materials for food packaging applications*", *US Pat.*, **6,358,576**, 19.03.2002, Appl. 12.02.1998, to International Paper Co.

Agag, T. and T. Takeichi, "*Polybenzoxazine-montmorillonite hybrid nanocomposites: synthesis and characterization*", *Polymer*, **41**, 7083-90 (2000).

Ago, H, K. P. Petrisch, S. P. Shaffer, A. H. Windle and R. H. Friend, "*Composites of carbon nanotubes and conjugated polymers for photovoltaic devices*", *Adv. Mater.*, **11**, 1281-85 (1999).

Ajayan P. M., L. S. Schadler, and P. V. Braun, "*Nanocomposite science and technology*", Wiley-VCH, Weinheim (2003).

Ajayan, P. M., L. S. Schadler, C. Giannaris, and A. Rubio, "*Single-Walled Carbon Nanotube-Polymer Composites: Strength and Weakness*", *Adv. Mater.*, **12**, 750-753 (2000).

Ajji A. and L. A. Utracki, "*Compatibilization of Polymer Blends*", *Prog. Rubber & Plastics Technol.*, **13**, 153-188 (1997).

Ajji, A., "*Interphase and Compatibilization by Addition of a Compatibilizer*", in *Polymer Blends Handbook*, L. A. Utracki (Ed.), Kluwer Academic Publishers, Dordrecht (2002).

Akelah, A. and A. S. Moet, "*Polymer-clay nanocomposites: Free radical grafting of polystyrene onto organophilic montmorillonite interlayers*", *J. Mater Sci.*, **31** 3589-96 (1996).

Akelah, A., N. S. El-Deen, A. Hiltner, E. Baer and A. Moet, "*Organophilic rubber-montmorillonite nanocomposites*", *Materials Letters*, **22**, 97-102 (1995).

Akita, H. and H. Kobayashi, "*Studies on molecular composite. iv. Block copolymer as a single-component nano composite consisting of poly(p-phenylene benzobisthiazole) and thermoplastic aromatic polyamide*", *Polym. J.* (Japan), **31**, 890-894 (1999).

Akkapeddi, M. K., "*Glass fiber reinforced polyamide-6 nanocomposites*", NRCC/IMI symposium on Polymer Composites, Quebec, QC, Canada, 6-8.10.1999, *Polym. Compos*,. **21**, 576-585 (2000).

Akovali, G., C. A. Bernardo, J. Leidner, L. A. Utracki and M. Xanthos, Eds., "*Reprocessing of Commingled Polymers and Recycling of Polymer Blends*", NATO ASI vol. 351, Kluwer Academic Publishers, Dordrecht (1998).

Aldrich Library of FTIR Spectra, Edition II, Sigma-Aldrich Co. (1997).

Alexandre, M. and Ph. Dubois, "*Polymer-layered silicate nanocomposites: preparation, properties and uses of a new class of materials*", *Mater. Sci. Eng.: R*, **28**, 1-63 (2000).

Alexandre, M., G. Beyer, C. Henris, R. Cloots, A. Rulmont, R. Jérôme, and P. Dubois, "*'One-pot' preparation of polymer/clay nanocomposites starting from Na+ montmorillonite. 1. melt intercalation of ethylene-vinyl acetate copolymer*", *Chem. Mater.*, **13**, 3830-32 (2001).

Alexandre, M., P. Dubois, R. Jerome, M. Garcia-Marti, T Sun, J. M. Garces, D. M. Millar, and A Kuperman, "*Polyolefin nanocomposites*", *US Pat.*, **6,465,543**, 2002b.10.15, Appl. 2000.12.21, to Dow Chem. Co.

Alexandre, M., P. Dubois, T. Sun, J. M. Garces and R. Jérôme, "*Polyethylene-layered silicate nanocomposites prepared by the polymerization-filling technique: synthesis and mechanical properties*", *Polymer*, **43**, 2123-32 (2002a).

Anastasiadis, S. H., K. Karatasos, G. Vlachos, E. Manias and E. P. Giannelis, "*Nanoscopic-confinement effects on local dynamics*", *Phys. Rev. Lett.*, **84**, 915-918 (2000).

Anderson, K. W., J. E. White, C.-J. Chou, and C. A. Polansky, *Polymer-organoclay-composites and their preparation, to the WO* **98/29491A1**, 09.07.1998, Appl. 22.12.1997, Dow Chem. Co.

Andrews, R., D. Jacques, M. Minot, and T. Rantell, "*Fabrication of carbon multiwall nanotube/polymer composites by shear mixing*", *Macromol. Mater. Eng.*, **287**, 395-403 (2002).

Annabi-Bergaya, F., M. I. Cruz, L. Gatineau and J. J. Fripiat, "*Adsorption of alcohols by smectites*",(Parts I to IV), *Clay Minerals*, **14**, 249-258 (179); *ibid.*, **15**, 219-223; *ibid.*, **15**, 225-237 (1980); *ibid.*, **16**, 115-122 (1981).

Anonymous, "*Eastman uses Nylon/clay to boost PET properties*", *Europ. Chem. News*, **71**, 47 (1999b).

Anonymous, *Comments, Inorg. Chem.*, **20** (4-6), 327-371 (1999a).

Antonietti, M. and H. Sillescu, "*Self-diffusion of polystyrene chains in networks*", *Macromolecules*, **18**, 1162-66 (1985).

Aranda P., Y. Mosqueda, E. Pérez-Cappe, and E. Ruiz-Hitzky, "*Electrical characterization of poly(ethylene oxide)–clay nanocomposites prepared by microwave irradiation*", *J. Polym. Sci. Part B: Polym. Phys.*, **41**, 3249-3263 (2003).

Aranda, P. and E. Ruiz-Hitzky, "*Poly(ethylene oxide)/NH4+-smectite nanocomposites*", *Appl. Clay Sci.*, **15**, 119-135 (1999).

Aranda, P. and E. Ruiz-Hitzky, "*Poly(ethylene oxide)-silicate intercalation materials*", *Chem Mater*, **4**, 1395-1403 (1992).

Arroyo, M., M. A. López-Manchado and B. Herrero, "*Organo-montmorillonite as substitute of carbon black in natural rubber compounds*", *Polymer*, **44**, 2447-53 (2003).

Arzhakowa O. V., S. Yu. Ermusheva, L. M. Yarysheva, A. L. Volynskii, and N. F., Bakeev, "*General features of formation and structure of metal-filled polymer nanocomposites based on crazed polymer matrices*", *Polym. Sci.*, **45A**, 573-578 (2003).

Avrami, M., "*Kinetics of phase change. General theory*", *J. Chem. Phys.*, **7**, 1103-11 (1939).

Aymonier, C., U. Schlotterbeck, L. Antonietti, P. Zacharias, R. Thomann, J. C. Tiller, S. Mecking, "*Hybrids of Silver Nanoparticles with Amphiphilic Hyperbranched Macromolecules Exhibiting Antimicrobial Properties*", *Chem. Commun.*, **2002** (24), 3018-19.

Badesha, S. S., A. W. Henry, J. B. Maliborski, and C. O. Eddy, "*Polymer nanocomposites*", *US Pat.*, **5,840,796**, 24.11.1998, Appl. 09.05.1997, to Xerox Corp.

Bafna A., G. Beaucage, F. Mirabella, and S. Mehta, "*3D Hierarchical orientation in polymer–clay nanocomposite films*", *Polymer*, **44**, 1103-1115 (2003).

Balazs A. C., C. Singh, E. Zhulina, and Y. Lyatskaya, "*Modeling the phase behavior of polymer/clay nanocomposites*", *Acc. Chem. Res.*, **32**, 651-657 (1999).

Balazs, A. C., C. Singh, and E. Zhulina, "*Modeling the interactions between polymers and clay surfaces through self-consistent field theory*", *Macromolecules*, **31**, 8370-81 (1998).

Baljon, A. R. C., J. Y. Lee, and R. F. Loring, "*Molecular view of polymer flow into a strongly attractive slit*", *J. Chem. Phys.*, **111**, 9068-72 (1999).

Barbee, R. B. and J. C. Matayabas, Jr., "*Polyester nanocomposites for high barrier applications*", *US Pat.*, **6,034,163**, 07.03.2000, Appl. 22.12.1997, to Eastman Chemical Co.

Barbee, R. B., J. C. Matayabas, Jr. and J. W. Gilmer, "*A polymer/clay nanocomposite comprising a functionalized polymer or oligomer and a process for preparing same*", *WO* **00/34393**, 15.06.2000d, Priority 07.12.1998, to Eastman Chemical Co.

Barbee, R. B., J. C. Matayabas, Jr., J. W. Trexler and R. L. Piner, "N*anocomposites for high barrier applications*", *WO* **00/788855 A1**, 28.12.2000e, Priority 22.06.1999, to Eastman Chemical Co.

Barbee, R. B., J. W. Gilmer, J. C. Matayabas, Jr. and T. Lan, "*A polymer/clay nanocomposite comprising a clay mixture and a process for making same*", *WO* **00/34375**; *WO* **00/34376**, 15.06.2000a,b, Priority 07.12.1998, to Eastman Chemical Co.

Barbee, R. B., J. W. Gilmer, S. R. Turner, J. C. Matayabas, Jr., "*Polymer/clay nanocomposite comprising a functionalized polymer or oligomer and a process for preparing same*", *US Pat.*, **6,384,121**, 2002.05.07, Appl. 1999.12.01, to Eastman Chemical Co.

Barbee, R. B., M. A. Weaver, and J. C. Matayabas, Jr., "*A colorant composition, a polymer nanocomposite comprising the colorant composition and articles produced therefrom*", *WO* **00/34379**, 15.06.2000c, Priority 07.12.1998, to Eastman Chemical Co.

Barry, A. J., W. H. Daudt, J. J. Domicone, and J. W. Gilkey, "*Crystalline organo silsesquioxanes*", *J. Am. Chem. Soc.*, 77, 4248-52 (1955).

Beall, G. W., *Molecular modeling of nanocomposite systems, SPE Techn. Pap.*, **45**(2), (1999); available at: www.nanocor.com/tech_papers.

Beall, G. W., S. Tsipursky, A. Sorokin and A. Goldman, "*Intercalates and exfoliates formed with oligomers and polymers and composite materials containing same*"*US Pat.*, **5,760,121**, 02.06.1998, to AMCOL International Corp.

Beall, G. W., S. Tsipursky, A. Sorokin, and A. Goldman, "*Intercalates and exfoliates formed with oligomers and polymers and composite materials containing same*", *US Pat.*, **5,552,469**, 03.09.1996a, to AMCOL International Corp.

Beall, G. W., S. Tsipursky, A. Sorokin, and A. Goldman, "*Exfoliated layered materials and nanocomposites comprising matrix polymers and said exfoliated layered materials formed with water-insoluble oligomers and polymers"*, *US Pat.*, **5,698,624**, 1997.12.16, Appl. 1995.06.07, to AMCOL International Corp.

Beall, G. W., S. Tsipursky, A. Sorokin, and A. Goldman, "*Intercalates; exfoliates; process for manufacturing intercalates and exfoliates and composite materials containing same"*, *US Pat.*, **5,578,672**, 26.11.1996b, Appl. 07.07.1995, to AMCOL International Corp.

Beall, G. W., S. Tsipursky, A. Sorokin, and A. Goldman, "*Intercalates and exfoliates formed with oligomers and polymers and composite materials containing same"*, *US Pat.*, **5,877,248**, 02.03.1999b, Appl. 12.11.1997, to AMCOL International Corp.

Beall, G. W., S. Tsipursky, and K. R. Turk, "*Intercalates and exfoliates formed with organic pesticides compounds and compositions containing the same"*, *US Pat.*, **5,955,094**, 21.09.1999a, to AMCOL International Corp.

Beck Tan N. C., L. Balogh, S. F. Trevino, D. A. Tomalia and J. S. Lin, "*A small angle scattering study of dendrimer-copper sulfide nanocomposites"*, *Polymer*, **40**, 2537-45 (1999).

Becker, O., R. Varley and G. Simon, "*Morphology, thermal relaxations and mechanical properties of layered silicate nanocomposites based upon high-functionality epoxy resins"*, *Polymer*, **43**, 4365-73 (2002).

Becker, O., Y.-B. Cheng, R. Varley and G. Simon, "*Layered silicate nanocomposites based on various high-functionality epoxy resins: The influence of cure temperature on morphology, mechanical properties and free volume"*, *Macromolecules*, **36**, 1616-25 (2003).

Bellamy, L. J., *The Infra-Red Spectra of Complex Molecules*, 3rd Ed., Chapman and Hall, 1975, pp. 16-18.

Bellemare, S. C., M. N. Bureau, J. I. Dickson, and J. Denault, "*Fatigue-induced damage in PA-6/clay nanocomposite"*, *SPE Techn. Pap.*, **48** (2002).

Benedek, G., P. Milani, and V. G. Ralchenko, Eds., *Nanostructured Carbon for Advanced Applications*, NATO Advanced Study Institute on Nanostructured Carbon for Advanced Applications, 2000, Erice, Italy; Kluwer Academic Publishers, Dordrecht (2001).

Bernholc, J., D. Brenner, M. Buongiorno Nardelli, V. Meunier and C. Roland, "*Mechanical and electrical properties of nanotubes"*, *Annu. Rev. Mater. Res.*, **32**, 347-75 (2002).

Bevis, M. J., "*Physical properties enhancements by composition and microstructure control*", Chapter 13 in *Structure Development During Processing*, A. Cunha and S. Fakirov, Eds., NATO ASI vol. XXX, Kluwer Academic Publishers, Dordrecht (2000).

Beyer, G., "*Nanocomposites. 2. A new flame protection system for polymers"*, *Gummi Fasem Kunststoffe,* 2001(5), 321; *Intl. Polym. Sci. Technol.*, **28**, T/1-T/5 (2001).

Bharadwaj R. K., "*Modeling the Barrier Properties of Polymer-Layered Silicate Nanocomposites*", *Macromolecules*, **34**, 9189-9192 (2001).

Bharadwaj R. K., A. R. Mehrabi, C. Hamilton, C. Trujillo, M. Murga, R. Fan, A. Chavira and A. K. Thompson, "*Structure-property relationships in crosslinked polyester-clay nanocomposites"*, *Polymer*, **43**, 3699-3705 (2002).

Bharadwaj, R. K., R. J. Berry and B. L. Farmer "*Molecular dynamics simulation study of norbornene–POSS polymers"*, *Polymer*, **41**, 7209-21 (2000).

Biasci, L., M. Aglietto, G. Ruggieri and F. Ciardelli, "*Functionalization of montmorillonite by methyl methacrylate polymers containing side ammonium cations*", *Polymer*, **35**, 3296-3304 (1994).

Biercuk, M. J., M. C. Llaguno, M. Radosavljevic, J. K. Hyun, and A. T. Johnson, "*Carbon nanotube composites for thermal management*", *Appl. Phys. Letters*, **80**, 2767-69 (2002).

Binder K., J. Horbach, W. Kob, W. Paul, and F. Varnik, "Molecular dynamics simulations", *J. Phys. Condens. Matter*, **16**, S429-S453 (2004).

Binder, K., Ed., "*Monte Carlo and Molecular Dynamics Simulations in Polymer Science*", Oxford University Press, New York (1995).

Binsbergen, F. L., "*Heterogeneous nucleation in the crystallization of polyolefins*", *J. Polym. Sci., Polym. Phys. Ed.*, **11**, 117-135 (1973).

Bishop C. E., and S. G. Niyogi, "*Compound compatible with inorganic solids, and with homopolymers and copolymers of propylene and of ethylene*", *US Pat.*, **6,548,705**, 2003.04.14, Appl. 2000.06.16, to Basell Poliolefine Italia S.p.A.

Bishop C. E., and S. G. Niyogi, "*Intercalated clay useful for making an α-olefin polymer material nanocomposite*", US Pat., **6,500,892**, 2002.12.31, Appl. 2000.06.16, to Basell Poliolefine Italia S.p.A.

Biswas M., S. S. Ray, Y. Liu, "*Water dispersible conducting nanocomposites of poly(n-vinylcarbazole), polypyrrole and polyaniline with nanodimensional manganese(iv) oxide*", *Synthetic Metals*, **105**, 99-105 (1999).

Biswas, M. and S. S. Ray, "*Preparation and evaluation of composites from montmorillonite and some heterocyclic polymers. I. Poly(n-vinylcarbazole)-montmorillonite nanocomposite system*", *Polymer*, **39**, 6423-28 (1998).

Blumstein, A., "*Études des polymerization en couche adsorbée*", *Bull. Soc. Chim., France*, 1961, 899-905.

Blumstein, A., "*Polymerization of adsorbed monolayers: II. Thermal degradation of the inserted polymers*", *J. Polym. Sci.*, **A3**, 2665-72 (1965).

Bonardelli, P., G. Moggi, and A. Turturro, "*Glass transition temperature of copolymer and terpolymer fluoroelastomers*", *Polymer*, **27**, 905-909 (1986).

Bondi, A., "*Physical Properties of Molecular Crystals, Liquids and Glasses*", John Wiley & Sons, New York (1968).

Bora, M., J. N. Ganguli and D. K. Dutta, "*Thermal and spectroscopic studies on the decomposition of [Ni{di(2-aminoethyl)amine}2]- and [Ni(2,2':6',2''-terpyridine)2]- montmorillonite intercalated composites*", *Thermochimica Acta*, **346**, 169-175 (2000).

Boryniec, S. and W. Przygocki, "*Polymer combustion processes. 3. Flame retardants for polymeric materials*", *Polimery*, **10**, 23 (1999); *Intl. Polym. Sci. Technol.*, **27**, T/94-T103 (1999).

Bourbigot S., D. L. Vanderhart, J. W. Gilman, W. H. Awad, R. D. Davis, A. B. Morgan, C. A. Wilkie, "*Investigation of nanodispersion in polystyrene–montmorillonite nanocomposites by solid-state NMR*", *J. Polym. Sci. Part B: Polym. Phys.*, **41**, 3188-3213 (2003).

Brace, W. F. and J. B. Walsh, "*Some direct measurements of the surface energy of quartz and orthoclase*", *Am. Mineral.*, **47**, 1111-22 (1962).

Bradley, W. F., "*Molecular associations between montmorillonite and some polyfunctional organic liquids*", *J. Am. Chem. Soc.*, **67**, 975-981 (1945).

Brady, J. F. and G. Bossis, "*The rheology of concentrated suspensions of spheres in simple shear flow by numerical simulations*", *J. Fluid Mech.*, **155**, 105-129 (1985).

Brandrup, J., E. H. Immergut and E. A. Grulke, Eds., *Polymer Handbook*, 4th edition, John Wiley & Sons, New York (1999).

Brassat, B., "*Process for production of polyamide mouldings*", *US Pat.*, **3,883,469,** 13.05.1975, to Bayer A.-G.

Brindley, G. W. and M. Rustom, "*Adsorption and retention of an organic material by montmorillonite in the presence of water*", *Am. Mineralogist*, **43**, 627-640 (1958).

Brinker, C. J. and G. W. Scherer, *Sol-Gel Science: The Physics and Chemistry of Sol-Gel Processing*, Academic Press, New York (1990).

Brittain, W. and X. Huang, "*Synthesis and characterization of nanocomposites by emulsion polymerization*", *PCT* **WO 02/28911 A2,** 11.04.2002, Appl. 01.10.2001 (priority 02.10.2000), to University of Akron.

Brochard, F., J. Jouffroy and P. Levinson, "*Polymer-polymer diffusion in melts*", *Macromolecules*, **16,** 1638-41 (1983).

Brodnyan, J., "*The concentration dependence of the Newtonian viscosity of prolate ellipsoids*", *Trans. Soc. Rheol.*, **3,** 61-68 (1959).

Brown G. D., S. D. Smith, and J. J. Watkins, "*Preparation of platinum and silver nanoclusters via templating reactions in CO2-swollen diblock copolymers*", *ACS Polym. Mat. Sci. Eng.*, **82,** 292-293 (2000)

Brown, A. B. D., S. M. Clarke and A. R. Rennie, "*Shear induced alignment of kaolinite. Studies using diffraction technique*", *Prog. Colloid Polym. Sci.*, **110,** 80-82 (1998).

Brown, D., P. Mélé, S. Marceau, and N. D. Albérola, "*A molecular dynamics study of a model nanoparticle embedded in a polymer matrix*", *Macromolecules*, **36,** 1395-1406 (2003).

Brown, G., R. Greene-Kelly, and K. Norrish, "*Organic derivatives of montmorillonite*", *Clay Minerals Bull.*, **1,** 214 (1952).

Brown, S. B., "*Reactive Compatibilization of Polymer Blends*", in *Polymer Blends Handbook*, L. A. Utracki (Ed.), Kluwer Academic Publishers, Dordrecht (2002).

Brune, D. A. and J. Bicerano, "*Micromechanics of nanocomposites*", *Polymer*, **43,** 369-387 (2002).

Bruno, T. J., A. Lawandowska, F. Tsvetkov and H. J. M. Hanley, "*Determination of heats of adsorption on a synthetic clay by gas-solid chromatography using a wall coated open tubular column approach*", *J. Chromatogr.*, **A844,** 191-199 (1999).

Bruque, S., L. Moreno-Real, T. Mozas and A. Rodriguez-Garcia, "*Interlayer complexes with lanthanides-montmorillonite with amines*", *Clay Miner.*, **17,** 201-208 (1982).

Buffett, J. B., "*Attapulgite clay thickening agent and method for making same*", *US Pat.*, **3,205,082,** 1965.09.07, Appl. 1962.11.29, to Minerals & Chemicals Philipp Corp.

Bureau, M. N., J. Denault, K. C. Cole and G. D. Enright, "*The role of crystallinity and reinforcement in the mechanical behavior of polyamide-6/clay nanocomposites*", Paper presented at the international symposium Polymer Nanocomposites 2001, Boucherville, QC, Canada, 2001.09.17-19; *Polym. Eng. Sci.*, **42,** 1897-1906 (2002).

Bureau, M., J. Denault and F. Głowacz, "*Mechanical behavior and crack propagation in injection molded polyamide-6/clay nanocomposites*", *SPE Techn. Pap.*, **47,** 2125-29 (2001).

Burns, R. A., "*Clay thixotrope for polyester resins and use thereof*", *US Pat.*, **3,795,650**, 1974.03.05, Appl. 1972.10.16, to Engelhard Minerals & Chemicals Corp.

Burnside, S. D. and E. P. Giannelis, "*Synthesis and properties of new poly(dimethyl siloxane) nanocomposites*", *Chem. Mater.*, 7, 1597-1600 (1995).

Burnside, S. D. and E. P. Giannelis, "*Nanostructure and properties of polysiloxane-layered silicate nanocomposites*", *J. Polym. Sci.: Part B: Polym. Phys.*, **38**, 1595-1604 (2000).

Burstein E, "*A major milestone in nanoscale material science: the 2002 Benjamin Franklin Medal in Physics presentedto Sumio Iijima*", *J. Franklin Institute*, **340**, 221-242 (2003).

Butzloff, P., N. A. d'Souza, T. D. Golden and D. Garrett, "*Epoxy + montmorillonite nanocomposite: effect of composition on reaction kinetics*", *Polym. Eng. Sci.*, **41**, 1794-02 (2001).

Carr, D. D., Ed., *Industrial Minerals and Rocks*, Society for Mining, Metallurgy, and Exploration, Inc., Littleton, Colorado (1994).

Carrado, K. A., "*Synthetic organo- and polymer-clays: preparation, characterization and materials applications*", *Appl. Clay Sci.*, **17**, 1-23 (2000).

Carrado, K. A., L. Q. Xu, R. Csencsits, and J. V. Muntean, "*Use of organo- and alkoxy silanes in the synthesis of grafted and pristine clays*", *Chem. Mater.*, **13**, 3766-73 (2001).

Carrado, K. A., P. Thiyagarajan and D. L. Elder, "*Polyvinyl alcohol-clay complexes formed by direct synthesis*", *Clays Clay Miner.*, **44**, 506-514 (1996).

Carter, L. W., J. G. Hendrics and D. S. Billey, "*Elastomer reinforced with a modified clay*", *US Pat.*, **2,531,396**, 28.11.1950, Appl. 29.03.1947, to National Lead Co.

Castro, C., A. Millan, and F. Palacio, "*Nickel oxide magnetic nanocomposites in an imine polymer matrix*", *J. Mater. Chem.*, **10**, 1945-47 (2000).

Chaiko, D. J., "*Process for the prepaation of organoclays*", **6, 521, 678**, 18.02.2003, Appl. 21.11.2000, to Argonne Natl. Lab.; US Pat. Appl., **0162877**, 28.08.2003a.

Chaiko, D. J., "*Composite materials with improved physllosilicate dispersion*", US Pat. Appl., **0176537**, 18.09.2003b, Apl. 18.03.2002; *US Pat. Appl.*, **0014823**, 22.01.2004, Appl 02.06.2003, to Argonne Natl. Lab.

Chaiko, D. J., "*Liquid crystalline composites containing phyllosilicates*", US Pat. Appl. **0068450**, 10.04.2003c, Appl. 09.10.2001, to Argonne Natl. Lab.

Chamberlin, R. V., "*Mean-field cluster model for the critical behaviour of ferromagnets*", *Nature*, **408**, 337-339 (2000).

Chan, T. V., G. D. Shyu and A. V. Isayev, "*Master curve approach to polymer crystallization kinetics*", *Polym. Eng. Sci.*, **35**, 733-740 (1995).

Chang Y.-W., Y.-C. Yang, S.-H. Ryu, and C.-W. Nah, "*Preparation and properties of EPDM/organomontmorillonite hybrid nanocomposites*", *Polym. Int.*, **51**, 319-324 (2002b).

Chang, B. H., Z. Q. Liu, L. F. Sun, D. S. Tang, W. Y. Zhou, G. Wang, L. X. Qian, S. S. Xie, J. H. Fen, and M. X. Wan, "*Conductivity and Magnetic Susceptibility of Nanotube/Polypyrrole Nanocomposites*", *J. Low Temp. Phys.*, **119**, 41-48 (2000).

Chang, J.-H. and D.-K. Park, "*Nanocomposites of poly(ethylene terephthalate-co-ethylene naphthalate) with organoclay*", *J. Polym. Sci. Part B: Polym. Phys.*, **39**, 2581-88 (2001b).

Chang, J.-H. and D.-K. Park, "*Polyimide nanocomposites: comparison of their properties with precursor polymer nanocomposites*", *Polym. Eng. Sci.*, **41**, 2226-30 (2001c).

Chang, J.-H. and D.-K. Park, "*Various organoclays based nanocomposites of poly(ethylene terephthalate-co-ethylene naphthalate)*", *Polym. Bull.*, **47**, 191-197 (2001a).

Chang, J.-H. and Y. U. An, "*Nanocomposites of polyurethane with various organoclays: Thermomechanical properties, morphology, and gas permeability*", *J. Polym. Sci. Part B: Polym. Phys.*, **40**, 670-677 (2002).

Chang, J.-H., B.-S. Seo, and D.-H. Hwang, "*An exfoliation of organoclay in thermotropic liquid crystalline polyester nanocomposites*", *Polymer*, **43**, 2969-74 (2002a).

Chang, J.-H., D.-K. Park, and K. J. Ihn, "*Montmorillonite-based nanocomposites of polybenzoxazole: synthesis and characterization*", *J. Polym. Sci.: Part B: Polym. Phys.*, **39**, 471-476 (2001a).

Chang, J.-H., K. M. Park, D. Cho, H. S. Yang, and K. J. Ihn, "*Preparation and characterization of polyimide nanocomposites with different organo-montmorillonites*", *Polym. Eng. Sci.*, **41**, 1514-20 (2001b).

Chang, J.-H., Y. U. An, and G. S. Sur, "*Poly(lactic acid) nanocomposites with various organoclays. I. Thermomechanical properties, morphology, and gas permeability*", *J. Polym. Sci. Part B: Polym. Phys.*, **41**, 94-103 (2003).

Chazeau, L., J. Y. Cavaille, G. Canova, R. Dendievel and B. Boutherin, "*Viscoelastic properties of plasticized PVC reinforced with cellulose whiskers*", *J. Appl. Polym. Sci.*, **71**, 1797-08 (1999).

Chen H., D. F. Schmidt, M. Pitsikalis, N. Hadjichristidis, Y. Zhang, U. Wiesner, and E. P. Giannelis, "*Poly(styrene-block-isoprene) nanocomposites: Kinetics of intercalation and effects of copolymer on intercalation behaviors*", *J. Polym. Sci. Part B: Polym. Phys.*, **41**, 3264-3271 (2003).

Chen W., and B. Qu, "*Structural characteristics and thermal properties of pe-g-ma/mgal-ldh exfoliation nanocomposites synthesized by solution intercalation*", *Chem. Mater.*, **15** (16), 3208-3213 (2003).

Chen W., L. Feng, and B. Qu, "*Preparation of nanocomposites by exfoliation of ZnAl Layered Double Hydroxides in nonpolar LLDPE solution*", *Chem. Mater.*, **16** (3), 368-370 (2004).

Chen, C.-G. and D. Curliss, "*Resin matrix composites: organoclay-aerospace epoxy nanocomposites, Part II*", *SAMPE J.*, **37(5)**, 11-18 (2001).

Chen, G., S. Liu, S. Chen, and Z. Qi, "*FTIR spectra, thermal properties, and dispersibility of a polystyrene/montmorillonite nanocomposite*", *Macromol. Chem. Phys.*, **202**, 1189-93 (2001b).

Chen, G., S. Liu, S. Zhang, and Z. Qi, "*Self-assembly in a polystyrene/montmorillonite nanocomposite*", *Macromol. Rapid Commun.* **21**, 746-749 (2000).

Chen, G., Y. Ma and Z. Qi, "*Preparation and morphological study of an exfoliated polystyrene/montmorillonite nanocomposite*", *Scripta Mater.*, **44**, 125-128 (2001a).

Chen, G.-H., D.-J. Wu, W.-G. Weng, and W.-L. Yan, "*Preparation of polymer / graphite conducting nanocomposite by intercalation polymerization*", *J Appl. Polym. Sci.*, **82**, 2506-13 (2001b).

Chen, J. H., Z. P. Huang, D. Z. Wang, S. X. Yang, J. G. Wen and Z. F. Ren, "*Electrochemical synthesis of polypyrrole/carbon nanotube nanoscale composites using well-aligned carbon nanotube arrays*", *Appl. Phys. A Mater. Sci. Proces.*, 73, 129-131 (2001a).

Chen, J., J. Yun and W. Guoquan, "*Nanoparticles offer performance boost in commodity materials*", *Modern Plast.*, **79** (10), 64-65 (2002).

Chen, L., K. Liu, and C. Z. Yang, *Preparation of Ultrafine Metal Particles/ Polyurethane Composites, Polym. Bull.*, **37**, 377-383 (1996).

Chen, L., S.-C. Wong, and S. Pisharath, "*Fracture properties of nanoclay-filled polypropylene*", *J. Appl. Polym. Sci.*, **88**, 3298-05 (2003).

Chen, T.-K., Y.-I. Tien and K.-H., Wei, *Synthesis and characterization of novel segmented polyurethane/clay nanocomposites, Polymer*, **41**, 1345-53 (2000).

Chen, T.-K., Y.-I. Tien, K. H. Wei, *Synthesis and characterization of novel segmented polyurethane/clay nanocomposite via poly(ε-caprolactone)/clay, J. Polym. Sci. A: Polym. Chem.*, **37**, 2225-33 (1999b).

Chen, Y., L. T. Chadderton, J. Fitz Gerald, and J. S. Williams, "*Solid state formation of carbon and boron nitride nanotubes*", NANOTUBE-99 International Workshop on the Science and Application of Nanotubes, East Lansing, Michigan, USA, July 24-27 (1999a).

Cheng, B., W. Q. Jiang, Y. R. Zhu and Z. Y. Chen, "*Preparation of ZnS-PSA nanocomposites by in situ simultaneous polymerization-precipitation of ZnS nanoparticles using γ-radiation*", *Chem. Lett.*, **9**, 935-936 (1999).

Cheng, L. P., D. J. Lin and K. C. Yang, "*Formation of mica-intercalated nylon-6 nanocomposite membranes by phase inversion method*", *J. Membrane Sci.*, **172**, 157-166 (2000).

Chico, L., V. H. Crespi, L. X. Benedict, S. G. Loui and M. L. Cohen, "*Pure carbon nanoscale devices: nanotube heterojunctions*", *Phys. Rev. Letters*, **76**, 971-974 (1996).

Chin, I.-J., T. Thurn-Albrecht, H.-Ch. Kim, T. P. Russell and J. Wang, "*On exfoliation of montmorillonite in epoxy*", *Polymer*, **42**, 5947-52 (2001).

Cho, J. W. and D. R. Paul, "*PA-6 nanocomposites by melt compounding*", *Polymer*, **42**, 1083-94 (2001). .

Cho, J. W., D. R. Paul, H. R. Dennis and D. L. Hunter, "*Nylon-6-based nanocomposites by melt processing*", Symposium on "Commercialization of Nanostructured Materials", Wyndham Miami Beach, Miami, FL, 2000.04.06-07.

Cho, J. W., J. Logsdon, S. Omachinski, G.-Q. Qian, T. Lan, T. W. Womer and W. S. Smith, "*Nanocomposites: A single screw mixing study of nanoclay-filled polypropylene*", *SPE Techn. Pap.*, **48**, (2002), SPE ANTEC San Francisco, CA, 2002.05.05-09.

Choe C. R., W. H. Jo, J. K. Kim, S. W. Kim, M. B. Ko, S. H. Lim, B. S. Jung, M. S. Lee, P. M. Sung, "*Preparation of clay-dispersed polymer nanocomposite*", *WO* **02/20646**, 2002.03.14, Appl. 2000.09.06, to Korea Inst. Sci. Technol.

Choe C. R., W. H. Jo, J. K. Kim, S. W. Kim, M. B. Ko, S. H. Lim, B. S. Jung, M. S. Lee, and M. Park, "*Preparation of clay-dispersed polymer nanocomposite*", *Pat.*, **WO0220646** 2002.03.14, Appl. 2000.09.06, to Korea Institute of Science and Technology (KAIST).

Choi Y. S., Y. K. Kim, and I. J. Chung, "*Method for synthesis of exfoliated polymer/silicate nanocomposite*", US Pat. Appl., 20030050382, 2003.03.13, Appl., 2001.09.11, to Korea Advanced Institute of Science and Technology.

Choi, Y. S., M. H. Choi, K. H. Wang, S. O. Kim, Y. K. Kim, and I. J. Chung, "*Synthesis of exfoliated PMMA/Na-MMT nanocomposites via soap-free emulsion polymerization*", *Macromolecules*, **34**, 8978-85 (2001).
Choo, D. and L. W. Jang, "*Preparation and characterization of PMAM-clay hybrid composite by emulsion polymerization*", *J. Appl. Polym. Sci.*, **61**, 1117-22 (1996).
Chopra, N. G., H. Luyken, V. H. Crespi, K. Cherrey, A. Zettl, and M. L. Cohen, "*Synthesis of BN nanotubes*", *Science,* **269**, 966-967 (1995).
Chou C.-J., and E. Garcia-Meitin, I, "*Nanocomposite*", **EP1155066**, 2001.11.21, Appl. 1999.11.04, to Dow Chem. Co.
Chow, T. S., "*The glass transition of nanoscale polymeric films*", *J. Phys.: Condens. Matter*, **14**, L333-L339 (2002).
Christiani, B. R. and MacRae Maxfield, "*Melt process formation of polymer nanocomposite of exfoliated layered material*", *US Pat.*, **5,747,560**, 05.05.1998, to AlliedSignal Inc.
Ciferri, A., "*Phase behavior of rigid and semirigid mesogens*", in *Liquid Crystallinity in Polymers*, A. Ciferri, Ed., VCH Pub., New York (1991).
Clarey, M., J. Edwards, S. J. Tsipursky, G. W. Beall, and D. D. Eisenhour, "*Method of manufacturing polymer-grade clay for use in nanocomposites*", *US Pat.*, **6,050,509**, 18.04.2000, to AMCOL International Corp.
Clearfield A., A. Oskarsson, and C. Oskarsson, "*On the mechanism of ion exchange in crystalline zirconium phosphates; 6*", *Ion Exch. Membr.*, **1**, 91-107 (1972).
Clearfield A., W. L. Duax, A. S. Medina, G. O. Smith, and J. R. Thomas, "*On the mechanism of ion exchange in crystalline zirconium phosphates; 1*", *J. Phys. Chem.*, **73**, 3424-3430 (1969).
Clegg, D. W. and A. A. Collyer, Eds., *Mechanical Properties of Reinforced Thermoplastics*, Elsevier Appl. Sci. Pub., London (1986).
Clocker, E. T., W. Paterek, N. D. Farel and M. J. Selsley, "*Conversion of clay to its colloidal form by hydrodynamic attrition*", *US Pat.* **3,951,850**, 20.04.1976, Appl. 22.06.1973.
Cohn, M. I., "*Reducing particle size of clay and rheology control of clay-water systems*", *US Pat.*, **3,348,778**, 24.10.1967, Appl. 04.03.1966, to Mineral Industries Corp.
Cole, K. C., *personal communication* (2003).
Colombini D., "*Apports de la modelisation mecanique a l'etude de matériaux polymeres multiphases*", PhD thesis INSA, Lyon, France (1999).
Condon, E. U. and H. Odishaw, Eds., *Handbook of Physics*, McGraw-Hill, New York (1967).
Conroy, J. L., J. W. Piche, P. J. Glatkowski, and D. H. Landis, "*Polymer nanocomposites and methods of preparation*", *US Pat. Appl.*, **20020143094**, 03.10.2002
Cooper, C. A., R. J. Young, and M. Halsall, "*Investigation into the deformation of carbon nanotubes and their composites through the use of Raman spectroscopy*", *Composites: Part A*, **32**, 401-411 (2001).
Cosgrove, T., T. G. Heath, J. S. Phipps and R. M. Richardson, "*Neutron reflectivity studies of polymer adsorbed on mica from solution*", *Macromolecules*, **24**, 94-98 (1991).
Cosgrove, T., T. G. Heath, K. Ryan and T. L. Crowley, "*Neutron scattering from adsorbed polymer layers*", *Macromolecules*, **20**, 2879-82 (1987).

Costa, R. O. R., W. L. Vasconcelos, R. Tamaki and R. M. Laine, "*Organic/inorganic nanocomposite star polymers via atom transfer radical polymerization of methyl methacrylate using octafunctional silsesquioxane cores*", *Macromolecules*, **34** (16), 5398-07 (2001).

Cotton, F. A., G. Wilkinson, C. A. Murillo, and M. Bochmann, *Advanced Inorganic Chemistry*, J. Wiley & Sons, New York (1999).

Cowan C. T. and D. White, "*The mechanism of exchange reactions occurring between sodium montmorillonite and various n-primary aliphatic amine salts*", *Trans. Faraday Soc.*, **54**, 691-697 (1958).

Crawford, R. J. and P. P. Benham, "*Some fatigue characteristics of thermoplastics*", *Polymer*, **16**, 908-14 (1975).

Cullity, B. D. and S. R. Stock, *Elements of X-ray Diffraction*, 3rd Ed., Chapters 5 and 12, Prentice Hall, Upper Saddle River, NJ (2001).

Cumings J. and A. Zettl, "*Mass-production of boron nitride double-wall nanotubes and nanococoons*", *Chem. Phys. Let.*, **316**, 211-216 (2000).

Curran, S. A., P. M. Ajayan, W. J. Blau, D. L. Carroll, J. N.Coleman, A. B. Dalton, A. P. Davey, A Drury, B. McCarthy, S. Maier and A. Strevens, "*Composite from poly(m-phenylenevinylene-co-2,5-dioctoxy-p-phenylenevinylene)(PmPV) and carbon nanotubes: a novel material for molecular optoelectronics*", *Adv. Mater.*, **10**, 1091-93 (1998).

Dąbrowski, F., S. Bourbigot, R. Delobel and M. Le Bras, *Kinetic modeling of the thermal decomposition of polyamide-6 nanocomposite, Europ. Polym. J.*, **36**, 273-284 (2000).

Dahman, S. J., "*Polymer-silicate nanocomposites: commercial opportunities via melt processing*", Conference *New Plastics Asia 2000*, Singapore, 07.03.2000.

Dai, L., "*Advanced syntheses and microfabrications of conjugated polymers, C60-containing polymers and carbon nanotubes for optoelectronic applications*", *Polym. Adv. Technol.*, **10**, 357-420 (1999).

Dai, L.-M. and A. W. H. Mau, "*Controlled synthesis and modification of carbon nanotubes and C60: carbon nanostructures for advanced polymeric composite materials*", *Adv. Mater.*, **13**, 899-913 (2001).

Davis R. D., J. W. Gilman, and D. L. VanderHart, "*Processing degradation of polyamide 6/montmorillonite clay nanocomposites and clay organic modifier*", *Polym. Degrad. Stab.*, **79**, 111-121 (2003).

Davis, C. H., L. J. Mathias, J. W. Gilman, D. A. Schiraldi, J. R. Shields, P. Trulove, T. E. Sutto, and H. C. Delong, "*Effects of melt-processing conditions on the quality of poly(ethylene terephthalate) montmorillonite clay nanocomposites*", *J Polym Sci Part B: Polym Phys.*, **40**, 2661-66 (2002).

de Kimpe, C., M. C. Gastuche, and G. W. Brindley, "*Ionic coordination in alumino-silicic gels in relation to clay mineral formation*", *Am. Mineral.* **46**, 1370-81 (1961).

de Siquira, A. V., C. Lobban, N. T. Skipper, G. D. Willams, A. K. Soper, R. Done, J. W. Dreyer, R. J. Humphreys and J. A. R. Bones, "*The structure of pore fluids in swelling clays at elevated pressures and temperatures*", *J. Phys.: Condens. Matter.*, **11**, 9179-88 (1999).

Debbaut, B. and H. Burhin, "*Large amplitude oscillatory shear and Fourier-transform rheology for a high density polyethylene: experiments and numerical simulation*", *J. Rheol.*, **46**, 1155-76 (2002).

Dee, G. T. and D. J., Walsh, "*Equations of state for polymer liquids*", *Macromolecules*, **21**, 811-815 (1988).

Deguchi, R., T. Nishio, and A. Okada, *"Polyamide composite material and method for preparing the same"*, *US Pat.*, **5,102,948**, 07.04.1992, Appl. 02.05.1990, to Ube Industries, Ltd. and Toyota Jidosha Kabushiki Kaisha, Aichi, Japan.

Deguchi, R., T. Nishio, and A. Okada, "*Polyamide composite material and method for preparing the same*"; "*Method for preparing a polyamide composite material*", *EP* **398 551 B1**, 15.11.1995, Appl. 02.05.1990, to Ube Industries, Ltd. and Toyota Jidosha Kabushiki Kaisha, Aichi, Japan.

Delevoye, L., J.-L. Robert and J. Grandjean, "*23Na 2D 3QMAS NMR and 29Si, 27Al MAS NMR investigation of Laponite and synthetic saponites of variable interlayer charge*", *Clay Minerals,* **38,** 63-69 (2003).

Delozier D. M., R. A. Orwoll, J. F. Cahoon, J. S. Ladislaw, J. G. Smith, Jr., and J. W. Connell, "*Polyimide nanocomposites prepared with a novel aromatic surfactant*", *High Performance Polym.,* **15**, 329-346 (2003).

Delozier, D. M., R. A. Orwoll, J. F. Cahoon, J. S. Ladislaw, J. G. Smith, Jr. and J. W. Connell, "*Polyimide nanocomposites prepared from high-temperature, reduced charge organoclays*", *Polymer*, **44**, 2231-41 (2003).

Delozier, D. M., R. A. Orwoll, J. F. Cahoon, N. J. Johnston, J. G. Smith, Jr. and J. W. Connell, "*Preparation and characterization of polyimide/organoclay nanocomposites*", *Polymer*, **43,** 813-822 (2002).

Deng, J.-G., X.-B. Ding, W.-C. Zhang, Y.-X. Peng, J.-H. Wang, X.-P. Long, P. Li and A. S. C. Chan, "*Carbon nanotube–polyaniline hybrid materials*", *Europ. Polym. J.*, 38, 2497-2501 (2002).

Deng, Q., C. A., Wilkie, R. B. Moore, and K. A. Mauritz, "*TGA-FTIR investigation of the thermal degradation of Nafion and Nafion/(silicon oxide)-based nanocomposites*", *Polymer*, **39**, 5961-72 (1998).

Dennis, H. R., "*Organoclay products containing a branched chain alkyl quaternary ammonium ion*", *US Pat.*, **5,739,087**, 14.04.1998, to Southern Clay Products, Inc.

Dennis, H. R., D. L. Hunter, D. Chang, S. Kim, J. L. White, J. W. Cho and D. R. Paul, "*Effect of melt processing conditions on the extent of exfoliation in organoclay-based nanocomposites*", *Polymer,* **42**, 9513-22 (2001).

Dennis, H. R., D. L. Hunter, J. W. Cho, D. R. Paul, D. Chang, S. Kim, J. L. White, "*Nanocomposites by Extruder Processing*", SPE ANTEC *Techn. Pap.*, **46,** 428-433 (2000).

Despotakis, A., "Nanotechnology", SRI Consulting Explorer Report (2003).

Despotakis, A., *Technology Map Nanotechnology*, SRI Explorer, 2001.

Deuel, H., "*Organic derivatives of clay minerals*", *Clay Minerals Bull.*, **1**, 205-213 (1952).

Di Y., S. Iannace, E. Di Maio, and L. Nicolais, "*Nanocomposites by melt intercalation based on polycaprolactone and organoclay*", *J. Polym. Sci. Part B: Polym. Phys.*, **41**, 670-678 (2003).

Diamond S. and E. B. Kinter, "*Characterization of montmorillonite saturated with short chain amine cations*", *Clays, Clay Minerals*, **10**, 163-173 (1963).

Dietsche, F., F. and R. Mülhaupt, *Thermal Properties and flammability of acrylic nanocomposites based upon organophilic layered silicates, Polym. Bull.*, **43,** 395-402 (1999).

Dietsche, F., Y. Thomann, R. Thomann, and R. Mülhaupt, *Translucent acrylic nanocomposites containing anisotropic laminated nanoparticles derived from intercalated layered silicates, J. Appl. Polym. Sci.,* **75,** 396-405 (2000).

Dirè S., R. Campostrini, and R. Ceccato, "*Pyrolysis chemistry of sol-gel-derived PDMS-zirconia nanocomposites. Influence of zirconium on polymer-to-ceramic conversion*", *Chem. Mater.*, **10**, 268-278 (1998).

Dirix, Y., C. Bastiaansen, W. Caseri, and P. Smith, "*Preparation, structure and properties of uniaxially oriented polyethylene-silver nanocomposites*", *J. Mater. Sci.*, **34**, 3859-66 (1999).

Dirix, Y., C. Darribere, W. Heffels, C. Bastiaansen, W. Caseri, and P. Smith, "*Optically anisotropic polyethylene-gold nanocomposites*", *Appl. Optics*, **38**, 6581-86 (1999).

Dobreva, A. and I. Gutzow, "*Activity of substrates in the catalyzed nucleation of glass-forming melts*", *J. Non-crystalline Solids*, **162**, 1-12; 13-25 (1993).

Doi M. and S. F. Edwards, "*Dynamics of concentrated polymer systems*", *J. Chem. Soc., Faraday Trans.* #2, **74**, 1802-17 (1978).

Domszy R., http://www.science.doe.gov (2004).

Donkai, N., H. Hoshino, K. Kajiwara and T. Miyamoto, "*Lyotropic mesophase of imogolite*", *Macromol. Chem.*, **194**, 559-590 (1993).

Dresselhaus, M. S., G. Dresselhaus and P. C. Eklund, *Science of Fullerenes and Carbon Nanotubes*, San Diego: Academic Press (1996).

Dresselhaus, M. S., G. Dresselhaus, and P. Avouris (Eds.), *Carbon Nanotubes: Synthesis, Structure, Properties, and Applications*, Applied Physics Series, Springer, Berlin (2001).

Drits V. A., "*Structural and chemical heterogeneity of layered silicates and clay minerals*", *Clay Minerals*, **38**, 403-432 (2003).

Du, J.-X., D.-Y. Wang, C. A. Wilkie, and J.-Q. Wang, "*An XPS investigation of thermal degradation and charring on poly(vinyl chloride)–clay nanocomposites*", *Polym. Degrad. Stabil.*, **79**, 319-324 (2003).

Du, J.-X., J. Zhu, C. A. Wilkie and J.-Q. Wang, "*An XPS investigation of thermal degradation and charring on PMMA clay nanocomposites*", *Polym. Degrad. Stab.*, 77, 377-381 (2002).

Dubief, D., E. Samain, and E. Dufresne, "*Polysaccharide microcrystals reinforced amorphous poly(beta-hydroxyoctanoate) nanocomposite materials*", *Macromolecules*, **32**, 5765-71 (1999).

Ebbesen, T. W. (Ed.), *Carbon Nanotubes*, CRC Press, Boca Raton, FL (1997).

Edström, K., S. Nordlinder, T. Gustafsson, J.-L. and E. Olsson, "*Vanadium oxide nanotubes as cathode materials in rechargeable lithium batteries*", International Workshop on the Science and Application of Nanotubes, Potsdam July 22-25 (2001).

Eidt, Jr., C. M., M. L. Gorbaty, C. W. Elspass, D. G. Peiffer, "*Thermoplastic elastomer-asphalt nanocomposite composition*", *US Pat.*, **5,652,284**, 29.07.1997, Appl. 09.11.1995, to Exxon Research & Engineering Company.

Eilliott, D. R. and G. W. Beall, "*Organophilic clay modified with silane compounds*", *US Pat.*, **4,874,728**, 1989.10.17, Appl. 1988.03.21, United Catalyst Inc.

Ellis, G., M. A. Gómez and C. Marco, "*Mapping the crystalline morphology of isotactic polypropylene by infrared microscopy*", *Int. J. Vibr. Spec.*, **5**, 4-7 (2001).

Ellsworth, M. W., "*Organoclay-polymer composites*", *US Pat.*, **5,962,553**, 05.10.1999, Appl. 03.10.1996, to Raychem Corporation.

Ellsworth, M. W., "*Organoclay-polymer composites*", *WO* **98/10012**, 12.03.1998, Appl. 03.10.1996, to Raychem Corporation.

Elspass, C. W. and D. G. Peiffer, "*Nanocomposite materials formed from inorganic layered materials dispersed in a polymer matrix*", *US Pat.*, **6,034,164**, 07.03.2000, Appl. 06.11.1998, to Exxon Research and Engineering Co.

Elspass, C. W., D. G. Peiffer, E. N. Kresge, D.-T. Hsieh, J. J. Chludzinski and K. S. Liang, "*Nanocomposite materials (LAW392)*", *US Pat.*, **5,883,173** 16.03.1999, Appl. 20.12.1996, to Exxon Res. & Eng. Co.

Elspass, C. W., E. N. Kresge, D. G. Peiffer, D.-T. Hsieh, and J. J. Chludzinski, "*Polymer nanocomposite formation by emulsion polymerization*", *PCT Pat. Appl.*, **WO9700910**, 09.01.1997, Priority 1995, to Exxon Res. & Eng. Co.

English, R. J., H. S. Gulati, R. D. Jenkins and S. A. Khan, "*Solution rheology of a hydrophobically modified alkali-soluble polymer*", *J. Rheol.*, **41**, 427-444 (1997).

Fan J.-H., M.-X. Wan, D.-B. Zhu, B.-H. Chang, Z.-W. Pan and S.-H. Xie, "*Synthesis, characterizations, and physical properties of carbon nanotubes coated by conducting polypyrrole*", *J. Appl. Polym. Sci.*, **74**, 2605-10 (1999).

Fan, J., S. Liu, G. Chen, and Z. Qi, "*SEM study of a polystyrene/clay nanocomposite*", *J. Appl. Polym. Sci.*, **83**, 66-69 (2002).

Farmer V. C. and J. D. Russel, "*Interlayer complexes in Layer silicates*", *Trans. Faraday Soc.*, **67**, 2737-49 (1971).

Farmer, V. C., "*The characterization of adsorption bands in clays by infrared spectroscopy*", *Soil Sci.*, **112**, 62-68 (1971).

Farrow, T. C., C, A. Rasmussen, W. R. Menking, D. H. Durham and P. W. Carroll, "*Organoclay compositions and method of preparation*", *US Pat.*, **6,036,765**, 01.04.1998, to Southern Clay Products.

Favier, V., G. R. Canova, S. C. Shrivastava and J. Y. Cavaille, "*Mechanical percolation in cellulose whisker nanocomposites*", *Polym. Eng. Sci.*, **37**, 1732-39 (1997).

Feher, F. J, K. D. Wyndham, R. K. Baldwin, D. Soulivong, J. D. Lichtenhan and J. W. Ziller, "*Methods for effecting monofunctionalization of (CH2=CH)8Si8O12*"*Chem. Commun.*, 1289-90 (1999).

Feng, B., Y. Su, J.-H. Song and K.-C. Kong, "*Electro-polymerization of polyaniline /montmorillonite nanocomposite*", *J. Mater. Sci., Letters*, **20**, 293-294 (2001).

Feng, W., A. Ait-Kadi, and B. Riedl, "*Polymerization compounding: epoxy-montmorillonite nanocomposites*", *Polym. Eng. Sci.*, **42**, 1827-35 (2002).

Fermeglia M, M. Ferrone and S. Pricl, "*Computer simulation of nylon-6/ organoclay nanocomposites: prediction of the binding energy*", *Fluid Phase Equilib.*, **212**, 315-329 (2003).

Ferraro, R. W., C. R. Landis, G. W. Beall, S. Tsipursky, A. Sorokin, and A. Goldman, "*Compositions and methods for manufacturing waxes filled with intercalates and exfoliates formed with oligomers and polymers*", *US Pat.*, **5,837,763**, 17.11.1998, to AMCOL International Corp.

Ferreiro, V., G. Schmit, C. C. Han, and A. Karim, Chapter 14 in "*Polymer Nanocomposites: Synthesis, Characterization, and Modeling*", R.Krishnamoorti, and R. A. Vaia, Eds., ACS Symposium Series 804, American Chemical Society, Washington, DC (2001).

Ferris J. P., "*Montmorillonite catalysis of 30-50 mer oligonucleotides: laboratory demonstration of potential steps in the origin of the RNA world*", *Origins Life Evol. Biosphere*, **32**, 311-332 (2002).

Ferry, J. D., *Viscoelastic Properties of Polymers*, 3rd edition, John Wiley & Sons, New York (1980).

Fibiger R. F., C.-J. Chou, L. Schilhab, and E. Garcia-Meitin I, "*Thermoplastic olefin nanocomposite*", WO0148080, 2001.07.05, Appl. 1999.12.29. This application is under a United States Government contract with The Department of Commerce (NIST)-Advanced Technology Program Project#70NANB7H3028.

Fibiger R. F., M. A. Barger, K. W. Suh, H. C. Tung, W.-B. Liang, G. A. Mackey and J. A. Schomaker, "*Nanocomposite Articles And Process For Making*", *WO* **0047657**, 2000.08.17, Appl. 1999.02.12, to Dow Chemical Co.

Finlayson, C. M. and J. W. Jordan, "*Organophilic clay having enhanced dispersability*", *US Pat.*, **4,105,578**, 08.08.1978, Appl. 23.02.1977, to NL Ind., Inc.

Finlayson, C. M. and W. S. Mardis, "*Organophilic clay complexes, their preparation and compositions comprising said complexes*", *US Pat.*, **4,412,018**, 25.10.1983, to NL Ind., Inc.

Finlayson, C. M., "*Viscosity increasing additive for non-aqueous fluid systems*", *US Pat.*, **4,208,218**, 17.06.1980, to NL Ind., Inc.

Fischer, E. W., H. J. Sterzel and G. Wegner, "*Investigation of the structure of solution grown crystals of lactide copolymers by means of chemical reactions*", *Kolloid-Z. u. Z. Polymere*, **251**, 980-990 (1973).

Fischer, H. R. and L. H. Gielgens, "*Nanocomposite material*", **PCT/NL99/00006** priority 08.01.1999, to Nederlandse Organisatie voor Toegepast Natuurweten Schappelijk Onderzoek.

Fischer, H. R. and L. H. Gielgens, "*Nanocomposite material*", *US Pat.*, **6,365,661**, 02.04.2002a, Appl. 20.10.2000, to Nederlandse Organisatie voor Toegepast (TNO).

Fischer, H. R. and L. H. Gielgens, "*Nanocomposite material*", *US Pat.*, US Pat. **6,372,837**, 16.04.2002b, Appl 31.08.2000, to Nederlandse Organisatie voor Toegepast (TNO).

Fischer, H. R. and L. H. Gielgens, "*Nanocomposite material*", **WO 99/35186**, 15.07.1999, ***PCT*/NL99/00007**, 08.01.1999, to Nederlandse Organisatie voor Toegepast (TNO).

Fischer, H. R., "*Nanocomposite material*", *WO* **99/07790**, 1999.02.18, Appl. 1997.08.08, to Nederlandse Organisatie voor Toegepast (TNO).

Fischer, H. R., L. F. Batenburg, H. A. Meinema, M. P. Hogerheide, and C. H. A. Rentrp, "*Nanocomposite coatings*", WO **01/04050** A1, 18.01.2001, Appl., 13.07.1999, to Nederlandse Centrale Organisatie voor Toegepast-Natuurwetenschappelijk Onderzoek TNO.

Fischer, H. R., L. H. Gielgens, and T. P. M. Koster, "*Nanocomposites from polymers and layered minerals*", *Acta Polym.*, **50**, 122-126 (1999).

Fisher F. and C. Brinson, "*Carbon Nanotubes Literature Review*", Last updated: February 21, 2001; www.mech.northwestern.edu/fac/brinson/nano/LitReview/

Fleer, G., M. A. Cohen-Stuart, J. M. H. M. Scheutjens, T. Cosgrove and B. Vincent, *Polymers at Interfaces*, Chapman and Hall, London (1993).

Fleischer, G., "*The chain length dependence of self-diffusion in melts of polyethylene and polystyrene*", *Colloid Polym. Sci.*, **265**, 89-95 (1987).

Fleurot, O. and D. D. Edie, "*Steady and transient rheological behavior of mesophase pitches*", *J. Rheol.*, **42**, 781-793 (1998).

Flory, P. J., D. Y. Yoon, and K. A. Dill, "*The interphase in lamellar semicrystalline polymers*", *Macromolecules*, **17**, 862-868 (1984).

Flory, P. J., R. A., Orwoll, and A., Vrij, "*Statistical thermodynamics of chain molecule liquids. I. An equation of state for normal paraffin hydrocarbons*", *J. Am. Chem. Soc.*, **86**, 3507-14 (1964).

Fornes T. D., and D. R. Paul, "*Crystallization behavior of nylon 6 nanocomposites*", *Polymer*, **44**, 3945-3961 (2003).

Fornes T. D., D. L. Hunter, and D. R. Paul, "*Nylon-6 nanocomposites from alkylammonium-modified clay: The role of alkyl tails on exfoliation*" *Macromolecules*, **37**, 1793-98 (2004).

Fornes T. D., P. J. Yoon, and D. R. Paul, "*Polymer matrix degradation and color formation in melt processed nylon 6/clay nanocomposites*", *Polymer*, **44**, 7545-7556 (2003).

Fornes, T. D., P. J. Yoon, H. Keskkula and D. R. Paul, "*Nylon 6 nanocomposites: the effect of matrix molecular weight*", *Polymer*, **42**, 9929-40 (2001).

Forrest, J. A. and K. Dalnoki-Veress, "*The glass transition in thin polymer film*", *Adv. Colloid Interf. Sci.*, **94**, 167-196 (2001).

Forte, C., M. Geppi, S. Giamberini, G. Ruggeri, C. A. Veracini and B. Mendez, "*Structure determination of clay/methyl methacrylate copolymer interlayer complexes by means of 13C solid state NMR.*", *Polymer*, **39**, 2651-56 (1998).

Fournaris, K. G., M. A. Karakassides, D. Petridis and K. Yiannakopoulou, "*Clay-polyvinylpyridine nanocomposites*", *Chem. Mater.*, **11**, 2372-81 (1999).

Frankland S. J. V., A. Caglar, D. W. Brenner, and M. Griebel, "*Molecular simulation of the influence of chemical cross-links on the shear strength of carbon nanotube-polymer interfaces*", *J. Phys. Chem. B*, **106**, 3046-3048 (2002).

Frankland S. J. V., V. M. Harik, G. M. Odegard, D. W. Brenner, and T. S. Gates, "*The stress–strain behavior of polymer–nanotube composites from molecular dynamics simulation*", *Comp. Sci. Tech.*, 63, 1655-1661 (2003).

Freudenberg, K. and P. Maitland, "*Der Quebracho Gerbstoff*", *Ann. Chem. Liebigs*, **510**, 193-205 (1934).

Fricke, H., "*A mathematical treatment of the electric conductivity and capacity of disperse systems. 1. The electrical conductivity of a suspension of homogeneous spheroids*", *Phys. Rev.*, **24**, 575-587 (1924).

Frisch, H. L. and J. E. Mark, "*Nanocomposites prepared by threading polymer chains through zeolites, mesoporous silica, or silica nanotubes*", *Chem. Mater.*, 8, 1735-38 (1996).

Frisch, H. L. and R. Simha, "*The viscosity of colloidal suspensions and macromolecular solutions*", in *Rheology*, Vol. 1., F. R. Eirich, Ed., Academic Press, New York (1956).

Frisk, P. and J. Laurent, "*Nanocomposite polymer container*", *US Pat.*, **5,876,812**, 02.03.1999a, Appl. 09.07.1996a,Tetra Laval Holdings & Finance, SA.

Frisk, P. and J. Laurent, "*Nanocomposite polymer container*", *US Pat.*, **5,972,448**, 26.10.1999b, Appl. 27.01.1999b, to Tetra Laval Holdings & Finance, SA.

Frisk, P. and J. Laurent, "*Nanocomposite polymer container*", *WO* **98/01346**, 1998.01.15, Appl. 09.07.1996, to Tetra Laval Holdings & Finance, SA.

Frisk, P., "*Polyolefin material integrated with nanophase particles*", *US Pat.*, **6,117,541**, 12.09.2000, Appl. 02.07.1997, to Tetra Laval Holdings & Finance, SA.

Frisk, P., "*Transparent high barrier multilayer structure*", *US Pat.*, **6,265,038**, 24.07.2001, Appl. 27.07.1999, to Tetra Laval Holdings & Finance, SA.

Fröhlich J., R. Thomann, and R. Mülhaupt, "*Synthesis and characterization of anhydride-cured epoxy nanocomposites containing layered silicates modified with phenolic alkylimidazoline-amide cations*", *Macromol. Mater. Eng.*, **289**, 13-19 (2004).

Fröhlich J., R. Thomann, and R. Mülhaupt, "*Toughened epoxy hybrid nanocomposites containing both an organophilic layered silicate filler and a compatibilized liquid rubber*", *Macromolecules*, **36**, 7205-7211 (2003).

Fu, X.-A. and S. Qutubuddin, "*Polymer-clay nanocomposites: exfoliation of organophilic montmorillonite Nanolayers in polystyrene*", *Polymer*, **42**, 807-813 (2001).

Fu, X.-A. and S. Qutubuddin, "*Synthesis of polystyrene-clay nanocomposites*", *Mater. Lett.*, **42**, 12-15 (2000).

Fujimoto Y., S. S. Ray, M. Okamoto, A. Ogami, K. Yamada, and K. Ueda, "*Well-controlled biodegradable nanocomposite foams: from microcellular to nanocellular*", *Macromol. Rapid Commun.*, **24**, 457-461 (2003).

Fujiwara, S. and T. Sakamoto, "*Method for manufacturing a clay-polyamide composite*", *Japan Kokai*, **109,998**, 1976, to Unitika Ltd.

Fukatani, J., K. Iwasa, N. Ueda and K. Shibayama, "*Polyolefin resin composite, thermoplastic resin composite, and process for producing thermoplastic resin composite*", *Europ. Pat.* **1193290** (**WO0061676**), 2002.04.03, Appl. 2000.04.10, to Sekisui Chemical Co. Ltd.

Fukui, O., K. Tsutsui, T. Akagawa, I. Sakai, T. Nomura, T. Nishio, T. Yokoi and N. Kawamura, "*Thermoplastic resin composition*" *US Pat.*, **5,091,462**, 25.02.1992, to Ube Industries Ltd.

Furukawa, T. and G. W. Brindley, "*Adsorption and oxidation of benzidine and aniline by montmorillonite and hectorite*", *Clays Clay Miner.*, **21**, 279-287 (1973).

Gabriel, J.-C. P. and P. Davidson, "*New trends in colloidal liquid crystals based on mineral moieties*", *Adv. Mater.*, **12**, 9-20 (2000).

Gałęski, A., "*Nucleation in polypropylene*", in *Polypropylene. Structure, Blends and Composites. I. Structure and Morphology*, J. Karger-Kocsis, Ed., Ch. 4, Chapman & Hall, London (1995).

Galgali G., C. Ramesh and A. Lele, "*A rheological study on the kinetics of hybrid formation in polypropylene nanocomposites*", *Macromolecules*, **34**, 852-858 (2001).

Garcia-Rejon, A. and Y. Simard, *unpublished results* (2002).

Gardoliński, J. E., L. C. M. Carrera, M. P. Cantdo and F. Wypych, "*Layered polymer-kaolinite nanocomposites*", *J Mater. Sci.*, **35**, 3113-19 (2000).

Gardoliński, J. E., P. Peralta-Zamora and F. Wypych, "*Preparation and characterization of a kaolinite-1-methyl-2-pyrrolidone intercalation compound*', *J Colloid Interface Sci.*, **211**, 137-141 (1999).

Garrett, W. G. and G. F. Walker, "*Swelling of some vermiculite-organic complexes in water*", *Clays, Clay Minerals*, **9**, 557-567 (1962).

Gavillet J., «*Étude par microscopie électronique en transmission de la nucléation et de la croissance des nanotubes de carbone monofeuillets*», PhD thesis, Université Pierre et Marie Curie-Paris VI (2001).

Gee, M. L. and J. N., Israelachvili, "*Interactions of surfactant monolayers across hydrocarbon liquids*", *J. Chem. Soc. Faraday Trans.*, **86**, 4049-58 (1990).

Gelfer, M., H. H. Song, L. Liu, C. Avila-Orta, L. Yang, M. Si, B. S. Hsiao, B. Chu, M. Rafailovich and A. H. Tsou, "*Manipulating the microstructure and rheology in polymer-organoclay composites*", *Polym. Eng. Sci.*, **42**, 1841-51 (2002).

George, E. R., and R. L. Ballard, "*Nanocomposite reinforced polymer blend and method for blending thereof*", *US Pat. Appl.*, **20030099798 A1**, 2003.05.29, Appl. 2001.11.29.

Geprägs M., V. Rauschenberger, V. Warzelhan, P. Wolf, T. Heitz, M. Klatt, M. Weber, M. Fischer, S. Grutke, T. Plesnivy, V. Rauschenberger and J. Wünsch, "*Thermoplastic polystyrene molding compositions*", *Ger. Pat.*, **DE19714548, EP0973828 (WO9845362), JP2001518954T**, 1998.10.15, Appl. 1997.04.09, to BASF A.G.

Giannelis, E. P. and P. B. Messersmith, "*Method of preparing layered silicate-epoxy nanocomposites*", *US Pat.*, **5,554,670**, 10.09.1996, to Cornell Research Foundation, Inc.

Giannelis, E. P., "*Polymer-layered silicate nanocomposites: synthesis, properties and applications*", *Appl. Organometal. Chem.*, **12**, 675-680 (1998).

Giannelis, E. P., R. Krishnamoorti and E. Manias, "*Polymer-silicate nanocomposites: model systems for confined polymer and polymer brushes*", *Adv. Polym. Sci.*, **138**, 107-147 (1999).

Giaquinta, D. M., L. Soderholm, S. E. Yuchs and S. R. Wasserman, "*The speciation of uranium in smectic clay*", *Radiochim. Acta*, **76**, 113-121 (1997).

Giese, R. F. and C. J. van Oss, *Colloid and surface properties of clays and related materials*, Marcel Dekker, New York (2002).

Gieseking, J. E., "*Mechanism of cation exchange in the montmorillonite-beidellite-nontronite type of clay minerals*", *Soil Sci.*, **47**, 1-14 (1939).

Gilman J. W., "*Recent advances in flame retardant polymer nanocomposites*", *Polymer nanocomposites-2001*, International symposium, NRCC-IMI, Boucherville, QC, Canada, 14 to 16.11.2001.

Gilman, J. W., C. L. Jackson, A. B. Morgan, R. Harris, Jr., E. Manias, E. P. Giannelis, M. Wuthenow, D. Hilton and S. H. Phillips, "*Flammability properties of polymer-layered-silicate nanocomposites. polypropylene and polystyrene nanocomposites*", *Chem. Mater.*, **12**, 1866-73 (2000a).

Gilman, J. W., *Flammability and thermal stability studies of polymer layered-silicate (clay) nanocomposites, Appl. Clay Sci.*, **15**, 31-49 (1999).

Gilman, J. W., T. Kashiwagi and J. D. Lichtenhan, "*Nanocomposites: a Revolutionary New Flame Retardant Approach*", 42nd International SAMPE Symposium, 4-8.05.1997; *SAMPE J.*, **33**, 40-46 (1997).

Gilman, J. W., T. Kashiwagi, A. B. Morgan, R. H. Harris Jr., L. Brassell, H. W. Award, R. D. Davis, L. Chyall, T. Sutto, P. C. Trulove and H. DeLong, "*Layered Silicates, Nano-silicas, and POSS Nano-additives as Flame Retardants in Commodity Polymers*", Proceedings of additives 2001, March, 2001.

Gilman, J. W., T. Kashiwagi, A. B. Morgan, R. H. Harris, Jr., L. Brassell, M. Vanlandingham, and C. L. Jackson, "*Flammability of polymer clay nanocomposites consortium: Year One Annual Report*", US Dept. Commerce, Technology Administration, National Institute of Standards and Technology, 55 pg. Report NISTIR **#6531**, Gaithersburg (2000b).

Gilman, J. W., T. Kashiwagi, E. P. Giannelis, E. Manias, S. Lomakin, J. D. Lichtenhan and P. Jones, *Fire Retardancy of Polymers. The use of Intumescence*, **Vol. 203**, M. Le Bras, G. Camino, S. Bourbigot and R. Delobel, Eds, Royal Chem. Soc, Cambridge (UK) (1998b).

Gilman, J. W., T. Kashiwagi, J. E. T. Brown and S. Lomakin, "*Flammability studies of polymer layered silicate nanocomposites*", *SAMPE J.* **43** 1053-66 (1998a).

Gilman, J. W., T. Kashiwagi, M. Nyden, J. E. T. Brown, C. L. Jackson, S. Lomakin, E. P. Giannelis and E. Manias, in *Chemistry and Technology of polymer Additives*, S. AI-Malaika, A. Golovoy and C. A. Wilkie, Eds., Blackwell Scientific, Oxford (1999).

Gilmer, J. W., J. Ch. Matayabas, Jr. and R. B. Barbee, *"Polymer/clay nanocomposite and process for making same"*, *WO* **00/34378**, 15.06.2000d, Priority 07.12.1998, to Eastman Chemical Co.

Gilmer, J. W., J. Ch. Matayabas, Jr., G. W. Connell, J. T. Turner, and R. L. Piner, *"Process for preparing an exfoliated, high I.V. polymer nanocomposite with an oligomer resin precursor and an article produced therefrom"*, *WO* **00/ 34377**, 15.06.2000c, Priority 07.12.1998, to Eastman Chemical Co.

Gilmer, J. W., J. Ch. Matayabas, Jr., R. B. Barbee, and T. Lan, *"A polymer/clay nanocomposite having improved gas barrier comprising a clay material with a mixture of two or more organic cations and a process for preparing same"*, *WO* **00/34180**; *WO* **00/34380**, 15.06.2000a,b, Priority 07.12.1998, to Eastman Chemical Co.

Gin, D. L., W. M. Fischer, D. H. Gray, and R. C. Smith, *"Highly ordered nanocomposites via a monomer self-assembly in situ condensation approach"*, *US Pat.*, **5,849,215**, 15.12.1998, Appl. 08.01.1997, to The Regents of the University of California.

Ginzburg V. V. and A. C. Balazs, *"Calculating phase diagrams for nanocomposites: the effect of adding end-functionalized chains to polymer/clay mixtures"*, *Adv. Mater.*, **12**, 1805-09 (2000).

Ginzburg V. V. and A. C. Balazs, *"Calculating phase diagrams of polymer-platelet mixtures using density functional theory: implications for polymer/clay composites"*, *Macromolecules*, **32**, 5681-88 (1999).

Ginzburg V. V., C. Singh, and A. C. Balazs, *"Theoretical phase diagrams of polymer/clay composites: the role of grafted organic modifiers"*, *Macromolecules*, **33**, 1089-99 (2000).

Ginzburg V. V., O. V. Gendelman, and L. I. Manevitch, *"Simple "Kink"Model of Melt Intercalation in Polymer-Clay Nanocomposites"*, *Phys. Rev. Lett.*, **86**, 5073-5075 (2001).

Giza, E., H. Ito, T. Kikutani and N. Okui, *"Structural control of polyamide-6/clay nanocomposite fibers by in-line drawing process"*, *J. Polym. Eng.*, **20**, 403-25 (2000).

Gloaguen, J. M. and J. M. Lefebvre, *"Plastic deformation behavior of thermoplastic/clay nanocomposites"*, *Polymer*, **42**, 5841-47 (2001).

Goda, H. and C. W. Frank, *Fluorescence Studies of the Nanocomposites of Segmented-Polyurethane and Silica, ACS Polymeric Materials Science and Engineering*, 77, 532-533 (1997).

Golden, J. H., F. J. DiSalvo, J. M. J. Frechet, J. Silox, M. Thomas and J. Elman., *"Subnanometer-diameter wires isolated in a polymer matrix by fast polymerization"*, *Science*, **273**, 782-784 (1996).

Goldman, A. Y., J. A. Montes, A. Barajas, G. Beall, and D. D. Eisenhour, *"Effect of aging on mineral-filled nanocomposites"*, SPE Techn. Pap., **44**, 2415-25 (1998).

Goldsmith, H. L. and S. G. Mason, *"The microrheology of dispersions"*, in *Rheology*, vol. 4., F. R. Eirich, Ed., Academic Press, New York (1967).

Gole J. L., Z. L. Wang, Z. R. Dai, J. Stout, and Mark White, *"Silica-based nanospheres, nanowires, nanosubstrates, nanotubes, and nanofiber arrays"*, *Colloid Polym. Sci.*, **281**, 673-685 (2003).

Gong, X., J. Liu, S. Baskaran, R. D. Voise and J. S. Young, "*Surfactant-assisted processing of carbon nanotube/polymer composites*", *Chem. Mater.*, **12**, 1049-52 (2000).

Gonsalves, K. E., G. Carlson, X., Chen, S. K. Gayen, R. Perez and M. José-Yacáman, "*Synthesis and nonlinear optical characterization of nanostructures gold/polymer composites*", *Nanostructured Mater, 7, 293-303* (1996).

Gonzales-Carreño, T., J. A. Raussel-Colom and J. M. Serratosa, "*Complexes vermiculite-alkylammonium*", Proceed. 3rd Europ. Clay Conf., Oslo (1977).

Gonzales, A., K. L. Nichols, C. E. Powell and B. P. Thill, "*Organoclay compositions*", *US Pat.*, **5,780,376**, 14.07.1998a, to Southern Clay Products, Inc.

Gonzales, G., M. A. Santa Ana and E. Benavente, "*Lithium chemical diffusion coefficients in poly(ethylene oxide)-molybdenum sulfide nanocomposites*", *J. Phys. Chem. Solids,* **58**, 1457-66 (1997).

Gonzales, G., M. A. Santa Ana and E. Benavente, "*Mixed conductivity of lithium diffusion in polyethylene oxide molybdenum disulfide nanocomposites*", *Electrochim. Acta*, **43**, 1327-32 (1998b).

Gorrasi G., L. Tammaro, M. Tortora, V. Vittoria, D. Kaempfer, P. Reichert, and R. Mülhaupt, "*Transport properties of organic vapors in nanocomposites of isotactic polypropylene*", *J. Polym. Sci. Part B: Polym. Phys.*, **41**, 1798-1805 (2003).

Gorrasi G., M. Tortora, V. Vittoria, E. Pollet, M. Alexandre, and P. Dubois, "*Physical properties of poly(ε-caprolactone) layered silicate nanocomposites prepared by controlled grafting polymerization*", *J. Polym. Sci. Part B: Polym. Phys.*, **42**, 1466-1475 (2004).

Gorrasi, G., M. Tortora, V. Vittoria, E. Pollet, B. Lepoittevin, M. Alexandre and P. Dubois, "*Vapor barrier properties of polycaprolactone montmorillonite nanocomposites: effect of clay dispersion*", *Polymer*, **44**, 2271-79 (2003).

Gorrasi, G., M. Tortora, V. Vittoria, G. Galli, and E. Chiellini, "*Transport and mechanical properties of poly(ε-caprolactone) and a modified montmorillonite poly(ε-caprolactone) nanocomposites*", *J. Polym. Sci. Part B: Polym. Phys.*, **40**, 1118-24 (2002).

Gotoh, Y., Y. Ohkoshi and M. Nagura, "*Preparation and optical absorption property of poly(acrylic acid)/copper sulfide nanocomposite films*", *Polym. J.*, **33**, 303-305 (2001).

Goward, G. R., F. Leroux and L. F. Nazar, "*Poly(pyrrole) and poly(thiophene)/ vanadium oxide interleaved nanocomposites: Positive electrodes for lithium batteries*", *Electrochim. Acta*, **43**, 1307-13 (1998).

Graff, G., "*Nanocomposite suppliers see promise for custom solutions*", *Modern Plast.*, **76(6)**, 38-40 (1999).

Granick, S., "*Motions and relaxations of confined liquids*", *Science*, **253**, 1374-79 (1991).

Greene-Kelly, R., "*Sorption of aromatic organic compounds by montmorillonite*", *Trans Faraday Soc.*, **51**, 412-424 (1955).

Greenland, D. J., "*Adsorption of polyvinyl alcohols by montmorillonite*", *J. Colloid Sci.*, **18**, 647-664 (1963).

Grest, G. S. and M. Murat, "*Computer simulations of tethered chains*", in *Monte Carlo and Molecular Dynamics Simulations in Polymer Science*, K. Binder, Ed., Oxford University Press, Oxford (1995).

Grim, R. E., *Clay Mineralogy*, 2nd edition, McGraw-Hill, London (1968).

Grimshaw, R. W., *The Chemistry and Physics of Clays*, 4th edition, J. Wiley & Sons, New York (1980).

Grutke, S., G. E. Mc Kee, and C. Mehler, "*Thermoplastic materials containing nanocomposites and an additional elastomer*", *US Pat.*, **6,458,879**, 2002.10.01, Appl. 1999.11.18, to BASF A.-G.

Gu, A. and F.-C. Chang, "*A novel preparation of polyimide/clay hybrid films with low coefficient of thermal expansion*", *J. Appl. Polym. Sci.*, **79**, 289-294 (2001).

Gu, A., S.-W. Kuo, and F.-C. Chang, "*Syntheses and properties of PI/clay hybrids*", *J. Appl. Polym. Sci.*, **79**, 1902-10 (2001).

Guimarães, J. L., C. J. da Cunha, and F. Wypych, "*Intercalation of hexylamine into Hydrated Kaolinite Phenyl phosphonate*", *J Colloid Interface Sci.*, **218**, 211-216 (1999).

Guimarães, J. L., P. Peralta-Zamora and F. Wypych, "*Covalent grafting of phenyl phosphonate groups onto the interlamellar aluminol surface of kaolinite*", *J Colloid Interface Sci.*, **206**, 281-287 (1998).

Hackett, E., E. Manias and E. P. Giannelis, "*Molecular dynamics simulation of organically modified layered silicates*", *J. Chem. Phys.*, **108**, 7410-15 (1998).

Haddad T. S. and J. D. Lichtenhan, "*Hybrid organic-inorganic thermoplastics: styryl-based polyhedral oligomeric silsesquioxane polymers*", *Macromolecules*, **29**, 7302-7304 (1996).

Haggenmueller, R., H. H. Gommans, A. G. Rinzler, J. E. Fischer and K. I. Winey, "*Aligned single-wall carbon nanotubes in composites by melt processing methods*", *Chem. Phys. Letters*, **330**, 219-225 (2000).

Hahn, H. W., Ed., "*Synthesis, functional properties and applications of nanostructures*", MRS symposium held April 17-20, 2001, San Francisco, CA, 2001, Materials Research Society, Warrendale, PA, (2002).

Hajii, P., L. David, J. F. Gerard, J. P. Pascault and G. Vigier, "*Synthesis, structure, and morphology of polymer-silica hybrid nanocomposites based on hydroxyethyl methacrylate*", *J. Polym. Sci.: Polymer Phys. Ed.*, **37**, 3172-87 (1999).

Hambir S., N. Bulakh, P. Kodgire, R. Kalgaonkar and J. P. Jog, "*PP/clay nanocomposites: A study of crystallization and dynamic mechanical behavior*", *J. Polym. Sci. Part B: Polym. Phys.*, **39**, 446-450 (2001).

Hambir, S., N. Bulakh and J. P. Jog, "*Polypropylene/clay nanocomposites: Effect of compatibilizer on the thermal, crystallization and dynamic mechanical behavior*", Paper presented at the international symposium Polymer Nanocomposites 2001, Boucherville, QC, Canada, 2001.09.17-19; *Polym. Eng. Sci.*, **42**, 1800-07 (2002).

Hamilton, C. C., "*Insecticide composition*", *US Pat.*, **2,035,546**, 31.03.1936, to Endowment Foundation, New Brunswick, N. J.

Hammami, A., J. Spruiell and A. K. Mehrotra, "*Quiescent nonisothermal crystallization kinetics of isotactic polypropylenes*", *Polym. Eng. Sci.*, **35**, 797-804 (1995).

Han B., A. Cheng, G. Ji, S.-S. Wu, and J. Shen, "*Effect of Organophilic Montmorillonite on Polyurethane/ montmorillonite Nanocomposites*", *J. Appl. Polym. Sci.*, **91**, 2536-2542 (2004).

Hand, R. J., P. F. James and A. B. Seddon (Eds.), *Proceeding of 9th International Workshop on Glasses, Ceramics, Hybrids and Nanocomposites from Gels* (Sol-Gel '97), Sheffield, 31 August-5 Sept 1997, Kluwer Acad. Pub., Dordrecht (1998).

Hang, W., Y. Bando, K. Kurashima and T. Sato, "*Synthesis of boron nitride nanotubes from carbon nanotubes by a substitution reaction*", *Applied Physics Letters*, **73**, 3085-87 (1998).

Hansen, C. M., "*Solubility parameters and coatings*", *Europ. Coat. J.*, **5/94**, 305-317 (1994).

Hansen, C. M., "*Solubility Parameters*", Ch. 35 in *Paint and Coating Testing Manual*, J. V. Koleske, Ed., ASTM Manual Series: MNL 17, ASTM, Philadelphia (1995).

Hansen, C. M., "*The three dimensional solubility parameter-key to paint component affinities*", *J. Paint Technol.*, **39**, 104-117 (1967).

Hansen, C. M., *Hansen Solubility Parameters: A User's Handbook*, CRC Press, Boca Raton FL (2000).

Harada, Y., M. Kimura, K. Hashimoto, and K. Abe, "*Gas barrier film containing inorganic stratified particles and a production method thereof*", *UC Pat.*, **5,981,029**, 09.11.1999, Appl. 05.04.1996, to Toray Industries, Inc.

Harada, Y., M. Kimura, K. Hashimoto, and K. Abe, "*Polyester-based gas barrier film containing inorganic stratified particles*", *UC Pat.*, **6,083,605**, 04.07.2000, Appl. 05.04.1996, to Toray Industries, Inc.

Harben, P. W. and R. L. Bates, "*Industrial Minerals and World Deposits*", *Industrial Material Division, Metal Bulletin Plc*, London, UK (1990).

Harik, V. M., "*Ranges of applicability for the continuum beam model in the mechanics of carbon nanotubes and nanorods*", *Solid State Commun.*, **120**, 331-35 (2001).

Harris, P. J. F., *Carbon Nanotubes and Related Structures: New Materials for the Twenty-First Century*, Cambridge University Press, Cambridge (1999).

Hartmann, B. and M. A., Haque, "*Equation of state for polymer liquids*", *J. Appl. Polm. Sci.*, **30**, 1553-63 (1985).

Hasegawa, N., H. Okamoto, M. Kato, A. Usuki and N. Sato, "*Nylon-6/Na–montmorillonite nanocomposites prepared by compounding Nylon-6 with Na–montmorillonite slurry*", *Polymer*, **44**, 2933-37 (2003).

Hasegawa, N., H. Okamoto, M. Kato, and A. Usuki, "*Preparation and mechanical properties of polypropylene-clay hybrids based on modified polypropylene and organophilic clay*", *J Appl. Polym. Sci.*, 78, 1918-22 (2000b).

Hasegawa, N., H. Okamoto, M. Kawasumi and A. Usuki, "*Preparation and mechanical properties of polystyrene-clay hybrids*", *J. Appl. Polym. Sci.* **74**, 3359-64 (1999).

Hasegawa, N., H. Okamoto, M. Kawasumi, A. Usuki and A. Okada, "*Resin composite*", *US Pat.*, **6,117,932**, 2000c.09.12, Appl. 1998.09.17, to Kabushiki Kaisha Toyota Chuo Kenkyusho.

Hasegawa, N., H. Okamoto, M. Kawasumi, M. Kato, A. Tsukigase, A. Usuki, "*Polyolefin-clay hybrids based on modified polyolefins and organophilic clay*", *Macromol. Mater. Eng.*, **280/281**, 76-79 (2000a).

Hasegawa, N., M. Kawasumi, M. Kato, A. Usuki, and A. Okada, "*Preparation and mechanical properties of polypropylene-clay hybrids using a maleic anhydride-modified polypropylene oligomer*", *J. Appl. Polym. Sci.*, **67**, 87-92 (1998).

Hatharasinghe, H. L. M., M. V. Smalley, J. Swenson, and G. D. Williams, "*Neutron Scattering Study of Vermiculite-PEO Mixtures*", *J Phys. Chem. B*, **102**, 6804-08 (1998).

Hauser, E. A., "*Modified gel-forming clay and process of producing same*", *US Pat.*, **2,531,427**, 1950.11.28, Appl. 1946.05.03.

Havni, I. and Z. S. Petrovic, *Structure and Properties of Polyurethane Nanocomposite*, Polyurethanes Expo '98. Conference proceedings, Dallas, TX.

Heinemann, J., P. Reichert, R. Thomann and R, Mülhaupt, "*Polyolefin nanocomposites formed by melt compounding and transition metal catalyzed homo-and copolymerization in the presence of layered silicates*", *Macromol. Rapid Commun.*, **20**, 423-430 (1999).

Helfand, E. and Y. Tagami, "*Theory of the interface between immiscible polymers*", *Polym. Letters, 9, 741 (1971); J. Chem. Phys.*, **56**, 3592-01 (1972).

Hendricks, S. B., "*Base exchange of the clay mineral montmorillonite for organic cations and its dependence upon adsorption due to van der Waals forces*", *J. Phys Chem.*, **45**, 65-81 (1941).

Hentschke, R., "*Molecular modeling of adsorption and ordering at solid interfaces*", *Macromol. Theory Simul.*, **6**, 287-316 (1997).

Hesselink, F. Th., "*On the theory of the stabilization of dispersions by adsorbed macromolecules-I, J. Phys. Chem.*, **75**, 65-71 (1971).

Hesselink, F. Th., A. Vrij and J. Th. G. Overbeek, "*On the theory of the stabilization of dispersions by adsorbed macromolecules-II*", *J. Phys. Chem.*, **75**, 2094-2103 (1971).

Hettinger W. P., "*Contribution to catalytic cracking in the petroleum industry*", *Appl. Clay Sci.*, **5**, 445-468 (1991).

Hill, D. E., Y. Lin, A. M. Rao, L. F. Allard and Y.-P. Sun, "*Functionalization of carbon nanotubes with polystyrene*", *Macromolecules*, **35**, 9466-71 (2002).

Hill, T. L. and R. V. Chamberlin, "*Extension of the thermodynamics of small systems to open metastable states: an example*", *Proc. Natl. Acad. Sci. USA*, **95**, 12,779-82 (1998).

Hill, T. L., "*Extension of nanothermodynamics to include a one-dimensional surface excess*", *Nano Letters*, **1**, 159-160 (2001b).

Hill, T. L., "*Perspective: nanothermodynamics*", *Nano Letters*, **1**, 111-112; 273-275 (2001a).

Hill, T. L., *Thermodynamics of Small Systems*, Dover: New York, 1994.

Hirai, T., M. Miyamoto, and I. Komasawa, "*Composite nano-CdS-polyurethane transparent films*", *J. Mater. Chem.*, **9**, 1217-19 (1999).

Ho, D. L., R. M. Briber, and C. J. Glinka, "*Characterization of organically modified clays using scattering and microscopy techniques*", *Chem. Mater.*, **13**, 1923-31 (2001).

Hoffman, B., C. Dietrich, R. Thomann, C. Friedrich, and R. Mülhaupt, "*Morphology and rheology of polystyrene nanocomposites based upon organoclay*", *Macromol. Rapid Commun.* **21**, 57-61 (2000a).

Hoffman, J. D., G. T. Davies, and J. I. Lauritzen, Jr., "*The rate of crystallization of linear polymers with chain folding*", in *Treatise on Solid State Chemistry*, N. B. Hannay, Ed., Plenum Press, New York (1976).

Hoffmann B., J. Kressler, G. Stöppelmann, Chr. Friedrich and G.-M. Kim, "*Rheology of nanocomposites based on layered silicates and polyamide-12*", *Colloid Polym. Sci.*, **278**, 629-636 (2000b).

Hofmann, A. W., "*Beiträge zur Kenntnifs der flüchtigen organischen Basen; X. Uebergang der flüchtigen Basen in eine Reihe nichtflüchtiger Alkaloïde*", *Liebigs Ann. Chem.*, **78**, 253 (1851).

Horn, R. G. and J. N. Israelachvili, "*Molecular organization and viscosity of a thin film of molten polymer between two surfaces as probed by force measurements*", *Macromolecules*, **21**, 2836-41 (1988).
Hoshino, H., M. Yamana, N. Donkai, V. Sinigersky, K. Kajiwara, T. Miyamoto and H. Inagaki, "*Lyotropic mesophase formation in HPC/imogolite mixture*", *Polym. Bull.*, **29**, 607-612 (1992b).
Hoshino, H., T. Ito, N. Donkai, H. Urakawa and K. Kajiwara, "*Lyotropic mesophase formation in PVA/imogolite mixture*", *Polym. Bull.*, **29**, 453-460 (1992a).
Hostenson, E. T., Z. Ren and T. -W. Chou, "*Advances in the science and technology of carbon nanotubes and their composites: a review*", *Compos. Sci. Technol.*, **61**, 1899 (2001).
Hsu S. L.-C., U. Wang, J.-S. King, J.-L. Jeng, "*Photosensitive poly(amic acid)/ organoclay nanocomposites*", *Polymer*, **44**, 5533-5540 (2003).
Hsu, S. L.-C. and K.-C. Chang, "*Synthesis and properties of polybenzoxazole-clay nanocomposites*", *Polymer* **43**, 4097-01 (2002).
Hsueh H.-B., and C.-Y. Chen, "*Preparation and properties of LDHs/polyimide nanocomposites*", *Polymer,* **44**, 1151-1161 (2003).
Hsueh H.-B., and C.-Y. Chen, "*Preparation and properties of LDHs/epoxy nanocomposites*", *Polymer,* **44**, 5275-5283 (2003)
Hu H., M. Pan, X. Li, X. Shi, and L. Zhang, "*Preparation and characterization of poly(vinyl chloride)/organoclay nanocomposites by in situ intercalation*", *Polym. Int.*, **53**, 225-231 (2004).
Hu, H.-W. and S. Granick, "*Viscoelastic dynamics of confined polymer melts*", *Science*, **258**, 1339-42 (1992).
Hu, X.-S., W.-h. Zhang, M.-Y. Si, M. Gelfer, B. Hsiao, M. Rafailovich, J. Sokolov, V. Zaitsev, and S. Schwarz, "*Dynamics of Polymers in Organosilicate Nanocomposites*", *Macromolecules*, **36**, 823-829 (2003).
Hu, Y., L. Song, J. Xu, L. Yang, Z. Chen and W. Fan, "*Synthesis of polyurethane/ clay intercalated nanocomposites*", *Colloid Polym. Sci.*, **279**, 819-822 (2001).
Huang, C., L. Chen, and C. Yang, "*Study of Cu2+-poly(itaconic acid-co-acrylic acid) complex and copper-polymer nanocomposites*", *Polym. Bull.*, **41**, 585-592 (1998).
Huang, J. C., Z.-K. Zhu, J. Yin, X.-F. Qian, and Y.-Y. Sun, "*Poly(etherimide)/ montmorillonite nanocomposites prepared by melt intercalation: morphology, solvent resistance properties and thermal properties*", *Polymer* **42**, 873-877 (2001b).
Huang, J.-C., Z.-K. Zhu, X.-D. Ma, X.-F. Qian, and J. Yin, "*Preparation and properties of montmorillonite/organo-soluble polyimide hybrid materials prepared by one-step approach*", *J. Mater. Sci.*, **36**, 871-875 (2001a).
Huang, X., S. Lewis, W. J. Brittain and R. A. Vaia, "*Synthesis of polycarbonate-layered silicate nanocomposites via cyclic oligomers*", *Macromolecules*, **33**, 2000-2004 (2000).
Huang, Z. P., J. W. Xu, Z. F. Ren, J. H. Wang, M. P. Siegal and P. N. Provencio, "*Growth of highly oriented carbon nanotubes by plasma-enhanced hot filament chemical vapor deposition*", *Appl. Phys. Letters*, **73**, 3845-47 (1998).
Hudson, S. D., "*Polyolefin nanocomposites*", *US Pat.*, **5,910,523**, 08.06.1999, Appl. 01.12.1997.
Hui, C. Y. and D. Shia, "*Simple formulae for the effective moduli of unidirectionally aligned composites*", *Polym. Eng. Sci.*, **38**, 774-82 (1998).

Hybrid Plastics™ on-line catalogue, 2003: http://www.hybridplastics.com/
Hyun, Y. H., S. T. Lim, H. J. Choi, and M. S. Jhon, "*Rheology of poly(ethylene oxide)/organoclay nanocomposites*", *Macromolecules*, **34**, 8084-93 (2001).
Iijima, S. and T. Ichlhashi, "*Single-shell carbon nanotubes of 1-nm diameter*", *Nature*, **363**, 603-605 (1993).
Iijima, S., "*Carbon nanotubes: past, present and future*", *Physica B*, **323**, 1-5 (2002).
Iijima, S., "*Helical microtubules of graphitic carbon*", *Nature*, **354**, 56-58 (1991).
Iijima, S., C. J. Brabec, A. Maiti and J. Bernholc, "*Structural flexibility of carbon nanotubes*", *J. Chem. Phys.*, **104**, 2089-92 (1996).
Imai Y., Y. Inukai, and H. Tateyama, "*Properties of poly(ethylene terephthalate)/ layered silicate nanocomposites prepared by two-step polymerization procedure*", *Polym. J.*, **35**, 230-235 (2003).
Imai, Y., N. Satoshi, E. Abe, H. Tateyama, A. Abiko, A. Yamaguchi, T. Aoyama, and H. Taguchi, "*High-modulus poly(ethylene terephthalate)/expandable fluorine mica nanocomposites with a novel reactive compatibilizer*", *Chem. Mater.*, **14**, 477-479 (2002).
Incarnato L., P. Scarfato, G. M. Russo, L. Di Maio, P. Iannelli, and D. Acierno, "*Preparation and characterization of new melt compounded copolyamide nanocomposites*", Polymer, 44 4625-4634 (2003).
Inoue, H., K. Tamura, T. Ebata and M. Noguchi, "*Highly-rigid, flame-resistant polyamide composite material*", *US Pat.*, **6,294,599**, 25.09.2002, Appl. 08.07.1998, to Showa Denko, K. K.
Inoue, T., "*Morphology of polymer blends*", Ch. 8 in *Polymer Blends Handbook*, L. A. Utracki, Ed., Kluwer, Dordrecht (2002).
Ishida, H., "*General approach to nanocomposite preparation*", *US Pat.*, **6,271,297**, 07.08.2001; **WO 00/69957**, 23.11.2000, Appl. 13.05.1999, to Case Western Reserve University (Cleveland, OH).
Ishida, H., S. Campbell, and J. Blackwell, "*General approach to nanocomposite preparation*", *Chem. Mater.*, **12**, 1260-67 (2000).
Ishikawa, T., A. Nakayama and T. Ishida, "*Intercalation complexes of clay minerals with fullerene and polymers as materials for electrodes for secondary lithium batteries*", *Jap. Pat.*, **09132695**, 1997; to Agency of Industrial Sciences and Technology, Japan.
Ishizaka, M. and T. Fujii, "*Mica-silicone composites*", *US Pat.*, **4,244,911**, 1981.01.13, to Toshiba Silicone Co., Ltd.
Israelachvili, J. N., *Intermolecular and Surface Forces with Applications to Colloidal and Biological Systems*, Academic Press, New York (1985).
Israelachvili, J. N., M. Tirrell, J. Klein and Y. Almog, "*Forces between two layers of adsorbed polystyrene immersed in cyclohexane below and above the Θ temperature*", *Macromolecules*, **17**, 204-209 (1984).
Itagaki, T., A. Matsumura, M. Kato, A. Usuki, K. Kuroda, "*Preparation of kaolinite–nylon-6 composites by blending nylon-6 and a kaolinite–nylon-6 intercalation compound*", *J. Mater. Sci. Lett.*, **20**, 1483-84 (2001).
Ivanov, Y., T. G. M. van de Ven and S. G. Mason, "*Damped oscillations in the viscosity of suspensions of rigid rods*", *J. Rheol.*, **26**, 213-230; 231-244 (1982).
Iwasaki, T., H. Hayashi, K. Torii, T. Sekimoto, T. Fujisaki, M. Ikegami, and I. Ishida, "*Organophilic clay*", *US Pat.*, **5,376,604**, 27.12.1994, Appl. 07.07.1992, to CO-OP Chemical Co., Ltd.
Jain R. K. and R., Simha, "*On the statistical thermodynamics of multicomponent fluids: equation of state*", *Macromolecules*, **13**, 1501-08 (1980).

Jain R. K. and R., Simha, "*Statistical thermodynamics of multicomponent fluids*", *Macromolecules*, **17**, 2663-68 (1984).
Jain T. S. and J. J. de Pablo, "*Monte Carlo simulation of free-standing polymer films near the glass transition temperature*", *Macromolecules*, **35**, 2167-76 (2002).
Jang, L. W., C. M. Kang, and D. C. Lee, "*A new hybrid nanocomposite prepared by emulsion copolymerization of ABS in the presence of clay*", *J Polym Sci B: Polym Phys.*, **39**, 719-727, (2001).
Jarjayes, O., P. H. Fries and G. Bidan, "*New nanocomposites of polypyrrole including γ-Fe2O3 particles: Electrical and magnetic characterizations*", *Synth. Metals*, **69**, 343-344 (1995).
Jayan P. M., "*Carbon nanotubes: novel architecture in nanometer space*", Prog. Crystal Growth Charact., 34, 37-51 (1997).
Jeon H. S., J. K. Rameshwaram, and G. Kim, "*Structure–property relationships in exfoliated polyisoprene/clay nanocomposites*", *J. Polym. Sci. Part B: Polym. Phys.*, **42**, 1000-1009 (2004).
Jeong, H. G., H. T. Jung, S. W. Lee and S. D. Hudson, "*Morphology of polymer/ silicate nanocomposites*", *Polym. Bull.*, **41**, 107-113 (1998)
Ji, X. L., J. K. Jing, W. Jiang, and B. Z. Jiang, "*Tensile modulus of polymer nanocomposites*", *Polym. Eng. Sci.*, **42**, 983-993 (2002).
Jia, Z., Z. Wang, C. Xu, Ji Liang, B. Wei, D. Wu and S. Zhu, "*Study on poly(methyl methacrylate)/carbon nanotube composites*", *Mater. Sci. Eng. A*, **271**, 395-400 (1999).
Jie W., L. Yubao, C. Weiqun, and Z. Yi, "*A study on nano-composite of hydroxyapatite and polyamide*", *J. Mater. Sci.*, **38**, 3303-3306 (2003).
Jin, L., C. Bower and O. Zhou, "*Alignment of carbon nanotubes in a polymer matrix by mechanical stretching*", *Appl. Phys. Letters*, **73**, 1197-99 (1998).
Jin, Y. H., H.-J. Park, S.-S. Im, S.-Y. Kwak, and S. J. Kwak, "*Polyethylene/clay nanocomposite by in-situ exfoliation of montmorillonite during Ziegler-Natta polymerization of ethylene*", *Macromol. Rapid Commun.*, **23**, 135-140 (2002).
Jinnai, H., M. V. Smalley, T. Hashimoto and S. Koizumi, "*Neutron scattering study of vermiculite-poly(vinyl methyl ether) mixtures*", *Langmuir*, **12**, 1199-1203 (1996).
Jogun, S. and C. F. Zukoski, "*Rheology of dense suspensions of plate like particles*", *J. Rheol.*, **40**, 1211-32 (1996).
Johnson, K. L., K. Kendall and A. D. Roberts, "*Surface energy and the contact of elastic solids*", *Proc. Roy. Soc., London*, **324A**, 301-313 (1971).
Jommersbach, K., "*RTP Company breaks ground with first commercial Nylon/ organo-clay compound*", press release (1999).
Jones, B. F. and E. Galan, "*Sepiolite and Palygorskite*", Ch. 16 in *Hydrous Phyllosilicates*, Rev. Mineralogy, 19, Mineralogical Society of America, Washington, D.C. (1988).
Jordan, F. W., "*Organophilic clay with dual modifiers, and method for its manufacture*", *US Pat.*, **5,576,257**, 19.11.1996, to T. O. W. Inc.
Jordan, F. W., "*Organophilic clay and method for its manufacture*", *US Pat.*, **5,334,241**, 02.08.1994, to T. O. W. Inc.
Jordan, J. W., "*Lubricants*", *US Pat.*, **2,531,440**, 1950.11.28, Appl. 1947.03.29, to National Lead Co.

Jordan, J. W., "*Organophilic bentonites, I. Swelling in organic liquids*", *J. Phys. & Colloid Chem.*, **53**, 294-306 (1949).

Jordan, J. W., B. J. Hook, and C. M. Finlayson, "*Organophilic bentonites, II. Organic liquid gels*", *J. Phys. & Colloid Chem.*, **54**, 1196-08 (1950).

Jordan, R., N. West, A. Ulman, Y.-M. Chou and O. Nuyken, "*Nanocomposites by Surface-Initiated Living Cationic Polymerization of 2-Oxazolines on Functionalized Gold Nanoparticles*", *Macromolecules*, **34**, 1606-11 (2001).

Kajiwara, K., N. Donkai, Y. Hiragi and H. Inagaki, "*Lyotropic mesophase of imogolite*", *Macromol. Chem.*, **187**, 2883-95 (1986).

Kamal, M. R. and E. Chu, "*Isothermal and nonisothermal crystallization of polyethylene*", *Polym. Eng. Sci.*, **23**, 27-31 (1983).

Kamal, M. R., N. K. Borse and A. Garcia-Rejon, "*The effect of pressure on the crystallization behavior and kinetics of polyamide-6/clay nanocomposites*", Paper presented at the international symposium Polymer Nanocomposites 2001, Boucherville, QC, Canada, 2001.09.17-19; *Polym. Eng. Sci.*, **42**, 1883-96 (2002).

Kamena, K., "*Emerging nanocomposite technologies for barrier and thermal improvements in PET containers*", Nova-Pack Americas '99. Conference proceedings, Orlando, Fl., 1-2 Feb, 1999, p.1-6.

Kamigaito, O., Y. Fukushima and H. Doi, "*Composite material composed of clay mineral and organic high polymer and method for producing the same*", *US Pat.*, **4,472,538**, 18.09.1984, to Kabushiki Kaisha Toyota Chuo Kenkyusho.

Karande, S. V., C.-J., Chou, J. H., Solc and K. W., Suh, "*Dispersions of delaminated particles in polymer foams*", *US Pat.*, **5,717,000**, 1998.02.10, Appl. 1997.02.20, to Dow Chem. Co.

Katahira, S., K. Kojima, T. Tamura, and K. Yasue, "*High dispersion of synthetic mica in polyamide 6 matrix through hydrolysis of caprolactam with acid*", *Kobunshi Ronbunshu*, **55**, 90-95 (1998b).

Katahira, S., K. Yasue and M. Inagaki, "*Studies on production of nanocomposites of polyamide 6 with expandable synthetic mica* (parts III and IV*)*", *Kobunshi Ronbunshu*, **55**, 401-406, 477-482 (1998c, d).

Katahira, S., T. Tamura and K. Yasue, "*High dispersion of synthetic mica in polyamide 6 matrix through hydrolysis of caprolactam under high pressure*", *Kobunshi Ronbunshu*, **55**, 83-89 (1998a).

Katdare, S. P., V. Ramaswamy and A. V. Ramaswamy, "*Preparation of alumina pillared montmorillonite clays employing ultrasonics and their catalytic properties*", *Eurasian ChemTech J.*, **2**, 53-58 (2000).

Kato M., M. Matsushita, and K. Fukumori, "*Development of a new production method for a polypropylene-clay nanocomposite*", Proceedings "*Polymer Nanocomposites 2003*". *Polym. Eng. Sci.*, **44**, 1205-11 (2004).

Kato, M., A. Usuki and A. Okada, "*Synthesis of polypropylene oligomer-clay intercalation compounds*", *J. Appl. Polym. Sci.*, **66**, 1781-85 (1997).

Kausch, H. H. and M. Tirrell, "*Polymer interdiffusion*", *Ann. Rev. Mater. Sci.*, **19**, 341-377 (1989).

Kawasumi, M., M. Kohzaki, Y. Kojima, A. Okada, and O. Kamigaito, "*Process for producing composite material*", *US Pat.*, **4,810,734**, 07.03.1989, to Toyota Chuo Kenkyusho.

Kawasumi, M., N. Hasegawa, A. Usuki, and A. Okada, "*Nematic liquid crystal/ clay mineral composites*", *Mat. Sci. Eng.*, **C6**, 135-143 (1998).

Kawasumi, M., N. Hasegawa, M. Kato, A. Usuki, and A. Okada, "*Preparation and mechanical properties of polypropylene-clay hybrids*", *Macromolecules*, **30**, 6333-38 (1997).

Ke, Y. C., C. F. Long and Z. N. Qi, "*Crystallization, properties, and crystal and nanoscale morphology of PET⁻clay nanocomposites*", *J. Appl. Polym. Sci.*, **71**, 1139-46 (1999).

Keller, W. D., in *Kirk-Othmer Encyclopedia of Chemical Technology*, M. Grayson and D. Eckroth, Eds., 3rd Edition, Vol. 6, Wiley-Interscience Pub., New York (1979).

Khanna, Y. P. and W. P. Kuhn, "*Measurement of crystalline index in Nylons by DSC*", *J. Polym. Sci.: Polym. Phys.*, **35**, 2219-31 (1997).

Khanna, Y. P., P. K. Han, and E. D. Day, "*New developments in the melt rheology of nylons. I: Effect of moisture and molecular weight*", *Polym. Eng. Sci.*, **36**, 1745-54 (1996).

Kilburn D., D. Bamford, G. Dlubek, J. Pionteck, and M. A. Alam, "*The size and number of local free volumes in polypropylenes: A positron lifetime and PVT study*", *J. Polym. Sci. Part B: Polym. Phys.*, **41**, 3089-3093 (2003).

Kim K., L. A. Utracki, and M. R. Kamal, "*Numerical simulation of polymer nanocomposites using a self-consistent mean-field model*", *J. Chem. Phys.*, (accepted 2004).

Kim, B. K., J. W. Seo and H. M. Jeong, "*Morphology and properties of waterborne polyurethane/clay nanocomposites*", *Europ. Polym. J.*, **39**, 85-91 (2003).

Kim, B.-H., J.-H. Jung, S.-H. Hong, J.-S. Joo, A. J. Epstein, K. Mizoguchi, J. W. Kim and H. J. Choi, "*Nanocomposite of polyaniline and Na^+-montmorillonite clay*", *Macromolecules,* **35**, 1419-23 (2002).

Kim, G.-M., D.-H. Lee, B. Hoffmann, J. Kressler and G. Stöppelmann, "*Influence of nanofillers on the deformation process in layered silicate/polyamide-12 nanocomposites*", *Polymer*, **42** 1095-1100 (2001b).

Kim, J. W., M. H. Noh, H. J. Choi, D. C. Lee and M. S. Jhon, "*Synthesis and electrorheological characteristics of SAN-clay composite suspensions*", *Polymer,* **41**, 1229-31 (2000).

Kim, J.-H., K. Emoto, M. Iijima, Y. Nagasaki, T. Aoyagi, T. Okano, Y. Sakurai, and K. Kataoka, "*Core-stabilized polymeric micelle as potential drug carrier: increased solubilization of taxol*", *Polym. Adv. Technol.*, **10**, 647-654 (1999).

Kim, J.-K., R. Ahmed and S. J. Lee, "*Synthesis and linear viscoelastic behavior of poly(amic acid)-organoclay hybrid*", *J. Appl. Polym. Sci.*, **80**, 592-603 (2001c).

Kim, K.-N., H.-S. Kim and J.-W. Lee, "*Mixing characteristics and mechanical properties of polypropylene-clay composites*", *ANTEC, SPE Techn. Pap.*, **46**, 3782-86 (2000a).

Kim, S. W., W. H. Jo, M. S. Lee, M. B. Ko and J. Y. Jho, "*Preparation of clay-dispersed poly(styrene-co-acrylonitrile) nanocomposites using poly(ε-caprolactone) as a compatibilizer*", *Polymer*, **42**, 9837-42 (2001a).

Kim, T. H., L. W. Jang, D. C. Lee, H. J. Choi, and M. S. Jhon, "*Synthesis and Rheology of Intercalated Polystyrene/Na+-montmorillonite Nanocomposites*", *Macromol. Rapid Commun.*, **23**, 191-195 (2002).

Kim, T. H., S. T. Lim, D. C. Lee, H. J. Choi, and M. S. Jhon, "*Preparation and rheological characterization of intercalated polystyrene/organophilic montmorillonite nanocomposite*", *J. Appl. Polym. Sci.*, **87**, 2106-12 (2003).

Kinloch, I. A., S. A. Roberts and A. H. Windle, "*A rheological study of concentrated aqueous nanotube dispersions*", *Polymer*, **43**, 7483-91 (2002).

Kiss, G. and R. S. Porter, "*Rheology of concentrated solutions of helical polypeptides*", *J. Polym. Sci., Polym. Phys.*, **18**, 361-388 (1980).

Klein, J., D. Fletcher and L. J. Fetters, "*Diffusional behavior of entangled star polymers*", *Nature*, **304**, 526-527 (1983).

Klein, J., Y. Kamiyama, H. Yoshizawa, J. N. Israelachvili, G. N. Fredrickson, P. Pincus and L. J. Fetters, "*Lubricating forces between surfaces bearing polymer brushes*", *Macromolecules*, **26**, 5552-60 (1993).

Kloster, G. M., T. J. Albritton P. W. Brazis, R. C. Kannewurf and D. F. Shriver, "*Synthesis, characterization, and transport properties of new mixed ionic-electronic conducting V2O5-polymer electrolyte xerogel nanocomposites*", *Chem. Mater.*, **8**, 2418-20 (1996).

Knight, W. C., "*Bentonite*", *Eng. Min. J.*, **66**, 491 (1896).

Knudson, Jr., M. I. and T. R. Jones, "*Process for manufacturing organoclays having enhanced gelling properties*", *US Pat.*, **5,160,454**, 03.11.1992b, to Southern Clay Products, Inc.

Knudson, Jr., M. I. and T. R. Jones, "*Process for manufacturing organoclays having enhanced gelling properties*", *US Pat.*, **4,569,923**, 11.02.86, to Southern Clay Products, Inc.

Knudson, Jr., M. I. and T. R. Jones, "*Process for manufacturing organoclays having enhanced gelling properties*", *US Pat.*, **5,110,501**, 05.05.1992a, to Southern Clay Products, Inc.

Kodgire, P., R. Kalgaonkar, S. Hambir, N. Bulakh and J. P. Jog, "*PP/clay nanocomposites: Effect of clay treatment on morphology and dynamic mechanical properties*", *J. Appl. Polym. Sci.*, **81**, 1786-92 (2001).

Köhn, R. D., Z. Pan, J. Sun and C.-F. Liang, "*Ring-opening polymerization of D,L-lactide with bis(trimethyl triazacyclohexane) praseodymium triflate*", *Catalysis Commun.*, **4**, 33-37 (2003).

Kojima Y., A. Usuki, M. Kawasumi, A. Okada, T. Kurauchi, and O. Kamigaito, "*Synthesis of Nylon-6-clay hybrid by montmorillonite intercalated with e-caprolactam*', *J Polym. Sci., Part A: Polym. Chem.*, **31**, 983-986 (1993c).

Kojima, Y., A. Usuki, M. Kawasumi, A. Okada, T. Kurauchi, O. Kamigaito and K. Kaji, "*Novel preferred orientation in injection-molded Nylon-6-clay hybrid*', *J Polym. Sci., Part B: Polym. Phys.*, **33**, 1039-45 (1995).

Kojima, Y., A. Usuki, M. Kawasumi, A. Okada, T. Kurauchi, O. Kamigaito and K. Kaji, "*Fine structure of Nylon-6-clay hybrid*", *J Polym. Sci.: Part B: Polym. Phys.*, **32**, 625-30 (1994).

Kojima, Y., A. Usuki, M. Kawasumi, A. Okada, Y. Fukushima, T. Karauchi and O. Kamigaito, "*Mechanical properties of nylon-6 clay hybrid*", *J. Mater. Res.*, **8**, 1185-89 (1993b).

Kojima, Y., K. Fukumori, A. Usuki, A. Okada and T. Karauchi, "*Gas permeabilities in rubber-clay hybrid*", *J. Mater. Sci., Letters*, **12**, 889-890 (1993a).

Komarneni, S., Ed., "*Nanophase and nanocomposite materials IV*", MRS symposium, Boston, MA November 26-29, 2001, Materials Research Society, Warrendale, PA, (2002).

Komarneni, S., Ed., "*Nanophase and Nanocomposite Materials III*", MRS Symposium, Proceedings Volume 581, Boston, MA, November 29-December 2, 1999, Materials Research Society, Warrendale, PA (1999).

Komarneni, S., J. C. Parker and, H. Wollenberger (Eds.), *Nanophase and Nanocomposite Materials II*, MRS Symposium, 1999.12.2-5, Boston, MA, USA.

Komarneni, S., J. C. Parker, H. Hahn, Eds., *Nanophase and Nanocomposite Materials III*, MRS symposium, Boston, MA, November 29-December 2, 1999, Materials Research Society, Warrendale, PA (2000).

Komori, Y., H. Enoto, R. Takenawa, S. Hayashi, S. Yoshiyuki and K. Kuroda, "*Modification of the interlayer surface of kaolinite with methoxy groups*", *Langmuir,* **16**, 5506-08 (2000).

Komori, Y., Y. Sugahara and K. Kuroda, *"Intercalation of alkylamines and water into kaolinite with methanol kaolinite as an intermediate", Appl. Clay Sci.,* **15**, 241-252 (1999a).

Komori, Y., Y. Sugahara, and K. Kuroda, "*Direct intercalation of poly(vinylpyrrolidone) into kaolinite by a refined guest displacement method", Chem. Mater.*, **11**, 3-6 (1999b).

Koo, C. M., H. T. Ham, M. H. Choi, S. O. Kim, and I. J. Chung, "*Characteristics of polyvinylpyrrolidone-layered silicate nanocomposites prepared by attrition ball milling", Polymer*, **44**, 681-689 (2003).

Korbee, R. A. and A. A. Van Geenen, "*Process for the preparation of a polyamide nanocomposite composition", US Pat.*, **6,350,805**, 26.02.2002, Appl. 11. 12.1997, to DSM N. V.

Kornmann, X., H. Lindberg and L. A. Berglund, "*Synthesis of epoxy-clay nanocomposites: influence of the nature of the clay on structure", Polymer*, **42**, 1303-10 (2001).

Kornmann, X., H. Lindberg and L. A. Berglund, "*Synthesis of Epoxy-Clay Nanocomposites", SPE Techn. Pap.*, **45**, 1623-28 (1999).

Kornmann, X., L. A. Berglund, J. Sterte, and E. P. Giannelis, "*Nanocomposites based on montmorillonite and unsaturated polyester", Polym. Eng. Sci.*, **38**, 1351-58 (1998).

Kornmann, X., R. Thomann, R. Mülhaupt, J. Finter, and L. A. Berglund, "*High Performance Epoxy-Layered Silicate Nanocomposites*", Proceedings Nanocomposites-2001 International Symposium, Boucherville, QC, Canada, 2001.11.16-18; *Polym. Eng. Sci.*, **42**, 1815-26 (2002).

Kornmann, X., *Synthesis and Characterization of Thermoset-Layered Silicate Nanocomposites, PhD thesis,* Luleå University of Technology (2001).

Kosiński, L. E. R., G. P. Rajendran, and R. R. Reitz, "*Polyurethane fibers and films containing clay" WO* **9749847**, 31.12.1997, to E.I. Du Pont De Nemours and Co.

Kotaka, T. and H. Watanabe, "*Rheological and morphological properties of heterophase block copolymer solutions*", in *Current Topics in Polymer Science*, Vol. II, R. M. Ottenbrite, L. A. Utracki and S. Inoue, Hanser Verlag, Munich (1987).

Kotsilkova, R., "*Rheology–Structure Relationship of Polymer/Layered Silicate Hybrids", Mech. Time-Dependent Mater.*, **6**, 283-300 (2002).

Kourogi, M., T. Noma and N. Suzuki Noriyuki, "*Resin compositions and process for producing the same*", **WO0022042**, 2000.04.20, to Kaneka Corp.

Kovar, R. F. and R. W. Lusignea, in *Ultrastructure Processing of Advanced Ceramics*, J. D. McKenzie and D. R. Ulrich, Eds., Wiley Interscience, New York (1988).

Kresge, E. N. and D. J. Lohse, "*Composite tire innerliners and inner tubes*", *US Pat.*, **5,576,373**, 19.11.1996, Appl. 07.06.1995, to Exxon Chemical Patents Inc.

Kresge, E. N. and D. J. Lohse, "*Tire inner-liners comprising a solid rubber and a complex of a reactive rubber and layered silicate clay*", *US Pat.*, **5,665,183**, 09.09.1997, Appl. 07.06.1995, to Exxon Chemical Patents Inc.

Krieger, I. M. and T. J. Daugherty, "*A mechanism for non-Newtonian flow in suspensions of rigid spheres*", *Trans. Soc. Rheol.*, **3**, 137-152 (1959).

Krishnamoorti R. and E. P. Giannelis, "*Rheology of end-tethered polymer layered silicate nanocomposites*", *Macromolecules*, **30**, 4097-02 (1997).

Krishnamoorti, R. and R. A. Vaia, Eds., "*Polymer nanocomposites: synthesis, characterization, and modeling*", Proceedings of ACS Polymer Nanocomposites Symposium (San Francisco, CA, 2000), American Chemical Society Washington, DC (2002).

Krishnamoorti, R., R. A. Vaia, and E. P. Giannelis, "*Structure and dynamics of polymer-layered silicate nanocomposites*", *Chem. Mater.*, **8**, 1728-34 (1996).

Krook, M., A.-C. Albertsson, U. W. Gedde, and M. S. Hedenqvist, "*Barrier and mechanical properties of montmorillonite/polyesteramide nanocomposites*", *Polym. Eng. Sci.*, **42**, 1238-46 (2002).

Krumeich, F., H. J. Muhr, M. Niederberger, F. Bieri and R. Nesper, "*The cross-sectional structure of vanadium oxide nanotubes studied by transmission electron microscopy and electron spectroscopic imaging*", *Z. Anorg. Allg. Chem.*, **626**, 2208-16 (2000).

Kuchta, F.-D., P. J. Lemstra, A. Keller, L. F. Batenburg and H. R. Fischer, "*Polymer crystallization studied in confined dimensions using nanocomposites from polymers and layered minerals*", MRS Symposium *Organic/Inorganic Hybrid Materials-2000*, San Francisco, CA, Proceedings, **628**, CC11.12.1-7 (2000).

Kuhn, W. and H. Kuhn, "*Die Abhängigkeit der Viscosität vom Strömungsgefälle bei hochverdünnten Suspensionen und Lösungen*", *Helv. Chim. Acta*, **28**, 97-127 (1945).

Kumar S., J. P. Jog, and U. Natarajan, "*Preparation and characterization of poly(methyl methacrylate)–clay nanocomposites via melt intercalation: the effect of organoclay on the structure and thermal properties*", *J. Appl. Polym. Sci.*, **89**, 1186-1194 (2003).

Kumar, R. V., Yu. Koltypin, Y. S. Cohen, Y. Cohen, D. Aurbach, O. Palchik, I. Felner, and A. Gedanken, "*Preparation of amorphous magnetite nanoparticles embedded in polyvinyl alcohol using ultrasound radiation*", *J. Mater. Chem.*, **10**, 1125-29 (2000).

Kumar, S., T. D. Dang, F. E. Arnold, A. R. Bhattacharyya, B. G. Min, X. Zhang, R. A. Vaia, C. Park, W. W Adams, R. H. Hauge, R. E. Smalley, S. Ramesh, and P. A. Willis, "*Synthesis, structure, and properties of PBO/SWNT composites*", *Macromolecules*, **35**, 9039-43 (2002).

Kuo, W.-F., M.-S. Lee, and C.-S. Liao, "*Method for forming nanocomposites*", *US Pat.*, **6,252,020**, 26.06.2001, Appl. 05.04.1999, Priority 29.10.1998, to Industrial Technology Res. Inst., Hsinchu, TW.

Kuritzkes, A. M., "*Nail enamel composition containing quarternary ammonium cation modified montmorillonite clays*", *US Pat.*, **3,422,185**, 14.01.1969, to the Mearl Corp.

Kurokawa, Y., H. Yasuda and A. Oya, "*Preparation of a nanocomposite of polypropylene and smectite*", *J Mater. Sci. Lett.*, **15**, 1481-83 (1996).

Kurokawa, Y., H. Yasuda, M Kashiwagi and A. Oya, "*Structure and properties of a montmorillonite/polypropylene nanocomposite*", *J. Mater. Sci. Letters*, **16**, 1670-72 (1997).

Kuwabara, S., K. Nagata, Y. Kazuhisa, Y. Imashiro and E. Sasaki, "*Stock solution composition for use in production of hard polyurethane foam*", *US Pat.*, **5,779,775**, 14.07.1998, Appl. 08.04.1997, to Nisshinbo Industries, Inc.

Kuznetsov D. V. and A. C. Balazs, "*Phase behavior of end-functionalized polymers confined between two surfaces*", *J. Chem. Phys.*, **113**, 2479-83 (2000b).

Kuznetsov, D. V. and A. C. Balazs, "*Scaling theory for end-functionalized polymers confined between two surfaces: Predictions for fabricating polymer/clay nanocomposites*", *J. Chem. Phys.*, **112**, 4365-75 (2000a).

Kwiatkowski, J. and A. K. Whittaker, "*Molecular motion in nanocomposites of poly(ethylene oxide) and montmorillonite*", *J. Polym. Sci. Part B: Polym. Phys.*, **39**, 1678-85 (2001).

La Mantia, F. P., Ed., *Recycling of PVC & Mixed Plastic Waste*, ChemTec Pub., Toronto (1996).

La Mer, V. K. and T. W. Healy, "*Adsorption-flocculation reaction of macromolecules at the solid-liquid interface*", *Rev. Pure Appl. Chem.*, **13**, 112-133 (1963).

Ladika, M., R. F., Fibiger, C.-J. Chou, and A. C. Balazs, "*Functionalized polymer nanocomposites*", *Pat.*, **WO0185831**, 2001-11-15, Appl. 2000.05.05, to Dow Chem. Co.

Lagaly, G. and S. Ziesmer, "*Colloid chemistry of clay minerals: the coagulation of montmorillonite dispersions*", *Adv. Colloid Interface Sci.*, **100-102**, 105-128 (2003).

Lagaly, G., K. Beneke and A. Weiss, "*Magadiite and H-magadiite: I. Sodium magadiite and some of its derivatives*", *Am. Mineralog.*, **60**, 642-649 (1975).

Lahr, B. and J. Sandler, "*Carbon nanotubes-high-strength reinforcing compounds for composites*", *Kunststoffe Plast Europe*, **90**, 42-43; 94-96 (2000).

Laine, R., *Organic/Inorganic Hybrid Materials—2000*, MRS symposium, San Francisco, CA, April 24-28, 2000, Materials Research Society, Warrendale, PA, (2001).

Lan T., R. B. Barbee, J. W. Gilmer, and J. C. Matayabas Jr., "*A polymer/clay nanocomposite having improved gas barrier comprising a clay material with a mixture of two or more organic cations and a process for preparing same*", **WO0034180; AU758250**, 2003.03.20, Appl. 1999.11.30, to Univ. South Carolina Res. Foundation.

Lan, T. and E. K. Westphal, "*Nanocomposites formed by onium ion-intercalated clay and rigid anhydride-cured epoxy resins*", *US Pat.*, **6,251,980**, 26.06.2001a, Appl. 19.03.1999, to AMCOL International Corp.

Lan, T. and G.-Q. Quian, "*Preparation of high performance polypropylene nanocomposites*", Paper presented at the *Additives-2000* conference, Clearwater Beach, FL, 2000.04.10-12.

Lan, T. and T. J. Pinnavaia, "*Clay-reinforced epoxy nanocomposites*", *Chem. Mat.*, **6**, 2216-19 (1994).

Lan, T., "*Nanocomposites formed by onium ion-intercalated clay and rigid anhydride-cured epoxy resins*", *EP.*, **1 038 913 A1**, 27.09.2000, Appl. 17.03.2000, to AMCOL International Corp.

Lan, T., E. K. Westphal, V. Psihogios and Y. Liang, "*Layered compositions with multi-charged onium ions as exchange cations, and their application to prepare monomer, oligomer, and polymer intercalates and nanocomposites prepared with the layered compositions of the intercalates"*, *US Pat.*, **6,262,162**, 17.07.2001b, Appl. 19.03.1999, to AMCOL International Corp.

Lan, T., G. W. Beall, and S. Tsipursky, *"Intercalates formed by co-intercalation of monomer, oligomer or polymer intercalants and surface modifier intercalants and layered materials and nanocomposites prepared with the intercalates"*, *US Pat.*, **6,057,396**, 02.05.2000, to AMCOL International Corp.

Lan, T., G. W. Beall, and S. Tsipursky, *"Intercalates formed by co-intercalation of monomer, oligomer or polymer intercalants and surface modifier intercalants and layered materials and nanocomposites prepared with the intercalates"*, *Europ. Pat. Appl.*, **846,661** A2, 06.10.1998, to AMCOL International Corp.

Lan, T., H. T. Cruz and A. S. Tomlin, *"Intercalates formed by co-intercalation of onium ion spacing/coupling agents and monomer, oligomer or polymer MXD6 nylon intercalants and nanocomposites prepared with the intercalates"*, *US Pat.*, **6,232,388**, 15.05.2001b, Appl. 19.03.1999, to AMCOL International Corp.

Lan, T., J. W. Gilmer, J. Ch. Matayabas, Jr. and R. B. Barbee, *"Polymer/clay intercalates, exfoliates; and nanocomposites having improved gas permeability comprising a clay material with a mixture of two or more organic cations and a process for preparing same"*, *US Pat.*, **6,387,996**, 14.05.2002a, Appl. 01.12.1999, to AMCOL International Corp.

Lan, T., P. D. Kaviratna and T. J. Pinnavaia, *"Mechanism of clay tactoid exfoliation in epoxy-clay nanocomposites"*, *Chem. Mater.*, 7, 2144-50 (1995).

Lan, T., V. Psihogios, S. Bagrodia, L. T. Germinario and J. W. Gilmer, *"Layered clay intercalates and exfoliates having low quartz content"*, *US Pat. Appl.*, **0028870**, 2002b.03.07, Appl. 2001.06.29, to AMCOL International Corp.

Lan, T., V. Psihogios, S. Bagrodia, L. T. Germinario and J. W. Gilmer, *"Intercalates and exfoliates thereof having an improved level of extractable material"*, *US Pat. Appl.*, **0037953**, 2002c.03.28, Appl. 2001.05.30, to AMCOL International Corp.

Larson, R. G. and M. Doi, *"Mesoescopic domain theory for textures liquid-crystalline polymers"*, *J. Rheol.*, **35**, 539-563 (1991).

Laude, T., A. Marraud, Y. Matsui, and B. Jouffrey, *"Long ropes of boron nitride nanotubes grown by a continuous laser heating"*, *Appl. Phys. Lett.*, **76**, 3239 –41 (2000).

Laus, M., M. Camerani, M. Lelli, K. Sparnacci, F. Sandrolini and O. Francescangeli, *"Hybrid nanocomposites based on polystyrene and a reactive organophilic clay"*, *J. Mater. Sci.*, **33**, 2883-88 (1998).

LeBaron, P. C. and T. J. Pinnavaia, *"Clay nanolayer reinforcement of a silicone elastomer"*, *Chem. Mater.*, **13**, 3760-65 (2001).

Lee E. C.-C., and D. F. Mielewski, *"Method for producing a well-exfoliated and dispersed polymer silicate nanocomposite by ultrasonication"*, *US Pat. Appl.*, **20030134942 A1**, 2003.07.17, Appl. 2002.05.30.

Lee J. Y., Z. Shou, and A. C. Balazs, *"Predicting the Morphologies of Confined Copolymer/Nanoparticle Mixtures"*, *Macromolecules*, **36**, 7730-7739 (2003).

Lee L. J., K. W. Koelling, D. L. Tomasko, X. Han, and C. Zeng, *"Polymer nanocomposite foams"*, *US Pat. Appl.*, **20030205832** A1, 2003.11.06, Appl. 2002.05.02, to The Ohio State University Research Foundation.

Lee S. Y., and S. J. Kim, *"Dehydration behavior of hexadecyltrimethylammonium-exchanged smectites"*, *Clay Minerals*, **38**, 225-232 (2003).

Lee S.-S., and J. Kim, "*Preparation of the Polymer–Clay Nanocomposites in Exfoliated State by Interface Stabilization*", *J. Polym. Sci. Part B: Polym. Phys.*, **42**, 246-252 (2004).

Lee, D. C. and L. W. Jang, "*Preparation and characterization of PMMA-clay hybrid composite by emulsion polymerization*", *J. Appl. Polym. Sci.*, **61**, 1117-22 (1996).

Lee, D. K. and Y. S. Kang, "*Fabrication of Thin Films-Preparation and Characterization of Iron Oxide Nanoparticle/Poly(maleic monoester) Nanocomposite Films*", *Molecular Cryst. Liquid Cryst.*, **316**, 153-156 (1998).

Lee, J. W., Y. T. Lim, and O. O. Park, "*Thermal characteristics of organoclay and their effects upon the formation of polypropylene/organoclay nanocomposites*", *Polym. Bull.*, **45**, 191-198 (2000a).

Lee, J. Y., A. R. C. Baljon, and R. F. Loring, "*Spontaneous swelling of layered nanostructures by a polymer melt*", *J. Chem. Phys.*, **111**, 9754-60 (1999).

Lee, J. Y., A. R. C. Baljon, D. Y. Sogah, and R. F. Loring, "*Molecular dynamics study of the intercalation of diblock copolymers into layered silicates*", *J. Chem. Phys.*, **112**, 9112-19 (2000).

Lee, J. Y., A. R. C. Baljon, R. F. Loring, and A. Z. Panagiotopoulos, "*Simulation of polymer melt intercalation in layered nanocomposites*", *J. Chem. Phys.*, **109**, 10321-330 (1998a).

Lee, S.-R., H.-M. Park, H. Lim, T. Kang, X. Li, W.-J. Cho, and C.-S. Ha, "*Microstructure, tensile properties and biodegradability of aliphatic polyester/clay nanocomposites*", *Polymer*, **43**, 2495-2500 (2002b).

Lee, T. H., N. Yao, T. J. Chen and W. K. Hsu, "*Fullerene-like carbon particles in petrol soot*", *Carbon*, **40**, 2275-79 (2002a).

Lemmon, J. P. and M. M. Lerner, "*Preparation and Characterization of Nanocomposites of Polyethers and Molybdenum Disulfide*", *Chem. Mater.*, **6**, 207-210 (1994).

Lemmon, J. P. and M. M. Lerner, "*Preparation of nanocomposites of poly(ethylene oxide) with TiS2*", *Solid State Commun.*, **94**, 533-537 (1995).

Lentz, C. W., "*Trimethylsilyl derivatives of silicate minerals*", *Inorg. Chem.*, **3**, 574-579 (1964).

Lepoittevin, B, M. Devalckenaere, M. Alexandre, N. Pantoustier, C. Calberg, R. Jerôme, and Ph. Dubois, "*Poly(ε-caprolactone)/clay nanocomposites by in-situ intercalative polymerization catalyzed by dibutyltin dimethoxide*", *Macromolecules*, **35**, 8385-90 (2002a).

Lepoittevin, B., M. Devalckenaere, N. Pantoustier, M. Alexandre, D. Kubies, C. Calberg, R. Jérôme, and P. Dubois, "*Poly(ε-caprolactone)/clay nanocomposites by in-situ intercalative polymerization catalyzed by dibutyltin dimethoxide*", *Polymer*, **43**, 4017-23 (2002b).

Leroux, F. and J.-P. Besse, "*Polymer interleaved layered double hydroxide: a new emerging class of nanocomposites*", *Chem. Mater.*, **13** (10), 3507-15 (2001).

Leśniak, E., "*Silseskwioksany*", *Polimery*, **46**, 516-521; 582-589 (2001).

Levy, R. and C. W. Francis, "*Interlayer adsorption of polyvinylpyrrolidone on montmorillonite*", *J. Colloid Interface Sci.*, **50**, 442-450 (1975).

Li J., C. Zhou, G. Wang, and D. Zhao, "*Study on rheological behavior of polypropylene/clay nanocomposites*", *J. Appl. Polym. Sci.*, **89**, 3609-3617 (2003).

Li, D.-M., D. G. Peiffer, C. W. Elspass and H.-C. Wang, "*Rubber toughened thermoplastic resin nano composites*", *US Pat.*, **6,060,549**, 09.05.2000, Appl. 20.05.1997, to Exxon Chemical Patents, Inc.

Li, G. Z., L. Wang, H. Toghiani, T. L. Daulton and C. U. Pittman, Jr., "*Viscoelastic and mechanical properties of vinyl ester (VE)/multifunctional polyhedral oligomeric silsesquioxane (POSS) nanocomposites and multifunctional POSS-styrene copolymers*", *Polymer*, **43**, 4167-76 (2002).

Li, L.-S., J. Hu, W. Yang and A. P. Alivisatos, "*Band gap variation of size-and shape-controlled colloidal CdSe quantum rods*", *Nano Letters*, **1**, 349-351 (2001a).

Li, X., T. Kang, W.-J. Cho, J.-K. Lee, and C.-S. Ha, "*Preparation and Characterization of Poly(butyleneterephthalate)/Organoclay Nanocomposites*", *Macromol. Rapid Commun.*, **22**, 1306-12 (2001b).

Liang Z.-M., J. Yin, J.-H. Wu, Z.-X. Qiu, and F.-F. He, "*Polyimide/montmorillonite nanocomposites with photolithographic properties*", *Eur. Polym. J.*, **40**, 307-314 (2004).

Liang, Z.-M., J. Yin and H.-J. Xu, "*Polyimide/montmorillonite nanocomposites based on thermally stable, rigid-rod aromatic amine modifiers*", *Polymer*, **44**, 1391-99 (2003).

Liao, C.-S. and L. K. Lin, "*Preparation of conductive polymeric nanocomposite*", *US Pat.*, **6,136,909**, 24.10.2000b, Appl. 20.05.1999, to Industrial Technology Research Institute (Hsinchu, TW).

Liao, C.-S., W.-F. Kuo and L. Lin, "*Preparation of thermoplastic nanocomposite*", *US Pat.* **6,136,908**, 24.10.2000a, Appl. 17.06.1999, to Industrial Technology Research Institute (Hsinchu, TW).

Lichtenhan J. D., "*Polyhedral oligomeric silsesquioxanes: building blocks for silsesquioxane-based polymers and hybrid materials*", *Comments Inorg. Chem.*, **17**, 115-130 (1995).

Lichtenhan, J. D. and J. J. Schwab, "*Nanostructured' Chemicals: simplifying nanoreinforcement via blending or polymerization*", Proceedings of Polymer Nanocomposites 2001, NRCC/IMI, Boucherville, QC, Canada 2001.09.17-19.

Lichtenhan, J. D., in *Polymer Materials Encyclopedia*, J. C. Salamone, Ed., Vol. **10**, (1996).

Lichtenhan, J. D., J. J. Schwabb, J. Feher, and D. Soulivong, "*Method of functionalizing polycyclic silicones and the resulting compounds*", *US Pat.*, **5,942,638**, 1999.08.24, Appl. 1998.06.05, to the USA Secretary of the Air Force, Washington, DC.

Lichtenhan, J. D., N. Q. Vu, J. A. Carter, J. W. Gilman, F. J. Feher, "*Silsesquioxane-Siloxane Copolymers from Polyhedral Silsesquioxanes*", *Macromolecules*, **26**, 2141-42 (1993).

Lichtenhan, J. D., *personal communication* (2000).

Lichtenhan, J. D., Y. A. Otonari, and M. J. Carr, "*Linear hybrid polymer building blocks: methacrylate-functionalized polyhedral oligomeric silsesquioxane monomers and polymers*", *Macromolecules*, **28**, 8435-37 (1995).

Lim, Y. T. and O. O. Park, "*Phase morphology and rheological behavior of polymer/layered silicate nanocomposites*", *Rheol Acta.*, **40**, 220-229 (2001).

Lim, Y. T. and O. Park, "*Rheological evidence for the microstructure of intercalated polymer/layered silicate nanocomposites*", *Macromol. Rapid. Commun.*, **21**, 231-235 (2000).

Lin, J.-M., J.-H. Wu, Z.-F. Yang, and M.-L. Pu, "*Synthesis and properties of poly(acrylic acid)/montmorillonite superabsorbent composites*", *Polym. Polym. Compos.*, **9**, 469-471 (2001).

Linares, J. and F. Huertas, "*Kaolinite: synthesis at room temperature*", *Science*, **171**, 896-897 (1971).

Liu X., Q. Wu, L. A. Berglund, H. Lindberg, J. Fan, and Z. Qi, "*Polyamide 6/Clay Nanocomposites Using a Cointercalation Organophilic Clay via Melt Compounding*", *J. Appl. Polym. Sci.*, **88**, 953-958 (2003).

Liu Z., K. Chen, and D. Yan, "*Crystallization, morphology, and dynamic mechanical properties of poly(trimethylene terephthalate)/clay nanocomposites*", *Eur. Polym. J.*, **39**, 2359-2366 (2003).

Liu Z., P. Zhou, and D. Yan, "*Preparation and properties of nylon-1010/ montmorillonite nanocomposites by melt intercalation*", *J. Appl. Polym. Sci.*, **91**, 1834-1841 (2004).

Liu, H., X. Ge, Y. Zhu, X. Xu, Z. Zhang, and M. Zhang, "*Synthesis and characterization of polyacrylamide-nickel amorphous nanocomposites by γ-irradiation*", *Mater. Letters*, **46**, 205-208 (2000).

Liu, H.-I., D. W. Kim, A. Blumstein, J. Kumar, and S. K. Tripathy, "*Nanocomposite Derived from Intercalative Spontaneous Polymerization of 2-Ethynyl pyridine within Layered Aluminosilicate: Montmorillonite*", *Chem. Mater.*, **13**, 2756-58 (2001).

Liu, L. and Q. Wu, "*PP/clay nanocomposites prepared by grafting-melt intercalation*", *Polymer*, **42**, 10013-16 (2001).

Liu, L., Z. Qi and X. Zhu, "*Studies on Nylon 6/clay nanocomposites by melt-intercalation process*", *J. Appl. Polym. Sci.*, **71**, 1133-38 (1999).

Liu, X and Q. Wu, "*Polyamide 66/clay nanocomposites via melt intercalation*", *Macromol. Mater. Eng.*, **287**, 180-186 (2002a).

Liu, X. and Q. Wu, "*Phase transition in nylon-6/clay nanocomposites on annealing*", *Polymer*, **43**, 1933-36 (2002b).

Liu, X. and Q. Wu, "*Phase transition in polyamide-66/montmorillonite nanocomposites on annealing*", *J. Polym. Sci.: Part B: Polym. Phys.*, **41**, 63-67 (2003).

Liu, Y-J., J. L. Schindler, D. C. DeGroot, C. R. Kannewurf, W. Hirpo, and M. G. Kanatzidis, "*Synthesis, structure, and reactions of poly(ethylene oxide)/V2O5 intercalative nanocomposites*", *Chem. Mater.*, **8**, 525-534 (1996).

Loo L. S., and K. K. Gleason, "*Fourier transform infrared investigation of the deformation behavior of montmorillonite in Nylon-6/nanoclay nanocomposite*", *Macromolecules*, **36**, 2587-2590 (2003).

López-Manchado, M. A., M. Arroyo, B. Herrero, and J. Biagiotti, "*Vulcanization kinetics of natural rubber-organoclay nanocomposites*", *J. Appl. Polym. Sci.*, **89**, 1-15 (2003).

Lorah, D. P. and R. V. Slone, "*Emulsion polymerization methods involving lightly modified clay and compositions comprising same*", *US Pat. Appl.*, **20020055581**, 2002b.05.09, Appl. 2001.09.17, to Rohm and Haas Co.

Lorah, D. P. and R. V. Slone, "*High acid aqueous nanocomposite dispersions*", *US Pat. Appl.*, **20020055580**, 2002a.05.09, Appl. 2001.09.17, to Rohm and Haas Co.

Lorah, D. P. and R. V. Slone, "*Nanocomposite compositions and methods for making and using same*", *US Pat. Appl.*, **20020058740**, 2002c.05.16, Appl. 2001.09.17, to Rohm and Haas Co.

Lorah, D. P., R. V. Slone, and T. G. Madle, "*Hydrophobically modified clay polymer nanocomposites*", *US Pat. Appl.*, **20020058739**, May 16, 2002.05.16, Appl. 2001.09.17, to Rohm and Haas Co.

Lordi, V. and N. Yao, "*Molecular mechanics of binding in carbon-nanotube-polymer composites*", *J. Mater. Res.*, **15**, 2770-79 (2000).

Lordi, V., S. X. C. Ma and N. Yao, "*Towards probing pentagons on carbon nanotube tips*", *Surface Sci.*, **421**, L150-L155 (1999).

Lourie O. and H. D. Wagner, "*Buckling and collapse of embedded carbon nanotubes.*"*Phys. Rev. Letters*, **81**, 1638-41 (1998b).

Lourie O. and H. D. Wagner, "*Transmission electron microscopy observations of fracture of single-wall carbon nanotubes under axial tension.*"*Appl. Phys. Letters*, 73, 3527-9 (1998a).

Lourie, O. and H. D. Wagner, "*Evidence of stress transfer and formation of fracture clusters in carbon nanotube-based composites*", *Compos. Sci. Technol.*, **59**, 975-977 (1999).

Lourie, O. and H. D. Wagner, "*Evidence of stress transfer and formation of fracture clusters in carbon nanotube-based composites*", *J. Mater. Res.*, **13**, 2418-21 (1998c).

Lozano, K. and E. V. Barrera, "*Nanofiber-reinforced thermoplastic composites. I. Thermoanalytical and mechanical analyses*", *J. Appl. Polym. Sci.*, **79**, 125-133 (2001).

Lü, J., Y. Ke, Z.-N. Qi, X.-S. Yi, "*Study on intercalation and exfoliation behavior of organoclays in epoxy resin*", *J. Polym. Sci. B: Polym. Phys.*, **39**, 115-120 (2001).

Luciani, A. and L. A., Utracki, "*The extensional flow mixer*", *Intl. Polymer Process.*, **11**, 299-309 (1996).

Luciani, A., M. F. Champagne, and L. A. Utracki, "*Interfacial Tension in Polymer Blends, Part 1: Theory*", *Polymer Networks & Blends*, **6**, 41-50 (1996); "Part 2: Measurements", *ibid.*, **6**, 51-62 (1996).

Luciani, A., M. F. Champagne, and L. A. Utracki, "*Interfacial tension determination by retraction of an ellipsoid*", *J. Polym. Sci. B, Polym. Phys. Ed.*, **35**, 1393-03 (1997).

Luciani, A., M. F. Champagne, and L. A. Utracki, "*Interfacial tension in polymer blends*", *Makromol Chem., Macromol. Symp.*, **126**, 307-321 (1998).

Luengo G., F.-J. Schmitt, R. Hill, and J. N. Israelachvili, "*Thin film rheology and tribology of confined polymer melts: contrasts with bulk properties*", *Macromolecules*, **30**, 2482-94 (1997).

Lyon, L. A., Ed., "*Anisotropic nanoparticles-synthesis, characterization and applications*", MRS symposium, Boston, MA, November 27-29, 2000, Materials Research Society, Warrendale, PA, (2001).

Lysek B. A., C. E. Powell and L. A. Goettler, "*Polymer nanocomposite composition*", WO9941299 (A1), EP1054922 (A1), EP1054922 (A4), CA2320988, 1999.08.19, Appl. 1998.02.13, to SOLUTIA Inc.

Lysek B. A., L. A. Goettler, and S. S. Joardar, "*Process to prepare a polymer nanocomposite composition*", CA2320909, WO9941060 (A1), EP1089866 (A1),1999.08.19, Appl. 1998.02.13, to SOLUTIA Inc.

Ma J., Z.-Z. Yu, Q.-X. Zhang, X.-L. Xie, Y.-W. Mai, and I. Luck, "*A novel method for preparation of disorderly exfoliated epoxy/clay nanocomposite*", *Chem. Mater.*, **16** (5), 757-759 (2004).

Ma, J., L.-H. Shi, M.-S. Yang, B.-Y. Li, H.-Y. Liu, and J, Xu, "*Synthesis and structure of polymethylsilsesquioxane-clay nanocomposite via in situ intercalative polymerization*", *J. Appl. Polym. Sci.*, **86**, 3708-11 (2002).

Ma, J., S.-F. Zhang, Z.-N. Qi, "*Synthesis and characterization of elastomeric polyurethane/clay nanocomposites*", *J. Appl. Polym. Sci.*, 82, 1444-48 (2001).

MacEwan, D. M. C., "*Complexes of clays with organic compounds*", *Trans. Faraday Soc.*, **44**, 349-367 (1948).

Macintyre J. E. (Ed.), *Dictionary of Inorganic Compounds*, Vol. 1-3, Chapman & Hall, London (1992).

Maeda, S. and S. P. Armes, "*Preparation and characterisation of polypyrrole-tin(IV) oxide colloidal nanocomposites*", *Chem. Mater.*, 7, 171-178 (1995).

Maeda, S. and S. P. Armes, "*Surface area measurements on conducting polymer-inorganic oxide nanocomposites*", *Synth. Metals*, **69**, 499-506; **73**, 151-155 (1995).

Maes A., L. Van Leemput, A. Cremers, and J. Uytterhoevenm, "*Electron density distribution as a parameter in understanding organic cation exchange in montmorillonite*", *J. Colloid Interface Sci.*, 77, 14-20 (1980).

Malfoy C., A. Pantet, P. Monnet, and D. Righi, "*Effects of the nature of the exchangeable cation and clay concentration on the rheological properties of smectites suspensions*", *Clays Clay Minerals*, **51**, 656-663 (2003).

Mallick P. K., and Y. Zhou, "*Yield and fatigue behavior of polypropylene and polyamide-6 nanocomposites*", *J. Mater. Sci.*, **38**, 3183-3190 (2003).

Mallick, P. K., *Fiber-Reinforced Composites*, Marcel Dekker Pub., New York (1993).

Maloney, W. T., "*Clay product and process of preparing some*", *US Pat.* **2,158,987**, 16.05.1939, Appl. 1934, to Georgia Kaolin Co.

Mamalis A. G., L. O. G. Vogtländer and A. Markopoulos, "*Nanotechnology and nanostructured materials: trends in carbon nanotubes*", *Precision Eng.*, **28**, 16-30 (2004).

Manas-Zloczower, I. and Z. Tadmor, Eds., *Mixing and Compounding of Polymers*, Hanser Pub., Munich (1994).

Manias, E. and V. Kuppa, "*Molecular simulations of ultra-confined polymers. Polystyrene intercalated in layered-silicates*" *in* "*Polymer nanocomposites*", R. Vaia and R. Krishnamoorti, Eds., ACS Symp. Ser., **804**, Ch. 15 (2002).

Manias, E., A. Touny, L. Wu, K. Strawhecker, B. Lu, and T. C. Chung "*Polypropylene/Montmorillonite Nanocomposites. Review of the Synthetic Routes and Materials Properties*", *Chem. Mater.*, **13**, 3516-23 (2001).

Marchand, D. and K. Jayaraman, "*Rheological probing of structure in polypropylene/clay nanocomposites*", *SPE Techn. Pap.*, **47**, 2130-34 (2001).

Marcolli, C. and G. Calzaferri, "*Monosubstituted octasilasesquioxanes*", *Appl. Organometallic Chem.*, **13**, 213-226 (1999).

Mark, J. E., "*Hybrid organic-inorganic composites containing mixed-oxide ceramics and high-temperature polymers.*" *J. Macromol. Sci., Pure Appl. Chem.* **A** 33, 2005-12 (1996); "*Ceramic reinforced polymers and polymer-modified ceramics.*" *Polym. Eng. Sci.*, **36**, 2905 - 20 (1996).

Mark, J. E., C. Y.-C. Lee and P. A. Bianconi (Eds.), *Hybrid Organic-Inorganic Composites*, American Chemical Society, Washington, DC (1995).

Marrucci, G., "*Remarks on the viscosity of polymeric liquid crystals*", in *Advances in Rheology*, B. Mena, A. Garcia-Rejon and C. Rangel-Nafaile, Eds., Univ. National Autónoma de México, p. 441-448 (1984).

Marrucci, G., "*Rheology of nematic polymers*", in *Liquid Crystallinity in Polymers*, A. Ciferri, Ed., VCH Pub., New York (1991).
Martin, C. E., "*Membrane-based synthesis of nanomaterials*", *Chem. Mater.*, 8, 1739-46 (1996).
Maruyama, B., and K. Alam, "*Carbon nanotubes and nanofibers in composite materials*", *SAMPE J.*, **38**, 59-70 (2002).
Mason, E. A. and Viehland, L. A., "*Statistical-mechanical theory of membrane transport for multicomponent system: passive transport through open membranes*", *J. Chem. Phys.*, **68**, 3562-73 (1978).
Massam, J., Z. Wang, T. J. Pinnavaia, T. Lan and G. Beall, "*Clay nanolayer reinforcement in epoxy-clay nanocomposites: a comparison of commercial and laboratory purified clays*", *ACS Polym. Mat. Sci. Eng.*, 78, 274-275 (1998).
Masson, P. and L. A. Utracki, *unpublished work* (1987).
Masuda, T., Y. Ohta and S. Onogi, "*Viscoelastic properties of multibranch polystyrenes*", in *Current Topics in Polymer Science*, Vol. II, R. M. Ottenbrite, L. A. Utracki and S. Inoue, Hanser, Munich (1987).
Matayabas Jr., J. C., S. R. Turner, B. J. Sublett, G. W. Connell, J. W. Gilmer, and R. B. Barbee, "*High I.V. polyester compositions containing platelet particles*", *US Pat.*, **6,084,019**, 04.07.2000, Appl. 22.12.1997, to Eastman Chemical Corp.
Mathot, V. B. F., Ed., "*Calorimetry and Thermal Analysis of Polymers*", Hanser, Munich (1994).
Mathur, R., D. R. Sharma, S. R. Vadera, S. R. Gupta, B. B. Sharma and N. Kumar, "*Room temperature synthesis of nanocomposites of Mn-Zn ferrites in a polymer matrix*", *Nanostructured Mater.*, **11**, 677-686 (1999).
Mauritz, K. A., I. D. Stefanithis, S. V. Davis, R. W. Scheetz, R. K. Pope, G. L. Wilkes and H. H. Huang, "*Microstructural evolution of a silicon oxide phase in a perfluorosulphonic acid ionomer by an in situ sol-gel reaction*", *J. Appl. Polym. Sci.*, **55**, 181-190 (1995).
Mauritz, K. A., P. R. Start, S. K. Young, and D.A. Reuschle, "*Organic/inorganic nanocomposite materials via polymer in situ sol-gel processes*", Polymer Nanocomposites 2001, Montreal, Quebec, Canada, September 17-19 (2001).
Maxfield, MacRae, B. R. Christiani, and V. R. Sastri, "*Polymer nanocomposites comprising a polymer and an exfoliated particulate material derivatized with organosilanes, organo titanates, and organo zirconates dispersed therein and process of preparing same*", *US Pat.*, **5,514,734**, 07.05.1996, Appl. 10.07.1995, to AlliedSignal Inc.
Maxfield, MacRae, B. R. Christiani, S. N. Murthy, and H. Tuller, "*Nanocomposites of gamma phase polymers containing inorganic particulate material*", *US Pat.*, **5,385,776**, 31.01.1995, Appl. 16.11.1992, to AlliedSignal.
McAtee, Jr., J. L., "*Cation exchange of organic compounds on montmorillonite in organic media*", *Clays, Clay Minerals*, **9**, 444-450 (1962).
McCarthy, B., J. N. Coleman, S. A. Curran, A. B. Dalton, A. P. Davey, Z. Konya, A. Fonseca, J. B. Nagy, W. J. Blau, "*Observation of site selective binding in a polymer nanotube composite*", *J. Mater. Sci. Letters*, **19**, 2239-41 (2000).
McCauley, J. R., *US Pat.*, **4,818,737**, 04.04.1989, to UOP (withdrawn).
McNeil, D., "*Tar and pitch*", in *Kirk-Othmer Encyclopedia of Chemical Technology*, Vol. **22**, J. Wiley & Sons, New York (1983).
Medellin-Rodriguez, F. J., C. Burger, B. S. Hsiao, B. Chu, R. Vaia and S. Phillips, "*Time-resolved shear behavior of end-tethered Nylon-6 clay nanocomposites followed by non-isothermal crystallization*", *Polymer*, **42**, 9015-21 (2001).

Mei, L., J. Ming, Z. Lei and W. Chi, "*Novel surfactant-free stable colloidal nanoparticles made of randomly carboxylated polystyrene ionomers*", *Macromolecules*, **30**, 2201-03 (1997).

Meincke, O., B. Hoffmann, C. Dietrich, and C. Friedrich, "*Viscoelastic properties of polystyrene nanocomposites based on layered silicates*", *Macromol. Chem. Phys.*, **204**, 823-830 (2003).

Menéndez, A., M. Bárcena, E. Jaimez, J. Garcia and J. Rodríguez, "*Intercalation of n-alkylamines by γ-titanium phosphate. Synthesis of new materials by thermal treatment of the intercalation compounds*", *Chem. Mater.*, **5**, 1078-84 (1993).

Merkel, T. C., B. D. Freeman, R. J. Spontak, Z. He, I. Pinnau, P. Meakin, A. J. Hill, "*Ultrapermeable, reverse-selective nanocomposite membranes*", *Science*, **296**, 519-522 (2002).

Messersmith, P. B. and E. P. Giannelis, "*Synthesis and characterization of layered silicate-epoxy nanocomposites*", *Chem. Mat.*, **6**, 1719-25 (1994).

Metzner, A. B. and G. M. Prilutski, "*Rheological properties of polymeric liquid crystals*", *J. Rheol.*, **30**, 661-691 (1986).

Mi, Y.-L., G. Xue, and X.-R. Wang, "*Glass transition of nano-sized single chain globules*", *Polymer*, **43**, 6701-05 (2002).

Michalczyk, M. J., K. G. Sharp and C. W. Stewart, *Fluoropolymer nanocomposites, US Pat.*, **5,726,247**, 10.03.1998, Appl. 14.06.1996, to E. I. du Pont de Nemours and Company.

Milewski, J. V. and H. S. Katz, Eds., *Handbook of Reinforcements for Plastics*, Van Nostrand Reinhold Co., New York (1987).

Millar, D. M. and J. M. Garces, "*Process of modifying the porosity of aluminosilcates and silicas and mesoporous composition derived therefrom*", *US Pat.* **6,017,508**, 2000.01.25, Appl. 1996.10.08, to Dow Chem. Co.

Mitchell, C. A., J. L. Bahr, S. Arepalli, J. M. Tour, and R. Krishnamoorti, "*Dispersion of Functionalized Carbon Nanotubes in Polystyrene*", *Macromolecules*, **35**, 8825-30 (2002).

Mitchell, I. V., Ed., *Pillared Layered Structures: Current Trends and Applications*, Elsevier, London (1990).

Mitsunaga M., Y. Ito, M. Okamoto, S. Sinha Ray, K.Hironaka, "*Intercalated polycarbonate/clay nanocomposites: nanostructure control and foam processing*"*Macromol. Mater. Eng.*, **288**(7), 543-548 (2003).

Miyanaga, S., Y. Doi, K. Nishimura, I. Nishi and Y. Sumida, "*Process for antistatic treatment of resin and antistatic resin composition*", *US Pat.*, **5,879,589**, 09.03.1999, Kao Corp.

Moet, A. and A. Akelah, "*Polymer-clay nanocomposites: polystyrene grafted onto montmorillonite interlayers*", *Mater. Lett.*, **18**, 97-102 (1993).

Moet, A., "*Fatigue failure*", in *Failure of plastics*, W. Brostow and R. D. Corneliussen, Eds., Hanser Pub., Munich (1986).

Moldenaers, P. and J. Mewis, "*Parallel superposition measurements on polymeric liquid crystals*", in *Theoretical and Applied Rheology*, P. Moldenaers and R. Keunings, Eds., Elsevier Sci. Pub., Amsterdam (1992).

Morgan, A. B. and J. D. Harris, "*Effects of organoclay Soxhlet extraction on mechanical properties, flammability properties and organoclay dispersion of polypropylene nanocomposites*", *Polymer*, **44**, 2313-20 (2003).

Morgan, A. B., J. W. Gilman and C. L. Jackson, "*Characterization of the dispersion of clay in polyetherimide nanocomposite*", *Macromolecules*, **34**, 2735-38 (2001).

Morlat S., B. Mailhot, D. Gonzalez, and J.-L. Gardette, "*Photo-oxidation of Polypropylene/Montmorillonite Nanocomposites. 1. Influence of Nanoclay and Compatibilizing Agent*", *Chem. Mater.*, **16**, 377-383 (2004).

Mortensen, J. L., "*Adsorption of hydrolyzed polyacrylonitrile on kaolinite*", *Clays, Clay Minerals*, **9**, 530-545 (1962).

Mortland, M. M., "*Clay organic complexes and interactions*", *Adv. Agron.*, **22**, 75-117 (1970).

Mortland, M. M., "*Urea complexes with montmorillonite*", *Clay Minerals.*, **6**, 143-156 (1966).

Mubarak, Y., P. J. Martin and E. Harkin-Jones, "*Effect of nucleating agents and pigments on crystallisation, morphology, and mechanical properties of polypropylene*", *Plast. Rubber Compos.*, **29**, 307-15 (2000).

Muhr H.-J., F. Krumeich, U. P. Schönholzer, F. Bieri, M. Niederberger, L. J. Gauckler and R. Nesper, "*Vanadium oxide nanotubes-a new flexible vanadate nanophase*", *Adv. Mater.*, **12**, 231-234 (2000).

Mukhopadhyay, A. and S. Granick, "*Micro- and nano-rheology*", *Current Opinion Colloid Interface Sci.*, **6**, 423-29 (2001).

Murase, S., A. Inoue, Y. Miyashita, N. Kimura and Y. Nishio, "*Structural characteristics and moisture sorption behavior of nylon-6/clay hybrid films*", *J. Polym. Sci. Part B: Polym. Phys.*, **40**, 479-87 (2002).

Murthy, N. S. and H. Minor, "*General procedures for evaluating amorphous scattering and crystallinity from X-ray diffraction scans of semicrystalline polymers*", *Polymer*, **31**, 996-1002 (1990).

Murugavel, R., M. Bhattacharjee, and H W. Roesky, "*Organosilanetriols: model compounds and potential precursors for metal-containing silicate assemblies*", *Appl. Organometallic Chem.*, **13**, 227-243 (1999).

Muzny, C. D., B. D. Butler, H. J. M. Hanley, F. Tsvetkov and D. G. Peiffer, "*Clay platelet dispersion in polymer matrix*", *Mater. Letters*, **28**, 379-384 (1996).

Nah, C., H. J. Ryu, S. H. Han, J. M. Rhee and M.-H. Lee, "*Fracture behavior of acrylonitrile-butadiene rubber/clay nanocomposites*", *Polym. Int.*, **50**, 1265-68 (2001).

Nahin, P. G. and P. S. Backlund, "*Organoclay-polyolefin compositions*", *US Pat.*, **3,084,117**, 02.04.1963, Appl. 04.04.1961, to Union Oil Co.

Nair, S. V., L. A. Goettler, B. A. Lysek, "*Toughening behavior of nanoscale and multiscale polymer composites*", Proceedings Nanocomposites-2001 International Symposium, Boucherville, QC, Canada, 2001.11.16-18; *Polym. Eng. Sci.*, **42**, 1872-82 (2002).

Nakada Y. and K. Kanaida, "*Method for manufacturing clay composite material*", *Japanese Pat.*, **JP2001354710**, 2001.12.25, to Nippon Shokubai Co Ltd.

Nakatani, A. I., Ed., *Filled and Nanocomposite Polymer Materials*, MRS symposium Boston, MA, November 27-30, 2000, Materials Research Society, Warrendale, PA (2001).

Nalwa, H. S., Ed., "*Handbook of nanostructured materials and nanotechnology*", 5 volumes, Academic Press, San Diego (2000).

Nam, P. H., P. Maiti, M. Okamoto, T. Kotaka, N. Hasegawa and A. Usuki, "*A hierarchical structure and properties of intercalated polypropylene/clay nanocomposites*", *Polymer*, **42**, 9633-40 (2001).

Nam, P. H., P. Maiti, M. Okamoto, T. Kotaka, T. Nakayama, M. Takada, M. Ohshima, A. Usuki, N. Hasegawa, H. Okamoto, "*Foam Processing and Cellular Structure of Polypropylene/Clay Nanocomposites*", *Polym. Eng. Sci.*, **42**, 1907-18 (2002).

Nanocor Technical data sheets T-10 and T-12 for "*Epoxy nanocomposites using Nanomer®*"I. 22 E and L28E"(2001b).

Nanocor's Technical Service Group, "*P-801 Guidelines for the Production of Polypropylene Nanocomposites*"(2001a).

Nanoscale Science and Engineering (NSE), *Program Solicitation for FY2001*, NSF 00-119, 31.07.2000.

Nelson, E. J. and E. T. Samulski, "*Preparation of CdS nanoparticles within an ordered polypeptite matrix*", *Mater. Sci. Eng.*, **C2**, 133-140 (1995).

Nesper, R., M. E. Spahr, M. Niederberger, P. Bitterli, "*Use and production of nanotubes containing a mixed valence vanadium*", *US Pat.*, **6,210,800**, 2001.04.03; Int. Patent Appl. **PCT/CH97/00470**; Priority 1999.08.04, to Eidg. Technische Hochschule, Zurich.

Newman A. C. D., Ed., "*Chemistry of clays and clay minerals*", *Mineralogical Soc. Monograph; No.* **6**, Burnt Mill, Harlow, Essex (1987).

Newman, A. C. D. (Ed.), "*Chemistry of Clays and Clay Minerals*", Mineralogical Soc. Monograph, 6, Burnt Mill (1987).

Nguyen, O., A. Repetto, M. Ortiz and R. A. Radovitzky, "*A cohesive model of fatigue crack growth*", *Int. J. Fract.* **110**, 351-69 (2001).

Nguyen, X. Q. and L. A., Utracki, "*Extensional Flow Mixer*", *U. S. Pat.*, **5,451,106**, 1995.09.19, Appl. 08 Aug. 1984, to National Research Council of Canada, Ottawa, Canada.

Nichols, K. L. and C.-J. Chou, "*Polymer composite comprising a inorganic layered material and a polymer matrix and a method for its preparation*", *US Pat.*, **5,952,093**, 14.09.1999, Appl., 20.02.1997, to The Dow Chem. Co.

Noh, M. H. and D. C. Lee, "*Comparison of characteristics of SAN-MMT nanocomposites prepared by emulsion and solution polymerization*", *J. Appl. Polym. Sci.*, **74**, 2811-19 (1999).

Noh, M. H., L. W. Jang and D. C. Lee, "*Intercalation of styrene-acrylonitrile copolymer in layered silicate by emulsion polymerization*", *Appl. Polym. Sci.* **74**, 179-188 (1999).

Norrish, K., "*The swelling of montmorillonite*", *Trans. Faraday Soc.*, **18**, 120-134 (1954).

Nowicki, W., "*Structure and entropy of a long polymer chain in the presence of nanoparticles*", *Macromolecules*, **35**, 1424-36 (2002).

Nyden, M. R. and J. W. Gilman, "*Molecular dynamics simulations of the thermal degradation of nano-confined polypropylene*", *Computational Theoretical Polym. Sci.*, **7**, 191-198 (1997).

Odegard G. M., T. S. Gates, L. M. Nicholson, and K. E. Wise, "*Equivalent continuum modeling of nano-structured materials*", *Comp. Sci. Tech.*, **62**, 1869-80 (2002).

Odegard, G. M., "*Nanotube Composite Properties Modeling*", Polymer Nanocomposites 2003, 2nd International symposium NRCC/IMI, Boucherville, 2003. 10. 06-08.

Odegard, G. M., T. S. Gates, L. M. Nicholson and K. E. Wise, "*Equivalent-continuum modeling with application to carbon nanotubes*", NASA/TM-2002-211454.

Odegard, G. M., V. M. Harik, K. E. Wise and T. S. Gates, "*Constitutive modeling of nanotube-reinforced polymer composite systems*", Am. Soc. Composites, 16th Annual Tech. Conf., Blacksburg, VA, 2001.09.09-12.

Ogawa, M., "*Formation of novel oriented transparent films of layered silica-surfactant nanocomposites*", *J. Am. Chem. Soc.*, **116**, 7941-42 (1994).

Ogawa, M., H. Shirai, K. Kuroda, and C. Kato, "*Solid-state intercalation of naphthalene and anthracene into alkylammonium-montmorillonites*", *Clays Clay Mineral.*, **40**, 485-90 (1992).

Ogawa, M., S. Okutomo and K. Kuroda, "*Control of interlayer microstructures of a layered silicate by surface modification with organochlorosilanes*", *J. Am. Chem. Soc.*, **120**, 7361-62 (1998).

Okada A. and A. Usuki, "*The chemistry of polymer-clay hybrids*", *Mater. Sci. Eng.*, **C3**, 109-115 (1995).

Okada, A., Y. Fukushima, M. Kawasumi, S. Inagaki, A. Usuki, S. Sugiyama, T. Kurauchi, and O. Kamigaito, "*Composite material and process for manufacturing same*", *US Pat.*, **4,739,007**, April 19, 1988, Kabushiki Kaisha Toyota Chuo Kenkyusho.

Okamoto K., S. S. Ray, and M. Okamoto, "*New Poly(butylene succinate)/Layered Silicate Nanocomposites. II. Effect of Organically Modified Layered Silicates on Structure, Properties, Melt Rheology, and Biodegradability*", *J. Polym. Sci. Part B: Polym. Phys.*, **41**, 3160-3172 (2003).

Okamoto M., "*Polymer/layered silicate nanocomposites*", *Rapra Review Reports*, **14** (7), Report 163, Rapra Technology Ltd., Shawbury, Shrewsbury, Shropshire, UK (2003).

Okamoto, M., P. H. Nam, P. Maiti, T. Kotaka, N. Hasegawa and A. Usuki, "*A house of cards structure in polypropylene/clay nanocomposites under elongational flow*", *Nano Lett.*, **1**, 295-298 (2001a).

Okamoto, M., P. H. Nam, P. Maiti, T. Kotaka, T. Nakayama, M. Takada, M. Ohshima, A. Usuki, N. Hasegawa and H. Okamoto, "*Biaxial flow-induced alignment of silicate layers in polypropylene/clay nanocomposites foam*", *Nano Lett.*, **1**, 503-505 (2001d).

Okamoto, M., P. H. Nam, T. Kotaka, N. Hasegawa and A. Usuki, "*Structure control and elongational rheology of polymer-layered silicate nanocomposites*", *Polymer*, unpublished preprint (2002).

Okamoto, M., S. Morita and T. Kotaka, "*Dispersed structure and ionic conductivity of smectic clay/polymer nanocomposites*", *Polymer*, **42**, 2685-88 (2001c).

Okamoto, M., S. Morita, H. Taguchi, Y. H. Kim, T. Kotaka and H. Tateyama, "*Synthesis and structure of smectic clay/poly(methyl methacrylate) nanocomposites by copolymerization with polar comonomers*", *Polymer*, **42**, 1201-06 (2001b).

Okamoto, M., S. Morita, H. Taguchi, Y. H. Kim, T. Kotaka and H. Tateyama, "*Synthesis and structure of smectic clay/poly(methyl methacrylate) and clay/polystyrene nanocomposites via in situ intercalative polymerization*", *Polymer*, **41**, 3887-90 (2000).

Ollivier, P. J., J. R. D. deBoard, P. J. Zapf, J. Zubieta, L. M. Meyer, C.-C. Wang, T. E. Mallouk and R. C. Haushalter, "*Hydrothermal synthesis and crystal structures of two novel vanadium oxides containing interlamellar transition metal complexes*", *J. Molecular Structure*, **470**, 49-60 (1998).

Olson, B. G., Z. L. Peng, J. D. McGervy, A. M. Jamieson, E. Manias, and E. P. Giannelis, "*Free volume in layered organosilicate-polystyrene nanocomposites*", *Materials Sci. Forum*, **255-257**, 336-340 (1997).

Onogi, S. and T. Asada, "*Rheology and rheo-optics of polymer liquid crystals*", in *Rheology*, Vol. 1., p. 127-147, G. Astarita, G. Marrucci and L. Nicolais, Eds., Plenum Press, New York (1980).

Oohara, Y. and N. Suzuki, "*Thermoplastic Resin Composition Containing Clay Composite and Process for Preparing the Same*"**EP0899308**, 1999.03.03, **WO9743343**, to Kanegafuchi Chemical Ind.

Ou C.-F., "*Nanocomposites of Poly(trimethylene terephthalate) with Organoclay*", *J. Appl. Polym. Sci.*, **89**, 3315-3322 (2003a); Ou C.-F., "*Crystallization behavior and thermal stability of poly(trimethylene terephthalate)/clay nanocomposites*", *J. Polym. Sci. Part B: Polym. Phys.*, **41**, 2902-2910 (2003b).

Ou C.-F., M. T. Ho, and J. R. Lin, "*Synthesis and characterization of Poly(ethylene terephthalate) nanocomposites with Organoclay*", *J. Appl. Polym. Sci.*, **91**, 140-145 (2003).

Oya, A., Y. Kurokawa and H. Yasuda, "*Factors controlling mechanical properties of clay mineral/polypropylene nanocomposites*", *J. Mater. Sci.*, **35**, 1045-50 (2000).

Ozawa, T., "*Kinetics of non-isothermal crystallization*", *Polymer*, **12**, 150-158 (1971).

Özdilek, C., K. Kazimierczak, D. van der Beek and S. J. Picken, "Preparation and properties of polyamide-6-boehmite nanocomposites", *Polymer*, **45**, 5207-5214 (2004).

Pan, C.-X. and Q.-L. Bao, "*Well-aligned carbon nanotubes from ethanol flame*", *J. Mater. Sci. Letters*, **21**, 1927-29, (2002).

Pantoustier, N., B. Lepoittevin, M. Alexandre, D. Kubis, C. Calberg, R. Jérôme, and P. Dubois, "*Biodegradable Polyester Layered Silicate Nanocomposites Based on Poly(ε-caprolactone)*", *Polym. Eng. Sci.*, **42**, 1928-37 (2002).

Pantoustier, N., M. Alexandre, P. Degée, C. Calberg, R. Jérôme, C. Henrist, R. Cloots, A. Rulmont and Ph. Dubois, "*Poly(ε-caprolactone) layered silicate nanocomposites: effect of clay surface modifiers on the melt intercalation process*", *e-Polymers*, **009**, 1-9 (2001).

Papazoglou E., R. Simha, and F. H. J., Maurer, "*Thermal expansivity of particulate composites: interlayer versus molecular model*", *Rheol. Acta*, **28**, 302-308 (1989).

Papirer, E. and H. Barard, "*Surface energy of silicas*", in *The Surface Properties of Silicas*, A. P. Legrand, Ed., John Wiley and Sons, New York (1998).

Parfitt, R. L. and D. J. Greenland, "*The adsorption of polyethyleneglycol on clay mineral*", *Clay Miner.*, **8**, 305-316 (1970).

Park, C. I., O O. Park, J. G. Lim and H. J. Kim, "*The fabrication of syndiotactic polystyrene/organophilic clay nanocomposites and their properties*", *Polymer*, **42**, 7465-75 (2001).

Paul, M.-A., M. Alexandre, P.Degée, C. Henrist, A. Rulmont and P. Dubois, "*New nanocomposite materials based on plasticized poly(*L*-lactide) and organo-modified montmorillonites: thermal and morphological study*", *Polymer*, **44**, 443-450 (2003).

Peigneur, P., A. Maes and A. Cremers, "*Ion exchange of the polyamide complexes of some transition metal ions in montmorillonite*", Proc. International Clay Conf., Oxford, Elsevier, Amsterdam (1978).

Penel-Pierron, L., C. Depecker, R. Séguéla, and J.-M. Lefebvre, "*Structural and mechanical behavior of nylon 6 films part I. Identification and stability of the crystalline phases*", *J. Polym. Sci. B: Polym. Phys.*, **39**, 484-95 (2001).

Penel-Pierron, L., R. Séguéla, J.-M. Lefebvre, V. Miri, C. Depecker, M. Jutigny, J. Pabiot, "*Structural and mechanical behavior of nylon-6 films. II. Uniaxial and biaxial drawing*", *J. Polym. Sci. B: Polym. Phys.*, **39**, 1224-36 (2001).

Pérez-Cardenas, F. C., L. F. Del Castillo and R. Vera-Graziano, "*Modified Avrami expression for polymer crystallization kinetics*", *J. Appl. Polym. Sci.*, **43**, 779-782 (1991).

Pérez-Maqueda L., F. Franco, M. A. Avilés, J. Poyato, and J. L. Pérez-Rodrigues, "*Effect of sonication on particle size distribution in natural muscovite and biotite*", *Clays Clay Minerals*, **51**, 701-708 (2003).

Perrin, F., personal communications (2002).

Petrarch Systems, *Silicon Compounds-Register and Review*, (1987).

Petrović, Z. S., I. Javni, A. Waddon and G. Bánhegyi, "*Structure and properties of polyurethane-silica nanocomposites*", *J. Appl. Polym. Sci.*, **76**, 133-151 (2000).

Phiriyawirut, P., R. Megaraphan and H. Ishida, "*Preparation and characterization of polybenzoxazine-clay immiscible nanocomposite, Mat. Res. Innovat.*, **4**, 187-196 (2001).

Pinnavaia T. J. and T. Lan, "*Sealant method of epoxy resin-clay composites*", *US Pat.*, **5,760,106**, 02.06.1998a, Appl. 13.09.1996, to Board of Trustees Michigan State University.

Pinnavaia T. J. and T. Lan, "*Hybrid nanocomposites comprising layered inorganic material and methods of preparation*", *US Pat.*, **5,853,886**, 29.12.1998c, Appl. 17.07.1996, to Claytec, Inc.

Pinnavaia T.J. and Z. Wang, "*Hybrid mixed ion clay structures for polymer nanocomposite formation*", *US Pat.*, **6,414,069**, 2002.07.02, Appl. 2000.03.03, to Board of Trustees of Michigan University.

Pinnavaia, T. J. and G. W. Beall, Eds., *Polymer-Clay Nanocomposites*, Wiley Series in Polymer Science, New York, (2000).

Pinnavaia, T. J. and T. Lan, "*Flexible resin-clay composite, method of preparation and use*", *US Pat.*, **5,801,216**, 01.09.1998b, Appl. 07.07.1997 to Board of Trustees operating Michigan State University.

Pinnavaia, T. J. and T. Lan, "*Hybrid organic-inorganic nanocomposites and methods of preparation*", *US Pat.*, **6,017,632**, 25.01.2000a, Appl. 20.08.1998, to Claytec, Inc.

Pinnavaia, T. J. and T. Lan, "*Methods of preparation of organic-inorganic hybrid nanocomposites*", *US Pat.*, **6,096,803**, 01.08.2000b, Appl. 20.08.1998, to Claytec, Inc.

Pinnavaia, T. J., T. Lan, Z. Wang, H.-Z. Shi and P. D. Kaviratna, "*Clay-reinforced epoxy nanocomposites*", *Chem. Mater.*, **8**, 1584-87 (1996).

Pironon J., M. Pelletier, P. de Donato, and R. Mosser-Ruck, "*Characterization of smectite and illite by FTIR spectroscopy of interlayer NH4+ cations*", *Clay Minerals,* **38**, 201-211 (2003).

Plueddemann, E. P., *Silane Coupling Agents*, Plenum Press, New York (1982).

Pluta, M., A. Gałeski, M. Alexandre, M.-A. Paul, and P. Dubois "*Polylactide/montmorillonite nanocomposites and microcomposites prepared by melt blending: Structure and some physical properties*", *J. Appl. Polym. Sci.*, **86**, 1497-06 (2002).

Polansky, C. A., J. E. White, J. M. Garces, A. Kuperman and D. Z. Ridley, *Polymer composite and method for its preparation, PCT,* **WO09954393A1**, 28.10.1999, Appl. 6.04.1999, to Dow Chem. Co.

Porter, D., *Group Interaction Modeling of Polymer Properties*, Marcel Dekker, Inc., New York (1995).

Porter, R. S. and J. F. Johnson, "*The rheology of liquid crystals*", *Rheology*, vol. 4., F. R. Eirich, Ed., Academic Press, New York (1967).

Pötschke, P., T. D. Fornes and D. R. Paul, "*Rheological behavior of multiwalled carbon nanotube/polycarbonate composites*", *Polymer*, **43**, 3247-55 (2002).

Powell, C. E., "*Process for treating smectite clays to facilitate exfoliation*", *US Pat.*, **6,271,298**, August 7, 2001b, Appl. April 27, 2000, to Southern Clay Products, Inc. (Gonzales, TX).

Powell, C. E., "*Process for treating smectite clays to facilitate exfoliation*", *US Pat. Appl.*, **20010056149**, 27.12.2001a, filed 04.05.2001, to Southern Clay Products, Inc.

Pozsgay A., T. Fráter, L. Százdi, P. Müller, I. Sajó, and B. Pukánszky, "*Gallery structure and exfoliation of organophilized montmorillonite: effect on composite properties*", *Eur. Polym. J.*, **40**, 27-36 (2004).

Pozsgay, A., L. Papp, T. Fráter and B. Pukánsky, "*Polypropylene/montmorillonite nanocomposiotes prepared by the delamination of the filler*", *Prog. Colloid Polym. Sci.*, **117**, 120-125 (2001).

Pramanik, M., S. K. Srivastava, B. K. Samantaray, and A. K. Bhowmick, "*Preparation and properties of ethylene vinyl acetate-clay hybrids*", *J. Mater. Sci. Letters*, **20**, 1377-80 (2001).

Pramanik, M., S. K. Srivastava, B. K. Samantaray, and A. K. Bhowmick, "*Synthesis and characterization of organosoluble thermoplastic elastomer/ clay nanocomposites*", *J. Polym. Sci.: Part B: Polym. Phys.*, **40**, 2065-72 (2002).

Premachandran, R., S. Banerjee, V. T. John, G. L. McPherson, J. A. Akkara, and D. L. Kaplan, "*The Enzymatic Synthesis of Thiol-Containing Polymers to Prepare Polymer-CdS Nanocomposites*", *Chem. Mater.*, **9**, 1342-47 (1997).

Privalko, V. P., T. Kawai and Yu. S. Lipatov, "*Crystallization of filled Nylon 6*", *Polym. J.*, **11**, 699-709 (1979).

Priya, L. and J. P. Jog, "*Intercalated poly(vinylidene fluoride)/clay nanocomposites: Structure and properties*", *J. Polym. Sci. Part B: Polym. Phys.*, **41**, 31-38 (2003).

Pucciariello R., V. Villani, S. Belviso, G. Gorrasi, M. Tortora, and V. Vittoria, "*Phase behavior of modified montmorillonite-poly(ε-caprolactone) nanocomposites*", *J. Polym. Sci. Part B: Polym. Phys.*, 42, 1321-1332 (2004).

Pullukat, T. J. and R. S. Shinomoto, "*High activity olefin polymerization catalysts*", *US Pat.*, **6,177,375**, 2001.01.23, to PQ Corporation.

Qian, D., E. C. Dickey, R. Andrews and T. Rantell, "*Load transfer and deformation mechanisms in carbon nanotube-polystyrene composites*", *Appl. Phys. Letters*, **76**, 2868-70 (2000).

Qian, G., J. W. Cho, and T. Lan, "*Preparation and properties of polyolefin nanocomposites*", Proceedings "Polyolefins 2001", Houston, TX, February 25-28, 2001.

Qian, X. F., C. Wang, X. M. Zhang, Y. Xie, Y. T. Qian, "*Solvothermal preparation of nanocrystalline tin chalcogenide*", *J. Phys. Chem. Solids*, **60**, 415-417 (2000).

Qian, X. F., X. M. Zhang, C. Wang, Y. Xie, Y. T. Qian, "*Benzene-thermal preparation of nanocrystalline chromium nitride*", *Mater. Res. Bull.*, **34**, 433-436 (1999a).

Qiao, Y. L. L. Cui, Y. Liu, S. Cui, J. S. Yang and D. A. Yang, "*Investigation into the preparation of the purified multiwalls carbon nanotubes-epoxy-polyethersulfone composites*", *J. Mater. Sci. Letters*, **21**, 1813-15 (2002).

Qiao, Z., Y. Xie, J. Xu, Y. Zhu and Y. Qian, "*Single-step confined growth of CdSe/polyacrylamide nanocomposites under γ-irradiation*", *Radiation Phys. Chem.*, **58**, 287-292 (2000a).

Qiao, Z., Y. Xie, J. Xu, Y. Zhu and Y. Qian, "*Synthesis of CdS/polyacrylonitrile nanocomposites by γ-irradiation*", *Mater. Res. Bull.*, **35**, 1355-60 (2000b).

Qiao, Z., Y. Xie, M. Chen, J. Xu, Y. Zhu, and Y. Qian, "*Synthesis of lead sulfide/(polyvinyl acetate) nanocomposites with controllable morphology*", *Chem. Phys. Letters*, **321**, 504-507 (2000c).

Qiao, Z., Y. Xie, Y. Zhu, and Y. Qian, "*Synthesis of PbS/poly(vinyl acetate) nanocomposites by γ-irradiation*", *Mater. Sci. Eng.-B-Solid State Mater. Adv. Technol.*, **77**, 144-146 (2000d).

Quach, A. and R. Simha, "*Pressure-volume-temperature properties and transitions of amorphous polymers; polystyrene and poly(orthomethylstyrene)*", *J. Appl. Phys.*, **42**, 4592-4606 (1971).

Quian, X. F., J. Yin, X. X. Guo, Y. F. Yang, Z. K. Zhu and J. Lu, "*Polymer-inorganic nanocomposites prepared by hydrothermal method*", *J. Mater. Sci. Letters*, **19**, 2235-37 (2000).

Qutubuddin, S., X.-A. Fu, and Y. Tajuddin, "*Synthesis of polystyrene-clay nanocomposites via emulsion polymerization using reactive surfactant*", *Polym. Bull.*, **48**, 143-149 (2002).

Ramsay, J. D. F. and P. Lindner, "*Small-angle neutron scattering investigations of the structure of thixotropic dispersions of smectic clay colloids*", *J Chem. Soc., Faraday Trans.*, **89**, 4207-14 (1993).

Ranade, A., N. A. D'Souza and B. Gnade, "*Exfoliated and intercalated polyamide-imide nanocomposites with montmorillonite*", *Polymer*, **43**, 3759-66 (2002).

Ratna D., O. Becker, R. Krishnamurthy, G. P. Simon, and R. J. Varley, "*Nanocomposites based on a combination of epoxy resin, hyperbranched epoxy and a layered silicate*", *Polymer*, **44**, 7449-7457 (2003).

Raussell-Colom J. A. and J. M. Serratosa, "*Reactions of clays with organic substances*", *in Chemistry of Clays and Clays Minerals*, Ch. 8, p. 371-422, A. C. D. Newman (Ed.), Mineralogical Society monograph, No. 6, Longman Scientific & Technical, Burnt Mill, Harlow, Essex, England (1987).

Rauwendaal, C., *Polymer Mixing*, Hanser Verlag, Munich (1998).

Ray S. S., and M. Okamoto, "New polylactide/layered silicate nanocomposites, 6a; Melt rheology and foam processing", *Macromol. Mater. Eng.*, **288**, 936-944 (2003).

Ray S. S., K. Yamada, M. Okamoto, Y. Fujimotoa A. Ogami, and K. Ueda, "*New polylactide/layered silicate nanocomposites. 5. Designing of materials with desired properties*", *Polymer*, **44**, 6633-6646 (2003b).

Ray, S. S. and M. Biswas, "*Preparation and evaluation of composites from montmorillonite and some heterocyclic polymers. II. A nanocomposite from n-vinylcarbazole and ferric chloride-impregnated montmorillonite polymerization system*", *J Appl. Polym. Sci.*, **73**, 2971-76 (1999).

Ray, S. S., K. Okamoto, K. Yamada, and M. Okamoto, "*Novel Porous Ceramic Material via Burning of Polylactide/Layered Silicate Nanocomposite*", *Nano Letters*, **2**, 423-425 (2002b).

Ray, S. S., K. Yamada, A. Ogami, M. Okamoto, and K. Ueda, "*New polylactide/layered silicate nanocomposite: nanoscale control over multiple properties*", *Macromol. Rapid Commun.*, **23**, 943-947 (2002c).

Ray, S. S., K. Yamada, M. Okamoto and K. Ueda, "*New polylactide-layered silicate nanocomposites. 2. Concurrent improvements of material properties, biodegradability and melt rheology*", *Polymer*, **44**, 857-866 (2003).

Ray, S. S., P. Maiti, M. Okamoto, K. Yamada, and K. Ueda, "*New polylactide/layered silicate nanocomposites. 1. Preparation, characterization, and properties*", *Macromolecules*, **35**, 3104-10 (2002a).

Reichert, P., H. Nitz, S. Klinke, R. Brandsch, R. Thomann and R. Mülhaupt, "*Poly(propylene)/organoclay nanocomposite formation: Influence of compatibilizer functionality and organoclay modification*", *Macromol. Mater. Eng.*, **275**, 8-17 (2000).

Ren, J., A. S. Silva, and R. Krishnamoorti, "*Linear viscoelasticity of disordered polystyrene-polyisoprene block copolymer based layered-silicate nanocomposites*", *Macromolecules*, **33**, 3739-46 (2000).

Reynaud, E., C. Gauthier and J. Perez, "*Nanophases in polymers*", *Rev. Metall./Cah. Inf. Tech.*, **96** 169-176 (1999).

Riess, G., J. M. Schwob, G. Guth, M. Roche and B. Laude, "*Ring opening polymerization of benzoxazines-A new route to phenolic resins*"in *Advances in Polymer Synthesis*, B. M. Culbertson and J. E. McGrath, Eds., Plenum Press, New York (1985).

Robello, D. R., N. Yamaguchi, T. N. Blanton, and C. L. Barnes, "*Exfoliated polystyrene-clay nanocomposite comprising star-shaped polymer*", *US Pat.*, **6,686,407**, 2004.02.03, Appl. 2002.05.24, to Eastman Kodak Co.

Robertson, D. H., D. W. Brenner, and J. W. Mintmire, "*Energetics of nanoscale graphitic tubules*", *Phys. Rev. B*, **45**, 12592-95 (1992).

Rodgers, P. A., "*Pressure-volume-temperature relationships for polymeric liquids*", *J. Appl. Polym. Sci.*, **50**, 1061-80, 2075-83 (1993).

Romero, D. B., M. Carrard, W. de Heer and L. Zuppiroli, "*Carbon nanotube/organic semiconducting polymer heterojunction*", *Adv. Mater.*, 8, 899-902 (1996).

Roovers, J. in *Encyclopedia of Polymer Science and Engineering*, 2nd ed., Kroschwitz, J. I., Ed., Wiley-Interscience Pub., New York (1985).

Rosenquist, N. R. and K. F. Miller, "*Composite comprising polymerized cyclic carbonate oligomer*", *US Pat.*, **4,786,710**, 22.11.1988, Appl. 26.05.1987, to General Electric Co.

Rosenquist, N. R. and K. F. Miller, "*Composition comprising bicyclic carbonate oligomer*", *US Pat.*, **4,701,538**, October 20.10.1987, Appl. 12.11.1985, to General Electric Co.

Ross, C. S. and S. B. Hendrics, *U. S. Geological Survey Prof. Paper* No. **205B** (1945).

Ross, C. S., *7th National Conference on Clays & Clay Minerals*, Pergamon Press, London (1960).

Ross, M. and J. Kaizerman, "*Clay/organic chemical compositions useful as additives to polymer, plastic and resin matrices to produce nanocomposites and nanocomposites containing such compositions*", *US Pat.*, **6,380,295**, 30.04.2002; **EP1055706,** 29.11.2000, Appl. 22.04.1998, to Rheox Inc.

Rossetti, F., *Meccanismi di degradazione termica e di ritardo all fiamma di nanocomposite polimerici,* Thesis, University of Torino (1999/2000).

Roy, R., "*The preparation and properties of synthetic clay minerals", C.N.R.S. Groupe Fr. Argiles C.R. Reun Etud.* **105**, 83-99 (1962).

Ruban, L., S. Lomakin and G. E. Zaikov, "*Polymer nanocomposites based on layered aluminium-silicates", Intern. J. Polym. Mater.*, **47**, 117-137 (2000).

Rueckes, T., K. Kim, E. Joselevich, G. Y. Tseng, C. Cheung, C. M. Lieber, "*Carbon nanotube-based nonvolatile random access memory for molecular computing", Science*, **289**, 94-97 (2000).

Ruiz-Hitzky, E. and B. Casal, "*Crown ether intercalations with phyllosilicates", Nature*, **276**, 596-597 (1978).

Ryul, J.-G., P.-S. Lee, H.-S. Kim and J.-W. Lee, "*Development of PP-based nanocomposites via in-situ copolymerization and melt intercalation with the power ultrasonic wave", ANTEC, SPE Techn. Pap.*, **47**, 2135-39 (2001).

Saito R., G. Dresslhaus, and M. S. Dresselhaus, "*Physical Properties of Carbon Nanotubes*" book in English and Japanese 258 pages, Imperial College Press (London) (1998).

Sakai, T., "*Reactive processing using twin-screw extruders*", Polymer Processing Society Regional Meeting, Taipei, TW, 2002.11.04-08.

Sakai, Y., T. Ikehara, T. Nishi, K. Nakajima and M. Hara, "*Nanorheology measurement on a single polymer chain", Appl. Phys. Letters*, **81**, 724-26 (2002).

Salahuddin, N. and A. Rehab, "*New nanocomposite materials based on polymerization of [oligo(oxyethylene) methacrylates] in the presence of montmorillonite clay", Polym. Int.*, **52**, 241-248 (2003).

Salahuddin, N. and M. Shehata, "*Polymethylmethacrylate-montmorillonite composites: preparation, characterization and properties", Polymer*, **42**, 8379-85 (2001).

Salem, A. J., "*A rheokinetic model of cyclic thermoplastic oligomer polymerization", SPE Techn. Pap.*, **37**, 748-751 (1991).

Samson, M. S. U., J. M. Van der Aalst, G. R. Meima, G.-S. J. Lee and J. M. Graces, "*Alkylation/transalkylation process with pretreatment of the alkylation/transalkylation feedstock", US Pat.*, **6,297,417**, 2001.10.02, Appl. 1999.02.18, to Dow Chem. Co.

Sanchez, I. C. and R. H., Lacombe, "*Statistical thermodynamics of polymer solutions", Macromolecules*, **11**, 1145-56 (1978).

Sánchez-Portal D., E. Artacho, J. M. Soler, Angel Rubio, and P. Ordejón, "*Ab initio structural, elastic, and vibrational properties of carbon nanotubes*", *Phys. Rev. B*, **59**, 12678-688 (1999).

Sandler, J., M. S. P. Shaffer, T. Prasse, W. Bauhofer, K. Schulte and A. H. Windle, "*Development of a dispersion process for carbon nanotubes in an epoxy matrix and the resulting electrical properties", Polymer*, **40**, 5967-71 (1999).

Sarangi, D., C. Godon, A. Granier, R. Moalic, A. Goullet, G. Turban and O. Chauvet, "*Carbon nanotubes and nanostructures grown from diamond-like carbon and polyethylene", App. Phys.*, **73A**, 765-768 (2001).

Sauer, J. A., A. D. McMaster and D. R. Morrow, "*Fatigue behavior of polystyrene and effect of mean stress", J. Macromol. Sci.-Phys.*, **B12**, 535-62 (1976).

Saujanya A., Y. Imai, and H. Tateyama, "*Structure development and isothermal crystallization behaviour of compatibilized PET/expandable fluorine mica hybrid nanocomposite*", *Polym. Bull.*, **51**, 85-92 (2003).

Saujanya, C. and S. Radhakrishnan, "*Structure development and crystallization behaviour of PP/nanoparticulate composite", Polymer*, **42**, 6723-31 (2001).

Saunders: W. S., Jr., "*Ionic Aliphatic Reactions*", p. 67-73, Prentice Hall. Englewood Cliffs (1965).

Schadler, L. S., S. C. Giannaris, and P. M. Ajayan, "*Load transfer in carbon nanotube epoxy composites*", *Appl. Phys. Letters*, 73, 3842-44 (1998).

Schmidt, G., A. I. Nakatani, and C. C. Han, "*Rheology and flow birefringence from viscoelastic polymer-clay solutions*", *Rheol. Acta*, **41**, 45-54 (2002a).

Schmidt, G., A. I. Nakatani, P. D. Butler, A. Karim and C. C. Han, "*Shear orientation of viscoelastic polymer-clay solutions probed by flow birefringence and SANS*", *Macromolecules*, **33**, 7219-22 (2000).

Schmidt, G., A. I. Nakatani, P. D. Butler, and C. C. Han, "*Small-angle neutron scattering from viscoelastic polymer-clay solutions*", *Macromolecules*, **35**, 4725-32 (2002b).

Schmidt-Rohr, K. and H. W. Spiess, *Multidimensional Solid-State "N"and Polymers*, Academic Press, London (1994).

Schön, F., R. Thomann and W. Gronski, "*Shear controlled morphology of rubber/organoclay nanocomposites and dynamic mechanical analysis*", *Macromol. Symp.* **189**, 105-110 (2002).

Schott, H., "*Deflocculation of swelling clays by nonionic and anionic detergents*", *J. Colloid Interface Sci.*, **26**, 133-139 (1968).

Schwab, J. J., W. A. Reinert, Sr., J. D. Lichtenhan, Y.-Z. An, S. H. Phillips and A. Lee, "*POSS™ nanostructured chemicals: true multifunctional polymer additives*", *Polym. Prepr.*, **42**(2), 48-49 (2001).

Schwerzel, R. E., K. B. Spahr, J. P. Kurmer, V. E. Wood and J. A. Jenkins, "*Nanocomposite photonic polymers. 1. Third-order nonlinear optical properties of capped cadmium sulfide nanocrystals in an ordered polydiacetylene host*", *J. Phys. Chem. A.*, **102**, 5622-26 (1998).

Scobbo, J. J., "*Conductive polyphenylene ether/polyamide blends for electrostatic painting applications*", *SPE Tech. Pap.*, **44**, 2468-72 (1998).

Scobbo, J. J., D. Grimes, T. Lemmen and V. Umamaheswaran, "*Conductive thermoplastic resin for electrostatically painted applications*", SAE Meeting, International Congress and Exposition Proceedings, (1998).

Sellinger, A., P. M. Weiss, A. Nguyen, Y. Lu, R. A. Assink, W. Gong and C. J. Brinker, "*Continuous self-assembly of organic–inorganic nanocomposite coatings that mimic nacre*", *Nature*, **394**, 256-260 (1998).

Sennett, M., E. Welsh, J. B. Wright, W. Z. Li, J. G. Wen and Z. F. Ren, "*Dispersion and alignment of carbon nanotubes in polycarbonate*", *Appl. Phys.*, **A76**, 111-113 (2003).

Serrano, F., G. W. Beall and H. Cruz, "*Exfoliated layered materials and nanocompositions comprising said exfoliated layered materials having water-insoluble oligomers or polymers adhered thereto*"; *Europ. Pat. Appl.*, **822,163A2**, 02.04.1998a, Priority 06.12.1996, to AMCOL International Corp.

Serrano, F., S. J. Engman, and G. W. Beall, "*Intercalates and exfoliates formed with non-EVAL monomers, oligomers and polymers; and EVAl composite materials containing same*"; *U.S. Pat.*, **5,844,032**, 01.12.1998; *Europ. Pat. Appl.*, 0 **846,723 A1**, 10.06.1998b, Priority. 06.12.1996, to AMCOL International Corp.

Severe G., A. J. Hsieh and B. E. Koene, "*Effect of layered silicates on thermal characteristics of polycarbonate nanocomposites*", *SPE Techn. Pap.*, **46**,1523-27 (2000).

Shaffer, M. S. P. and A. H. Windle, "*Analogies between polymer solutions and Carbon Nanotube dispersions*", *Macromolecules*, **32**, 6864-66 (1999a).

Shaffer, M. S. P. and A. H. Windle, "*Fabrication and characterization of carbon nanotube/ poly(vinyl alcohol) composites*", *Adv. Mater.*, **11**, 937-941 (1999b).

Shao, P. L., K. A. Mauritz and R. B. Moore, "*(Perfluorosulfonate ionomer)/(SiO2-TiO2) nanocomposites via polymer-in situ sol-gel chemistry: sequential alkoxide procedure*", *J. Polym. Sci.-B-Polym. Phys. Ed.*, **34**, 873-882 (1996).

Shen, Z., G. P. Simon and Y.-B. Cheng, "*Comparison of solution intercalation and melt intercalation of polymer-clay nanocomposites*", *Polymer*, **43**, 4251-60 (2002).

Sheng N., M. C. Boyce, D. M. Parks, G. C. Rutledge, J. I. Abes, and R. E. Cohen, "*Multiscale micromechanical modeling of polymer/clay nanocomposites and the effective clay particle*", *Polymer*, **45**, 487-506 (2004).

Shenton, W., T. Douglas, M. Young, G. Stubbs, S. Mann, "*Inorganic-organic nanotube composites from template mineralization of tobacco mosaic virus*"*Advanced Materials*, **11**, 253-256 (1999).

Shewring N. I. E., T. G. J. Jones, G. Maitland, and J. Yarwood, "*Fourier Transform Infrared Spectroscopic Techniques to Investigate Surface Hydration Processes on Bentonite*", *J. Colloid Interface Sci.*, **176**, 308-317 (1995).

Shi, H.-Z., T. Lan, and T. J. Pinnavaia, "*Interfacial effects on the reinforcement properties of polymer-organoclay nanocomposites: synthesis, and mechanism of formation*", *ACS Symposium Series*, **622**, 251-261 (1996).

Shia, D., C. Y. Hui, S. D. Burnside and E. P. Giannelis, "*An interface model for the prediction of Young's modulus of layered silicate elastomer nanocomposites*", *Polym. Compos.*, **19**, 608-617 (1998).

Shih, K. S., A. M., Adur, B. Frank and R. D. Solimeno, "*Multi-layer resin/paper laminate structure containing at least a polymer/nanoclay composite layer and packaging materials made thereof*", *WO Pat. Appl.*, **00/76862 A1**, 21.12.2000, Appl., 07.06.2000, to International Paper Co.

Shipway, A. N., E. Katz, I. Willner, "*Nanoparticle arrays on surfaces for electronic, optical, and sensor applications*", *Chem. Phys. Chem.*, **1**, 18-52 (2000).

Shockey, E. G., A. G. Bolf, P. F. Jones, J. J. Schwab, K. P. Chaffee, T. S. Haddad and J. D. Lichtenhan, "*Functionalized polyhedral oligosilsesquioxane (POSS) macromers: New graftable POSS-hydride, POSS-alpha olefin, POSS-epoxy, and POSS-chlorosilane macromers and POSS-siloxane triblocks*", *Appl. Organometallic Chem.*, **13**, 311-327 (1999).

Shrivastava, K. N., "*Specific heat of nanocrystals*", *Nano Lett.*, **2**, 21-24 (2002).

Sikka, M., L. N. Cerini, S. S. Ghosh, and K. I. Winey, "*Melt intercalation of polystyrene in layered silicates*"*J. Polym. Sci. Part B: Polym. Phys.*, **34**, 1443-49 (1996).

Simha R. and Somcynsky, T., "*On the statistical thermodynamics of spherical and chain molecule fluids*", *Macromolecules*, **2**, 342-350 (1969).

Simha, R. and P. Moulinié, "*Statistical thermodynamics of gas solubility in polymers*", in *Polymer Foams*, T. S. Lee, Ed., Technomic, Lancaster, PA (2000).

Simha, R., "*A treatment of the viscosity of concentrated suspensions*", *J. Appl. Phys.*, 23, 1020-24 (1952).

Simha, R., "*The influence of Brownian movement on the viscosity of solutions*", *J. Phys. Chem.*, **44**, 25-34 (1940).

Simha, R., L. A., Utracki and Garcia-Rejon, A., "*Pressure-volume-temperature relations of a poly-ε-caprolactam and its nanocomposite*", *Composite Interfaces*, 8, 345-353 (2001).

Simha, R., R. K., Jain and F. H. J., Maurer, "*Bulk modulus of particulate composites: a comparison between statistical and micro-mechanical approach*", *Rheol. Acta*, 25, 161-165 (1986).

Simha, R., R. K., Jain and S. C., Jain, "*Bulk modulus and thermal expansivity of melt polymer composites: statistical versus macro-mechanics*", *Polym. Composites*, 5, 3-10 (1984).

Simonik, J., A. Klendova and L. Kovarova, "*Polymer/clay nanocomposites; modified clay in polyvinylchloride (PVC) matrix*", Conference Nano '02, Brno, 2002.

Singh, A. and R. Haghighat, "*High-temperature polymer/inorganic nanocomposites*", *US Pat.*, **6,057,035**, 02.05.2000, Appl. 05.06.1998, to Triton Systems, Inc.

Singh, C. and A. C Balazs, "*Effect of polymer architecture on the miscibility of polymer/clay mixtures*", *Polym. Int.*, **49**, 469-471 (2000).

Sinn, C., "*Dynamic light scattering by rodlike particles: examination of the vanadium(V)-oxide system*", *Europ. Phys. J. B*, 7, 599-605 (1999).

Siripurapu S., J. R. Royer, J. M. DeSimone, S. A. Khan, and R. J. Spontak, "*Generation of diffusion-controlled microcellular polymer foams and foamed nanocomposites by supercritical carbon dioxide*", Proceedings *4th International Symposium on high pressure process technology and chemical engineering*, September 22-25, 2002, Venice-Italy.

Slaughter, M. and I. Milne, Eds., *Proceedings of the 7th National Conference of Clays and Clay Minerals*, Pergamon Press, Inc., New York (1960).

Smith, C. R., "*Base exchange reaction of bentonites and salts of organic bases*", *J. Am. Chem. Soc.*, **56**, 1561-63 (1934).

Sohn, J.-I., C. H. Lee, S. T. Lim, T. H. Kim, H. J. Choi, and M. S. Jhon, "*Viscoelasticity and relaxation characteristics of polystyrene/clay nanocomposite*", *J. Mater. Sci.*, **38**, 1849-52 (2003).

Solomon M. J., A. S. Almusallam, K. F. Seefeldt, A. Somwangthanaroj and P. Varadan, "*Rheology of polypropylene/clay hybrid materials*", *Macromolecules*, **34**, 1864-72 (2001).

Someya Y., T. Nakazato, N. Teramoto, and M. Shibata, "*Thermal and mechanical properties of poly(butylene succinate) nanocomposites with various organo-modified montmorillonites*", *J. Appl. Polym. Sci.*, **91**, 1463-1475 (2004).

Song W., I. A. Kinloch, and A. H. Windle, "*Nematic liquid crystallinity of multiwall carbon nanotubes*", *Science*, **302**, *1363 (*2003).

Song, C. H. and A. I. Isayev, "*Nanocomposites of PET and PET/HBA based LCP*", *SPE Techn. Pap.*, **44**, 1637-42 (1998); *J. Polym. Eng.*, **18**, 417-450 (1998).

Song, W., "*Comprehensive study of a new extensional flow mixer*", *SPE Techn. Pap.*, **46**, 270-275 (2000); SPE ANTEC 2000.05.07-11, Orlando FL.

Song, W., "*New extensional flow mixing technology: breakthrough mixer for industries*", Nanocomposite Compounding Workshop, Union Chemical Laboratories of the Industrial Technology Research Institute (UCL/ITRI), Hsinchu, Taiwan, 2002.11.01.

Spenleuhauer, G., D. Bazile, M. Veillard, C. Prud'Homme, and J. P.Michalon, "*Process for preparing nanoparticles*", *US Pat.*, **5,766,635**, 16.06.1998, to Rhone-Poulenc Rorer SA.

Srikhirin, T., A. Moet and J. B. Lando, "*Polydiacetylene-Inorganic Clay Nanocomposites*", *Polym. Adv. Technol.*, **9**, 491-503 (1998).

Stackhouse S., P. V. Coveney, and E. Sandre, "*Plane-Wave Density Functional Theoretic Study of Formation of Clay-Polymer Nanocomposite Materials by Self-Catalyzed in Situ Intercalative Polymerization*", *J. Am. Chem. Soc.*, **123**, 11764-74 (2001).

Starnes, Jr., W. H., *personal communication* (2002).

Stéphan, C., T. P., Nguyen, M. Lamy de la Chapelle, S. Lefrant, C. Journet and P. Bernier, "*Characterisation of singlewalled carbon nanotubes-PMMA composites*", *Synth. Met.*, **108**, 139-149 (2000).

Stokes, R. J. and D. F. Evans, *Fundamentals of Interfacial Engineering*, Wiley-VCH, New York (1997).

Su S., D. D. Jiang, and C. A. Wilkie, "*Novel polymerically-modified clays permit the preparation of intercalated and exfoliated nanocomposites of styrene and its copolymers by melt blending*", *Polym. Degrad. Stab.*, **83**, 333-346 (2004).

Su S., D. D. Jiang, and C. A. Wilkie, "*Poly(methyl methacrylate), polypropylene and polyethylene nanocomposite formation by melt blending using novel polymerically-modified clays*", *Polym. Degrad. Stab.*, **83**, 321-331 (2004).

Su, S.-J. and N. Kuramoto, "*Processable polyaniline-titanium dioxide nanocomposites: Effect of titanium dioxide on the conductivity*", *Synth. Metals*, **114**, 147-153 (2000).

Sue H.-J., K. T. Gam, N. Bestaoui, N. Spurr, and A. Clearfield, "*Epoxy nanocomposites based on the synthetic a-zirconium phosphate layer structure*", *Chem. Mater.*, **16**, 242-249 (2004).

Sugahara, Y., T. Sugiyama, T. Nagayama, K. Kuroda and C. Kato, "*Clay-organic nano-composite; preparation of a kaolinite-poly(vinylpyrrolidone) intercalation compound*", *J Ceramic Soc. Japan*, **100**, 413-416 (1992).

Suh, D. J. and O O. Park, "*Nanocomposite structure depending on the degree of surface treatment of layered silicate*", *J. Appl. Polym. Sci.*, **83**, 2143-47 (2002).

Suh, D. J., Y. T. Lim and O. O. Park, "*The property and formation mechanism of unsaturated polyester-layered silicate nanocomposite depending on the fabrication methods*", *Polymer*, **41**, 8557-63 (2000).

Sumpter, B. G., K. Fukui, M. D. Barnes, and D. W. Noid, "*Computational simulation and modeling of polymer-composite nanoparticles: new insights and directions for advanced materials*", *Mater Today*, **2(4)**, 3-6 (2000).

Sun, T., D. R. Wilson and J. M. Garces, "*Ion exchanged aluminum-magnesium silicate or fluorinated magnesium silicate aerogels and catalyst supports therefrom*", *US Pat.*, **6,376,421**, 2002.04.23, Appl. 2001.02.27, to Dow Chem. Co.

Sur, G. S., H. L. Sun, S. G. Lyu and J. E. Mark, "*Synthesis, structure, mechanical properties, and thermal stability of some polysulfone/organoclay nanocomposites*", *Polymer*, **42**, 9783-89 (2001).

Suzuki, N. and Y. Oohara, "*Thermoplastic resin composition containing silan-treated foliated phyllosilicate and method for producing the same*", *US Pat.*, **6,239,195 B1**, 2001.05.29, Appl. 1996.05.13, to Kanegafuchi Chemical Ind.

Suzuki, N. and Y. Oohara, "*Thermoplastic resin compositions containing clay composite and manufacturing thereof with excellent mechanical properties, heat resistance, and molding*", **WO9743343**, 1997, to Kaneka Corp.

Suzuki, N., "*Thermoplastic polyester resin containing silane/clay composite*", **WO9923162**, 1999, **EP1026203**, 2000.08.09, to Kaneka Corp.

Suzuki, N., T. Noma, and M. Kourogi, "*Resin compositions and process for producing them*", **WO2000022042**, 2000, to Kaneka Corp.

Tabtiang, A., S. Lumlong and R. A. Venables, "*Effects of shear and thermal history on the microstructure of solution polymerized poly(methyl methacrylate)–clay composites*", *Polym. Plast. Technol. Eng.*, **39** 293-303 (2000a).

Tabtiang, A., S. Lumlong and R. A. Venables, "*The influence of preparation method upon the structure and relaxation characteristics of poly(methyl methacrylate)/clay composites*", *Europ. Polym. J.*, **36**, 2559-68 (2000b).

Tadmor, Z., "*Method and apparatus for processing polymeric materials*", *US Pat.*, **4,194,841** (1980).

Tadmor, Z., "*Method for processing polymeric material*", *U.S. Pat.*, **4,142,805**, 1979.03.06.

Tadmor, Z., L. N. Valsamis, J. C. Yang, P. S. Mehta, O. Duran and J. C. Hinchcliffe, "*The corotating disk plastics processor*", *Polym. Eng. Rev.*, **3**, 29-62 (1983).

Tadmor, Z., P., Hold and L., Valsamis, "*A novel polymer processing machine theory and experimental results*", *SPE Tech. Papers*, **25**, 193-211 (1979).

Tai, N. H., *personal information* (2002).

Takahashi, T., *PhD Thesis*, Yamagata University, Yonezawa (1996).

Takeichi, T., R. Zeidam and T. Agag, "*Polybenzoxazine/clay hybrid nanocomposites: influence of preparation method on the curing behavior and properties of polybenzoxazines*", *Polymer,* **43**, 45-53 (2002).

Takekoshi, T., F. F. Khouri, J. R. Campbell, T. C. Jordan and K. H. Dai, "*Layered minerals and compositions comprising the same*", *US Pat.*, **5,530,052**, 1996.06.25, Appl. 1995.04.03, to General Electric Co.

Taki K., K. Tabata, T. Yanagimoto, and M. Ohshima, "*Visual observation of microcellular foaming of PP/clay nanocomposites*", Proceedings *Polymer Nanocomposites 2003 International Symposium*, 2003.10.06-08, Boucherville, QC, Canada.

Tanaka G., and L. A. Goettler, "*Predicting the binding energy for nylon 6,6/clay nanocomposites by molecular modeling*", *Polymer*, **43**, 541-553 (2002).

Tang B. Z., Y. Geng, J. W. Y. Lam, B. Li, X. Jing, X. Wang, F. Wang, A. B. Pakhomov and X. X. Zhang, "*Processible nanostructured materials with electrical conductivity and magnetic susceptibility: preparation and properties of maghemite/polyaniline nanocomposite films*", *Chem. Mater.*, **11**, 1581-89 (1999).

Tang, B. Z. and H.-Y. Xu, "*Preparation, alignment and optical properties of soluble poly(phenylacetylene)-wrapped carbon nanotubes*", *Macromolecules*, **32**, 2569-76 (1999).

Tanoue S., L. A. Utracki, A. Garcia-Rejon, J. Tatibouët and M. R. Kamal, "*Preparation of Polystyrene/Organoclay Nanocomposites by Melt Compounding Using a Twin Screw Extruder*", Japanese Society of Polymer Processing annual meeting 2003, Symposium Papers, November 3-4 2003, Kanazawa (Japan).

Tanoue, S., L. A. Utracki, A. Garcia-Rejon, M. R. Kamal, P. Sammut, M.-T. Ton-That, I. Pesneau and J. Lyngae-Jørgensen, "*Melt compounding of different grades polystyrene with organoclay: Rheological properties*", *Polym. Eng. Sci.*, **44**, 1061-76 (2004).

Tanoue, S., L. A. Utracki, A. Garcia-Rejon, J. Tatibouët, K. C. Cole and M. R. Kamal, "*Melt compounding of different grades of polystyrene with organoclay: Compounding and characterization*", *Polym. Eng. Sci.*, **44**, 1046-60 (2004).

Tans, S. J., M. H. Devoret, H. Dai, A. Thess, R. E. Smalley, L. J. Geerligs and C. Dekker, "*Individual single-wall carbon nanotubes as quantum wires*", *Nature*, **386**, 474-477 (1997).

Tateyama, H., K. Tsunematsu, K. Kimura, H. Hirosue, K. Jinnai and T. Furusawa, "*Method for producing fluorine mica*"*US Pat.*, **5,204,078**, 1993.04.20, to CO-OP Chemical Co., Ltd.

Taub, A., *GM-Basell Polyolefins Press release* of 28.08.2001.

Tenne, R., L. Margulis, M. Genut, and G. Hodes, "*Polyhedral and cylindrical structures of tungsten disulphide*", *Nature*, **360**, 444-445 (1992).

Tersoff, J. and R. S. Ruoff, "*Structural properties of a carbon-nanotube crystal*", *Phys. Rev. Lett.*, **73**, 676-79 (1994).

Theng, B. K. G., D. J. Greenland and J. P. Quirk "*Adsorption of alkylammonium cations by montmorillonite, Clay Minerals*, 7, 1-17 (1967).

Theng, B. K. G., *Organic Complexes of Montmorillonite*, PhD Thesis, Adelaide University (1964).

Theng, B. K. G., *The Chemistry of Clay-Organic Reactions*, J. Wiley & Sons, New York (1974).

Thiessen, P. A., "*Kennzeichnung submikroskopischer Grenzflächenbereiche verschiedennartiger Wirksamkeit*", *Z. Anorg. Chem.*, **253**, 161-169 (1947); n.b., paper written by June 1944.

Thostenson, E. T., Z. Ren and T.-W. Chou, "*Advances in the science and technology of carbon nanotubes and their composites: a review*", *Compos. Sci. Technol.*, **61**, 1899-12 (2001).

Tian D., S. Blacher and R. Jerome, "*Biodegradable and biocompatible inorganic-organic hybrid materials: 4. effect of acid content and water content on the incorporation of aliphatic polyesters into silica by the sol-gel process*", *Polymer*, **40**, 951-957 (1999).

Tjong, S. C., Y. Z. Meng, and A. S. Hay, "*Novel preparation and properties of polypropylene-vermiculite nanocomposites*", *Chem. Mater.*, **14**, 44-51 (2002).

Ton-That, M.-T., K. Cole, J. Denault, Y. Simard and G. Enright, "*The Role of Reactive Processing in the Manufacturing of Polymer Nanocomposites*", Proceedings of ECM, Inc., Nanocomposites 2002-The second world congress on Nanocomposites, San Diego CA, 2002.09.23-25.

Ton-That, T.-M., *personal communication* (2000).

Tortora, M., V. Vittoria, G. Galli, S. Ritrovati, and E. Chiellini, "*Transport properties of modified montmorillonite-poly(ε-caprolactone) nanocomposites*", *Macromol. Mater. Eng.*, **287**, 243-249 (2002).

Treacy, M. M. J., T. W. Ebbesen and J. M. Gibson "*Exceptionally high Young's modulus observed for individual carbon nanotubes*", *Nature*, **381**, 680-687 (1996).

Tsai, H.-L., "*Plastic superconducting polymer-NbSe2 nanocomposites*", *Chem. Mater.*, **9**, 875-878 (1997).

Tsai, M.-H. and W.-T. Whang, "*Dynamic mechanical properties of polyimide/poly(silsesquioxane)-like hybrid films*", *J. Appl. Polym. Sci.*, **81**, 2500-16 (2001a).

Tsai, M.-H. and W.-T. Whang, "*Low dielectric polyimide/poly(silsesquioxane)-like nanocomposite material*", *Polymer*, **42**, 4197-4207 (2001b).

Tsai, T.-Y., "*Polyethylene terephthalate-clay nanocomposites*", in *Polymer-Clay Nanocomposites*, p. 173-188, T. J. Pinnavaia and G. W. Beall, Eds., John Wiley & Sons, New York (2000).

Tseng, C.-R., H.-Y. Lee, F.-C. Chang, "*Crystallization kinetics and crystallization behavior of syndiotactic polystyrene/clay nanocomposites*", *J Polym Sci Part B: Polym Phys.*, **39**, 2097-2107 (2001).

Tsipursky, S., V. Dolinko, V. Psihogios and G. W. Beall, "*Viscous carrier compositions, including gels, formed with an organic liquid carrier, a layered material: polymer complex, and a di-, and/or tri-valent cation*", *US Pat.*, **5,998,528**, 07.12.1999, to AMCOL International Corp.

Tsuchita, A., C. Bolln, F. G. Sernetz, H. Frey and R. Mülhaupt, "*Ethene and propene copolymers containing silsesquioxane side groups*", *Macromolecules*, **30**, 2818-24 (1997).

Tucker III, C. L., "*Principles of mixing measurements*", *in Mixing in Polymer Processing*, C. Rauwendaal, Ed., Marcel Dekker, Inc., New York, N. Y. (1991).

Tunney, J. J. and C. Detellier, "*Aluminosilicate nanocomposite materials. poly(ethylene glycol)kaolinite intercalates*", *Chem. Mater.*, **8**, 927-935 (1996).

Turcsáyi, B., B. Pukánszky, and F., Tüdös, "*Composition dependence of tensile yield stress in filled polymers*", *J. Mater. Sci. Letters*, **7**, 160-162 (1988).

Turnbull, D. and J. C., Fisher, "*Rate of nucleation in condensed systems*", *J. Chem. Phys.*, **17**, 71-73 (1949).

Tyan H.-L., Y.-C. Liu and K.-H. Wei, "*Enhancement of imidization of poly(amic acid) through forming poly(amic acid)/organoclay nanocomposites*", *Polymer*, **40**, 4877-86 (1999b).

Tyan, H.-L., K.-H. Wei, T.-E. Hsieh "*Mechanical properties of clay-polyimide (BTDA-ODA) nanocomposites via ODA-modified organoclay*"*J. Polym. Sci. Part B: Polym. Phys.*, **38**, 2873-77 (2000).

Tyan, H.-L., Y.-C. Liu and K.-H. Wei, "*Thermally and mechanically enhanced clay/polyimide nanocomposite via reactive organoclay*", *Chem. Mater.*, **11**, 1942-47 (1999a).

Ube Industries, Ltd., Nylon Resin Dept., *High Performance Nylon Resin-Technical Data* (Molding applications), March, 2000.

Ugaz, V. M., W. R. Burghardt, W. Zhou and J. A. Kornfield, "*Transient molecular orientation and rheology in flow aligning thermotropic liquid crystalline polymers*", *J. Rheol.*, **45**, 1029-63 (2001).

Uribe-Arocha, P., C. Mehler, J. E. Puskas, and V. Altstädt, "*Effect of sample thickness on the mechanical properties of injection-molded polyamide-6 and polyamide-6 clay nanocomposites*", *Polymer*, **44**, 2441-46 (2003).

Uribe, J., M. R. Kamal, A. Garcia-Rejon, and L. A. Utracki, "*Melt Blending of polystyrene/organoclay nanocomposites*", *Proceedings of the Polymer Processing Society Annual Meeting*, Guimarães, Portugal, 2002.06.16-20.

Usuki, A., "*Synthesis of polymer-clay nanocomposites*", in Workshop & International Symposium on Polymer Nanocomposites Science and Technology, NRCC/IMI and McGill University, Boucherville, QC, Canada 2001.09.17-19 (held 2001.11.14-16).

Usuki, A., A. Koiwai, Y. Kojima, M. Kawasumi, A. Okada, Y. T. Kurauchi, and O. Kamigaito, "*Interaction of nylon-6-clay surface and mechanical properties of nylon 6-clay hybrid*", *J. Appl Polym. Sci.*, **55**, 119-123 (1995).

Usuki, A., A. Tukigase and M. Kato, "*Preparation and properties of EPDM-clay hybrids*", *Polymer*, **43**, 2185-89 (2002).

Usuki, A., K. Okamoto, A. Okada, and T. Kurauchi, *Kobunsi Ronbunshu*, **52**, 728 (1995).

Usuki, A., M. Kato and A. Okada, "*Composite clay material and method for producing the same, blend material and composite clay rubber using the same and production method thereof*", *US Pat.*, **5,973,053**, 26.10.1999, Appl. 05.06.1995, to Kabushiki Kaisha Toyota Chuo Kenkyusho.

Usuki, A., M. Kato and A. Okada, "*Composite clay material and method for producing the same, blend material and composite clay rubber using the same and production method thereof*", *US Pat.*, **6,103,817**, 15.08.2000a, Appl. 05.06.1995, to Kabushiki Kaisha Toyota Chuo Kenkyusho.

Usuki, A., M. Kato and A. Okada, "*Composite clay material and method for producing the same, blend material and composite clay rubber using the same and production method thereof*", *US Pat.*, **6,121,361**, 19.09.2000b, Appl. 05.06.1995, to Kabushiki Kaisha Toyota Chuo Kenkyusho.

Usuki, A., M. Kato, A. Okada, and T. Kurauchi, "*Synthesis of polypropylene-clay hybrid*", *J. Apply. Polym. Sci.*, **63**, 137-139 (1997).

Usuki, A., M. Kato, and A. Okada, "*Composite clay material and method of producing the same, blend material and composite clay rubber using the same and production method thereof*", *EP Appl.*, **0747322**, 11.12.1996, Appl. May 30, 1996, to Kabushiki Kaisha Toyota Chuo Kenkyusho.

Usuki, A., M. Kawasumi, Y. Kojima, Y. Fukushima, A. Okada, T. Kurauchi, and O. Kamigaito, "*Synthesis of PA-6-clay hybrid*", *J Mater. Res.*, **8**, 1179-84 (1993b).

Usuki, A., M. Kawasumi, Y. Kojima, Y. Okada, T. Kurauchi, and O. Kamigaito, "*Swelling behavior of montmorillonite cation exchanged for ω-amino acids by ε-caprolactam*", *J. Mater. Res.*, **8**, 1174-78 (1993a).

Usuki, A., *personal communication* (2000).

Usuki, A., T. Mizutani, Y. Fukushima, M. Fujimoto, K. Fukumori, Y. Kojima, N. Sato, N. Kurauchi, and O. Kamigaito, "*Composite material containing a layered silicate*", *US Pat.*, **4,889,885**, Dec. 26, 1989, Appl. March 4, 1988, Kabushiki Kaisha Toyota Chuo Kenkyusho.

Usuki, A., Y. Kojima, M. Kawasumi, A. Okada, T. Kurauchi, and O. Kamigaito, "*Characterization and Properties of PA-6-clay hybrid*", *ACS Polym. Preprint*, **31**, 651-652 (1990).

Utracki, L. A. and R. Simha, "*Pressure-Volume-Temperature Dependence of Polypropylene/Organoclay Nanocomposites*", *Macromolecules* (submitted, 2004).

Utracki L. A., and R. Simha, "*Statistical thermodynamics predictions of the solubility parameter*", *Polym. Int.*, **53**, 279-286 (2004).

Utracki L. A., "*Statistical Thermodynamics evaluation of solubility parameters and polymer miscibility*", *J. Polym. Sci. Part B: Poly. Phys.*, **42**, 2909-2915 (2004).

Utracki, L. A, *Polymer Alloys and Blends*, Hanser Verlag, Munich (1989).

Utracki, L. A. and A. Luciani, "*Mixing in extensional flow field*", *Appl. Rheology*, **10**, 10-21 (2000).

Utracki, L. A. and A., Luciani, "*Extensional Flow Mixer*", *Canad. Pat. Appl.*, 1997, to National Research Council of Canada, Ottawa, Canada.

Utracki, L. A. and Gerard Z. H. Shi, "*Compounding polymer blends*", Chapter 10 in *Polymer Blends Handbook*, Kluwer Academic Publishers, Dordrecht (2002).

Utracki, L. A. and J. Lyngaae-Jørgensen, "*Dynamic melt flow of nanocomposites based on poly-ε-caprolactam*", *Rheolog. Acta*, **41**, 394-407 (2002).

Utracki, L. A. and K. C. Cole, "*First International Symposium on Polymeric Nanocomposites*", National Research Council Canada, Industrial Materials Institute, Boucherville, QC, 2001.11.14-16, *Polym. Eng. Sci.*, **42** (9), 1799 (2002).

Utracki, L. A. and M. R. Kamal, "*Clay-containing polymeric nanocomposites*", *Arabian J. Sci. Eng.*, 27, 43-67 (2002a).

Utracki, L. A. and M. R. Kamal, "*The rheology of polymer alloys and blends*", Chapter 7 in *Polymer Blends Handbook*, Kluwer Academic Publishers, Dordrecht (2002b).

Utracki, L. A. and R. Simha, "*Analytical representation of solutions to lattice-hole theory*", *Macromol. Chem. Phys., Molecul. Theory Simul.*, **10**, 17-24 (2001a).

Utracki, L. A. and R. Simha, "*Free volume and viscosity of polymer-compressed gas mixtures during extrusion foaming*", *J. Polym. Sci., Part B: Polym. Phys.*, **39**, 342-362 (2001b).

Utracki, L. A., "*Correlation between PVT behavior and the zero–shear viscosity of liquid mixtures*", *J. Rheol.*, **30**, 829-841 (1986).

Utracki, L. A., "*Rheology of two–phase flows*", Ch. 15 in *Rheological Measurements*, A. A. Collyer and D. W. Clegg, Eds., Elsevier, London (1988).

Utracki, L. A., "*Temperature and pressure dependence of liquid viscosity*", *Canadian J. Chem. Eng.*, **61**, 753-758 (1983).

Utracki, L. A., "*The rheology of multiphase systems*", in *Rheological Fundamentals of Polymer Processing*, J. A. Covas, J. F. Agassant, A. C. Diogo, J. Vlachopoulos, and K. Walters, Eds., pg. 113-137, Kluwer Academic Publishers, Dordrecht (1995).

Utracki, L. A., "*Pressure-volume-temperature-viscosity relations in fluorinated polymers*", *J. Appl. Polym. Sci.*, **84**, 1101-05 (2002d).

Utracki, L. A., "*Role of polymer blends' technology in polymer recycling*", Ch. 17 in *Polymer Blends Handbook*, Utracki, L. A., Ed., Kluwer Academic Pub., Dordrecht (2002c).

Utracki, L. A., "*Thermodynamics of polymer blends*", Chapter 2 in *Polymer Blends Handbook*, Kluwer Academic Publishers, Dordrecht (2002b).

Utracki, L. A., *Commercial Polymer Blends*, Chapman & Hill, London (1998).

Utracki, L. A., Ed., *Encyclopaedic Dictionary of Commercial Polymer Blends*, ChemTec Publishing, Toronto, Canada (1994).

Utracki, L. A., Ed., *Polymer Blends Handbook*, Kluwer Academic Pub., Dordrecht (2002a).

Utracki, L. A., *Polymer Blends*, RAPRA Review Reports, 11(3), 1-170 (2000).

Utracki, L. A., R. Simha, and A. Garcia-Rejon, "*Pressure-volume-temperature relations in nanocomposites*", *Macromolecules*, **36**, 2114-21 (2003).

Uytterhoeven, J., "*Determination of the surface hydroxyl groups of kaolinite by organometallic compounds (CH3MgI and CH3Li)*", *Bull. Groupe Franc. Argiles*, **13**, 69-75 (1962).

Uytterhoeven, J., "*Organic derivatives of silicates and aluminosilicates*", *Silicates Inds.*, **25**, 403-409 (1960).

Vacatello, M., "*Monte Carlo simulations of polymer melts filled with solid nanoparticles*", *Macromolecules*, **34**, 1946 (2001a).

Vacatello, M., "*Monte Carlo simulations of the interface between polymer melts and solids. Effect of chain stiffness*", *Macromol. Theory Simul.*, **10**, 187-195 (2001b).

Vaia R. A., K. D. Jandt, E. J. Kramer, and E. P. Giannelis, "*Microstructural evolution of melt intercalated polymer-organically modified layered silicates nanocomposites*", *Chem. Mater.*, **8**, 2628-35 (1996).

Vaia, R. A. and Giannelis, E. P., "*Lattice model of polymer melt intercalation in organically-modified layered silicates*", *Macromolecules*, 30, 7990-99 (1997a).

Vaia, R. A. and Giannelis, E. P., "*Polymer melt intercalation in organically-modified layered silicates: model predictions and experiment*", *Macromolecules*. 30, 8000-09 (1997b).

Vaia, R. A., G. Price, P. N. Ruth, H. T. Nguyen and J. Lichtenhan, "*Polymer/layered silicate nanocomposites as high performance ablative materials*", *Appl. Clay Sci.*, **15**, 67-92 (1999).

Vaia, R. A., H. Ishii and E. P. Giannelis, "*Synthesis and properties of two-dimensional nanostructure by direct intercalation of polymer melts in layered silicates*", *Chem. Mater.*, **5**, 1694-96 (1993).

Vaia, R. A., K. D. Jandt, E. J. Kramer and E. P. Giannelis, "*Kinetics of polymer melt intercalation*", *Macromolecules*, **28**, 8080-85 (1995b).

Vaia, R. A., K. D. Jandt, E. J. Kramer, and E. P. Giannelis, "*Microstructural evolution of melt intercalated polymer-organically modified layered silicates nanocomposites*", *Chem. Mater.*, **8**, 2628-35 (1996).

Vaia, R. A., *Polymer Intercalation in Mica-Type Layered Silicates*, PhD thesis, Cornell University (1995).

Vaia, R. A., R. K. Teukolsky and E. P. Giannelis, "*Interlayer structure and molecular environment of alkylammonium layered silicates*", *Chem. Mater.*, **6**, 1017-22 (1994).

Vaia, R. A., S. Vasudevan, W. Krawiec, L. G. Scanlon and E. P. Giannelis, "*New polymer electrolyte nanocomposites: melt intercalation of poly(ethylene oxide) in mica-type silicates*", *Adv. Materials*, 7, 154-156 (1995a).

Vaia, R., H., Ishii and E., Giannelis, "*Synthesis and properties of two-dimensional nanostructures by direct intercalation of polymer melts in layered silicates*", *Chem. Mater.*, 5, 1694-96 (1993).

van de Ven, T. G. M., *Colloidal Hydrodynamics*, Academic Press, New York (1989).

van der Giessen, E. and A. Needleman, "*Micromechanics simulations of fracture*", *Annul. Rev. Mater. Res.*, **32**, 141-162 (2002).

van der Gucht, J., N. A. M. Besseling, and G. J. Fleer, "*Chain-length dependence of the polymer surface excess near the adsorption/depletion transition*", *Macromolecules*, **35**, 2810-16 (2002).

van der Waals, J. D., *PhD dissertation*, University of Leiden (1873).

van Krevelen, D. W., *Properties of Polymers*, 3rd Ed., Elsevier, Amsterdam (1993).

van Olphen, H. and C. T. Deeds, "*The stepwise hydration of clay-organic complexes*", *Nature*, **194**, 176-177 (1962).

van Olphen, H. and Fripiat, J. J., Eds., *Data Handbook for Clay Minerals and Other Non-Metallic Minerals*, Pergamon Press, Elmsford NY (1979).

van Olphen, H., *An Introduction to Clay Colloid Chemistry*, 2nd Ed., John Wiley & Sons, New York (1977).

van Opstal, L., R. Koningsveld, and L. A. Kleintjens, "*Description of partial miscibility in chain molecular mixtures*", *Macromolecules*, **24**, 161-167 (1991).

van Oss, C. J., *Interfacial Forces in Aqueous Media*, Marcel Dekker, New York (1994).

VanderHart, D. L., A. Asano, and J. W. Gilman, "*Solid-state NMR investigation of paramagnetic Nylon-6 clay nanocomposites. 1. Crystallinity, morphology, and the direct influence of Fe3+ on nuclear spins*", *Chem. Mater.*, **13**, 3781-95 (2001).

Vansant, E. F. and J. B. Uytterhoeven, "*Thermodynamics of the exchange of n-alkyl ammonium ions in Na-montmorillonite*", *Clay Clays Miner.*, **20**, 47-54 (1972).

Varlot, K, E. Reynaud, M. H. Kloppfer, G. Vigier and J. Varlet, "*Clay-reinforced polyamide: Preferential orientation of the montmorillonite sheets and the polyamide crystalline lamellae*", *J. Polym. Sci., Part B: Polym. Phys.*, **39**, 1360-70 (2001).

Vaughan, D. E. W., R. J., Lussier and J. S. Magee, Jr., "*Pillared interlayered clay materials useful as catalysts and sorbents*", *US Pat.*, **4,176,090**, 27.11.1979, to W. R. Grace & Co.

Velde, B., *Clays and Clay Minerals in Natural and Synthetic Systems*, Elsevier, New York, (1977).

Vesely, D., "*Microstructural characterization of polymer blends*", *Polym. Eng. Sci.*, **36**, 1586-93 (1996).

Viola, G. G. and D. G. Baird, "*Studies on the transient shear flow behavior of liquid crystalline polymers*", *J. Rheol.*, **30**, 601-628 (1986).

von Meerwall, E. D., "*Self-diffusion in polymer systems, measured with field-gradient spin echo NMR methods*", *Adv. Polym. Sci.*, **54**, 1-29 (1983).

von Werne, T. and T. E. Patten, "*Preparation of structurally well defined polymer-nanoparticle hybrids with controlled/living radical polymerization*", *J. Am. Chem. Soc.*, **121** 7409-10 (1999).

Voulgaris, D. and D. Petridis, "*Emulsifying effect of dimethyl dioctadecyl ammonium-hectorite in polystyrene/poly(ethyl methacrylate) blends*", *Polymer*, **43**, 2213-18 (2002).

Vu, Y. T. J. E. Mark, L. H. Pham, and M. Engelhardt, "*Clay nanolayer reinforcement of cis-1,4-polyisoprene and epoxidized natural rubber*", *J. Appl. Polym. Sci.*, **82**, 1391-03 (2001).

Wagener R., and T. J. G. Reisinger, "A rheological method to compare the degree of exfoliation of nanocomposites", *Polymer,* **44**, 7513-7518 (2003).

Walker, G. F., "*Macroscopic swelling of vermiculite crystals in water*", *Nature*, **187**, 312-313 (23 July 1960).

Wang J., D. Montville and K. E. Gonsalves, "*Synthesis of polycarbonate-co-poly(p-ethyl-phenol) and CdS nanocomposites*", *J. Appl. Polym. Sci.*, **72**, 1851-68 (1999)

Wang Qi, H. Xia and C. Zhang, "*Ultrasonic radiation process for preparing composite-material of polymer and nm-class inorganic particles*", *Chinese Pat.*, **CN1280993**, 2001.01.24, Appl. 1999.07.20, to Univ. Sichuan, Chengdu.

Wang S., M. Wang, Y. Lei and L. Zhang, "*Anchor effect"in poly(styrene maleic anhydride)/TiO2 nanocomposites*", *J. Mater. Sci. Letters*, **18**, 2009-12 (1999).

Wang S., Y. Hu, L. Song, J. Liu, Z. Chen, and W. Fan, "*Study on the Dynamic Self-Organization of Montmorillonite in Two Phases*", *J. Appl. Polym. Sci.*, **91**, 1457-1462 (2004).

Wang Y., C. Shen, and J. Chen, "*Nonisothermal cold crystallization kinetics of poly(ethylene terephthalate)/clay nanocomposites*", *Polym. J.*, **35**, 884-889 (2003b).

Wang Y., C. Shen, H. Li, and J. Chen, "*Nonisothermal melt crystallization kinetics of poly(ethylene terephthalate)/clay nanocomposites*", *J. Appl. Polym. Sci.*, **91**, 308-314 (2004).

Wang Z. M., T. C. Chung, J. W. Gilman, and E. Manias, "*Melt-processable syndiotactic polystyrene/montmorillonite nanocomposites*", *J. Polym. Sci. Part B: Polym. Phys.*, **41**, 3173-3187 (2003).

Wang, R., F. Qiu, H. Zhang and Y. Yang, "*Interactions between brush-coated clay sheets in a polymer matrix*", *J. Chem. Phys.*, **118**, 9447-56 (2003).

Wang, D. and C. A. Wilkie, "*Preparation of PVC-clay nanocomposites by solution blending*", *J. Vinyl Add. Technol.*, **8**, 238-245 (2002).

Wang, D.-Y. and C. A. Wilkie, "*In-situ reactive blending to prepare polystyrene–clay and polypropylene–clay nanocomposites*", *Polym. Degrad. Stab.*, **80**, 171-182 (2003).

Wang, D.-Y., D. Parlow, Q. Yao, and C. A. Wilkie, "*PVC-clay nanocomposites: preparation, thermal and mechanical properties*", *J. Vinyl Add. Technol.*, **7**, 203-213 (2001c).

Wang, H., C.-C. Zeng, M. Elkovitch, L. J. Lee, and K. W. Koelling, "*Processing and properties of polymeric nano-composites*", *Polym. Eng. Sci.*, **41**, 2036-46 (2001a).

Wang, J., J.-X. Du, J. Zhu, and C. A. Wilkie, "*An XPS study of the thermal degradation and flame retardant mechanism of polystyrene-clay nanocomposites*", *Polym. Degrad. Stab.*, **77**, 249-252 (2002a).

Wang, K. H., M. H. Choi, C. M. Koo, M. Xu, I. J. Chung, M. C. Jang, S. W. Choi, and H. H. Song, "*Morphology and physical properties of polyethylene/silicate nanocomposite prepared by melt intercalation*", *J. Polym. Sci., Part B: Polym. Phys.*, **40**, 1454-63 (2002).

Wang, K. H., M. H. Choi, C. M. Koo, Y. S. Choi and I. J. Chung, "*Synthesis and characterization of maleated polyethylene/clay nanocomposites*", *Polymer*, **42**, 9819-26 (2001b).

Wang, L., J. Schindler, C. R. Kannewurf, and M. G. Kanatzidis, "*Lamellar polymer-LiMoO3 nanocomposites via encapsulative precipitation*", *J. Mater. Chem.*, **7**, 1277-83 (1997).

Wang, L., M. Rocci-Lane, P. Brazis, C. R. Kannewurf, Y.-I. Kim, W. Lee, J. H. Choy, and M. G. Kanatzidis, "*α-RuCl3/Polymer Nanocomposites: The first group of intercalative nanocomposites with transition metal halides*", *J. Am. Chem. Soc.*, **122**, 6629-40 (2000b).

Wang, S., C. Long, X. Wang, Q. Li, and Z. Qi, "*Synthesis and properties of silicone rubber/organo-montmorillonite hybrid nanocomposites*", *J. Appl. Polym. Sci.*, **69**, 1557-61 (1998).

Wang, Y., L. Zhang, C. Tang and D. Yu, "*Preparation and characterization of rubber-clay nanocomposites*", *J. Appl. Polym. Sci.*, **78**, 1879-83 (2000b).

Wang, Z. and T. J. Pinnavaia, "*Nanolayer reinforcement of elastomeric polyurethane*", *Chem. Mater.*, **10**, 3769-71 (1998b).

Wang, Z. and T. J. Pinnavaia, "*Clay-polymer nanocomposites formed from acidic derivatives of montmorillonite and an epoxy resin*", *Chem. Mater.*, **6**, 468-474 (1994).

Wang, Z. and T. J. Pinnavaia, "*Hybrid organic-inorganic nanocomposites: exfoliation of magadiite nanolayers in an elastomeric epoxy polymer*", *Chem. Mater.*, **10**, 1820-26 (1998a).

Wang, Z. L., Y. L. Liu, and Z. Zhang, Eds., "*Handbook of nanophase and nanostructured materials*", 4 volumes, Kluwer Acad. Pub., Dordrecht (2003a).

Wang, Z., J. Massam and T. J. Pinnavaia, "*Epoxy-clay nanocomposites*", in *Polymer-Clay Nanocomposites*, T. J. Pinnavaia and G. W. Beall, Eds., J. Wiley & Sons, Ltd., New York (2000a).

Wang, Z., Tie Lan, and T. J. Pinnavaia., "*Hybrid organic-inorganic nanocomposites formed from an epoxy polymer and a layered silicic acid (magadiite)*", *Chem. Mater.*, 8, 2200-04 (1996).

Wapner P. G. and W. C. Forsman, "*Fourier transform method in linear viscoelastic analysis: the vibrating reed*", *Trans. Soc. Rheol.*, **15**, 603-626 (1971).

Wapner, P. G., *PhD thesis*, University of Pennsylvania (1971).

Warkentin B. P. and R. D. Miller, "*Conditions affecting formation of the montmorillonite-polyacrylic acid bond*", *Soil Sci.*, **85**, 14-18 (1948).

Wasserman S. R., L. Soderholm and U. Staub, "*Effect of surface modification on the interlayer chemistry of iron in a smectite clay*", *Chem. Matter.*, **10**, 559-566 (1998).

Weaver C. E. and L. D. Pollard, *The Chemistry of Clay Minerals*, Elsevier Scientific Pub. Co., Amsterdam (1973).

Weber, K. E. and H. Mukamal, "*Rubber composition comprising phyllosilicate minerals, silanes, and quaternary ammonium salts*", *US Pat.*, **4,431,755**, 14.02.1984, to Standard Oil Co.

Weiss A. and I. Kantner, "*Uber eine einfache Möglichkeit zur Abschätzung der Schichtladung glimmerartiger Schichtsilikate*", *Z. Naturforsch.*, **15b**, 804-807 (1960).

Weiss, A., "*Organic derivatives of mica-type layered silicates*", *Angew. Chem. Int. Ed. Engl.*, **2**, 134-143 (1963).

Weiss, A., *Organic Geochemistry*, Eglinton-Murphu, Ed., Springer-V., Berlin (1969).

Wen J. and G. L. Wilkes, "*Organic-inorganic hybrid network materials by the sol-gel approach*", *Chem. Mater.*, 8, 1667-81 (1996).

Wilhelm, M., "*Fourier-transform rheology*", *Macromol. Mater. Eng.*, **287**, 83-105 (2002).

Wilson, Jr., O. C., T. Olorunyolemi, A. Jaworski, L. Borum, D. Young, A. Siriwat, E. Dickens, C. Oriakh and M. Lerner, "*Surface and interfacial properties of polymer-intercalated double hydroxide nanocomposites*", *Appl. Clay Sci.*, **16**, 265-279 (1999).

Wilson, P. S., "*Structural foam composite having nano-particle reinforcement and method making the same*", *WO* **00/37242**, 2000.06.29, Appl. 1998.12.21, to Magna Intl.

Wittingham, S. and A. Jacobson (Eds.), *Intercalation Chemistry*, Academic Press, New York (1982).

Wolf, D., A. Fuchs, U. Wagenknecht, B. Kretzschmar, D. Jehnichen, L. Hdussler, *Nanocomposites of polyolefin clay hybrids*, Proceedings of the "Eurofiller '99", Lyon-Villeurbanne, 6-9 Sept. (1999).

Wood, J. R., Q. Zhao and H. D. Wagner, "*Orientation of carbon nanotubes in polymers and its detection by Raman spectroscopy*", *Composites Part A*, **32**, 391-399 (2001).

Wu Q., X. Liu, and L. A. Berglund, "*FT-IR spectroscopic study of hydrogen bonding in PA6/clay nanocomposites*", *Polymer*, **43**, 2445-2449 (2002).

Wu T.-M. and C.-S. Liao, "*Polymorphism in nylon-6/clay nanocomposites*", *Macromol. Chem. Phys.*, **201**, 2820-25 (2000).

Wu, C.-G., D. C. DeGroot, H. O. Marcy, J. L. Schindler, C. R. Kannewurf, Y.-J. Liu, W. Hirpo and M. G. Kanzidis, "*Redox intercalative polymerization of aniline in V2O5 xerogel. The post intercalative intralamellar polymer growth in polyaniline/metal oxide nanocomposites is facilitated by molecular oxygen*", *Chem. Mater.*, **8**, 1992-2004 (1996).

Wu, H.-D., C.-R. Tseng, and F.-C. Chang, "*Chain conformation and crystallization behavior of the syndiotactic polystyrene nanocomposites studied using Fourier transform infrared analysis*", *Macromolecules*, **34**, 2992-99 (2001).

Wu, Q., Liu, X. and L. A. Berglund, "*FTIR spectroscopic study of hydrogen bonding in PA6/clay nanocomposites*", *Polymer*, **43**, 2445-49 (2002).

Wu, Q., X. Liu and L. A. Berglund, "*An unusual crystallization behavior on polyamide-6/montmorillonite nanocomposites*", *Macromol. Rapid Commun.*, **22**, 1438-40 (2001a).

Wu, T.-M., S.-F. Hsu, and J.-Y. Wu, "*Crystalline forms in melt-crystallized syndiotactic polystyrene/clay nanocomposites*", *Polym. Eng. Sci.*, **42**, 2295-05 (2002a).

Wu, Z., C. Zhou and N. Zhu, "*The nucleating effect of montmorillonite on crystallization of nylon 1212/montmorillonite nanocomposite*", *Polymer Testing*, **21**, 479-83 (2002b).

Wypart, R., G. S. Rav and W. E. Cormier, "*Aluminosilicate stabilized halogenated polymers*", *US Pat.*, **6,414,071**, 2002.07.02 PQ Corporation.

Xiao, P., M. Xiao and K. Gong, "*Preparation of exfoliated graphite/polystyrene composite by polymerization-filling technique*", *Polymer*, **42**, 4813-17 (2001).

Xie, H. and R. Simha, "*Theory of solubility of gases in polymers*", *Polym. Intern.*, **44**, 348-355 (1997).

Xie, H., E. Nies and A. Stroeks, "*Some considerations on equation of state and phase relations: polymer solutions and blends*", *Polym. Eng. Sci.*, **32**, 1654-64 (1992).

Xie, W., Z. Gao, W. Pan, R. Vaia, D. Hunter and A. Singh, "*Characterization of organically modified montmorillonite by thermal techniques*", *ACS Polym. Mater. Sci. Eng. Proc.*, **82**, 284-285 (2000).

Xu, H.-Y., S.-W. Kuo, J.-S. Lee, and F.-C. Chang, "*Preparations, thermal properties, and Tg increase mechanism of inorganic/organic hybrid polymers based on polyhedral oligomeric silsesquioxanes*", *Macromolecules*, **35**, 8788-93, (2002a).

Xu, W., G. Liang, W. Wang, S. Tang, P. He, and W.-P. Pan, "*PP-PP-g-MAH-Org-MMT nanocomposites*", *J. Appl. Polym. Sci.*, **88**, 3093-99; 3225-31 (2003).

Xu, W., M. Ge, and P. He, "*Nonisothermal crystallization kinetics of polyoxymethylene/ montmorillonite nanocomposite*", *J. Appl. Polym. Sci.*, **82**, 2281-89 (2001).

Xu, W., M. Ge, and P. He, "*Nonisothermal crystallization kinetics of polypropylene /montmorillonite nanocomposites*", *J. Polym. Sci. Part B: Polym. Phys.*, **40**, 408-414 (2002).

Yablokov, Yu. S., A. I. Prokof'ev, and I. M. Papisov, "*Preparation of polymer-iron nanocomposites by reduction of iron(ii) in aqueous solutions of polymers*", *Vysokomol. Soed., 41, 1055 (1999); Polym. Sci.-Ser. B.*, **41**, 161-163 (1999).

Yakobson, B. I., C. J. Brabec and J. Bernholc, "*Structural mechanics of carbon nanotubes: From continuum elasticity to atomistic fracture*", *J. Computer-Aided Materials Design*, **3**, 173-176 (1996); "*Nanomechanics of carbon tubes: instabilities beyond linear range.*"*Phys. Rev. Letters*, **76**, 2511-14 (1996).

Yakobson, B. I., M. P. Campbell, C. J. Brabec and J. Bernholc., "*Tensile strength, atomistics of fracture, and C-chain unraveling in carbon nanotubes*", *Computational Mater. Sci.*, 8, 341-348 (1997).

Yamaguchi, T., K. Kitajima, E. Sakai, and M. Daimon, "*Synthesis and properties of alumina-pillared fluorine micas having cation exchangeability*", *Clay Minerals*, **38**, 41-47 (2003).

Yan L., and J. W. Stucki, "*Effects of structural Fe oxidation state on the coupling of interlayer water and structural Si-O stretching vibrations in montmorillonite*", *Langmuir*, **15**, 4648-4657 (1999).

Yan L., C. B. Roth, and P. F. Low, "*Changes in the Si-O vibrations of smectite layers accompanying the sorption of interlayer water*", *Langmuir*, **12**, 4421-4429 (1996).

Yan, J., R. Bajpai, E. Iannotti, M. Popovic and R. Mueller, "*Lactic acid fermentation from enzyme-thinned starch with immobilized Lactobacillus amylovorus*", *Chem. Biochem. Eng.*,**15**, 59-63 (2001).

Yang, Y. Z.-K. Zhu, J. Yin, X.-Y. Wang and Z.-E. Qi, "*Preparation and properties of hybrids of organo-soluble polyimide and montmorillonite with various chemical surface modification methods*", *Polymer*, **40**, 4407-11 (1999).

Yano, K., A. Usuki and A. Okada, "*Synthesis and properties of polyimide-clay hybrid films*", *J Polym. Sci., Part A: Polym. Chem.*, **35**, 2289-94 (1997).

Yano, K., A. Usuki, A. Okada, T. Kurauchi and O. Kamigaito, "*Synthesis and properties of polyimide-clay hybrid*", *J Polym. Sci.: Part A: Polym. Chem.*, **31**, 2493-98 (1993).

Yao, H., J. Zhu, A. B. Morgan, and C. A. Wilkie, "*Crown ether modified clays and their polystyrene nanocomposites*", Proceedings Polymer Nanocomposites 2001, Boucherville, QC, Canada, 17-19.09. 2001.

Yao, K. J., M. Song, D. J. Hourston and D. Z. Luo, "*Polymer/layered clay nanocomposites: 2 polyurethane nanocomposites*", *Polymer*, **43**, 1017-20 (2002).

Yasmin A., J. L. Abot, and I. M. Daniel, "*Processing of clay/epoxy nanocomposites by shear mixing*", *Scripta Materialia*, **49**, 81-86 (2003).

Yasue, K., T. Tamura, S. Katahira, and M. Watanabe, "*Reinforced polyamide resin composition and process for producing the same*", *US Pat.* **5,414,042,** 09.05.1995, Appl. 28.12.1993, to Unitika Ltd.

Yoo, Y. J., C. H. Sohn, H. S. Kim and J. W. Lee, "*Development of polypropylene-clay nanocomposite through the ultrasonic melt intercalation*", Polymer Processing Society (PPS) Regional Meeting, paper D-8, Bangkok, 1999.12.01-03.

Yoon P. J., D. L. Hunter, and D. R. Paul, "*Polycarbonate nanocomposites: Part 2. Degradation and color formation*", *Polymer*, **44**, 5341-5354 (2003).

Yoon, J. T., W. H. Jo, M. S. Lee and M. B. Ko, "*Effects of comonomers and shear on the melt intercalation of styrenics/clay nanocomposites*", *Polymer*, **42**, 329-336 (2001).

Yu Y.-H., C.-Y. Lin, and J.-M. Yeh, "*Poly(N-vinylcarbazole)–Clay Nanocomposite Materials Prepared by Photoinitiated Polymerization with Triarylsulfonium Salt Initiator*", *J. Appl. Polym. Sci.*, **91**, 1904-1912 (2004).

Yu, C.-H., S. Q. Newton, M. A. Norman, D. M. Miller, L. Schäfer and B. J. Teppen, "*Molecular dynamics simulations of the adsorption of methylene blue at clay mineral surfaces*", *Clay Clay Miner.*, **48**, 665-681 (2000b).

Yu, M.-F., O. Lourie, M. J. Dyer, K. Moloni, T. F. Kelly, and R. S. Ruoff, "*Strength and Breaking Mechanism of Multiwalled Carbon Nanotubes Under Tensile Load*", *Science*, **287**, 637-40 (2000a).

Yu, Y.-Y., C.-Y. Chen, and W.-C. Chen, "*Synthesis and characterization of organic–inorganic hybrid thin films from poly(acrylic) and monodispersed colloidal silica*", *Polymer*, **44**, 593-601 (2003).

Yu, Z.-Z., M. Yang, Q. Zhang, C. Zhao, and Y.-W. Mai, "*Dispersion and distribution of organically modified montmorillonite in nylon-66 matrix*", *J. Polym. Sci.: Part B: Polym. Phys.*, **41**, 1234-1243 (2003).

Zanetti, M., G. Camino, C. Zilg, and R. Mülhaupt, "*Thermal degradation and combustion behaviour of EVA/fluorohectorite nanocomposites*", *First MoDeSt Conference*, Palermo, 2000a.09.03-07.

Zanetti, M., S. Lomakin and G. Camino, "*Polymer layered silicate nanocomposites*", *Macromol. Mater. Eng.*, **279**, 1-9 (2000b).

Zapata, L., J. Castelein, J. P. Mercier and J. J. Fripiat, "*Derivés organiques des silicates*", *Bull. Soc. Chim.*, 54-63 (1972).

Zarraga, I. E., D. A. Hill and D. T. Leighton, Jr., "*The characterization of the total stress of concentrated suspensions of noncolloidal in Newtonian fluids*", *J. Rheol.*, **44**, 185-220 (2000).

Zax, D. B., D.-K. Yang, R. A. Santos, H. Hegemann, E. P. Giannelis and E. Manias, "*Dynamical heterogeneity in nanoconfined polystyrene chains*"*J. Chem. Phys.*, **112**, 2945-51 (2000).

Zeigler, J. M. and F. W. G. Fearon, Eds., *Silicon Based Polymer Science: A Comprehensive Resource*, ACS Advances in Chemistry Ser. No. 224, American Chemical Society: Washington, DC (1990).

Zeng, C.-C. and L. J. Lee, "*Poly(methyl methacrylate) and polystyrene/clay nanocomposites prepared by in-situ polymerization*", *Macromolecules*, **34**, 4098-03 (2001).

Zerda, A. S. and A. J. Lesser, "*Intercalated clay nanocomposites: Morphology, mechanics, and fracture behavior*", *J. Polym. Sci. Part B: Polym. Phys.*, **39**, 1137-46 (2001).

Zerda, A. S., T. C. Caskey, and A. J. Lesser, "*Highly Concentrated, Intercalated Silicate Nanocomposites: Synthesis and Characterization*", *Macromolecules*, **36**, 1603-08 (2003).

Zhang G., and D. Yan, "*Crystallization Kinetics and Melting Behavior of Nylon 10,10 in Nylon 10,10-Montmorillonite Nanocomposites*", *J. Appl. Polym. Sci.*, **88**, 2181-2188 (2003).

Zhang G., Y. Li, and D. Yan, "*Polymorphism in Nylon-11/Montmorillonite Nanocomposite*", *J. Polym. Sci. Part B: Polym. Phys.*, **42**, 253-259 (2004).

Zhang K., L. Wang, F. Wang, G. Wang, and Z. Li, "*Preparation and Characterization of Modified-Clay-Reinforced and Toughened Epoxy-Resin Nanocomposites*", *J. Appl. Polym. Sci.*, **91**, 2649-2652 (2004).

Zhang Q.-X., Z.-Z. Yu, M. Yang, J. Ma, and Y.-W. Mai, "*Multiple melting and crystallization of Nylon-66/Montmorillonite nanocomposites*", *J. Polym. Sci. Part B: Polym. Phys.*, **41**, 2861-2869 (2003).

Zhang W. A., D. Z. Chen, H. Y. Xu, X. F. Shen, and Y. E. Fang, "*Influence of four different types of organophilic clay on the morphology and thermal properties of polystyrene/clay nanocomposites prepared by using the γ-ray irradiation technique*", *Eur. Polym. J.*, **39**, 2323-2328 (2003).

Zhang X., R. Xu, Z. Wu, and C. Zhou, "*The synthesis and characterization of polyurethane/clay nanocomposites*", *Polym. Int.*, **52**, 790-794 (2003).

Zhang, C., Qi Wang, H. Xia and G. Qiu, "*Ultrasonically induced microemulsion polymerization of styrene*", *Europ. Polym. J.*, **38**, 1769-76 (2002a).

Zhang, J. and C. A. Wilkie, "*Preparation and flammability properties of polyethylene–clay nanocomposites*", *Polym. Degrad. Stab.* **80**, 163-169 (2003).

Zhang, L., Y. Wang, Y. Wang, Y. Sui and D. Yu, "*Morphology and mechanical properties of clay styrene-butadiene rubber nanocomposites*", *J. Appl. Polym. Sci.*, **78**, 1873-78 (2000b).

Zhang, M., Y. Bando, K. Wada, "*Silicon dioxide nanotubes prepared by anodic alumina as templates*", *J. Mater. Res.*, **15**, 387-392 (2000a).

Zhang, Q., Q. Fu, L. Jiang and Y. Lei, "*Preparation and properties of polypropylene/ montmorillonite layered nanocomposites*", *Polym. Int.*, **49**, 1561-64 (2000b).

Zhang, Q., Y. Wang, and Q. Fu, "*Shear-induced change of exfoliation and orientation in polypropylene/montmorillonite nanocomposites*", *J. Polym. Sci. Part B: Polym. Phys.*, **41**, 1-10 (2003).

Zhang, X., J. G. Lim and S. C. Yoon, "*Syndiotactic polystyrene nanocomposite and manufacturing method thereof*", *PCT*, **WO 02/06394**, 2002b.01.24, Appl. 2001.06.29, Priority 2000.07.18, to Samsung General Chem. Co., Ltd.

Zhao Q., and E. T. Samulski, "*Supercritical CO2-Mediated Intercalation of PEO in Clay*", *Macromolecules*, **36**, 6967-6969 (2003).

Zheng, L., A. J. Waddon, R. J. Farris, and E. B. Coughlin, "*X-ray characterizations of polyethylene polyhedral oligomeric silsesquioxane copolymers*", *Macromolecules*, **35**, 2375-79 (2002a).

Zheng, L., R. J. Farris, and E. B. Coughlin, "*Novel polyolefin nanocomposites: synthesis and characterization of metallocene-catalyzed polyolefin polyhedral oligomeric silsesquioxane copolymers*", *Macromolecules*, **34**, 8034-39 (2001).

Zheng, L., R. M. Kasi, R. J. Farris, and E. B. Coughlin, "*Synthesis and thermal properties of hybrid copolymers of syndiotactic polystyrene and polyhedral oligomeric silsesquioxane*", *J. Polym. Sci.: Part A: Polym. Chem.*, **40**, 885-891 (2002b).

Zhou, H. S., T. Wada, H. Sasabe and H. Komiyama, "*Synthesis and optical properties of nanocomposite silver-polydiacetylene*", *Synthetic Metals*,**81**, *129-132*, (1996).

Zhu, J., A. B. Morgan, F. J. Lamelas, and C. A. Wilkie, "*Fire properties of polystyrene-clay nanocomposites*", *Chem. Mater.*, **13**, 3774-80 (2001a).

Zhu, J., F. M. Uhl, A. B. Morgan, and C. A. Wilkie, "*Studies on the mechanism by which the formation of nanocomposites enhances thermal stability*", *Chem. Mater.*, **13**, 4649-54 (2001b).

Zhu, J., P. Start, K. A Mauritz and C. A Wilkie, "*Thermal stability and flame retardancy of poly(methyl methacrylate)-clay nanocomposites*", *Polym. Degrad. Stab.*, **77**, 253-258 (2002).

Zhu, Y., Y. Qian, X. Li, and M. Zhang, "*A nonaqueous solution route to synthesis of polyacrylamide-silver nanocomposites at room temperature*", *Nanostructured Mater.*, **10**, 673-678 (1998).

Zhu, Y., Y. Qian, X. Li, and M. Zhang, "*γ-Radiation synthesis and characterization of polyacrylamide-silver nanocomposites*", *Chem. Commun.*, **12**, 1081-82 (1997).

Zhulina, E., C. Singh, and A. C. Balazs, "*Attraction between surfaces in a polymer melt containing telechelic chains: guidelines for controlling the surface separation in intercalated polymer-clay composites*", *Langmuir*, **15**, 3935-43 (1999).

Zilg, C., R. Mülhaupt, and J. Finter, "*Morphology and toughness/stiffness balance of nanocomposites based upon anhydride-cured epoxy resins and layered silicates*", *Macromol. Chem. Phys.*, **200**, 661-670 (1999a).

Zilg, C., R. Mülhaupt, and J. Finter, "*Organophilic phyllosilicates*", *WO **99/ 42518**, 1999.08.26;* priority 1998.02.20, to Ciba Specialty Chemicals/Vantico Inc.

Zilg, C., R. Mülhaupt, and J. Finter, "*Organophilic phyllosilicates*", *US Pat.*, **6,197,849**, 2001.03.06, priority 1998.02.20, to Ciba Specialty Chemicals/ Vantico Inc.

Zilg, C., R. Thomann, R. Mülhaupt and J. Finter, "*Polyurethane nanocomposites containing laminated anisotropic nanoparticles derived from organophilic layered silicates*", *Adv. Mat.*, **11**, 49-52 (1999b).

Part 9

Index

Clay-Containing Polymeric Nanocomposites

Index

C

D

E

F

G

H

I

J

K

L

M

N

O

P

Q

R

S

T

U

V

W

X

Y

Z